Biomathematical and Biomechanical Modeling of the Circulatory and Ventilatory Systems

Volume 2

For further volumes:
http://www.springer.com/series/10155

Marc Thiriet

Control of Cell Fate in the Circulatory and Ventilatory Systems

Springer

Marc Thiriet
Project-team INRIA-UPMC-CNRS REO
Laboratoire Jacques-Louis Lions, CNRS UMR 7598
Université Pierre et Marie Curie
Place Jussieu 4
75252 Paris Cedex 05
France
Marc.Thiriet@inria.fr

ISSN 2193-1682
ISBN 978-1-4614-0328-9 e-ISBN 978-1-4614-0329-6
DOI 10.1007/978-1-4614-0329-6
Springer New York Dordrecht Heidelberg London

Library of Congress Control Number: 2011940323

Printed on acid-free paper

Springer is part of Springer Science+Business Media (www.springer.com)

Contents

Introduction

"*Praeterea nisi erit minimum, paruissima quaeque corpora constabunt ex partibus infinitis... [... In addition, in the absence of limit to smallness, the smallest elements are constituted of countless parts...]*" (Lucretius) [1]

Cellular mechanisms that regulate cell fate are targeted by mathematical modeling to predict outcomes as well as approximately quantify contributions of major influence molecules and environmental physical and mechanical factors. Among influence agents, mechanical forces and derived stresses can prime molecule release and gene transcription in cells subjected to these stresses such as vascular endothelial and smooth muscle cells between flowing blood and vessel wall. Cells in beating heart as well as inflating–deflating arteries, airways, and lung air spaces bear time- and space-dependent mechanical stress fields. These fields participate in the regulation of cell fate (growth, differentiation, and migration) and tissue remodeling. Cells of vessel and airway walls control the local caliber of blood vessels and conduits of the respiratory tract.

Complexity arises from the large number of involved quantities that are related by non-linear relationships. Therefore, kinetic and transport equations with associated rates as well as kinetics and transport coefficients that govern cell signaling and tissue remodeling are strongly coupled. For example, the cell division cycle is controlled by a huge set of regulators, mainly cyclin-dependent kinases that have many activators and inhibitors. The genome must be precisely replicated and accurately partitioned into its 2 new progeny. Progression through the cell cycle is thus governed by interactions between proteins as well as between genes and proteins.

Volume 1 "Cell and Tissue Organization in the Circulatory and Ventilatory Systems" briefly presents in 9 chapters the cell types of the cardiovascular and respiratory systems and focuses on cellular constituents, processes, and principal elements of interaction with a cell's neighborhood that participate in cell's behavior, whatever the cell type. The cardiovascular and ventilatory systems, like any physiological system, have a complicated three-dimensional structure and composition. Their time-dependent behavior is regulated. With the preliminary set of data yielded by Volume 1, Volume 2 "Control of Cell Fate in the Circulatory and Ventilatory Systems", composed of 6 chapters, describes the major, strongly regulated, cellular events. The regulation of cell fate can be local or regional, and can originate from distant body sites. Before discussing characterization of cell events, Volume 2 reports in Chapter 1 on cells of the nervous and endocrine systems that are responsible for remote control. Chapters 2 to 5 detail the primary events of cell life down to death. The cell division cycle and cell growth are treated in Chapter 2. Involved growth factors are given in Chapter 3. Chapter 4 is devoted to cell decision between survival and death. Chapter 5 focuses on the circadian cycle, as circadian clocks coordinate behavior with daily light-dark cycles by driving rhythmic gene transcription. Chapter 6 describes cell migration for tissue adaptation and repair that strongly relies on the cell cytoskeleton and cell–matrix adhesion.

Signal transduction starts at the cell surface, where chemical messengers bind and/or stimulate specific proteic and lipidic sensors and receptors that trigger cellular responses. In addition, physical agents and mechanical stresses can efficiently deform these sensors and receptors, open their proper activation domains, and deliver messages into the cell. These sensors and receptors are described in Volume 3 "Signaling at the Cell Surface in the Circulatory and Ventilatory Systems". Initiated signaling cascades are usually composed of multiple nodes that correspond to major mediators. Intracellular effectors of signaling cascades are characterized in Volume 4 "Intracellular Signaling Mediators in the Circulatory and Ventilatory Systems". Signaling axes prime the release of substances from intracellular stores as well as the gene expression to synthesize messengers for the cell itself and its neighbors, whereas endocrine cells and neurons are responsible for remote control. Intracellular cascades of chemical reactions with signaling nodes, hubs, and modules are represented by mathematical models.

Volume 5 "Tissue Functioning and Remodeling in the Circulatory and Ventilatory Systems" is devoted to functioning and remodeling of vascular tissues (blood, heart, and blood and lymph vessels) and tissues of the respiratory tract. These regulated processes can also be described using mathematical models.

Volume 6 "Circulatory and Ventilatory Conduits in Normal and Pathological Conditions" mainly focuses on macroscopic aspects of the cardiovascular and respiratory systems. Diseases of walls of blood vessels and airways disturb blood and air flows. Conversely, local flow disturbances can contribute to trigger pathophysiological processes. Nowadays, minimally invasive therapies use natural ducts to implant medical devices. Development of these medical devices and techniques relies, at least partly, on numerical tests and medical simulators.

Therefore, Volumes 1 and 2 deal with the microcopic scale (cell); Volumes 3 and 4 with the nanoscopic scale (signaling mediators and cascades of chemical reactions); Volume 5 with the mesocopic scale (constituents of biological tissues of apparatus of blood circulation and body's respiration); and Volume 6 with the macroscopic scale (structure and function in normal conditions and diseases).

Volume 7 "Blood and Air Transport in the Circulatory and Ventilatory Systems" copes with mechanics of air and blood flows in relatively short, curved, deformable conduits. Consequently, blood and air flows driven by anatomical pumps are three-dimensional, developing, and time-dependent, but mostly laminar. In addition, inhaled air is transported in the respiratory tract by two different mechanisms: convection and diffusion. Numerical simulations, at least in some conditions, rely on the coupling of flow mechanics to wall dynamics. Volume 7 not only deals with the simulation of movements of fluids through biological conduits, but also their experimental validation.

The three basic natural sciences — biology, chemistry, and physics — interact with mathematics to explain the functioning of physiological systems. In this multidisciplinary series "Biomathematical and Biomechanical Modeling of the Circulatory and Ventilatory Systems", Volume 8 provides a necessary set of glossaries to rapidly get information on elements of physiological systems that should be incorporated in models or eliminated. Specific vocabulary used in each field of knowledge can indeed limit easy access to this field to researchers of other disciplines.

Common abbreviations such as "a.k.a." that stands for "also known as" are used throughout the text to lighten sentences. Latin-derived shortened expressions are also widely utilized: "e.g." (exempli gratia) and "i.e." (id est) mean "for example" and "in other words", respectively. Rules adopted for substance aliases as well as alias meaning and other notations are given as appendices in this book.

1

Remote Control

The remote control of any body's tissue is mainly carried out by the nervous and endocrine systems. The first information system exchanges electrochemical waves, the so-called action potentials, with the body's tissues via neurons that contact cells of innervated organs.[1] The second information system sends chemical messengers, the so-called hormones, via the blood circulation that irrigates the body's cells. The endocrine system is a set of ductless, endocrine glands that secrete hormones into the blood stream to target cells that possess hormone cognate receptors on their surfaces.[2]

Kidneys represent additional controlling organs. They produce urine and, consequently, regulate electrolyte and hydrogen ion (acid–base) balance as well as the body's fluid volume, hence blood pressure. They also belong to the endocrine system, as they synthesize hormones, such as renin and erythropoietin.

Adipose tissue that is either located close to organs or remotely is an important constituent of the endocrine system, as it releases multiple hormones and growth factors. Adipose tissue is composed of loose connective tissue with adipocytes that derive from lipoblasts. Its main role is to store energy in the form of available lipids. Two types of adipose tissue exist: white and brown adipose tissue. White adipose tissue acts as a thermal insulator, helping to maintain body temperature. Brown adipose tissue that is particularly related to skeletal muscles is aimed at compensating for fluctuating availability of energy.

---

[1] The nervous system transmits electrochemical signals down to synapses between 2 neurons or between neurons and target cells, where these cues are converted into chemical signals, as neurotransmitters are released into synaptic gaps (Sect. 1.1.5.2 and Vol. 3 – Chap. 2. Membrane Ion Carriers). In the postsynaptic membrane, ligand-gated ion channels convert chemical signals back into electrochemical signals.

[2] Exocrine hormones, or ectohormones, are secreted into a duct, from which they can reach blood circulation.

1.1 Nervous System

The nervous system is a network of specialized cells — neurons and assisting glial cells — that communicate information from peripheral and central sensors. These detectors survey the state of the organism and its environment and transmit information to processing centers via *afferent nerves*. These processing centers treat received information that arise from many sources to cause adaptive responses sent by *efferent nerves* to suitable peripheral organs to coordinate the activity of all the body's regions.

The nervous system can be split into 2 main compartments: peripheral and central nervous system. Processing centers localize to the central nervous system. Neurons relay impulses between and within the 2 compartments. Communication between neurons is done via neurotransmitters.

Neurotransmitters that are released from the axon terminal of an upstream neuron target their cognate receptors on the cell surface of a downstream neuron that are ligand-gated ion channels. Once activated, the latter change their conformation and transduce a chemical cue into an electrochemical signal, as they control gating of ion channels (allosteric transition) [2]. A rapid, 2-step, domino-like mechanism takes place between adjacent subunits of multimeric ion channels.

At least some neuron types control nervous activity using 2 neurotransmitters (*cotransmission*). In particular, dopaminergic neurons of the brain deliver messages either rapidly with glutamate or slowly with dopamine.

1.1.1 Central and Peripheral Nervous Systems

1.1.1.1 Central Nervous System

The central nervous system includes the spinal cord and brain. According to the developmental organization of the central nervous system, the brain consists of the brainstem and prosencephalon (Table 1.1). The *prosencephalon*, or forebrain, comprises the telencephalon, or cerebrum, and diencephalon, or interbrain, whereas the brainstem that connects the spinal cord to the brain consists of the mesencephalon, or midbrain, and rhombencephalon, or hindbrain. The *telencephalon* is constituted by the *basal ganglia* with the striatum (caudate nucleus and putamen), pallidum, substantia nigra, and subthalamic nucleus, *rhinencephalon* with the olfactory bulb, tract, tubercle, striae, and anterior nucleus and piriform cortex, *amygdala*, *hippocampus*, and *neocortex*, as well as lateral ventricles. The *diencephalon* is composed of the *prethalamus*, or subthalamus, *thalamus*, *hypothalamus*, *epithalamus*, and *pretectum*, as well as *pituitary* and *pineal* glands and third ventricle. The *mesencephalon* encompasses the *tectum*, *pretectum*, and *cerebral peduncles*, as well as the mesencephalic duct or cerebral aqueduct. The *rhombencephalon* that contains the fourth ventricle can be subdivided into the *metencephalon* and *myelencephalon*. The metencephalon is composed of the *pons* and the *cerebellum*; the myelencephalon corresponds to the *medulla oblongata*.

Table 1.1. Anatomical compartments of the central nervous system (brain) from organogenesis (V_L, V_3, V_4: lateral, third, and fourth cerebral ventricles). Epiphysis and hypophysis are also called pineal and pituitary glands, respectively. The cerebral cortex is constituted by the neocortex (gray matter), or isocortex, with 6 layers (from outer layer I to inner layer VI) and allocortex with the hippocampus and olfactory cortex, which contain less than 6 cell layers (type-1 classification). Type-2 classification includes the neo-, archi- (hippocampus), and paleocortex (anterior olfactory nucleus, anterior perforated substance, prepyriform area and peri-amygdalar area) that are the cortical parts of the limbic system. The limbic system, devoted to olfaction, behavior, and long-term memory, is mainly composed of the amygdala, hippocampus, parahippocampal and cingulate gyri, hypothalamus, thalamus, fornix, and other parts of the limbic cortex.

Compartment	Components
Prosencephalon	Forebrain: diencephalon, telencephalon
Diencephalon	Interbrain: subthalamus, thalamus, hypothalamus, epithalamus, pretectum, epiphysis, hypophysis, V_3
Telencephalon	Cerebrum: neocortex, basal ganglia, rhinencephalon, amygdala, hippocampus, V_L
Rhinencephalon	Olfactory bulb, tract, tubercle, striae, anterior nucleus, piriform cortex
Brainstem	Mesencephalon, rhombencephalon
Mesencephalon	Midbrain: cerebral peduncles, tectum, pretectum
Rhombencephalon	Hindbrain: metencephalon, myelencephalon, V_4
Metencephalon	Cerebellum, pons
Myelencephalon	Medulla oblongata

Neurons of the neocortex are organized into functional columns of neurons spanning from the pia mater (inner meninge layer) to white matter (inner brain region with mostly myelinated axons). Specific synapses ensure that adjacent neurons convey information independently. Synapses are preferentially formed between radially aligned sister excitatory neurons that arise from the same mother cell in the developing neocortex and give rise to unidirectional synapses with each other [3]. In fact, neurons communicate via both relatively slower, chemical, and quicker, electrical synapses, i.e., via release of vesicular content and gap junctions. Both types of communications are used for learning and memory by neurons of the central nervous system [4].

Limbic, cortical, and midbrain structures are responsible for the response to behavioral events, which in particular need rapid changes in sympathetic tone. At rest, the background sympathetic tone is set by the hypothalamus, the rostral ventrolateral medulla, the nucleus of the solitary tract, and the spinal cord, via the barosensitive sympathetic efferents. The background sympathetic activity can be either increased or decreased for the short- and long-term stabilization of the blood pressure.

The paraventricular nucleus and the dorsomedial nucleus of the hypothalamus are integrative centers for the circulatory control. The dorsomedial nucleus contributes to the cardiovascular responses to environmental stresses. The paraventricular nucleus controls the body's homeostasis. It is thus involved in the regulation of the body's fluid (particularly osmolality and blood volume), metabolism, and temperature.

The central nervous system is directly regulated by circulating factors. Hormones and cytokines act via circumventricular organs (subfornical organ, organum vasculosum lamina terminalis and area postrema). Endothelial cells release mediators (e.g., nitric oxide and prostaglandins) that cross the blood–brain barrier (Vol. 5 – Chap. 7. Vessel Wall) and stimulate the hypothalamus, rostral ventrolateral medulla,[3] and nucleus of the solitary tract. Changes in ion levels are detected in the hypothalamus. Brainstem chemoreceptors sense the level of blood gases. The renin–angiotensin system also directly acts on the central system. Circulating angiotensin-2 activates angiotensin-2 receptor-1 (AT_1) on endothelial cells, releasing nitric oxide.[4]

1.1.1.2 Peripheral Nervous Systems

The peripheral nervous system resides outside the skull and vertebral column that lodge the brain and spinal cord, respectively, which together make up the central nervous system. According to its functioning mode, the peripheral nervous system is divided into the somatic and autonomic nervous systems. The former is under conscious control, but not the latter. The somatic nervous system coordinates the body's movements and transmits received external stimuli. The autonomic nervous system is split into the stress-related sympathetic and parasympathetic systems as well as the enteric system (Tables 1.2 and 1.3). Released neurotransmitters bind to their cognate receptors on target cells (Table 1.4).

[3] The rostral ventrolateral medulla is a set of heterogeneous glutamatergic C1 neurons. The majority of C1 neurons synthesizes adrenaline. Neurons C1 are connected to the medulla oblongata and the pons. Barosensitive C1 neurons of this area also receive inputs from the brain and spinal cord. Input GABA from the caudal ventrolateral medulla is required in the baroreflex. Brainstem regions, such as the caudal pressor area, midline depressor area, subnuclei of the nucleus of the solitary tract, and gigantocellular depressor area, relay electrical or chemical stimulations aimed at triggering blood pressure adaptation. The rostral as well as the caudal ventrolateral medulla also comprise interneurons that contain γ-aminobutyric acid. Many interneurons have a baseline activity.

[4] Nitric oxide diffuses across capillaries and potentiates the release of γ-aminobutyric acid.

Table 1.2. Autonomic nervous system with its 2 major components: (ortho)sympathetic and parasympathetic systems. Nervous centers of the sympathetic and parasympathetic systems localize to the thoracic and lumbar spine cord and brainstem and sacral spinal cord, respectively. Preganglionic, short, cholinergic neurons of the sympathetic system synapse with postganglionic, long, adrenergic (most often), neurons in the para- and prevertebral ganglia (and also cervical, celiac, and mesenteric ganglia), which then innervate organs, such as airways, lungs, heart, and blood vessels. The 2 pelvic sympathetic trunks in front of the sacrum in continuity with the abdominal chains consist of 4 to 5 sacral ganglia interconnected by interganglionic cords. They converge and end in the ganglion impar (or ganglion of Walther) on the front of the coccyx. Adrenal medulla serves as a sympathetic ganglion, in addition to an endocrine gland. Preganglionic, long, cholinergic neurons of the parasympathetic system synapse with postganglionic, short, cholinergic neurons in ganglia on or near thoracic (e.g., heart, trachea, and lungs), abdominal, and pelvic viscera.

Sympathetic	Parasympathetic
Cervical (3 pairs of ganglia) Cardiac and pulmonary nerves	Facial (Parasympathetic cranial ganglia) Nose innervation
Thoracic (12 pairs of ganglia) Aortic and lung nerves	Visceral Vagus nerve Cardiac and pulmonary plexi Innervation of pharynx, larynx, trachea, lung, heart, blood vessels
Abdominal (4 pairs of ganglia)	Pelvic (S2–S4) Sacral spinal nerves
Sacral (4 pairs of ganglia and ganglion impar)	

The autonomic nervous system comprises afferent and efferent neurons between the central nervous system and the body's organs; these neurons belong to both the sympathetic and parasympathetic components. The *cholinergic preganglionic neurons* innervate ganglia, glands, or nervous ganglionic networks, such as the cardiac ganglionic network. The peripheral ganglia and networks contain motoneurons which control smooth muscles and other visceral targets. The *sympathetic ganglionic neurons* of the cardiovascular system are noradrenergic. The afferent components for the cardiovascular system include baroreceptors, chemoreceptors, and renal afferents. Afferents associated with baroreceptors and other mechanoreceptors such as voloreceptors project to the *nucleus of the solitary tract*, a major integrative center for circulatory control [5]. Afferents transmitting chemical or physical parameter-related signals project to the spinal cord.

Table 1.3. Effects of the autonomic nervous system in the circulatory and ventilatory systems. The sympathetic nervous system is responsible for the body's response to stress and counteracts the parasympathetic system that promotes the maintenance of the body at rest. Effects depend not only on the neurotransmitter type (noradrenaline for the sympathetic and acetylcholine for the parasympathetic system), but also receptor type for a given neurotransmitter (↑: increase). Bathmotropes modify the degree of sarcolemmal excitability, i.e., the threshold potential of excitation; chronotropic neurotransmitters the heart frequency; dromotropes the conduction velocity in the atrioventricular node; and inotropes the strength of muscular contractions. Tonotropes reduce the size of the relaxing heart, thereby influencing heart distensibility and tone. Lusitropes cause a rapid myocardial relaxation. Myocardial contractility increases via an augmented influx of calcium ions in the cytosol of cardiomyocytes during stimulation by the sympathetic nervous system via catecholamines. In addition, catecholamines promote positive lusitropy (L+), the heart relaxing more rapidly via the phosphorylation of phospholamban using the cAMP–PKA pathway. Sympathetic stimulation is also positively bathmo- (B+), dromo- (D+), ino- (I+), and tonotropic (T+). Conversely, parasympathetic stimulation is negatively chrono- (C−), dromo- (D−), and inotropic (I−). Positive (e.g., adrenaline and dopamine) and negative (e.g., acetylcholine) chronotropes increase and decrease the heart frequency, respectively. Positive inotropes include adrenaline, noradrenaline, dopamine, eicosanoids, and prostaglandins.

	Sympathetic	Parasympathetic
Heart	B+, C+, D+, I+, T+, L+	B−, C−, D−, I−, T−
Blood vessel	Vasodilation of perfusion arteries (brain, heart, liver, muscle, etc.) Vasoconstriction of arterioles of kidneys and skin	Vasoconstriction Vasodilation in sexual organs
Kidney	Renin secretion ↑	
Adrenal gland	Catecholamine release	
Bronchus	Bronchodilation	Bronchoconstriction Secretion ↑

The peripheral nervous system is composed of sensory and motor neurons as well as connecting neurons that couple the body's tissues to the central nervous system. *Nervous ganglia* serve as relay centers between peripheral and central neurons. Sympathetic ganglia are located close to the spinal cord, on each side of the vertebral (or spinal) column. Parasympathetic ganglia are most often small terminal ganglia or intramural ganglia, as they lie near or within the organs, respectively. The 4 paired parasympathetic ganglia of the

Table 1.4. Major types of receptors in circulatory and ventilatory organs targeted by neurotransmitters of the autonomic nervous system. Muscarinic cholinergic receptors (mAChR) are G-protein-coupled receptors of acetylcholine. Adrenergic receptors, or adrenoceptors, are also G-protein-coupled receptors for noradrenaline and adrenaline.

Type	Main loci	Effects
	Muscarinic cholinergic receptors	
M_2 (Gi-coupled)	Heart	C−, D−, I−
M_3 (Gq-coupled)	Artery, vein (endothelial cell)	Vasodilation
	Airway (smooth myocyte)	Bronchoconstriction
	Adrenergic receptors	
$\alpha 1$ (Gq-coupled)	Heart	I+
	Blood vessel	Vasoconstriction
	Airway	Bronchoconstriction ($\alpha < \beta$)
$\alpha 2$ (Gi-coupled)	Blood vessel	Vasoconstriction
	Airway	Bronchoconstriction
$\beta 1$ (Gs-coupled)	Heart	B+, C+, D+, I+, L+
	Blood vessel	Vasodilation
	Airway	Bronchodilation
$\beta 2$ (Gs-coupled)	Heart	I+, L+
	Blood vessel	Vasodilation ($\beta 2 > \beta 1$)
	Airway	Bronchodilation ($\beta 2 > \beta 1$)
$\beta 3$ (Gs-coupled)	Heart	I−

head and neck supply all parasympathetic innervation to these regions. Nerves are classified into cranial and spinal nerves depending on their origin.

Ten among 12 *cranial nerves* originate from the brainstem. They mainly control the functions of the anatomic structures of the head. The nuclei of cranial nerves I and II reside in the forebrain and thalamus, respectively. Other cranial nerve nuclei are located in the midbrain, pons, and medulla oblongata. The medulla oblongata contains the cardiovascular and respiratory centers, thereby controlling breathing and blood circulation.

Spinal nerves originate from the spinal cord. In humans, 31 pairs of spinal nerves (8 cervical, 12 thoracic, 5 lumbar, 5 sacral, and 1 coccygeal). In the cervical region, the spinal nerve roots come out above the corresponding vertebrae. From the thoracic region to the coccygeal region, the spinal nerve roots come out below the corresponding vertebrae. In adults, because the spinal cord stops growing in infancy, but not the spinal column, the spinal cord ends at about L1–L2 level of lumbar vertebra. In the lumbar and sacral region, the spinal nerve roots travel within the cauda equina, which is filled with cerebrospinal fluid.

1.1.2 Neurons and Glial Cells of the Nervous Regulation

Cells that participate in the nervous regulation of the cardiovascular and ventilatory system include sensors and afferent, efferent, and possible relay neurons, as well as glial cells of the central nervous system that can interfere in exchanges between the circulating blood and neurons. Glial cells are major neuronal progenitors, as they self-renew and give rise to neurons (Sect. 1.1.5.5). They also guide radial migration of postmitotic neurons.

The nervous control of blood circulation and ventilation operates via: (1) *cholinergic parasympathetic neurons* that innervate the body's organs and (2) 3 main classes of *sympathetic efferents*, baro-, thermo-, and glucosensitive that innervate heart, blood vessels, airways, kidneys, and adrenal medulla.[5]

The *sympathetic preganglionic neurons* receive synaptic contributions from the spinal cord, medulla oblongata, and hypothalamus. Barosensitive sympathetic efferents are mainly fed by the rostral ventrolateral medulla. The main inputs for the cutaneous circulation come from the rostral ventromedial medulla and the medullary raphe. The rostral and the caudal ventrolateral medullas extends to the ventral respiratory column. The caudal ventrolateral medulla strongly and negatively influences the rostral ventrolateral medulla. In addition, the adrenaline secretion is partially controlled by the rostral ventrolateral medulla.

A feedback loop which includes atrial voloreceptors, the nucleus of the solitary tract, the paraventricular nucleus of the hypothalamus, and the renal sympathetic nerves, regulates sodium reabsorption by the kidney. Peripheral and central osmoreceptors control the renal sympathetic activity, together with the arterial baroreceptors, which mostly operate via the rostral ventrolateral medulla.

Temperature sensations are transmitted from thermoreceptors to the central thermoregulatory command center, i.e., the preoptic area. The thermosensory pathway that triggers responses to elevated environmental temperature

---

[5] Thermosensitive efferents consist of cutaneous vasoconstrictors. They are activated by hypothermia, hyperventilation, and emotions. Glucosensitive efferents control adrenaline release from adrenal medulla. They are activated by hypoglycemia and exercise. Barosensitive efferents, the largest class, have a rest activity. They discharge in bursts synchronized with the the cardiac and respiratory frequency. The barosensitive efferents are responsible for the short-term control of the blood pressure. The barosensitive efferents are subjected to various stimuli, lung stretch-sensitive afferents activated by the ventilation, activated muscle receptors during exercise, activated visceral or cutaneous nociceptors, activated peripheral and central chemoreceptors. The chemoreflex corresponds to the activation of peripheral chemoreceptors by hypoxia and hypercapnia, which increases the activity of barosensitive sympathetic efferents.

involves glutamatergic neurons in the dorsal region of the lateral parabrachial nucleus [6]. These neurons transmit cutaneous warm signals from spinal somatosensory neurons directly to the preoptic area. Efferent nerves then provoke cutaneous vasodilation. Another adjacent population of glutamatergic neurons in the external lateral part of the parabrachial nucleus conveys cutaneous cool signals from somatosensory neurons in the spinal dorsal horn to a midline zone of the preoptic area.

Astrocytes, electrically non-excitable cells, yield structural and metabolic support to neighboring neuronal networks (Sect. 1.1.6).[6] They link neurons to the vasculature, thereby regulating the cerebrovascular tone and matching blood supply to local metabolic demands. In the brainstem, chemosensitive astrocytes contribute to the control of the body's respiration. They indeed sense, monitor, and integrate CO_2 and H^+ concentrations in the blood and brain parenchyma and relay information to chemosensitive neurons of the brainstem [7]. Fall in extracellular pH provokes an immediate and sustained increase in intracellular calcium in astrocytes near the ventral surface of the medulla oblongata that contact the adjacent pia mater and locally penetrating arterioles.[7] This calcium response to acidification primes a release of adenosine triphosphate. This signal is transmitted to adjoining, chemoresponsive neurons of the retrotrapezoid nucleus. Increased neuronal activity raises discharge of the phrenic nerve that innervates the diaphragm to adjust lung ventilation. Therefore, chemoresponsive neurons and astrocytes cooperate to fulfill the central chemosensory function.

1.1.3 Chemosensors

Chemosensors or chemoreceptors transduce chemical signals into action potentials. Chemoreceptors affect the breathing rate via: (1) central chemoreceptors, i.e., populations of highly specialized neurons, located in the ventrolateral surface of medulla oblongata such as chemoresponsive neurons of the retrotrapezoid nucleus that mainly detect changes in pH of cerebrospinal fluid, but also hypercapnia; and (2) peripheral aortic and carotid chemoreceptors (aortic and carotid bodies) that sense changes in blood concentration of oxygen and carbon dioxide, but not of hydrogen ion (pH), and all of these 3 concentrations, respectively. Therefore, hypoxemia (low blood O_2 concentration) is a specific stimulus to arterial chemoreceptors.

Arterial chemoreflex generally relates to that of the carotid body, which has been more studied. Glomus cells in the carotid body depolarize in response to hypoxemia and release multiple neurotransmitters that activate impulses in afferent fibers going to the medulla oblongata via the carotid sinus

[6] A single astrocyte may enwrap several neuronal somas (neuron body that contains the cell nucleus; Sect. 1.1.5) and contact thousands of synapses, thereby regulating information processing [7].

[7] Astrocytes can contact surface pial arteries that rest on glia limitans as well as penetrating arterioles and capillaries that are enwrapped by astrocytic endfeet [7].

nerve. Chemosensor afferents project in the pons and medulla to the primary sites for respiratory and cardiovascular signal receptors as well as sympathetic integration. Nevertheless, the hypothalamus modulates the ventilatory and cardiovascular responses to peripheral chemoreflex activation, especially the paraventricular nucleus of the hypothalamus (PVN), where ionotropic γ-aminobutyric acid receptor-A (Cl^- channel $GABA_A$) impedes chemoreflex that originates from the carotid body. The paraventricular nucleus of the hypothalamus sends efferent neurons to the rostral ventrolateral medulla (RVLM) that relays signals for sympathetic activity.

The inspiratory center in the medulla sends nervous impulses to respiratory muscles via the phrenic nerve to increase breathing rate and lung volume during inhalation. Chemoreceptor-mediated elevation in sympathetic nerve activity is progressively dampened when tidal volume increases because of inhibitory feedback from pulmonary volume receptors.

Central and peripheral chemoreflexes not only control alveolar ventilation to improve O_2 uptake and CO_2 removal, but also blood flow, as gas exchanges require matched ventilation and lung perfusion. A nervous impulse is sent to the cardiovascular centers in the medulla, which then yield feedback to the sympathetic ganglia, increasing the action potential frequency of sinoatrial node and sympathetic output to the vasculature. Arterial chemoreceptors activated by hypoxemia increase the sympathetic activity to systemic vasculature to compensate for vasodilation caused by hypoxemia on blood vessels and to redistribute blood flow to essential organs. Sympathetic-mediated vasoconstriction exerts particularly on muscular, splanchnic, and renal vascular beds. Moreover, sympathetic activity in adrenal glands augments noradrenaline release, whereas sympathetic activity to brown adipose tissue decays to reduce body temperature and oxygen consumption during hypoxemia [8]. Cardiac sympathetic activity elevates heart frequency and contractility. However, parasympathetic and sympathetic nerves that innervate heart and cerebral vessels are concomitantly activated to limit sympathetic vasoconstriction of coronary and cerebral vessels during hypoxemia and to modulate cardiac chronotropic and inotropic responses. The sympathetic response also depends on ventilation frequency. The higher the breathing rate, the stronger the chemoreflex responsiveness.

Parasympathetic and sympathetic nerve activity rapidly increase upon exposure to hypoxemia. When hypoxemia is sustained, parasympathetic and sympathetic nerve activity rises and is maintained, whereas the ventilatory rate increases transiently and then progressively declines [8]. When hypoxemia disappears, parasympathetic and sympathetic nerve activity gradually return to normal level, being maintained during a given period after blood gases and ventilation return to standard state.

1.1.3.1 Carotid Body

The carotid body is a sensory chemoreceptor organ located in the vicinity of the carotid bifurcation. It is composed of cell clusters formed by chemoreceptors and support cells. Sensory nerve endings of the carotid sinus nerve penetrate these clusters and synapse with chemoreceptor cells. The carotid body operates in the regulation of ventilation during hypoxia (lowered O_2 concentration), hypercapnia (elevated CO_2 concentration), and acidosis (augmented H^+ concentration or decayed pH).

However, in normal humans at rest, hyperoxia that inhibits peripheral chemoreceptor activity decreases sympathetic activity. A sympathetic tone may thus exist even in the absence of hypoxia [9]. Upon detection, chemical stimuli (hypoxia, hypercapnia, and acidosis on glomus cells of the carotid body) provoke an increase in release rate of neurotransmitters by chemoreceptors that heighten action potential frequency in the carotid sinus nerve.

Multiple neurohumoral factors, such as catecholamines, angiotensin-2, nitric oxide, endothelin, and natriuretic peptides, influence chemoreceptor activity. Glomus cells express oxygen-sensitive, Ca^{++}-insensitive, outward voltage-gated K^+ channels ($K_{V_{O_2}}$) constituted by $K_V4.1$ and -4.3 subunits. At least in rabbits, these cells also produce a $K_V3.4$ channel subunit that is insensitive to oxygen. Once $K_{V_{O_2}}$ channels are repressed by hypoxia, attenuated K^+ current contributes to cell membrane depolarization at rest. This initial depolarization subsequently activates voltage-gated Ca^{++} channels. Calcium influx allows release of neurotransmitters, such as acetylcholine, adenosine triphosphate, dopamine, serotonin, and substance-P that are produced by chemoreceptor cells to increase discharge in the carotid sinus nerve.

Nitric oxide is synthesized by constitutive nitric oxide synthases NOS1 and NOS3 in nerve fibers and vascular endothelium in the carotid body. Its synthesis requires oxygen. Nitric oxide impedes chemosensor discharge in normoxia by activating calcium-dependent, voltage-gated, O_2-sensitive K^+ channels (K_{Ca}) via a cGMP pathway in glomus cells in rabbits [8].[8] Potassium flux through K_{Ca} channels prevents glomus cell membrane depolarization at rest. Nitric oxide thus indirectly precludes Ca_V1 channel activity in rabbit glomus cells, in addition to direct inhibition via a cGMP-independent mechanism. On the other hand, angiotensin-2 activates Ca_V1 channels in vascular smooth muscle and possibly in glomus cells.

Inwardly rectifying K^+ current carried by $K_V11.1$ channel in rabbit glomus cells can also influence the resting membrane potential [8].

Nitric oxide in the paraventricular nucleus of the hypothalamus suppresses sympathetic output. In the paraventricular nucleus of hypothalamus, NOS

[8] Potassium channels participate in the response to hypoxia. Regulation of K^+ channels by O_2 has been observed in rabbit type-1 carotid body cells, pulmonary vascular smooth muscle cells, airway neuroepithelial body cells, and neurons. Oxygen regulates K^+ conductance of delayed rectifier, ATP-dependent, and Ca^{++}-activated (BK) K^+ channels.

inhibitor increases sympathetic and phrenic nerve responses to stimulation of the carotid body [9].

Endothelin-1 that is synthesized in the carotid body excites this chemosensor organ. Endothelin receptor antagonist lowers elevated carotid body discharge induced by chronic intermittent hypoxia in hypoxic as well as normoxic conditions [8].

1.1.3.2 Chemoreflex in Animal Models of Hypertension and Heart Failure

This section introduces signaling cascades (Vol. 3 – Chap. 1. Signal Transduction) that start by stimulation or inhibition of the activity of receptors and ion channels, as well as other mediators located in the plasma membrane that limits the cell (Vol. 1 – Chap. 7. Plasma Membrane).

Arterial chemoreflex contributes to sympathetic hyperactivity in hypertension and heart failure (Vol. 6 – Chaps. 6. Heart Pathologies and 7. Vascular Diseases). Central chemoreflex heightens in heart failure. Sensitivity of the arterial chemoreflex is enhanced in some types of neurogenic hypertension and in heart failure, thus augmenting sympathetic output to resistance vessels and kidney. Yet, the ventilatory response is not altered in many patients with heart failure, but the ventilatory response to hypoxia can be enhanced in advanced stages of heart failure [8].

Angiotensin-2 concentration rises during heart failure. Angiotensin-2 increases chemoreceptor activity via its AT_1 receptor on glomus cells. Augmentation of angiotensin-2 level and concomitant upregulation of AT_1 receptor in the carotid body contribute to exacerbated chemoreflex and sympathetic nerve activation in heart failure. Angiotensin-2 enhances hypoxia-induced renal sympathetic nerve activity (considered as an index of sympathetic output) in normal rabbits, but not in those with tachycardia-induced heart failure in which concentration of angiotensin-2 is already elevated and thus has no further effect on O_2 sensitivity of K_V channels [10]. Conversely, AT_1 receptor antagonist reduces hypoxia-induced renal sympathetic activity in rabbits with heart failure, but not normal rabbits, as the number of AT_1 receptors in the carotid body of rabbits with heart failure is greater than that in normal rabbits, which do not have angiotensin-2-induced heightened chemoreceptor sensitivity.

Angiotensin-2 enhances the sensitivity to oxygen of outward voltage-gated K^+ channels in glomus cells via its AT_1 receptor. Antagonists of AT_1 are able to reverse augmented hypoxic sensitivity, i.e., elevated hypoxia-mediated inhibition of $K_{V_{O_2}}$ channels ($K_V4.1$ and -4.3). Oxygen-insensitive $K_V3.4$ expression, but not that of $K_V4.3$, is attenuated by hypoxia as well as in heart failure [9]. Chemoreceptor sensitivity depends on reduced nicotinamide adenine dinucleotide phosphate oxidase (NADPH oxidase) that can interact with angiotensin-2 receptor and reactive oxygen species (Sect. 4.2.2). Receptor AT_1

actually promotes activation of NADPH oxidase, hence superoxide anion production that enhances $K_{V_{O_2}}$ channel sensitivity to hypoxia. Oxidative stress can then mediate the excitatory effect of angiotensin-2 in the carotid body. Yet, AT_1 blockade does not alter $K_{V_{O_2}}$ channel function in glomus cells in normoxia in rabbits with heart failure. Reduced nitric oxide production in the carotid body participates in enhanced basal activity in glomus cells in rabbits with heart failure under normoxic conditions, as K_{Ca} channel is markedly repressed and attenuated K^+ current contributes to cell membrane depolarization [8]. Concomitant elevation in angiotensin-2 level and reduction in nitric oxide concentration in the carotid body augment arterial chemoreceptor sensitivity.

Angiotensin-2 influences activity of both central and peripheral chemoreceptors. Angiotensin-2 targets the *nucleus tractus solitarius* (NTS) in the medulla (Fig. 1.1). In rats, activation of sympathetic nerve activity by carotid body chemoreceptors is enhanced by simultaneous excitation of cardiac sympathetic nerves [8]. Receptor AT_1 antagonist in the nucleus tractus solitarius abolishes this facilitatory effect of cardiac sympathetic afferent stimulation. Furthermore, downregulation of nitric oxide synthase in the *paraventricular nucleus* of the hypothalamus also leads to elevated sympathetic nerve activity caused by chemoreceptor activation.

Essential hypertensive patients and spontaneously hypertensive rats have a hypertrophied carotid body. They display hyperventilation under resting conditions and enhanced respiratory and sympathetic activity under hypoxia. In addition, patients with sleep apnea can develop hypertension with elevated levels at rest of muscular sympathetic activity that persists in periods without apnea [8]. Patients with sleep apnea have an enhanced peripheral chemoreflex response to hypoxemia, whereas central chemoreflex remains normal. Chronic intermittent hypoxia that causes hypertension generates carotid body chemoreceptor discharge in normoxia and hypoxia and persistent sympathetic activation, but not chronic intermittent hypercapnia. The underlying mechanism relies on superoxide anion production in the carotid body due to alterations in both mitochondrial and plasmalemmal redox systems. Upregulation of both endothelin-1 and superoxide anion in the carotid body in response to chronic intermittent hypoxia is related to activation of transcription factors hypoxia-inducible factor-1 and Activator protein-1.

Endothelin-1 induces vasoconstriction, hence reducing blood flow and O_2 delivery in the carotid body. Sympathetic innervation of the carotid body and circulating catecholamines can also excite glomus cells via vasoconstriction triggered by α1-adrenergic receptors. On the other hand, α2-adrenergic receptors in the carotid body inhibit noradrenaline release and chemoreceptor discharge. Increased and decreased levels of angiotensin-2 and nitric oxide, respectively, can also play a role on carotid body functioning in hypertension.

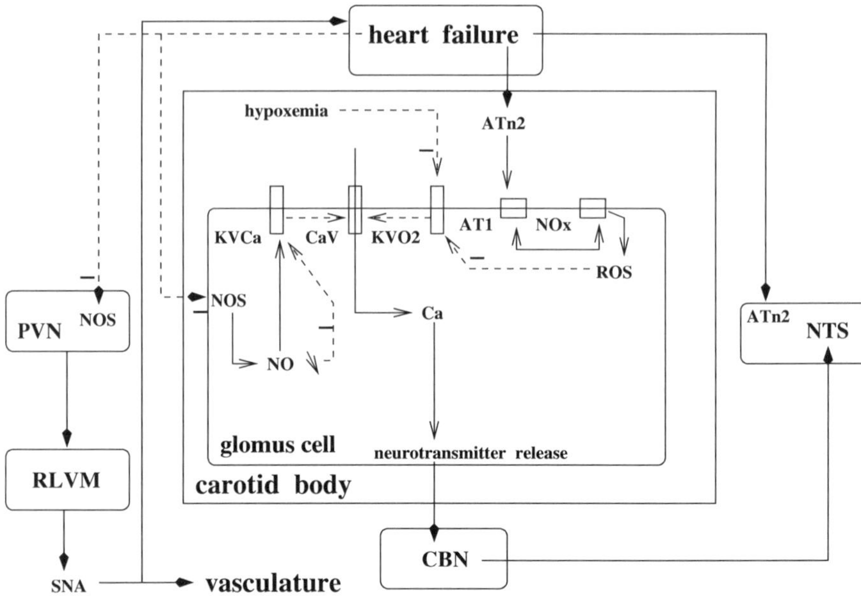

Figure 1.1. Heart failure and chemoreceptor activity (Source: [8]). Decreased activity of nitric oxide synthase (NOS)–NO axis and concomitant increased signaling by the angiotensin-2 (ATn2)–AT_1 receptor–reactive oxygen species (ROS) pathway in glomus cells of the carotid body enhance the sensitivity of K^+ channels to hypoxic inhibition, hence allowing initial membrane depolarization that activates Ca_V1 channels. Calcium influx allows release of neurotransmitters to increase discharge in the carotid body nerve (CBN). Chemosensor afferents project to nucleus tractus solitarius (NTS) in the medulla. The paraventricular nucleus (PVN) of the hypothalamus that modulates the ventilatory and cardiovascular responses to peripheral chemoreflex activation sends efferent neurons to the rostral ventrolateral medulla (RVLM) that relays signals for sympathetic nerve activity (SNA). Messengers NO and ATn2 target calcium-dependent voltage-gated K^+ channels (KVCa) and oxygen-sensitive, Ca^{++}-insensitive, outward voltage-gated K^+ channels (KVO2), respectively. Overexpression of Atn2 and NO underexpression hamper $K_{V_{O_2}}$ and K_{Ca} channel function, respectively. Reactive oxygen species are produced by reduced nicotinamide adenine dinucleotide phosphate oxidase (NADPH oxidase). Both Atn2 and NO extends their activity in the central nervous system. Increased ATn2 level in NTS in the medulla and decreased NO level in PVN of the hypothalamus further enhance activation in the carotid body of SNA in heart failure.

1.1.3.3 Interactions between Chemoreceptors and Baroreceptors

Both chemoreceptors and baroreceptors that respond to mechanical stimuli control sympathetic nerve activity. Peripheral chemoreceptors strongly interact with baroreceptors. Their respective afferent inputs to the central nervous system exert a mutual inhibition on sympathetic output. Baroreceptors suppress chemoreflex activation of sympathetic signal and, conversely, chemoreceptors preclude baroreflex inhibition of sympathetic output.

1.1.4 Baroreceptors

Baroreceptors, or baroceptors, detect blood pressure via wall circumferential stretch. Baroreceptors are classified into 2 categories: high-pressure arterial and low-pressure cardiopulmonary baroreceptors. Arterial baroreceptors are located in the aortic arch and carotid sinuses. Signals from the carotid and aortic baroreceptors are sent via the glossopharyngeal nerve (cranial nerve IX) and vagus nerve (cranial nerve X), respectively. When blood pressure falls (wall stretch decays), the baroreceptor firing rate decreases and, conversely, when blood pressure rises, the firing rate increases. Below a threshold of about 10 kPa, baroreceptors stop firing signals. However, chemoreceptor activity becomes stronger, especially below 8 kPa. In addition, baroreceptors quickly respond only to short-term changes and reset to a new reference in the case of permanent abnormal blood pressure. Cardiopulmonary baroreceptors are situated in veins and heart atria. They regulate blood volume. Low-pressure baroreceptors modify the secretion rate of hormones that influence the hydroelectrolytic balance.

Barosensitive C1 neurons not only correspond to the sympathetic preganglionic neurons of the cardiovascular system, with the exception of cutaneous arterioles, but also send axons to the midbrain, pons, and medulla [5]. Other groups of C1 neurons control adrenaline release from chromaffin cells. Certain C1 neurons innervate the hypothalamus. The activity of barosensitive C1 neurons also depends on ionotropic and metabotropic inputs. Barosensitive neurons are indeed regulated by several transmitters, including γ-aminobutyric acid, glutamate, acetylcholine, serotonin, corticotropin-releasing factor, oxytocin, substance-P, vasopressin, and orexin. Certain inputs can correspond to the circadian rhythm.

1.1.5 Neurons

Neurons or nerve cells are electrochemically excitable cells that generate, receive, process, and transmit information. Afferent neurons convey information from tissues to the central nervous system. Efferent neurons transmit signals from the central nervous system to the effectors. Interneurons connect neurons. Excitatory and inhibitory neurons excite and inhibit their targets, respectively. Modulatory neurons are responsible for neuromodulation.

1.1.5.1 Neuron Structure

Neurons are composed of a *soma* (or cell body with nucleus; typical size 10–25 μm), *dendrites* (thin cellular extensions with many branches close to

Table 1.5. Examples of neurotransmitters and their activity on blood circulation and respiration (GABA: γ-aminobutyric acid; NO: nitric oxide).

	Main functions
Acetylcholine	Transmission of nervous cues Decrease in heart frequency Release of NO and vasodilation Contraction of skeletal muscles Bronchoconstriction, increase in bronchial secretion
Adenosine	Vasodilation
Dopamine	Regulation of muscle tone Control of venous return
γ-Aminobutyrate acid	Inhibitory neurotransmitter
Glutamic acid (Glutamate)	Stimulatory neurotransmitter Conversion into GABA Precursor of proline, ornithine, arginine, and polyamines
Glycine	Inhibitory neurotransmitter Synthesis of many compounds
Neuropeptide-Y	Regulation of energy Augmentation of noradrenaline-induced vasoconstriction
Noradrenaline	Smooth muscle contraction Stimulation of cardiac activity
Serotonin	Vasoconstriction Temperature regulation Sensory perception

upstream axon terminals), and a single *axon* (long cellular projection with its neuronal plasma membrane, the so-called axolemma; thickness $\sim$1 µm). The majority of neurons receive input on the soma and dendrites and transmit output via axon. Neurons communicate via *synapses* (synaptic transmission) that allow the propagation of electrochemical signals (depolarization wave). Synapses contain neurotransmitters (Table 1.5).

Two types of synapses comprise *dendrite spines* (input signals) and *axon terminal buttons* (or knobs; output signals). Dendrites indeed form protrusions associated with upstream axon terminals, the so-called dendritic spines. Each spine typically receives input from a single synapse of a presynaptic axon. Most spines have a bulbous spine head and a thin neck that connects the spine head to the dendrite shaft.

The surface of the soma contains voltage-gated ion channels. These channels contribute to signal transmission from dendrites. The soma is connected to the axon by the *axon hillock*. Axon hillock possesses a high density of voltage-gated sodium channels, as it serves to initiate spike (action potential). This transition region between soma and axon integrates inhibitory and excitatory postsynaptic potentials from numerous inputs. Afterward, action potential propagates down the axon to reach synaptic knobs, where it triggers neurotransmitter release to excite the postsynaptic neuron. The *axon terminal region* branches into several axon terminals. These axon terminals then end with axon terminal buttons that have voltage-gated calcium channels.

Many neuron axons have *insulating sheaths of myelin* formed by glial cells, such as oligodendrocytes and Schwann cells in the central and peripheral nervous system, respectively. In peripheral nerves, the insulating myelin sheath allows nervous impulses to skip down the axon from node to node, thereby enhancing electrical conductivity. Schwann cells wrap themselves multiple times around an axonal segment to prevent ion entry into and exit from the axon, hence signal alteration, especially decay, as well as to ensure fast signal transmission.[9] *Nodes of Ranvier* are regularly spaced patches of non-insulated axon. These nodes of Ranvier have ion channels to boost signals.

In myelinated fibers of the central and peripheral nervous systems, glial-ensheathing cells interact with axons at specialized adhesive junctions, the *paranodal septate-like junctions* on each sides of nodes of Ranvier. Paranodal junctions hence separate the nodes of Ranvier from internodal regions. Axonal transmembrane glycoprotein paranodin[10] and contactin form a complex in the axolemma to stabilize the axoglial adhesion zone. Neurofascin is a neural cell adhesion molecule of the L1 family of the immunoglobulin superclass that serves as a glial receptor for the paranodin–contactin axonal complex at the axoglial junction. Oligodendroglial neurofascin isoform 155 kDa (NF155) binds myelinating glial cells specifically to the axolemmal paranodin–contactin complex [12]. Paranodin (CAsPr)-related substance contactin-associated protein-like molecule CAsPr2 (or CntnAP2) is concentrated in juxtaparanodal regions, where it links indirectly to delayed rectifier potassium channels $K_V1.2$ (Vol. 3 – Chap. 3. Main Classes of Ion Channels and Pumps).[11] Both CAsPr1 and CAsPr2 connect to the cytoskeleton via erythrocyte membrane protein band

[9] Axons signal using neuregulin to support Schwann cells in their formation of the myelin sheath that wraps around axons for their insulation over an axonal lenght of about 100 µm. Both under- and overmyelination must be avoided. Rounds of myelin insulation stops when the adequate amount for optimal nerve conductivity is reached due interaction of adaptor Disc large DLg1 with phosphatase and tensin homolog deleted on chromosome 10 (PTen) [11].

[10] A.k.a. contactin-associated protein CAsPr. Paranodin is a member of the NCP group (NCP: neurexin-4–CAsPr–paranodin) of the neurexin family.

[11] Both $K_V1.1$ and $K_V1.2$ are clustered in juxtaparanodes.

EPB4.1[12] and protein band-4.1-like-3[13] (EPB4.1L3) [13].[14] The latter is enriched in paranodal and juxtaparanodal regions. It associates with CAsPr1 in paranodes and CAsPr2 in juxtaparanodes.[15]

Amyloid precursor protein (APP) is implicated in interneuronal adhesion. It binds to different extracellular matrix proteins, such as heparin and collagen. Amyloid precursor protein and amyloid precursor-like proteins APLP1 and APLP2 share homology domains. They are processed by the same peptidase.

1.1.5.2 Neurotransmission – Synapses

Physiological systems (nervous, endocrine, cardiovascular, respiratory, and immune systems) interact via various substances that are synthesized in these systems, such as aminergic neurotransmitters, peptides, steroid hormones, prostaglandins, cytokines, and growth factors. Physiological systems express receptors for these substances. Neuromodulators impinge the release of neurotransmitters (acetylcholine, catecholamines, indolamines, excitatory amino acids such as glutamate, and inhibitory amino acids such as γ-aminobutyric acid) or neuropeptides. Chemokines and cytokines (interleukins, tumor-necrosis factors, and interferons) that are constitutively expressed in glial cells and neurons serve as neuromodulators.

Axonal growth relies on preprogrammed data, guidance molecules, and guidecells, similarly to guidebooks, travelling paths, and guideposts. Guidance molecules include netrins, Slits, and sonic Hedgehog (Vol. 3 – Chap. 10. Morphogen Receptors). Supporting cells that assist in axon growth and synapse localization and genesis use ephrins, netrins, neuregulins, and Wnt morphogens, among others, as well as a set of chemoattractant and -repellent receptors and adhesion molecules [15].

Spikes and Spikelets

An *action potential* is a short-lasting change (rapid rise and fall) in electrical membrane potential. Action potentials in neurons are also termed *impulses* and *spikes*. *Spikelets*, or fast prepotentials, are fast voltage fluctuations of small amplitude with a spike-like waveform that differ with excitatory postsynaptic potentials. They were initially described in the hippocampus. In fact, they

[12] A.k.a. protein band-4.1 Rh-linked (4.1R).

[13] A.k.a. protein band-4.1B.

[14] The intracellular region of paranodin possesses a short GNP motif (GNP: glycophorin-C, neurexin-4, and paranodin) that anchors the cortical cytoskeleton to the plasma membrane via protein-4.1R and -4.1B.

[15] CAsPr2 Protein as well as $K_V1.1$ and $K_V1.2$ channels require glycosylphosphatidylinositol-anchored cell adhesion molecule of the immunoglobulin superclass Transient axonal glycoprotein TAG1 expressed in both neurons and glial cells to accumulate at juxtaparanodes [14].

Table 1.6. Trans-synaptic regulators (DLg: Disc large homolog; GluR, GluN: ionotropic glutamate receptor; LRRTM: leucine-rich repeat-containing transmembrane protein; Nlgn: neuroligin; Nrxn: neurexin).

Type	Isotypes	Location	Partners
Neurexin	Nrxn1–Nrxn3	Presynaptic	Nlgn, LRRTM
Neuroligin	Nlgn1–Nlgn3	Postsynaptic	Nrxn
	Nlgn4X–Nlgn4Y		
LRRTM	LRRTM1–LRRTM4	Postsynaptic	Nrxn, DLg4,
			GluR (AMPA), GluN (NMDA)

are observed in various neuron types. They can result from action potential transmission between principal neurons and interneurons or between axons of principal neurons. Like action potentials, spikelets are all-or-none signals, but with different kinetics and amplitude. They occur in isolation or burst. They support about 30% of all action potentials. They can synchronize neuron activity during spontaneous high-frequency oscillations.[16]

Synaptic and Trans-Synaptic Regulators

Neurons communicate by synapses. Many trans-synaptic regulators contribute to synapse formation between pre- and postsynaptic cells (Table 1.6).

Neurexins, Neuroligins, and LRRTMs

Neurexins are presynaptic type-1 membrane proteins that assist in neuronal connection at the synapse, specify the synaptic function, and mediate signaling across the synapse. Neurexins function as receptors for trans-synaptic neuroligin-1, a type-1 membrane protein of the postsynaptic membrane that mediates synapse formation between neurons and also specifies the synaptic function.[17] Neurexin partners include not only neuroligins, but also Leu-rich repeat-containing transmembrane proteins (LRRTM1–LRRTM4) that are postsynaptic contributors of the formation of excitatory synapses. Protein LRRTM2 interacts with both ionotropic AMPA-type and NMDA-type glutamate receptors and scaffold Disc large homolog DLg4 [17,18].[18] Binding of neurexins to LRRTM2 can produce cell-adhesion junctions and promotes presynaptic differentiation.

[16] In addition, spikelets intervene in hippocampal activity during spatial exploration [16]. Localization in the environment modulates spikelet firing in cortical neuron networks to possibly contribute to information processing.

[17] In humans, the neurexin family encompasseses Nrxn1 to Nrxn3; the neuroligin family Nlgn1 to Nlgn3, and X- (Nlgn4X) and Y linked (Nlgn4Y) neuroligins-4.

[18] Neuroligins bind neurexins that contain or lack an insert in splice site 4, but LRRTM2 only binds neurexins that lack an insert in splice site 4 [17]. Protein LRRTM2 links to both neurexin-1α and -1β.

Scaffold Proteins

Scaffold proteins regulate the clustering of plasmalemmal receptors, organize synaptic signaling complexes, participate in the transfer of receptors and ion channels, and coordinate cytoskeletal dynamics. The postsynaptic density (PSD) of excitatory synapses at the cortex and plasma membrane of dendritic spines is a molecular complex composed of scaffold proteins, glutamate receptors (GluNs and GluRs), K^+ channels, signaling effectors and their regulators, cell adhesion molecules, as well as components of the cytoskeleton, their associated nanomotors and regulators in some synapses of brain neurons.[19] This complex can be decomposed into several modules.[20]

Many of the postsynaptic density proteins contain PDZ (postsynaptic density protein [PSD], disc large homolog [DLg], and zonula occludens protein [ZO]) domains. PDZ Domain-containing scaffold proteins include DLg4 (or PSD95), glutamate receptor-interacting proteins GRIP1 and GRIP2, amyloid-β precursor-binding protein APBa1, and partitioning defective homolog Par3 [19]. Members of the Src Homology domain SH3 and multiple ankyrin repeat-containing (SHANK) family form a 2-layer protein network below the postsynaptic membrane that is bridged by guanylate kinase-associated protein (GKAP). Scaffold DLg4 tethers to the plasma membrane and forms filamentous structures roughly equally spaced and perpendicular to the membrane.

MicroRNAs

MicroRNAs (Vol. 1 – Chap. 5. Protein Synthesis) participate in synapto- and dendritogenesis as well as in the control of synapse function and adaptivity to sensory experience in adults (Table 1.7). A subset of neuronal miRs, such as miR134 and miR138, regulate mRNA translation in the synaptodendritic compartment [20]. Conversely, the neural activity controls miR transcription and processing, subcellular localization, as well as functioning of microRNA-associated proteic complexes. Both miR124 and miR132 exert growth-promoting effect by regulating the activity of small Rho GTPases, that are regulators of the dendritic actin cytoskeleton [20].

[19] Signaling effectors comprise kinases, such as calcium–calmodulin-dependent protein kinase CamK2 and SRC family Tyr kinases (e.g., Src and Fyn; Vol. 4 – Chap. 3. Cytosolic Protein Tyr Kinases), phosphatases, nitric oxide synthase, A-kinase-anchoring proteins, microtubule-associated protein MAP1a, members of the Rho guanine nucleotide-exchange factor (GEF) family (e.g., RhoGEF11 and RhoGEF25), and synaptic Ras GTPase-activating protein SynGAP1 [19].

[20] These modules encompass DLg4, MAGUK(membrane-associated guanylate kinase), GRIP, syntenin–PAR3, and DH–PH (Dbl homology, pleckstrin homology; Rho regulatory) [19].

Table 1.7. MicroRNAs at neuronal synapses (Sources: [20–22]; ActL6a: actin-like protein-6A [a.k.a. 53-kDa BRG1-associated factor BAF53a and actin-related protein ARP4 and ARPnβ]; BDNF: brain-derived neurotrophic factor; CamK: calmodulin-dependent kinase; GluR: glutamate receptor; Itg: integrin; LIMK: Lin-1, Isl-1, and Mec-3 kinase [supports actin polymerization and spine growth via phosphorylation of actin-depolymerizing factor cofilin]; LyPLa1: lysophospholipase-1 [a.k.a. acyl-protein thioesterase APT1; catalyzes depalmitoylation of proteins]; MeCP2: methyl CpG-binding domain [MBD]-containing protein-2 [binds to methylated DNA and represses gene transcription]; MEF: myocyte enhancer factor; NAcc: nucleus accumbens-associated protein-1, BEN and BTB (POZ) domain-containing protein (BTBD14b); nAChR: nicotinic acetylcholine receptor; Nrp: neuropilin; PHLPP1: PH domain and Leu-rich repeat-containing protein phosphatase-1 [a.k.a. suprachiasmatic nucleus circadian oscillatory protein (SCOP)]; Pum: Pumilio homolog [accessory RNA-binding protein]; RFX4: regulatory factor-X4). MicroRNA124 is highly expressed in neurons, where it represses the translation of BDNF and Nrp2; CREB, Sox9, and REST (RE1-silencing) transcription factors; axon guidance molecules Sema6a and Sema6c; as well as NAcc1, Itg β_1, MAP2K4, urokinase, and dopamine D_3 and μ opioid receptors.

MicroRNA	Target
	Dendritic growth and arborization
miR124	ActL6a
miR132	RhoGAP32
miR134	Pum2
	Synapse formation and dendritic spine maturation
miR134	LIMK1
miR138	LyPLa1
	Synapse signaling and remodeling
miR1	nAChR subunits, MEF2
miR284	GluR subunits
	Circadian rhythm and alcohol tolerance
miR9	Ca^{++}-activated K^+ channel BK
miR132	RFX4
miR219	PHLPP1
	Neuronal function
miR132	MeCP2
miR219	CamK2γ

Metallic Ions

In synaptic terminals, the electrochemical wave, the so-called *action potential*, activates voltage-gated calcium channels. These channels are close to synaptic vesicles for short time scale release. The subsequent entry of extracellular calcium triggers the fusion of neurotransmitter filled synaptic vesicles with the plasma membrane of the neuron terminal. The fast calcium sensor synaptotagmin-2 favors bursts of neurotransmitter release synchronous with low-frequency action-potential firing.

Zinc ion abounds in many (but not all) glutamatergic nerve terminals. The cytosolic Zn^{++} concentration is in the picomolar range, but can rise to micromolar levels in the proximity of axon terminals [23]. Zinc ion enters and exits out of neurons using many carriers (Table 1.8; Vol. 3 – Chap. 4. Membrane Compound Carriers). It participates in the control of synaptic excitability. It modulates both glutamatergic and gabaergic neurotransmission [23].[21] Many proteins contribute to homeostasis of neurotoxic Zn^{++},[22] such as Zn^{++} transporters, Zn^{++}-importing proteins, and buffers (metallothioneins), as well as cell organelles (mitochondria, vesicles, and lysosomes). The production of Zn^{++} transporters that act as H^{+}–Zn^{++} exchangers in the cytoplasmic and Golgi compartments is coupled to changes in cytosolic Zn^{++} concentration.

Magnesium ion intervenes in the function of polyphosphate compounds, such as nucleotide ATP and nucleic acids DNA and RNA. This metallic ion is a cofactor for numerous enzymes. The regulation of the intracellular free Mg^{++} concentration by neurotransmitters such as glutamate contributes to the control of neuronal excitability. Intracellular free Mg^{++} concentration rises from its resting concentration of 0.5 mmol to beyond 1.0 mmol upon stimulation of ^{N}methyl ^{D}aspartate (NMDA)-type ionotropic glutamate receptors (GluN) by glutamate [24].[23] Magnesium *muffling*, i.e., the sum of mechanisms that damp changes in the intracellular free Mg^{++} concentration (buffering, sequestration, and extrusion) is mainly due to Na^{+}–Mg^{++} antiporter [25]. Adenosine triphosphate is a major intracellular Mg^{++} buffer. Calcium-binding protein CaBP1, a neuron-specific regulator of calcium channels, switches between a Mg^{++}-bound to Ca^{++}-bound state in response to Ca^{++} signaling to modulate the activity of Ca^{++} targets [26].

Manganese ions are cofactors of numerous enzymes (oxidoreductases, transferases, hydrolases, lyases, isomerases, ligases) as well as membrane proteins, such as lectins and integrins. In the human brain, Mn^{++} is bound

[21] Zinc ion influences the activity of ^{N}methyl ^{D}aspartate (NMDA) and γ-aminobutyric acid $GABA_A$ receptors. Many glutamatergic vesicles in the neocortex, amygdala, and hippocampus contain high concentrations of Zn^{++} ions.

[22] Zinc ion promotes neuron apoptosis. It supports the production and secretion of nerve growth factor, an agonist of the low-affinity nerve growth factor receptor (LNGFR), also called neurotrophin receptor $P75^{NTR}$ and tumor-necrosis factor receptor superfamily member TNFRSF16, that is associated with death executor nerve growth factor receptor-associated protein NGFRAP1. On the other hand, Zn^{++} fosters the activation of extracellular signal-regulated kinases ERK1 and ERK2 that stimulate the transcriptional regulator Early growth response factor EGR1 that regulates the expression of synaptobrevin-2 involved in synaptic exocytosis.

[23] The first component of this increase depends mainly on extracellular Ca^{++} rather than extracellular Mg^{++} ions. The second component that is independent of extracellular Ca^{++} requires extracellular Mg^{++} and is amplified by extracellular Na^{+} removal [24].

Table 1.8. Zinc carriers and buffers (Source: [23]). Zinc ion enters mainly neurons through voltage-gated Ca^{++} channels (Ca_V1) that regulate Ca^{++} and Zn^{++} influx and Ca^{++}- and Zn^{++}-permeable α-amino 3-hydroxy 5-methyl 4-isoxazole propionic acid (AMPA)-type GluR2 receptors. Zn^{++} Transporter ZnT1 controls Zn^{++} efflux. Na^+–Zn^{++} exchanger can transfer Zn^{++} in or out of neurons according to the Na^+ gradient. Zn^{++}-importing proteins (ZIP) may act as H^+– or HCO_3^-–Zn^{++} cotransporters for Zn^{++} influx. Ubiquitous Zn^{++} transporters (ZnT1–ZnT10) mediate Zn^{++} transfer from the cytosol to the lumen of intracellular organelles or out of the neuron. Except ZnT1 on the plasma membrane of neurons and glia, they lodge on organelles, such as the Golgi stack and secretory vesicles. In neurons, vesicular Zn^{++} transporters at the Golgi apparatus and other secretory vesicles act as H^+–Zn^{++} exchangers. Mitochondria sequester Zn^{++} via Ca^{++} uniporter and a Ca^{++} uniporter-independent mechanism. NMDA-Type ionotropic glutamate receptors (GluN) possess several modulatory Zn^{++}-binding sites. Zinc ion has a potentiation effect on inhibitory glycinergic synapses. ZnT3+ glutamatergic synapses can receive Zn^{++} from neighboring Zn^{++}+ excitatory synapses.

Event	Mediators
Influx	Ca_V1, AMPA-type GluR2, Na^+–Zn^{++} exchanger, H^+–Zn^{++}, HCO_3^-–Zn^{++} cotransporters (ZIP at the plasma membrane and organelles)
Outflux	ZnT1 Na^+–Zn^{++} exchanger
Sequestration	Metallothioneins Mitochondrion (Ca^{++} uniporter) Vesicle (H^+–Zn^{++} exchanger [ZnT])
Vesicles	H^+–Zn^{++} exchangers

to manganese metalloproteins, especially glutamine synthetase in astrocytes. Manganese ion can enter neurons through Ca^{++} channels and is rapidly transported along microtubules. Cobalt is a constituent of cobalamin (vitamin-B12). Both Mn^{++} and Co^{++} are Ca^{++}-channel blockers [27].

Mechanical Forces

Synaptic transmitter is stored in synaptic vesicles that cluster at the synaptic region. In addition to chemical-based signaling between neurons, vesicle clustering at the neuromuscular presynaptic terminal depends on mechanical tension generated by the actomyosin contractile filaments within axons [28]. ^{F}Actin serves as a scaffold for synaptic vesicles at the presynaptic terminal, where they can be subjected to mechanical forces.

Vesicular Filling and Storage

Vesicular storage and exocytosis of neurotransmitters is essential for chemical transmission in neurons. Many classes of transporters participate in the uptake of neurotransmitters into neuronal synaptic vesicles (Vol. 3 – Chap. 4. Membrane Compound Carriers): (1) vesicular acetylcholine transporters; (2) vesicular glutamate transporters; (3) vesicular inhibitory amino acid transporters; (4) vesicular monoamine transporters; and (5) vesicular nucleotide transporters. These vesicular transporters support the active accumulation of their respective neurotransmitters using an electrochemical gradient of protons across the membrane generated by vacuolar H^+ ATPase.

Adenosine triphosphate is a major chemical transmitter stored in secretory vesicles and subsequently secreted for nucleotide-based signaling used in glioneuronal interactions, mechanosensory transduction, and control of autonomic functions, in addition to the regulation of the vasomotor tone, angiogenesis, and platelet aggregation. The vesicular nucleotide ATP transporter (VNuT) in synaptic vesicles is the solute carrier superclass member SLC17a9 of the SLC17 phosphate transporter family [29].[24]

Vesicular acetylcholine transporter (VAChT) packages acetylcholine (ACh) into synaptic vesicles (Michaelis constant 0.6–1.0 mmol) [30]. VAChT expression quantity regulates filling of vesicles.[25] VAChT comprises 12 transmembrane segments around a central transport channel. It is located in cholinergic cell bodies, fibers, and terminals, as well as in non-neuronal cells. It limits the ACh release rate from nerve terminal. Vesicular-type H^+ ATPases (vATPases of endosomes, lysosomes, and secretory vesicles) yield vesicular protons for VAChT-mediated transfer of cytosolic ACh that leads to a 100-fold concentration gradient.

Fast Conduction by Action Potentials and Neurotransmitter Release

Information transfer is the final result of a series of cellular and molecular steps, in which an action potential increases the intracellular Ca^{++} concentration via Ca_V channels and triggers full or partial fusion of synaptic vesicles with the plasma membrane of the presynaptic cell at neuronal synapses and release of neurotransmitters.

In general, calcium-induced vesicular fusion empties the vesicular contents into the synaptic cleft, a tiny extracellular gap between tightly apposed pre-

[24] The transporter SLC17a9 of the anion transporter family is mainly synthesized in the brain and adrenal glands. It is associated with neuronal synaptic vesicles and adrenal chromaffin granules. It actively takes up ADP, ATP, and GTP by using a membrane potential as the driving force.

[25] The VACHT gene corresponds to the first intron of choline acetyltransferase gene.

and postsynaptic cells. Within a millisecond after release, secreted neurotransmitters bind to and activate receptors on the postsynaptic cell and trigger a downstream action potential.

Multiple Inputs and Outputs

A given axon can synapse with different targets (divergence from a single source) as well as possess many synapses with a single cell (convergence from a single source). These synapses are most often heterogeneous. The magnitude of a neural output depends on many factors: (1) the Ca^{++} concentration in the presynaptic terminal for a given input that depends on functioning of Ca^{++} channel and presence of Ca^{++} buffers; (2) the number and size of local synapses; and (3) the number of synaptic vesicles and amount of neurotransmitter released per synapse.

Neurons encode as well as decode and interpret information to and from various sources, respectively. Neurons work collectively and combine sources of information. Information from multiple neurons can converge on a given neuron (convergence from multiple sources). Consequently, neurons can have a coordinated activity to yield salience of an event and initiate faster responses. Neuronal interactions enhance or degrade their response according to the type of interaction (*excitatory* or *inhibitory* input) and spatiotemporal relationship among stimuli. For example, sensory information that originates from many different modalities (auditory, tactile, visual, etc.) can be integrated by a given neuron as well as neuronal network (multisensory integration) [31].

The neuron operates as a signal integrator. A neuron can combine and transform received inputs with a given firing rate and amplitude within a given time window that is converted into a suitable output. Additive and subtractive operations (linear combination of signals) carried out on driving (I_d) and distinct modulatory (I_m) synaptic inputs correspond to shifts in the input–output relationship in the direction given by the operation type (addition or subtraction) without change in shape [32]. Multiplicative and divisive operations (non-linear integration) correspond to variations in the slope of the input–output relationship (gain changes). A change in gain affects the sensitivity of a neuron to modifications in driving inputs, but not its selectivity. The maximum output firing rate is scaled accordingly. Inhibition, short-term synaptic remodeling (plasticity), synaptic noise (voltage fluctuations),[26] and changes in somatic and dendritic conductances[27] enable neurons to perform additive and multiplicative operations on their inputs.

[26] Voltage fluctuations enable synaptic inputs to cross the threshold and trigger a spike even when the mean voltage is below the spike threshold, thereby extending the range of excitations over which a neuron can signal [32].

[27] Hyperpolarization and increased somatic membrane conductance introduce subtractive shifts because additional excitation is required to reach the threshold.

Table 1.9. Modes of neurotransmitter release and their agents. Neuronal signal transmission results from squirting of stored packets of neurotransmitters into the synaptic cleft. The synaptic vesicle pool comprises both rapidly and slowly releasing vesicles that correspond to synchronous and asynchronous modes of neurotransmitter secretion.

Mode	Agents
Synchronous	Synaptotagmins
	^{N}ethylmaleimide-sensitive factor attachment receptor
Asynchronous	Double C2-like domain-containing proteins
	^{N}ethylmaleimide-sensitive factor attachment receptor

Modes of Neurotransmitter Release

Synaptic vesicle release can be (Table 1.9): (1) *synchronous*, being tightly coupled to action potentials; (2) *asynchronous*, occurring immediately after action potentials; or (3) *spontaneous*, independent of action potentials. Vesicular Ca^{++} sensors synaptotagmins (synaptotagmin-1, -2, and -9) operate in synchronous release. Cytosolic Ca^{++} sensors Double C2-like domain-containing proteins (DoC2α–DoC2β) act in spontaneous synaptic vesicle release, like synaptotagmin-1; but with a higher Ca^{++} sensitivity [33].[28]

In the interval between action potentials, i.e., in the absence of calcium influx and at low cytosolic calcium concentrations, neurons also secrete neurotransmitters via asynchronous release that is mediated by another single, slower, asynchronous calcium sensor [34]. The kinetics of calcium binding to and unbinding from calcium sensors govern the asynchronous release. In synapses with predominantly asynchronous release during high-frequency action-potential trains, asynchronous release out-competes synchronous release of neurotransmitters.

Slow Conduction

Nervous conduction is ensured by action potentials that drive fast regulation after synaptic transmission down to the final effector cell activation. However, slow regulation by the autonomic nervous system copes with membrane raft (specialized nanodomain)-mediated conduction based on ceramide that triggers an increase in intracellular calcium concentration followed by activation of neuronal nitric oxide synthase (NOS1) and then to cyclic guanosine monophosphate [35]. This sequence of second messengers is activated in cascade from raft to raft along the nerve fiber without action potential.

[28] Double C2 domain-containing protein promotes membrane fusion in response to very low Ca^{++} increases. Synaptotagmin and DoC2 bind to ^{N}ethylmaleimide-sensitive factor attachment receptor (SNARE) complexes and compete for SNARE connection during membrane fusion.

Ceramide-based conduction of excitation especially operates in digestive tract motility.

Synaptic Remodeling (Plasticity)

The ability of neurons to modify their structure and function according to experienced stimuli supports learning. Synaptic adaptivity, or remodeling (or plasticity)[29] is related to the modulation of neurotransmitter release. Some synapses transmit strongly to action potentials but weaken with repeated activation. Others transmit weakly at first but strengthen with sustained activity. Therefore, synapses are classified into *facilitating* and *depressing* types. Facilitating synapses have a low probability of initial transmitter release and exhibit facilitation of release during trains of stimuli. Depressing synapses possess a high probability of initial release and depression of release during trains of action potentials.

Neuronal activity can generate persistent forms of synaptic adaptivity to retain new information, such as long-term potentiation (LTP) and long-term depression (LTD). These 2 processes (LTP and LTD) are balanced by the action of a collection of molecules, such as acetylcholine, γ-aminobutyric acid, catecholamines, hormones, and cytokines that control the magnitude of LTP and LTD, as they can modulate the activity of ion channels and signaling kinases and phosphatases [36]. Furthermore, saturation must be prevented by neurons to continue to discriminate events and store information.[30]

---

[29] In rheology, plasticity describes the behavior of a material that bears an irreversible deformation in response to applied forces. Permanent changes occur between interacting constituents of the material. In biology, plasticity refers to an adaptative response. Phenotypic plasticity is the ability of an organism to modify its phenotype in response to environmental changes. Neural plasticity brands reversible variations in connectivity between neurons as well as between neurons and glial cells and/or input–ouput relationships that result from the history of sensed events. Whereas in biology plasticity is a concept associated with the notion of reversibility for adaptation and learning, in physics it is related to an irreversible fate. Therefore, this noun is avoided or employed in parenthesis in association with the term remodeling or adaptivity for immediate understanding.

[30] The activation of GluN glutamate receptors can cause a persistent reduction in LTP induction and an enhancement of LTD [36]. Prior activation of group-1 metabotropic glutamate receptors facilitates both the induction and persistence of long-term potentiation. Stimulation of protein synthesis of synaptic remodeling-related proteins by strong high-frequency stimulation in one set of synapses can facilitate LTP acquisition in another independent input pathway, although this second set of synapses experience a delayed, weak, high-frequency stimulation, which normally remains ineffective (cooperative upregulation). Molecules of synapses of the second pathway are tagged by the weak stimulation and then are able to capture proteins newly synthesized in the first pathway in response to stronger stimulation (tagging and capture mechanism). Modifications of functioning of postsynaptic ion channels and retrograde signaling

Synaptic transmission and remodeling are normal when presenilins are inactivated in postsynaptic neurons, but impaired when they are inactivated in presynaptic neurons [37]. Presenilins contribute to the control of release of the neurotransmitter glutamate, as they regulate liberation of calcium from its intracellular stores.

Synaptic adaptivity that results from the modulation of the strength of interneuronal connectivities contributes to learning, memory, and complex behavior. It relies on cytoskeletal reorganization in neurons, in addition to diverse synaptic component species and their density in the postsynaptic compartment. Actin filament depolymerizing protein cofilin controls dendritic spine morphology [38].

Neurotransmitter Clearance

Neurotransmitters must be quickly removed from the cleft to reset the synapse for subsequent action. Transport proteins rapidly pump neurotransmitters back into upstream neurons. Neurotransmitter reuptake pumps take energy from preexisting transmembrane gradients of sodium ions, which nerve cells maintain at concentrations much lower inside the cell than outside. Such plasmalemmal proteins are thus defined as ion-coupled transporters, which alternate between 2 conformations, inward- and outward-facing, exposing binding sites for the pumped substrate and its driver ions to the corresponding sides of the membrane.

1.1.5.3 Neural Activity and Blood Flow

The human brain comprises only 2% of the body's mass, but, at rest, it consumes 20% of the produced energy [39]. The energy is mostly used to reverse ion influxes associated with information transmission. An adequate supply of blood glucose[31] and oxygen is necessary to avoid death of neurons and glial cells. The neurovascular coupling heightens the local blood flow to regions of neuronal activity (functional hyperemia; Sect. 1.1.6.3)).

Local blood flow elevation of approximately 40% can actually be detected as soon as neurons start to act. Blood flow rushes into active regions of the central nervous system, where neurons start to fire to bring nutrients, especially oxygen and glucose, to these signaling neurons. Blood flow supplies nutrients not only to working neurons that need energy, but also neurons that will soon be functioning. Blood flow indeed rises in areas that process information in reponse to given stimuli as well as prior to stimulus occurrence. In addition, blood flow rate elevation is greater than that needed by

can also participate in this regulation. Inhibition of LTP by priming synaptic activity depends on the activation of GluNs, A_2 adenosine receptors, P38MAPK (mitogen-activated protein kinase), and protein phosphatases PPM1a, PP2, and PP3 [36].

[31] γλυκυς: sweet.

energy demand. Furthermore, blood flow contributes to neuronal processing and may direct a neuronal circuit for activity. Flowing blood can thus behave as a neuromodulator.

Glial cells couple neurons to the cerebral microvasculature. Astrocytes possess a multitude of fine processes (300–600 dendrites; >100 synapses). Therefore, a single astrocyte can sense and integrate the activity of hundreds of neurons. Each astrocyte extends at least one process with endfeet enwrapping blood capillaries. Therefore, astrocytes can adjust local and regional cerebral blood flow to local and regional energy metabolism. Increase in intracellular Ca^{++} concentration in astrocytes causes either vasodilation or -constriction according to the nature of the signal that triggers Ca^{++} influx. Ca^{++} waves can be transmitted in an endfeet population (i.e., of a given astrocyte as well as different astrocytes) through gap junctions that abound in astrocyte endfeet adjacent to the vasculature.

Hemoneural interaction takes place at the semipermeable blood–brain barrier (Vol. 5 – Chap. 7. Vessel Wall) that is constituted by vascular endothelial cells connected to both neurons and glial cells. Endothelial cells can influence neuronal behavior directly or via glial cells such as astrocytes that operate as sensors. Change in blood flow may provoke the release of signaling molecules and gliotransmitters from astrocytes.

On the other hand, certain substances conveyed in blood can modulate neuron activity. Nitric oxide that easily crosses the blood–brain barrier is able to excite or sometimes dampen neuronal activity. Blood flow can also control neuronal functioning via thermal and mechanical effects.

1.1.5.4 Nerve Pathfinding and Neuron Migration

The development of functional neuronal circuits relies on controlled growth of nerve fibers toward their targets (axon and dendrite pathfinding). At the tip of growing nerves, the *growth cone* with lamellipodia and filopodia (Chap. 6) explores the environment and leads the nerve growth toward the direction prescribed by guidance cues (i.e., concentration gradient of chemoattractants and -repellents). Among mediators of the growth cone guidance, intracellular Ca^{++} concentration plays a major role, as it rises during stimulation by many extracellular guidance factors of nerve growth (e.g., ephrins, brain-derived neurotrophic factor, myelin-associated glycoprotein, netrin-1, Slit, and stromal cell-derived factor) [40]. Furthermore, a gradient of intracellular Ca^{++} concentration at the growth cone is able to trigger growth cone extension toward the side of higher Ca^{++} concentration. Calcium ions regulate intracellular signaling effectors of nerve growth, such as microtubule- and actin-binding proteins, kinases, phosphatases, and peptidases. Canonical transient receptor potential channels (Vol. 3 – Chap. 2. Membrane Ion Carriers) allow Ca^{++} entry, hence neurite extension. In addition, TRPC channels can activate voltage-gated Ca^{++} channels via ion flux and resulting membrane voltage change.

The development of the nervous system also relies on neuronal migration. This process is triggered by Ca^{++} waves and ceases with the end of Ca^{++} waves. During neuron migration (as well as growth cone exploration), stresses are generated between the moving neuron and the extracellular matrix. Mechanical sensing by canonical transient receptor potential channels contributes to stretch-activated Ca^{++} influx, hence, neuronal migration.

1.1.5.5 Neural Progenitors

Multipotent, self-renewing neural progenitors differentiate into neurons, astrocytes, or oligodendrocytes. Neural progenitor proliferation and differentiation that follow damage in the central nervous system can replenish the nervous tissue and at least partially restore the nervous function.

However, in many pathologies in which inflammation is strongly involved, astrocyte production dominates neural regeneration. Inflammation generates oxidative conditions. Under oxidative conditions, the fraction of cultured neuronal stem cells that differentiate into astrocytes[32] and neurons is larger and smaller, respectively, and the converse occurs under reducing conditions [41]. The redox state of neural progenitors thus affects their cell fate. Oxidative stress and inflammation skew differentiation toward astrogenesis by modulating activity of anti-aging Sirtuin-1 (Sect. 2.2). The amount of nicotinamide adenine dinucleotide NAD^+-dependent histone deacetylase Sirtuin-1 is greater in neuronal stem cells under oxidative condition. This condition promotes SIRT1 binding to transcription factor Hairy enhancer of Split HES1 that represses achaete-scute homolog promoter ASCL1, which, in turn, encodes the neurogenic AScL1 (or ASH1) transcription factor. The SIRT1–HES1 complex indeed binds to and deacetylates histones at ASCL1 and recruits corepressors to impede neuronal differentation. In a reducing environment, HES1 recruits transcription activators such as CREB-binding protein to AScL1 so that neural progenitors take a neuronal fate.

Increases in cyclic adenosine (cAMP) and guanosine (cGMP) monophosphates (Vol. 4 – Chap. 10. Signaling Pathways) mediate antagonistic effects of extracellular messengers. Messengers cAMP and cGMP activate specific phosphodiesterases as well as protein kinase-A and -G. They reciprocally inhibit each other. In cultured hippocampal neurons with undifferentiated neurites (any projection, i.e., axon or dendrite), localized cytosolic cAMP and cGMP initiate axon and dendrite formation, respectively [42]. Local elevation of cAMP in one neurite causes cAMP reduction in all other neurites of the same neuron.

[32] Astrocytes support neurons because they secrete signaling molecules and favor uptake and metabolism of neurotransmitters.

Table 1.10. Glial cells of the nervous system. The central nervous system includes the cerebrum, cerebellum, brainstem, and spinal cord. The peripheral nervous system comprises somatic (for coordinating the body motions) and autonomic nervous systems (with sympathetic, parasympathetic, and enteric compartments) with afferent sensory neurons, efferent motor neurons, and relay neurons (Source: Wikipedia; IP_3: inositol triphosphate).

Type	Features and activity
	Central nervous system
Astrocytes (astroglia)	Macroglial cell Neuron environment regulation, neurotransmitter recycling, blood–brain barrier functioning, vasomotor tone Ca^{++}–IP_3 signaling
Oligodendrocytes	Axon coating (myelin sheath) for electrical insulation
Ependymal cells (ependymocytes)	CNS cavity lining, cerebrospinal fluid production cilia-mediated CSF circulation
Radial glia cells	Neuronal progenitors, scaffold for neuron migration
	Peripheral nervous system
Schwann cells	Axon myelination, phagocytosis
Satellite cells	Neuron environment regulation

1.1.6 Glial Cells

Glial cells are non-neuronal cells of the nervous system that maintain neuron homeostasis (Table 1.10). Some of them form myelin. They guide neuron migration, regulate neuron repair, support synaptic remodeling (neuronal plasticity), assist in the control of cerebral blood flow, and participate in signal transmission by neurotransmitter release and clearance as well as secretion of gliotransmitters. Glial cells are also involved in material exchange between the flowing blood and neurons. Moreover, microglial cells are specialized macrophages that degrade pathogens and dead neurons.

1.1.6.1 Astrocytes

Astrocytes have a spongiform morphology. Astrocytes are the most abundant type of glial cells in the central nervous system. They account for about half of the cells in the cerebral cortex. They actively participate in brain function. Astrocytes that are tightly associated with neighboring neurons couple neuronal activity to vascular signals. They communicate with neurons via signals and metabolism.

Astrocytes contain glial fibrillary acidic protein (GFAP), a type-3 intermediate filaments that is capable of forming both homo- and heterodimers. It can also polymerize with other type-3 proteins and neurofilament protein NFL, but not type-1 and -2 intermediate filament proteins. Agent GFAP is involved in cell structure, communication, and movement, as well as functioning of the blood–brain barrier.

Like neurons, astrocytes are organized into networks [43]. They occupy distinct domains from adjoining astrocytes, with which they communicate via intercellular junctions. Whereas neuronal networks exchange information through synapses, astrocytes are interconnected by gap junctions (Vol. 1 – Chap. 7. Plasma Membrane).[33] These aqueous channels permit intercellular communications that participate in neuroglial and gliovascular signalings. Gap junctions are controlled by neurotransmitters. Molecules cross gap junction channels according to their molecular weight, size, and shape, as well as charge and binding capacity with connexins (gap-junction constituents).[34] Moreover, the transjunctional and transmembrane voltages of gap junctions influence the channel conductance, hence transfer of signaling molecules.[35] The permeability of connexin channels is regulated by several ligands of neuronal receptors, such as endothelin and kainate.

In addition, connexins can operate as hemichannels that allow exchanges between the cytoplasm and extracellular medium (Sect. 1.1.2). Astrocyte hemichannels are permeable to Ca^{++}, glutamate, ATP, glucose, and glutathione, among other small molecules [43].

---

[33] Among 21 connexins, 11 have been detected in the brain. Connexins Cx43 and Cx30 are the main types in astrocytes.

[34] Connexin-43-based channels are selective for second messengers (cAMP, IP_3, and Ca^{++}), amino acids (glutamate, aspartate, and taurine), nucleotides (ADP, ATP, CTP, and NAD), and metabolites (glucose, glucose-6-phosphate, and lactate). Molecules ATP, inositol trisphosphate, aspartate, glutamate, glucose, and lactate also move through Cx30-based channels [43]. Glutamate increases glucose transfer through gap junctions. Glutamate released from neurons increases glucose transfer in astrocytes via activation of α-amino 3-hydroxy 5-methyl 4-isoxazole propionic acid (AMPA) receptors. Astrocytes then supply neurons with metabolic substrates. Furthermore, intercellular Ca^{++} waves trigger glutamate from astrocytes, hence amplifiying the process. Sodium ions act as second messengers in this neurometabolic coupling. Sodium-coupled glutamate uptake in astrocytes activates Na^+–K^+ ATPase to trigger glucose uptake from blood and glycolysis, with resulting release of lactate from astrocytes. Lactate and glucose can be used as fuel by neurons.

[35] The behavior of homotypic Cx30 and Cx43 channels and heterotypic Cx43–Cx30 channels depends on voltage.

In response to a stimulus, a change in intracellular calcium concentration is transmitted with a delay of a few seconds from neurons to adjacent astrocytes via neurotransmitter release. The functional contribution of astrocytes includes neuronal response magnitude and duration [44]. Astrocyte activity can then modify local blood flow. Calcium signaling in astrocytes couples regional changes in neural activity to oxygen availability and cerebral blood flow by modifying arteriole vasomotor tone. At the onset of neural activity, dendrites rapidly consume oxygen and cerebral metabolism then shifts. When an additional amount of oxygen is required, calcium concentration augments in astrocyte endfeet that collectively circumscribe cerebral arteries. Calcium ion transients from mGluR stimulation activate cytosolic phospholipase-A2 that forms arachidonic acid, which is either converted to 20-hydroxyeicosatetraenoic acid in smooth muscle cells to induce vasoconstriction or vasodilator prostaglandin-E2 in astrocytes via cyclooxygenases COx1 and COx2.

In hypoxia, astrocyte glycolysis rises and synthesized lactates are released [45]. High levels of extracellular lactate attenuate prostaglandin-E2 uptake from the extracellular space by prostaglandin transporters. Prostaglandin-E2 accumulates in the extracellular medium of the brain region that requires additional oxygen and causes local vasodilation. Furthermore, extracellular adenosine content also increases to suppress astrocyte-mediated vasoconstriction.

In the central nervous system, astrocytes bolster communications between pre- and postsynaptic neurons. Astrocytes can promote excitatory, glutamatergic as well as inhibitory, gabaergic synapses using distinct signals [46]. Astrocytes can actually enhance axon length and branching as well as increase the local density of synapses. Thrombospondins secreted by astrocytes are necessary and sufficient to heighten glutamatergic synaptogenesis, but not gabaergic synaptogenesis.

Astrocytes take up neurotransmitters. In addition, in astrocytes, increase in intracellular calcium concentration can trigger the release of the *gliotransmitters* glutamate, adenosine triphosphate, and ^{D}serine, stored or not in vesicles, to modulate neuronal excitability and transmitter release [47]. However, calcium signaling in hippocampal astrocytes is not associated with the release of gliotransmitters and does not influence spontaneous and evoked excitatory synaptic transmission [48].

1.1.6.2 Mathematical Modeling of Calcium Dynamics in Astrocytes

Mathematical modeling contributes to the understanding of neuroglial interactions, i.e., signaling from neurons to astrocytes and vice versa, as well as information processing. The smallest second (intracellular) messenger is the Ca^{++} ion that operates in cell contraction, gene transcription, molecule exocytosis, and cell metabolism. This ubiquitous inorganic transmitter permits communication among cells of different types (Vol. 4 – Chap. 10. Signaling Pathways). In addition, it regulates the activity of many intra- and extracellular proteins.

Calcium concentration within the cell ($\sim$100 nmol) is much lower than that in the extracellular medium as well as, inside the cell, in the endoplasmic reticulum and mitochondria that serve as intracellular store ($\sim$1 mmol). Ca^{++} influx results from activation of voltage-gated and stretch-activated Ca^{++} channels that provoke a fast transfer. Long-lasting processes can then take the relay. In the cytosol, Ca^{++} ions are buffered by many Ca^{++}-binding proteins. Calcium ions can return back to its store or be exported to the surrounding space.

Calcium signaling is able to generate coordinated activity patterns over large length and time scales. Cytosolic calcium displays various spatiotemporal oscillatory profiles. Trains of spikes of given amplitude and duration at a given station propagate as Ca^{++} waves. Calcium dynamics can create bursts of oscillations.

Forms of Ca^{++} signaling are separated based on whether they appear in excitable (neurons and myocytes) or non-excitable cells. Calcium sparks occur within myocytes (cardio-, skeletal, and smooth myocytes) transiently and locally following electrical excitations for excitation–contraction coupling.

Inositol triphosphate-sensitive Ca^{++}-release channel (IP_3R) releases Ca^{++} from the endoplasmic reticulum (ER). A kinetic model of Ca^{++} release was carried out to describe the properties of trimeric IP_3R channel [49]. Each equivalent and independent subunit is involved in Ca^{++} conduction. Each subunit possesses a single IP_3- and 2 Ca^{++}-binding sites. One Ca^{++}-binding site activates, whereas the other inhibits. The binding and unbinding of IP_3 is the fastest stage, followed by the binding of Ca^{++} on the activation binding site (one order of magnitude less), itself followed by the binding of Ca^{++} on the inactivation binding site (another factor of 10 slower). Channel IP_3R opens when its subunits are in the open state, i.e., the repressing Ca^{++}-binding site is empty, but the 2 other binding sites are filled. Eight different configurations exist with transitions governed by second-order (association) and first-order (dissociation) rate constants. The 9-variable model thus relies on a set of 8 equations for distinct states of IP_3R and one equation for the cytosolic free Ca^{++} concentration given by the difference between inward and outward fluxes from and in distinct involved compartments.

Calcium release was also modeled from a small isolated cluster of IP_3Rs in which the Ca^{++} concentration is assumed to be uniform using a stochastic Hodgkin-Huxley-like formalism, assuming stationary Markov transitions between the receptor states, in particular open and closed configurations with given opening and closing rates [50]. The fraction of Ca^{++} released from its intracellular store is assumed to be a tiny part of stored Ca^{++}. The model is limited to 2 variables: the Ca^{++} dynamics and gate-shutting variable (2-state gate S) that corresponds to the sum of the fraction subunits in different states:

$$\begin{aligned}\frac{d}{dt}[\mathrm{Ca}^{++}] &= f([\mathrm{Ca}^{++}]) + g([\mathrm{Ca}^{++}]),\\ \frac{d}{dt}S &= h([\mathrm{Ca}^{++}]S).\end{aligned} \tag{1.1}$$

The model yields 3 types of behaviors given by nullclines of the system of 2 ordinary differential equations: excitability, oscillations, and bistability.

An alternative strategy relies on the Langevin approximation of the previous model [51]. Instead of taking into account all intermediary steps, a 2-step approach is considered with the same opening and closing rates for all subunits. Langevin equations with a Gaussian, white noise $n(t)$ are given by:

$$\begin{aligned}\frac{d}{dt}[\mathrm{Ca}^{++}] &= f(S^3, [\mathrm{Ca}^{++}] - [\mathrm{Ca}^{++}]_{ER}) - g([\mathrm{Ca}^{++}]),\\ \frac{d}{dt}S &= h(S) - n(t).\end{aligned} \tag{1.2}$$

Next generations of models will be aimed at coupling activity of astrocytes, neurons, and capillaries.

1.1.6.3 Messengers of Hemoneural Coupling

In the central nervous system, the blood flow that supplies energy sources (ATP); Table 1.11) is regulated by neurons and astrocytes, the latter being major contributors (Sect. 1.1.5.3). Neurons are most often closer to capillaries (gap 8–23 µm) than arterioles (distance 70–160 µm) [39]. Yet, arterioles yield a more important fraction of the hemodynamical resistance interplay (i.e., range of relative caliber variations) than capillaries. Targeting arterioles appears to be an efficient strategy.

Neurons either signal directly to mural cells of the local microvasculature or activate astrocytes to release vasoactive agents onto blood vessels via, in both cases, neurotransmitters, particularly glutamate (Table 1.12). Synaptic release of glutamate activates its receptors on astrocytes and neurons. Glutamate signals trigger the release of nitric oxide from neurons and arachidonic acid derivatives from astrocytes to locally regulate the cerebral blood flow (Table 1.12). Epoxy (EET) and dihydroxyeicosatrienoic (DHET) acids are vasoactive metabolites of arachidonic acid processed by members of the cytochrome-P450 infraphylum. Subtypes (5,6)-, (8,9)-, (11,12)-, and (14,15)-EETs function as thromboxane receptor (TP) antagonists. In vascular smooth muscle cells, arachidonic acid is converted into 20-hydroxyeicosatetraenoic acid (20HETE) by CyP4a11 (or ω-hydroxylase) that causes a vasoconstriction. Prostaglandins are mainly produced by cyclooxygenase COx1, which lodges in astrocyte endfeet, with a possible contribution by COx3 and COx2. Prostaglandin-E2 (among other prostaglandins) can relax vascular smooth muscle cells by binding to Gs-coupled prostaglandin EP_4 receptors that lead to

Table 1.11. Energy supply and consumption in nervous tissue (Source: [39]). Adenosine triphosphate is generated from glycolysis and mitochondrial oxidative phosphorylation in neurons and glial cells. It is used to restore ion gradients after the generation and propagation of action potentials. Ca^{++} ATPase exchanges internal Ca^{++} for external H^+ ions to regulate intracellular Ca^{++} concentration after an action potential (by Ca^{++} extrusion), therefore acidifying the cytosol and alkalinizing the extracellular space.

ATP source	ATP sink
	Neuron
Glycolysis	Ionotropic glutamate receptors
Oxidative phosphorylation	Voltage-gated Na^+ channels
(glucose, O_2 input)	(Na^+ influx)
	Ca^{++}–H^+ ATPase
	Metabolism
	Glial cell
Glycolysis	Excitatory amino acid transporters
Oxidative phosphorylation	(Glutamate and Na^+ influx)
	Metabolism
	Gliotransmitter ATP

cAMP synthesis, activation of protein kinase-A, and a reduction of phosphorylation of the myosin light chain. On the other hand, nitric oxide synthesized in endothelial cells and/or neurons prevents the production of both vasoconstricting 20HETE and vasodilating EETs. Changes in O_2 concentration modulate the production of nitric oxide, prostaglandins, epoxyeicosatrienoic acids, and 20-hydroxyeicosatetraenoic acid.

In addition, the metabolic messenger adenosine produced by ATP hydrolysis participates in the coupling of the cerebral blood flow to neuronal activation, hence contributing to functional hyperemia. The vasodilator adenosine connects to vascular adenosine receptors A_{2A} (Vol. 3 – Chap. 7. G-Protein-Coupled Receptors). Adenosine then mediates glutamate-induced vasodilation, as it hinders vasoconstriction mediated by 20-hydroxyeicosatetraenoic acid. Adenosine also binds to astrocytic A_{2B} receptors and promotes the propagation of calcium signal throughout astrocytic processes [54]. Calcium signal can elicit the release of adenosine triphosphate through connexin hemichannels.

Another catabolite, lactate, produced when pyruvate production by glycolysis outmatches pyruvate consumption by oxidative phosphorylation, also elevates blood flow [39]. In particular, low O_2 concentrations attenuate the capacity of mitochondrial oxidative phosphorylation to consume all pyruvate molecules produced by glycolysis. Lactate is exported by monocarboxylate transporters. Extracellular lactate precludes the reuptake of prostaglandin-E2 by the prostaglandin transporter, thereby promoting vasodilation.

Table 1.12. Glutamate-mediated control of the cerebral blood flow via the regulation of the tonus of vascular smooth muscle cells (Source: [39, 52, 53]; $\ominus \longrightarrow$: inhibition; AA: arachidonic acid; CyP: member of the cytochrome-P450 infraphylum [epoxygenase]; EET: epoxyeicosatrienoic acid; sGC: guanylate cyclase; GluN: NMDA (^{N}methyl ^{D}aspartate)-type ionotropic glutamate receptors; mGluR: metabotropic glutamate receptors; $K_{Ca}1$: Ca^{++}-activated K^+ (BK_{Ca}); NOS2: neuronal nitric oxide synthase; PG: prostaglandin; PLA2: phospholipase-A2; PR: prostanoid receptor; TP: thromboxane-A2 receptor; TRPV4: vanilloid transient receptor potential channel). Synaptic release of glutamate activates its receptors on astrocytes and neurons. Epoxy- (EET) and dihydroxy-eicosatrienoic (DHET) are vasoactive metabolites of arachidonic acid (AA) processed by cytochrome-P450 infraphylum members. The former (EETs) are antagonists of the thromboxane receptor (TP). In vascular smooth muscle cells, AA is converted into 20-hydroxyeicosatetraenoic acid (20HETE) by CyP4a11 that causes a vasoconstriction. Prostaglandins are mainly produced by cyclooxygenases (COx). Prostaglandin-E2 binds to EP_4 receptor and relaxes vascular smooth muscle cells. On the other hand, nitric oxide (NO) synthesized in endothelial cells and/or neurons prevents the production of both vasoconstricting 20HETE and vasodilating EETs. Potassium ion efflux from astrocyte endfeets and modest elevation in extracellular K^+ concentration activate inward rectifier K^+ channels (K_{IR}) on smooth muscle cells of intracerebral arterioles that suppress Ca^{++} entry, hence provoking a SMC relaxation.

Initiation	Neural tissue	Vascular wall
	Pathways from astrocytes	
Glutamate–mGluR	Ca^{++}–PLA2–AA–COx1/3(2)–PG	PG–PR (PGE2–EP_4), vasodilation
	Ca^{++}–PLA2–AA–CyP450–EET	EET–TP, vasodilation
	Ca^{++}–PLA2–AA	AA–CyP4a11–20HETE, vasoconstriction
	Ca^{++}–$K_{Ca}1$	K^+–K_{IR}, vasodilation
	NO$\ominus\longrightarrow$EET	NO$\ominus\longrightarrow$20HETE
	Pathways from neurons	
Glutamate–GluN	Ca^{++}–NOS1–NO	NO–GC–cGMP, vasodilation
	Ca^{++}–PLA2–AA–COx2–PG	PG–PR, vasodilation

Capillaries are surrounded by pericytes instead of smooth muscle cells. Pericytes possess a contractile cytoskeleton. They contract in response to noradrenaline and relax in response to glutamate [39]. In addition, ionotropic GABA receptors foster dilation of capillaries.

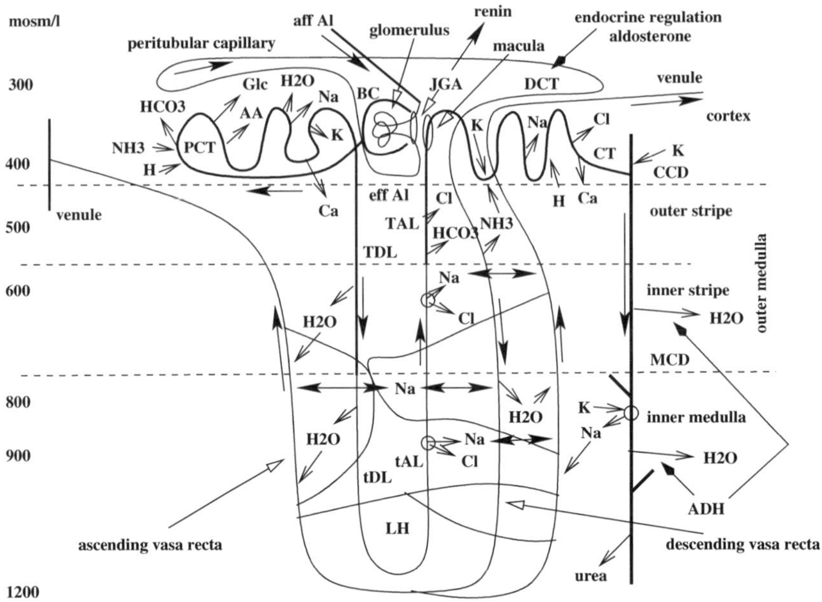

Figure 1.2. The juxtamedullar nephron with a long Henle loop (Source: [55]). The glomerulus with its afferent arteriole (aff Al) and efferent arteriole (eff Al), and with the Bowman capsule (BC). Water, electrolytes (Na, K, Cl, etc.), glucose (Glc), amino acids (AA), urea, and other filtered chemicals form the glomerular filtrate. The tubule is composed of several segments: (1) proximal convoluted tubule (PCT); (2) loop of Henle (LH), with its descending (thick [TDL] in the outer medulla and thin [tDL] in the inner medulla) and ascending (thin [tAL] in the inner medulla and the inner stripe of the outer medulla, thick [TAL] in the outer stripe of the outer medulla) limbs; (3) distal convoluted tubule (DCT); (4) connecting tubule (CT); and (5) cortical (CCD) and medullary (MCD) segments of the collecting duct. The juxtaglomerular apparatus (JGA) secretes renin. The macula densa regulates arteriolar resistance. Interstitial osmolarity of 300 mosm/l is observed at the level of cortical nephrons with short loops and ordinary peritubular capillaries. At the hairpin curve, the interstitium osmolarity is equal to 1200 mosm/l or more.

1.2 Nephron Cells

Fluid and electrolyte excretion by renal nephrons (νεφρος: kidney), i.e., urinary content and flow, regulates blood volume (volemia) and, subsequently, blood pressure. Urine is produced in the nephron (Fig. 1.2) by 3 regulated processes: filtration, reabsorption, and secretion.

Renal blood is first filtered from the *glomerulus*, a capillary ball-like region formed from an afferent arteriole and leading to a narrower efferent arteriole, by the *Bowman capsule* (glomerular filtration). The glomerular filter consists of 3 layers: (1) fenestrated endothelium, (2) glomerular basement membrane, and (3) interdigitated podocyte extensions, which completely enwrap the glomerular capillaries. The filtration barrier restricts the passage of molecules according to their size, shape, and charge. Water, electrolytes, glucose, amino acids, wastes (urea), and other filtered chemical species form the glomerular filtrate. The *juxtaglomerular apparatus* of the arteriole walls contains granular cells that secrete renin.

In cells of the thick ascending limb (TAL) of Henle's loop, apical Na^+–Ka^+–$2Cl^-$ cotransporters NKCC2 (Vol. 3 – Chap. 4. Membrane Compound Carriers) conduct Cl^- uptake. Chloride ions exits through basolateral channels. Renal outer medullary K^+ channel (ROMK; Vol. 3 – Chap. 3. Main Classes of Ion Channels and Pumps) is required for apical K^+ recycling.

Magnesium ion (Mg^{++}) is reabsorbed paracellularly. In the distal convoluted tubule (DCT), Mg^{++} is transported transcellularly by melastatin-related transient receptor potential channels TRPM6 (Vol. 3 – Chap. 2. Membrane Ion Carriers). Salt uptake in the distal convoluted tubule is mediated by Na^+–Cl^- cotransporters (NCC).

The connecting tubule (CT) and cortical collecting duct (CCD) are composed of 2 cell types. *Principal cells* absorb Na^+ and secrete K^+. *Intercalated cells* are implicated in: (1) acid–base transport due to H^+ ATPase and Cl^-–HCO_3^- exchanger; (2) K^+ reabsorption using apical H^+–K^+ ATPase under certain conditions; and (3) K^+ secretion via flow-dependent Ca^{++}- and stretch-activated BK channels ($K_{Ca}1$). In α intercalated cells of the collecting duct of the distal nephron, H^+ ions are secreted into the tubular lumen via H^+ ATPases and H^+–K^+ ATPases. Proton secretion is coupled to HCO_3^- reabsorption via chloride–bicarbonate exchanger SLC4a1 (or anion exchanger AE1). Chloride ion exits via a separate ion channel. Sodium reabsorption and potassium secretion in the distal nephron (i.e., distal convoluted tubule, connecting tubule, and collecting duct) are increased in response to augmented tubular urinary flow rates by enhancing the activity of apical (plasmalemmal) Na^+ channels and activating $K_{Ca}1$ channels. In the cortical collecting duct, WNK4 kinase stimulates ROMK endocytosis.

Epithelial Na^+ channel (ENaC) is expressed in the distal nephron and associated with cytoskeleton elements (ankyrin, α-spectrin, and actin). This mechanosensitive ion channel is activated by increased urinary flow rate that augments Na^+ reabsorption. Potassium secretion in the distal nephron depends on Na^+ absorption via ENaC channel. Sodium ion diffuses from urine across the apical membrane of the principal cell via ENaC and is extruded at the basolateral membrane in exchange of K^+ by the Na^+–K^+ ATPase. Potassium ion is then transported into urine via apical K^+ channels. The higher the urinary flow rate (in the physiological range), the larger the urinary K^+ excretion in the distal nephron.

Potassium channels $K_{Ca}1$ (BK) are composed of a pore-forming α- and a regulatory β subunit. Apical $K_{Ca}1$ channels are activated by plasmalemmal depolarization, elevation in intracellular calcium concentration, hyposmotic stress, and plasmalemmal stretch. Elevated urinary flow rate increases the circumferential stretch in the tubular wall, the intracellular Ca^{++} concentration, and ENaC-mediated Na^+ absorption, leading to apical membrane depolarization. The flow-stimulated (but not baseline) K^+ secretion in the distal nephron is prevented by specific inhibitors of BK, but not ROMK channels. Calcium binding to $K_{Ca}1$ channels is required, Ca^{++} shifting the voltage-dependent gating of the channel for activation within the physiological range of plasmalemmal potentials.[36]

The luminal surface of the distal nephron has a glycocalyx. The central cilium[37] senses and transduces a flow rise to an increase in intracellular calcium concentration. Polycystin-1 and -2, which are located not only in the primary cilium, but also in focal adhesions, are activated by stress transmitted to focal adhesions. Elevation in tubular volume could also activate release of Ca^{++} or ATP and then stimulate Ca^{++} sensors or purinergic receptors in both principal and intercalated cells.

In the kidney, friction generated by the urinary flow in the nephron participates in the control of transport of glucose, chloride, magnesium, potassium, sodium, as well as cytoskeleton organization, synthesis of matrix peptidases, and activity of specific transcription factors [56].

1.3 Adipocytes

1.3.1 Types of Adipose Tissues

Three types of adipose tissues (adeps: fat) exist in mammals: (1) *white adipose tissue* (WAT) that stores energy and is associated with obesity; (2) highly vascularized *brown adipose tissue* (BAT) that burns energy to generate heat (thermal energy) under the control of the sympathetic nervous system;[38] and (3) bone-marrow adipose tissue (BMAT). Brown adipose tissue is mainly found in neonates. White adipose tissue, the predominant type in adults, pertains to endocrine and inflammatory systems. Adipose tissue participates in hormonal control of body homeostasis. It contributes to the regulation of cell metabolism and inflammation (Table 1.13). Besides, scattered brown adipocytes lodge in any white adipose depot.

---

[36] In the cortical collecting duct, ROMK channels are found in principal cells whereas BK channels are detected in both principal and intercalated cells. When the urinary flow rates are low, K^+ secretion is mainly due to ROMK channels. Increased tubular flow rates activate BK channels.

[37] The principal cell has a central cilium. The intercalated cell is devoid of apical cilium, but its apical surface has numerous microvilli and microplicae.

[38] Brown and white adipocytes contain organelles that have different morphologies. Lipids accumulate in multiple, small droplets in brown adipocytes and a single, large lipid droplet in white adipocytes [57]. Mitochondria are large and numerous with laminar cristae in brown adipocytes, whereas white adipocytes are endowed with small and elongated mitochondria with randomly oriented cristae.

Table 1.13. Contribution of white adipose tissue to the regulation of cell metabolism and inflammation (Source: [59, 60, 444]; AMPK: adenosine monophosphate-activated protein kinase; BMP: bone morphogenetic protein; CCL: chemokine CC-motif ligand; IL: interleukin; IL1RA: IL1-receptor antagonist; PKA: protein kinase-A; PRDM: PR domain-containing protein; RBP4: retinol-binding protein-4; TNF: tumor-necrosis factor).

Function	Mediators
	Regulation of cell metabolism
Energy	Noradrenaline, adiponectin, leptin, IL1, IL1RA, IL6 lipoprotein lipase, triacylglycerides, PKA, AMPK, PRDM16
Adipocyte differentiation	Chemerin, TNFα, CCL2, IL1, IL1RA, IL6 angiotensin-2, BMP7, prostaglandins
Insulin sensitivity	Adiponectin, omentin, resistin, vaspin, visfatin, RBP4, TNFα, IL1, IL1RA, IL6
	Regulation of inflammation
Inflammation	Adiponectin, hepcidin, leptin, resistin, visfatin, TNFα, CCL2, CCL5, IL1, IL1RA, IL6, IL8, IL9
Vascular response, diapedesis	Adiponectin, apelin, resistin, CCL2, CCL5, IL8, IL6
Angiogenesis	Adiponectin, leptin, VEGF, TNFα, IL1, IL1RA, IL10

Adipocytes secrete hormones such as adiponectin, apelin, chemerin, hepcidine, leptin, omentin, resistin, vaspin, and visfatin, as well as inflammatory cytokines such as tumor-necrosis factor-α, chemokines such as CCL2 (or monocyte chemoattractant protein-1), and plasminogen activator protein. Adipose tissue contains adipocytes and preadipocytes, as well as fibroblasts, macrophages, leukocytes, and endothelial cells.

Some adipocytes in each population can reversibly transform into one another (*white-to-brown-to-white transdifferentiation*), i.e., they can change their function from heat production to energy (triglyceride) storage and conversely [57].[39]

[39] Moreover, adipocytes can undergo an *adipo-epithelial-adipo transformation*, such as that observed in mammary gland adipocytes during pregnancy, lactation, and postlactation [57].

Brown adipocytes devoted to heat production are able to uncouple respiration from ATP synthesis via specific Uncoupling protein-1 (Ucp1; the so-called *adaptive or facultative thermogenesis*).[40]

1.3.2 Localization of Subcutaneous and Visceral Adipose Compartments

White and brown adipose tissues are the predominant forms in adults and neonates, respectively. In humans, adipose tissue is organized into subcutaneous and visceral compartments. Subcutaneous adipose tissue forms a continuous sheet in nearly all parts of the body.

Brown adipose tissue especially localizes to the perivascular space around the major vessels (aorta and its main branches: carotids, subclavian, intercostal, and renal arteries). Brown adipocytes are thus able to quickly distribute produced heat via the blood stream to various body compartments. Moreover, a separate adipocyte type associated with a distinct basal membrane (and characterized by glycogen particles and numerous large mitochondria with several poorly differentiated cristae) are apposed to capillaries (1 in every 5–10 capillaries). They can be brown adipocyte precursors [57].

Local compartments of white adipose tissue with specialized functions include visceral, muscular, epicardial, perivascular, and perirenal adipose tissue [444]. Small amounts of white adipocytes are observed in organs, such as bone marrow, skin, parotid and parathyroid glands, thymus, lymph nodes, and pancreas.

The anterior subcutaneous depot that is mainly composed of brown adipose tissue is distributed within a pyramidal region, the apex of which lies deep in the interscapular area, that extends from the dorsal paravertebral skeletal muscles to the cervical and axillary regions [57]. The posterior subcutaneous depot that is predominantly made up of white adipose tissue resides in hind legs.[41] Visceral depots of adipose tissue localize to the thoracic and abdominal cavities, in zones delimited by serous membranes. The mediastinal depot corresponds mainly to brown adipose tissue, whereas the abdominopelvic depot is equally composed of white and brown adipose tissue.[42]

---

[40] Inward transfer of protons across the mitochondrial inner membrane by Ucp1 generates heat rather than ATP.

[41] At the base of posterior regions of inferior limbs, adipose tissue forms a single band that extends from the lumbar region to the inguinocrural, pubic, and gluteal regions. Subfascial depots are detected in the limbs. In the hind legs, 2 main depots are located in the thigh and popliteal fossa, respectively [57].

[42] The main mediastinal depot lies around the aorta and the proximal part of its thoracic branches. Abdominal depots are divided into retro- and intraperitoneal sites. The retroperitoneal depot extends longitudinally between the spine and the posterior abdominal wall in a paravertebral position and is separated from the perirenal depot by a peritoneal fold. Males also have epididymal depots. In

The ratio between white and brown adipose tissue in adipose depots that contain the majority of the body's lipidic stores varies with genetic background, sex, age, nutritional status, and environmental conditions [57].

1.3.3 Function of Subcutaneous and Visceral Adipose Tissue

Subcutaneous and visceral adipose tissue show functional differences [61]. Synthesis of angiotensinogen, complement factors, and fatty acid-binding protein-4 is greater in visceral than in subcutaneous fat. Leptin is mainly produced by subcutaneous adipose tissue. Visceral fat is the main contributor of blood IL6 concentration. Abdominal adipose tissue (but not other subcutaneous adipose tissues) drains directly into the portal circulation. In addition, infiltration rate of monocytes into visceral adipose tissue is higher than into subcutaneous adipose tissue.

Both visceral and subcutaneous adipose tissues are innervated by the autonomic nervous system and controlled by neuroendocrine feedback. Stimulation of the parasympathetic nervous system decreases lipolysis, whereas stimulation of the sympathetic nervous system reduces adipogenesis and activates lipolysis [61].

1.3.4 Adipocyte Differentiation

Embryonic mesoderm gives rise to mesenchymal stem cells that can differentiate into adipoblasts. Adipoblasts develop into committed white and brown preadipocytes and ultimately mature adipocytes.

Differentiation of preadipocytes into adipocytes is influenced by angiotensin-2 that can be generated by the renin–angiotensin system. Expression of type-1 and -2 angiotensin-2 receptors (AT_1 and AT_2) increases in the initial phase of differentiation and decays afterward [59]. Production of both leptin and resistin also significantly rises during preadipocyte–adipocyte transformation. Tumor-necrosis factor-α that elevates Wnt10b level (Vol. 3 – Chap. 10. Morphogen Receptors), promotes a macrophage-like phenotype of preadipocytes, but hinders normal differentiation of preadipocytes [60].

Pericytes in the walls of blood vessels that irrigate adipose tissue can be progenitor cells of white adipocytes. They express peroxisome proliferator-activated receptor-γ (nuclear receptor NR1c3), stem cell antigen-1, and transmembrane sialomucin CD34, in addition to markers of pericytes (smooth muscle actin, platelet-derived growth factor receptor-β, and protein neural/glial cell-2) [62].

Two master regulators of brown phenotype induction exist: (1) positive regulatory (PR) domain-containing protein[43] PRDM16, a zinc finger

females, perirenal, periovarian, parametrial, and perivesical adipose tissues form the abdominopelvic depot.

[43] The PR domain is an N-terminal module that was first detected in Retinoblastoma protein-interacting zinc finger protein (RIZ), now called PR domain-

transcription factor that controls the cell fate between myocyte and brown adipocyte, and (2) bone morphogenetic protein BMP7 that promotes differentiation of brown preadipocytes [63]. The latter activates a program of brown adipogenesis that includes: (1) early regulators of brown adipocyte fate PRDM16; (2) peroxisome proliferator-activated receptor-γ coactivator-1α (PGC1α); (3) Uncoupling protein UCP1; and (4) adipogenic transcription factors PPARγ and CCAAT–enhancer-binding proteins (C/EBP); as well as (5) mitochondrial biogenesis via mitogen-activated protein kinase MAPK14, or P38MAPKα, and PGC1 transcriptional coactivator. In addition, BMP7 triggers commitment of mesenchymal progenitor cells to a brown adipocyte lineage [63].

1.3.5 Adipocytes and Vascular Cells

Adipocyte precursors, CD34+ preadipocytes with a restricted potential and short lifespan as well as multipotent adipose-derived stem cells,[44] lodge in all adipose depots [64]. These cells have an important self-renewal capacity. They can differentiate into many mesenchymal cell types, in addition to functional adipocytes.

Adipocyte progenitors can give rise to macrophage-like cells [64]. Preadipocytes indeed can have a phagocytic activity against yeast and apoptotic bodies, especially during inflammation. Macrophage-specific properties decline after the commitment of preadipocytes into adipocytes. These features are thus confined to the progenitor state of the adipose lineage.

In adults, the development of a capillary network support remodeling of adipose tissue for optimal functioning as a metabolic and endocrine tissue. This functional interaction between adipose and vascular tissues is reinforced by the potent pro-angiogenic factors released by cells of the adipose lineage. In particular, adipocyte progenitor cells that are committed prenatally or early in postnatal life reside in walls of the adipose vasculature [65]. CD34+ Preadipocytes can spontaneously differentiate into vWF+ (von Willebrand factor-positive) cells that can participate in the formation of vascular-like structures [64]. Human mature adipocytes can dedifferentiate into cells with an angiogenic potential. CD34+, CD31− cells from human adipose tissue can differentiate into an endothelial type [66]. In addition, adipocytes and endothelial cells have a common progenitor cell [67].

Human mature adipocyte in culture loses lipid droplets and mature adipocyte markers, as it dedifferentiates, and acquire a proliferative morphology similarly to mesenchymal stem cells, as they express typical mesenchymal stem cell markers, such as stem cell antigen SCA1, stem cell growth factor

---

containing protein with ZNF domain PRDM2 and transcription repressor positive regulatory domain I-binding factor PRDI-BF1, now named PR domain-containing protein with ZNF domain PRDM1.

[44] Molecule CD34 is a hematopoietic progenitor cell antigen.

receptor SCFR, and Thy1 membrane glycoprotein (or CD90) [68]. These dedifferentiated adipocytes can then differentiate into several mesenchymal lineages (adipogenic, chondrogenic, and osteogenic). Cardiac differentiation of dedifferentiated adipocytes increases by inhibition of the bone morphogenetic protein or Wnt pathways. Noggin, a BMP inhibitor, is more efficient at promoting cardiomyocyte differentiation when applied to adipocytes, whereas Wnt inhibitor Dickkopf-1 is more potent in dedifferentiated adipocytes.

Adipose stromal cells and brown adipocytes can differentiate into spontaneously contracting cardiomyocyte-like cells [64]. Mouse white adipocytes isolated from subcutaneous adipose tissue can also dedifferentiate and then spontaneously differentiate into both autorhythmic and quiescent cardiomyocyte-like cells [68]. Among dedifferentiated adipocytes, 10 to 15% spontaneously differentiate into sarcomeric actin, GATA4+ or troponin-1, GATA4+ pseudo-cardiomyocytes. These cells possess the capability of triggering action potentials and Ca^{++} transients, hence coupling excitation to contraction. However, they have a limited self-renewal capacity.

The population of dedifferentiated adipocytes is more homogenous than that of adipose stromal cells. Dedifferentiated adipocyte population has a higher percentage of cells expressing SCFR and SCA1 than the adipose stromal cell pool [68]. Very few dedifferentiated adipocytes express α_L and α_M integrins, CD34, platelet–endothelial cell adhesion molecule PECAM1, phosphatase PTPRc, or α-smooth muscle actin. On the other hand, 5 to 15% of adipose stromal cells possess these markers.

1.3.6 Adipocyte Role in Metabolism

Adipocytes are large cells that store lipids for food-restriction periods, as they contain lipid droplets. The most common type of adipocyte is the *unilocular adipocyte* or white fat cell. Each cell contains a single lipid droplet surrounded by a thin rim of cytoplasm. The *multilocular adipocyte* or brown fat cell is a more specialized and localized type of adipocyte. It functions in thermogeneration, as it burns lipids to produce heat.

During the postprandial phase, blood triglycerides of very-low-density lipoproteins, chylomicrons, and their remnants are hydrolyzed by lipoprotein lipase into free fatty acids. Free fatty acids are stored in adipocytes after food intake and released during fasting to ensure sufficient energy.

In adipocytes, energy storage and homeostasis rely on triacylglycerides and, on energy demand, lipolysis and release of stored energy into the blood circulation in the form of non-esterified free fatty acids or glycerol.

Vitamin-A, or retinol, operates via its active metabolites, the visual chromophore, 11-cis-retinal, and the gene transcription activator, retinoic acid. Vitamin-A is transported in the blood circulation bound to retinol-binding protein. The retinol–retinol-binding protein complex binds the vitamin-A transporter and plasmalemmal signaling receptor, stimulated by retinoic acid gene-6 homolog (StRA6), for cell uptake and activation of STRA6 receptor [69].[45] Liganded STRA6 indeed undergoes a Tyr phosphorylation and subsequently recruits Janus kinase-2. The resulting signaling comprises the phosphorylation (activation) of JaK2 and STAT5 (signal transducer and activator of transcription) transcription factor. The JaK2–STAT5 axis causes the synthesis of SOCS3, a STAT-induced STAT inhibitor, and PPARγ, an intracellular receptor. In adipocytes, the latter supports triglycerides accumulation and the former prevents the signaling primed by the insulin receptor.

1.3.7 Energy Mobilization

Cold exposure causes sympathetic nerve fibers to release noradrenaline in brown adipose tissue. This neurotransmitter activates β3-adrenergic receptors on brown adipocytes to stimulate lipolysis and heat production. In white fat depots, prolonged exposure to cold or β3-adrenergic agonists causes tissue transdifferentiation to *beige adipose tissue* [70]. Prostaglandins actually trigger brown fat-specific gene expression in white adipocytes.

Mobilization of metabolic energy that results from hydrolysis of triacylglycerides is stimulated by β-adrenergic signals that lead to activation of cAMP-dependent protein kinase-A. The latter phosphorylates lipolytic enzymes such as hormone-sensitive lipase. On the other hand, triacylglyceride resynthesis is associated with high-energy consumption. Under depletion of ATP and accumulation of ADP, adenosine monophosphate-activated protein kinase (AMPK) restrains triacylglyceride hydrolysis. Enzyme AMPK phosphorylates hormone-sensitive lipase and can prevent further activation of hormone-sensitive lipase by protein kinase-A.

Hormone-sensitive lipase mobilizes adipocyte triglycerides and increases cellular AMP concentration. Its phosphorylation by PKA (activation) and AMPK (inactivation) are mutually exclusive. Protein kinase-A also phosphorylates (inactivates) AMPKα1 subunit (Ser173), hence preventing phosphorylation (activation) of AMPKα1 (Thr172) by LKB1 kinase [71].

Insulin is the main regulator of lipid content in adipocytes because it inhibits hormone-sensitive lipase and activates lipoprotein lipase, thereby promoting uptake of free fatty acids and triglyceride synthesis in adipocytes.

Prostaglandins synthesized by cyclooxygenase-2 triggers mesenchymal adipocyte progenitor cells (at least in mice) to differentiate into brown adipocytes rather than white adipocytes specialized in lipid storage. The brown phenotype aimed at wasting energy via an elevated heat production possesses an increased sensitivity to noradrenaline [70].

Prostaglandins PGE2 and PGI2 support beige adipose tissue development in white adipose pads. Prostaglandin-E2 acts via 4 different receptor types (EP1–EP4), whereas PGI2 binds to the prostacyclin receptor (IP_1; Vol. 3 – Chap. 7. G-Protein-Coupled Receptors). The intracellular signaling pathway

[45] Mutations of the STRA6 gene cause the Matthew-Wood syndrome.

stimulated by noradrenaline involves adenylate cyclase that produces cyclic adenosine monophosphate [70]. The signaling cascade also involves nuclear receptor NR1c (or peroxisome proliferator-activated receptor-γ). On the other hand, brown fat depots remain unresponsive to increased prostaglandin synthesis.

1.3.8 Adipocyte Production

Adipocytes produce numerous hormones, growth factors, and cytokines (Tables 1.14 and 1.15). A class of cytokines corresponds to *adipokines*, or adipocytokines. They are secreted by adipose tissues and other organs. Adipokines are devoted to auto-, para-, and endocrine communications. They mainly regulate adipocyte differentiation and metabolism. Adipokines reduce fatty acids in non-adipose tissue cells. The adipokine class includes adiponectin, apelin, chemerin, hepcidine, leptin, omentin, resistin, retinol-binding protein-4, vaspin, and visfatin.

Adipocytes also synthesize: (1) fibrinolytic cascade proteins such as plasminogen activator inhibitor-1;[46] (2) complement and complement-related proteins, such as adipsin (complement-D) and acylation-stimulating protein, in addition to adiponectin; and (3) vasoactive proteins (renin, angiotensinogen, and angiotensins-1 and -2. Angiotensin-2 regulates synthesis of leptin, plasminogen activator inhibitor-1, and interleukins-6 and -8 in human adipocytes [73]. Adipocytes also secrete adipocyte differentiation factor, nitric oxide, and prostaglandins.

Endocrine activity of adipose tissues is partly related to its renin–angiotensin system (Sect. 1.4.9.1 and Vol. 6 – Chap. 3. Cardiovascular Physiology). Angiotensin-converting enzyme converts angiotensin-1 to active angiotensin-2. Reduction in activity of angiotensin-converting enzyme decreases fat accumulation independently of food intake, but in association with increased energy expenditure related to increased metabolism of fatty acids in the liver and improved glucose clearance [74].

Visceral white adipose tissue controls local and systemic inflammation. Inflammatory cytokines and adhesion molecules synthesized by adipocytes comprise: (1) tumor-necrosis factor-α (Sect. 3.20); (2) interleukins (Sect. 3.22)

[46] Plasminogen activator inhibitor-1 is a marker of hypofibrinolysis. Visceral adipose tissues seem to have up to 5 times the number of PAI1-producing stromal cells compared with subcutaneous adipose tissues [72]. Circulating level in plasminogen activator inhibitor-1 is correlated with an accumulation of visceral fat.

Table 1.14. Adipocyte production (**Part 1**). Adipokins regulate food intake, and thereby energy homeostasis. Adiponectin inhibits hepatic glucose production and enhances glucose uptake in muscle. It increases fatty acid oxidation in both liver and muscle. It augments energy expenditure. Leptin production is stimulated by insulin and influenced by tumor-necrosis factor-α, estrogens, free fatty acids, and growth hormone, but not food uptake. Leptin increases hepatic lipid oxidation and lipolysis in skeletal myocytes and adipocytes. It inhibits AMP-activated protein kinase in the arcuate nucleus of hypothalamus (decreased hunger) and stimulates secretion of growth hormone and gonadotrophin-releasing hormone. Interleukin-6 upregulates vascular endothelial growth factor production by visceral and subcutaneous adipocytes. It also causes production of liver C-reactive protein and hemostasis proteins (plasminogen activator inhibitor-1, fibrinogen, and tissue plasminogen activator). Moreover, it relieves inhibition of microsomal triglyceride transfer protein that controls hepatic assembly of apolipoprotein-B-containing lipoproteins. Vaspin (or serpinA12) is a visceral adipose-specific serpin that has insulin-sensitizing effects. Plasminogen activator inhibitor-1 that regulates fibrinolysis and prevents plasminogen-induced migration of vascular smooth muscle cells is synthesized by visceral adipocytes subjected to insulin, glucocorticoids, TNFα, and free fatty acids.

Adipokines	Adiponectin, apelin, chemerin, hepcidine, leptin, omentin, resistin, retinol-binding protein-4, vaspin, visfatin
Lipogenic factors	Acylation-stimulating protein, cholesteryl ester transfer protein, lipoprotein and hormone-sensitive lipases, adipocyte fatty acid-binding protein-4, renitol-binding protein-4, perilipin
Angiogenic factors	Vascular endothelium growth factor, monobutyrin
Vasodilator	Apelin
Blood pressure regulators	Angiotensinogen, angiotensin-converting enzyme, angiotensin-2
Fibrinolysis	Plasminogen activator inhibitor-1

such as interleukin-6; and (3) chemokines, such as chemokine ligand CCL2,[47] CCL5,[48] and CXCL10[49] (Table 1.16).

[47] Chemokine CCL2, or monocyte chemoattractant protein MCP1, recruits monocytes, memory T lymphocytes, and dendritic cells. It is produced as a precursor. It binds chemokine receptors CCR2 and CCR4.

[48] Chemokine CCL5, also called Regulated upon activation, normal T-cell expressed, and secreted protein (RANTES), attracts basophils, eosinophils, and T lymphocytes.

[49] Chemokine CXCL10, or interferon-γ-inducible protein IP10, is also secreted by monocytes, endothelial cells, and fibroblasts. It is a chemoattractant for mono-

Table 1.15. Adipocyte production (**Part 2**).

Cytokines	Tumor-necrosis factor-α, interleukin-6
Inflammatory reactants	Serum amyloid A, pentraxin, lipocalin, C-reactive protein, ceruloplasmin, macrophage migration inhibitory factor (MIF)
Adhesion molecules	Macrophage chemoattractant protein-1 (CCL2), intercellular adhesion molecule-1
Matrix components	Collagen-4
Peptidases	Adipsin
Serpin	Vaspin
Miscellaneous	Osteonectin, stromolysin

Therefore, adipocytes participate in the control of: (1) blood pressure via angiotensinogen; (2) fibrinolysis via plasminogen activator inhibitor-1; (3) lipid metabolism, as they synthesize regulators of lipid (e.g., cholesteryl ester transfer protein and fatty acid-binding protein-4); (4) glucose metabolism via adiponectin and resistin; (5) feeding behavior via leptin; and (6) inflammation via secreted cytokines.

Two important transcription factors in adipocyte functioning include peroxisome proliferator-activated receptors, particularly PPARγ, and sterol regulatory element-binding proteins (SREBP). Transcription factor PPARγ activates genes that encode fatty acid transport protein, lipoprotein lipase, fatty acid-binding protein, adiponectin, and acylCoA synthase [61]. The main SREBP isoform SREBP1c[50] is activated by insulin in the postprandial phase to provoke lipogenesis by promoting fatty acid synthase, HMGCoA synthase, and low-density lipoprotein receptor and preadipocyte differentiation.

The adipokine of the C1q and tumor-necrosis factor superfamily[51] cartonectin[52] is secreted by adipocytes. Cartonectin stimulates the secretion of

cytes, macrophages, T and NK cells, as well as dendritic cells. It also promotes T-lymphocyte adhesion to endothelial cells. It is a ligand for chemokine receptor CXCR3.

[50] A.k.a. adipocyte determination and differentiation factor-1 (ADD1).

[51] Agent C1q is the recognition protein of the classical complement pathway. This protein connects innate to acquired immunity. It can engage many ligands via its globular domain (gC1q). This gC1q domain is also found in many non-complement proteins. It has a fold similar to that of ligands of the tumor-necrosis factor superfamily. The members of the C1qTNFSF superfamily are involved in immune defense, inflammation, apoptosis, autoimmunity, cell differentiation, and organogenesis [482].

[52] A.k.a. 26-kDa collagenous repeat-containing sequence protein (CoRS26) and complement C1q tumor-necrosis factor-related protein-3 (CTRP3).

Table 1.16. Pro- and anti-inflammatory factors synthesized and released by adipocytes (Source: [444]). Inflammatory agents include some growth factors, such as transforming growth factor-β (TGFβ), tumor-necrosis factor (TNF), interleukins (IL), interferons (Ifn), and chemokines. Tumor-necrosis factor increases its production as well as that of resistin, visfatin, IL6, nerve growth factor, and CCL2 chemokine. It also reduces adiponectin and leptin concentrations. Anti-inflammatory factors include anti-inflammatory cytokines, receptor antagonists (IL1RA), soluble receptors (sTNFR, IL1R2, and sIL1R), and adipokines. Anti-inflammatory cytokines and TGFβ repress inflammatory cytokine production. Adiponectin primarily is an anti-inflammatory agent that becomes inflammatory in the presence of lipopolysaccharides.

Inflammatory cytokines	Anti-inflammatory cytokines
Adipokines	
Adiponectin	Adiponectin
Leptin	
Resistin	
Visfatin	
Cytokines and their receptors	
TNF	TGFβ
IL1, IL6	IL4, IL10, IL1RA
Ifnβ, Ifnγ	sTNFR, IL1R2, sIL1R
Chemokines	
IL8	
IP10	
RANTES	
CCL2 (MCP1)	

adiponectin and resistin from murine adipocytes, but fails to provoke that of adiponectin or leptin from human adipocytes [75].

Adiponectin

Adiponectin increases insulin sensitivity, as in: (1) livers, it impedes gluconeogenesis and increases fatty acid oxidation and (2) skeletal muscles, it reduces triglyceride formation and also heightens fatty acid oxidation.

Adiponectin stimulates glucose utilization and fatty-acid oxidation via the activation of AMP-activated protein kinase [77]. Adiponectin also promotes phosphorylation of acetyl coenzyme-A carboxylase, glucose uptake, and lactate production in myocytes.

1.3.9 Adipose Tissue Growth

Adipose tissues have an important growth potential; it then requires angiogenesis for its expansion. However, excess inputs develop adipose tissues, increasing cell size and number, and augment the occurrence of type-2 diabetes mellitus and cardiovascular diseases. Lipid accumulation in adipocytes disturbs the adipokine secretion, impairs insulin signaling, and dysregulates cell functioning.

1.3.10 Dyslipidemia, Inflammation, and Obesity

Adipocytes and resident macrophages synergistically secrete tumor-necrosis factor-α and interleukin-6, particularly in obesity.[53] Obesity corresponds to a pro-inflammatory state in which adipocyte hyperplasia and hypertrophy alter adipokine secretion, hence disturbing insulin sensitivity. Obesity and insulin resistance increase cardiovascular risk by dyslipidemia, hypertension, and glucose dysmetabolism [78]. Certain compounds released by adipocyte actually alter glucose and lipid metabolism, blood pressure, coagulation, and fibrinolysis, and lead to inflammation.

Dyslipidemia in obesity is characterized by increased concentrations of VLDLs and of LDLs, and decreased levels of HDLs. Hepatic overproduction of VLDL is a consequence of hepatic steatosis. In insulin-resistant states of obesity,[54] the dyslipidemia is characterized by an increased concentration

---

[53] Saturated free fatty acids secreted by enlarged adipocytes bind to Toll-like receptor-4 of resident macrophages to cause nuclear factor-κB activation and tumor-necrosis factor-α synthesis and release [61]. (Polyunsaturated fatty acids prevent Toll-like receptor-4 effect.) Macrophage-derived tumor-necrosis factor-α activates adipocyte production of interleukin-6 as well as intracellular adhesion molecule-1 and macrophage chemoattractant protein-1 that promotes blood monocyte diapedesis into adipose tissue. These monocytes afterward mature into macrophages. This paracrine regulation establishes a gradual vicious cycle. Because adiponectin normally inhibits Toll-like receptor-activated nuclear factor-κB activity, low adiponectin levels re-enforce the macrophage–adipocyte interaction.

[54] Insulin resistance leads to: (1) impaired glucose uptake, particularly in myocytes, in hepatocytes, and in adipocytes, (2) impaired LDL receptor activity, with delayed VLDL clearance, and (3) inability to suppress hepatic glucose production and release of non-esterified fatty acids from hypertrophic adipocytes. The level of non-esterified fatty acids increases owing to decreased lipolysis, fatty acid oxidation and low levels of adiponectin (the latter favoring fatty acid oxidation), stress-induced adrenergic stimulation, and inflammation. Increased levels of non-esterified fatty acids cause lipotoxicity, impair endothelium-dependent regulation of the vasomotor tone, increase oxidative stress, and have cardiotoxic effects. Reduced lipoprotein lipase activity decays the clearance of triacylglycerol-rich lipoproteins. Impaired lipolysis of triacylglycerol-rich lipoproteins decreases the transfer of apolipoproteins and phospholipids from triacylglycerol-rich lipoproteins to HDL, thus reducing HDL concentration. Furthermore, delayed clearance

Table 1.17. The endocrine system and various types of hormones.

Production site	Hormones
Hypothalamus	CRH, dopamine, endorphins, GnRH, GHRH, adenylyl cyclase-activating polypeptide, oxytocin, neuropeptide-Y, somatostatin, TRH
Posterior pituitary	Vasopressin, lipotropin
Anterior pituitary	GH, ACTH, TSH, LH, FSH, prolactin, MSH, endorphins, lipotropin
Pineal gland	Melatonin
Thyroid gland	T_3, T_4, calcitonin
Parathyroid gland	PTH
Adrenal medulla	Adrenaline, noradrenaline, adrenomedullin
Adrenal cortex	Aldosterone, cortisol, DHEA
Heart	Natriuretic peptides, adrenomedullin
Adipose tissue	Leptin, adiponectin
Kidney	Renin, erythropoietin, thrombopoietin, calcitriol
Liver	IGF1, thrombopoietin, erythropoietin
Pancreas	Glucagon, insulin, somatostatin, VIP
Stomach	Gastrin, ghrelin, melatonin
Duodenum	Gastrin, CCK, GIP, secretin, motilin
Ileum	CCK, enteroglucagon, GIP
Ovary	Estradiol, progesterone, inhibin, activin
Testis	Testosterone, AMH, inhibin
Placenta	hCG, hPL, estrogen, progesterone

of smaller, denser LDLs after increased lipolysis by hepatic lipase [78]. These LDLs are more exposed to oxidation. They are mostly targeted by macrophage scavenger receptors rather than the normal LDL receptor.

1.4 Classical Endocrine System and Hormones

The endocrine system, composed of many scattered endocrine glands with feedback loops between them, secretes hormones for communication between the body's organs to coordinate their activities (Table 1.17). Hormone secretion is regulated by: (1) the nervous system; (2) hormones such as those released by the hypothalamic–pituitary axis; (3) plasma content in ions and nutrients such as glucose; and (4) environmental changes.

According to their structure and composition, hormones have been classified into: (1) amino acid-derived hormones, i.e., derivatives of tyrosine and

of triacylglycerol-rich lipoproteins facilitates CETP-mediated exchange between cholesterol esters in HDL and triacylglycerols in VLDL. Also, the degradation rate of ApoB100, which regulates VLDL secretion, is decreased in insulin resistance. In the cardiovascular system, insulin resistance is associated with inhibition of the PI3K pathway and overstimulation of the growth factor-like pathway.

tryptophan, such as catecholamines and thyroxine; (2) peptide hormones (polypeptide chain with at most 20 amino acids) such as vasopressin; (3) protein hormones (polypeptide chain with more than 20 amino acids), such as insulin and growth hormone; (4) glycoprotein hormones with carbohydrate side chains such as thyroid-stimulating hormone; (5) lipid, fatty acid- and phospholipid-derived hormones, such as steroid hormones and eicosanoids; (6) nucleotides that are composed of a 5-carbon sugar, phosphate, and nitrogenous base such as adenosine triphosphate; (7) retinoids, such as all-trans-retinoic and 9-cis-retinoic acid; and (8) small inorganic molecules such as nitric oxide.

Hormonal signaling consists of several stages: (1) hormone *synthesis* by producing cells that contains corresponding synthesis enzymes. These cells are subjected to feedback that depends on the concentration of circulating hormone; (2) eventual *maturation* of inactive precursors into active hormones and *storage*; (3) *secretion* triggered by chemical or electrochemical cues; (4) *transport* by blood flow to the target cells; (5) *reception* by cognate receptors; (6) *transduction* of the signal into an intracellular reaction that is characterized by an event cascade (chemical signaling pathway) with positive and negative feedback loops inside target cells; (7) *degradation* of hormones by catabolic enzymes.

Signal intensity depends on the concentration of received hormones as well as density and efficiency of their specific receptors. Signaling enzymes receive and transmit signals by switching from an inactive form to active state, producing messengers, and activating and inactivating signaling mediators (Vol. 3 – Chap. 1. Signal Transduction). Structure of signaling effectors is characterized by regulatory, partner-tethering, eventual membrane-targeting, and catalytic domains for signal reception and transmission. These domains can serve simultaneously or sequentially for signaling flexibility. Multiple domains allow registration of numerous inputs and controls that are integrated and processed to communicate various possible outputs. Regulatory and interaction domains are committed according to time and subcellular locus. Proteic mediators undergo one or several regulatory modifications on certain amino acid residues (e.g., Tyr and/or Ser/Thr phosphorylation, Lys acetylation, Lys and/or Arg methylation, Lys ubiquitination, Cys oxidation, and Cys nitrosylation) in a single or different sites (Vol. 1 – Chap. 5. Protein Synthesis). Signaling redundancy that relies on reactions controlled by distinct partners prevents failure of one signaling component. Partner affinity and reaction kinetic coefficients modulates signaling. Tight interactions can yield signaling specificity, but numerous low-affinity linkages ensure fidelity and quick response to environmental changes.

The hormones are released into the blood and circulate to reach their distant specific targets, binding to receptors. Two main types of hormones are synthesized in the endocrine system: (1) steroid and (2) non-steroid hormones. Non-steroid hormone receptors are plasmalemmal proteins that generally act via second messengers within the cell. The receptors of steroid (glucocorticoids, mineralocorticoids, androgens, and estrogens) and thyroid hormones are located within the cell.

Among endocrine glands, certain tissues belong to organs with other functions than hormone secretion, such as the digestive tract, kidneys, liver, gonads, and heart. The endocrine system also functions with the nervous system. The endocrine system accomplishes multiple tasks, including control of body homeostasis, tissue growth, reproduction, and responses to the surrounding medium. The endocrine system provides a chemical connection from the hypothalamus.

The hypothalamic–pituitary–adrenal axis is activated by numerous stimuli. Cues from various cerebral regions lead to the production by the paraventricular nucleus of the hypothalamus of vasopressin and corticotropin-releasing hormone (CRH) of the corticotropin-releasing factor family (CRF), which coordinates the hypothalamic–pituitary–adrenal axis. Corticotropin-releasing factor activates CRF receptors in anterior pituitary and releases adrenocorticotropic hormone. Adrenocorticotropic hormone targets its receptors in the adrenal cortex to increase the synthesis of glucocorticoids.

1.4.1 Hypothalamus

The hypothalamus produces corticotropin-releasing hormone, dopamine, endomorphins, gonadotropin-releasing hormone (GnRH), growth hormone-releasing hormone (GHRH), somatostatin, and thyrotropin-releasing hormone (TRH; Table 1.18).

Corticotropin-releasing hormone[55] is responsible for adrenocorticotropic hormone[56] (ACTH) secretion. Dopamine is a neurotransmitter, activating dopamine receptors, and a neurohormone which inhibits the release of prolactin. Dopamine acts on D, $\alpha 1$, $\beta 1$, and $\beta 2$ G-protein-coupled receptors. Once activated, D_1-like (D_1 and D_5) and D_2-like (D_2–D_4) groups of dopamine receptors (Table 1.19) increase and reduce cAMP amounts, respectively.

Gonadotropin-releasing hormone[57] activates the release of follicle-stimulating hormone (FSH) and luteinizing hormone (LH). Pulsatile GnRH secretion from the hypothalamus is required to initiate and maintain human sexual features. Release of GnRH is indeed modulated by excitatory and inhibitory control signals induced by regulators, such as kisspeptin and neurokinin-B

[55] Corticotropin-releasing hormone is also termed corticoliberin.

[56] Adrenocorticotropic hormone is also defined as corticotropin.

[57] A.k.a. luteinizing hormone-releasing hormone, gonadoliberin, luliberin, and gonadorelin.

Table 1.18. Hormones of the hypothalamus–hypophysis axis. The hypophysis, or pituitary gland, is an endocrine gland that protrudes from the base of the brain, below the hypothalamus at the base of the brain. It connects to the hypothalamus by the pituitary stalk (a.k.a. infundibular stem and infundibulum. Both anterior (adenohypophysis) and posterior (neurohypophysis) lobes are controlled by the hypothalamus (ACTH: adrenocorticotropic hormone; DA: dopamine; CRH: corticotropin-releasing hormone; FSH: follicle-stimulating hormone; GH: growth hormone; GHRH: growth hormone-releasing hormone; GnRH: gonadotropin-releasing hormone; LH: luteinizing hormone; MSH: melanocyte-stimulating hormone; PIF: prolactin-inhibiting factor (dopamine); PRH prolactin-releasing hormone; Prl: prolactin; Sst: somatostatin; TRH: thyrotropin-releasing hormone; TSH: thyroid-stimulating hormone $\oplus$: stimulation; $\ominus$: inhibition). The intermediate lobe is rudimentary in humans. Vasopressin (a.k.a. arginine vasopressin and antidiuretic hormone [ADH]) is also released from the supraoptic nucleus in the hypothalamus. Dopamine that prevents the secretion of prolactin is also called prolactin-inhibiting factor (PIF), prolactin-inhibiting hormone (PIH), and prolactostatin.

Hypothalamus	Anterior pituitary (adenohypophysis)	Intermediate pituitary	Posterior pituitary (neurohypophysis)
CRH ($\oplus$)	ACTH	MSH	Oxytocin
GHRH ($\oplus$)	GH	(or intermedin)	Vasopressin
Sst ($\ominus$)	GH		
GnRH ($\oplus$)	FSH, LH		
PRH ($\oplus$)	Prl		
DA ($\ominus$)	Prl		
TRH ($\oplus$)	TSH		

respectively.[58] In the hypothalamus, serotonin stimulates GnRH secretion using the Src–PLCγ1 pathway [79].[59]

Growth hormone-releasing hormone[60] stimulates the growth hormone (GH) secretion, in opposition to somatostatin.[61] Somatostatin also inhibits the release of thyroid-stimulating hormone (TSH)[62] and gastrointestinal hormones. Thyrotropin-releasing hormone[63] stimulates the release of thyroid-stimulating hormone and prolactin.

[58] Neurokinin-B that is encoded by gene TAC3 belongs to the tachykinin family. It targets its receptor NK3R encoded by gene TACR3. Tachykinins are synthesized mainly in the central and peripheral nervous systems. Other members include substance-P, neurokinin-A, and hemokinin-1. Kisspeptin receptor Kiss1R is also labeled as GPR54.

[59] Serotonergic nerve fibers and receptors lodge in both the supraoptic and paraventricular nuclei to contribute to the regulation of adeno- and neurohypophysical hormone secretion.

[60] Growth hormone-releasing hormone is also termed somatocrinin.

[61] Somatostatin is also designated as growth hormone-inhibiting hormone.

[62] Thyroid-stimulating hormone is also named thyrotropin.

[63] Thyrotropin-releasing hormone is also called thyroliberin or protirelin.

Endorphins, or endomorphins, are endogenous opioids. Endomorphin-1 and -2 have high affinity and selectivity for the μ opioid receptor (Vol. 3 – Chap. 7. G-Protein-Coupled Receptors). They reside in the central and peripheral nervous systems. They are involved in vigilance, arousal, stimulation of locomotion, stress responses, and pain perception, as well as autonomic, cognitive, neuroendocrine, and limbic homeostasis [80].

Hypothalamic nuclei also produce neuropeptide-Y that modulates the sympathetic nervous system in the control of blood pressure by a potentiation of noradrenaline-induced vasoconstriction via G-protein-coupled receptors, among other functions.

The hypothalamic neuropeptide adenylyl cyclase-activating polypeptide is a vasorelaxant. Oxytocin is made in cells in the supraoptic nucleus and paraventricular nucleus of the hypothalamus.

Orexin-A and -B, or hypocretin-1 and -2 because they are incretins synthesized in the hypothalamus, are neuropeptide hormones that regulate the energy expenditure and sleep/wakefulness and maintain readiness for activity. Orexin neurons are activated during wakefulness and inhibited during sleep. Orexin neurons regulate nuclei in the brainstem, which controls sleep and wakefulness. Both G-protein-coupled orexin-1 and -2 receptors are involved.[64] Orexin neurons also link the arcuate nucleus to regulate feeding. Orexin production is inhibited by leptin and activated by ghrelin and hypoglycemia (orexigenic activity). Orexins stimulate the sympathetic system. Intracerebroventricular injection of orexins increases blood pressure and heart rate [81].

Neuropeptide-B (NpB) and *neuropeptide-W* (NpW), found in anterior pituitary gland and hypothalamus, respectively, bind to G-protein-coupled receptors NpBWR1 (GPR7) and NpBWR2 (GPR8) in the central nervous system and adrenal gland. NpB and NpW modulate inflammatory pain and feeding behavior. They also regulate the release of hormones, such as corticosterone, prolactin, and growth hormone.

The corticotropin-releasing factor family includes corticotropin-releasing hormone, urotensin-1, sauvagine, and urocortins. Urocortins are encoded by 3 different genes (UCN1–UCN3). The corticotropin-releasing hormone is implicated in different responses to stress. The corticotropin-releasing hormone and urocortins (Ucn1–Ucn3) activate 2 members of the B1 family of G-protein-coupled receptors, CRHR1 and CRHR2.[65] These compounds modulate the functions of the central nervous system (appetite, addiction, hearing, and neurogenesis).[66] They also target the endocrine, cardiovascular, reproductive,

[64] Orexin-1 receptor (OX1R) has a greater affinity for orexin-A than for orexin-B. Orexin-2 receptor (OX2R) has similar affinity for orexin-A and -B.

[65] The B1 family of GPCRs includes receptors for growth hormone-releasing factor, secretin, calcitonin, vasoactive intestinal peptide, glucagon, glucagon-like peptide-1, and parathyroid hormone. Hormone CRH has a higher affinity for CRHR1 than for CRHR2. Urocortin binds CRHR2 with a much higher affinity than CRH.

[66] Urocortin is a potent anorexigenic peptide.

Table 1.19. Receptors of hormones of hypothalamo-pituitary axis and pineal gland, their main targeted G proteins, and order of ligand potency when available (Source: [82]).

Type	Main transducer	Ligand (potency order)
Hypothalamus hormones		
GnRHR1	Gq/11	Gonadotrophin-releasing hormone, GnRH1>GnRH2
GnRHR2	Gq/11	GnRH1<GnRH2
GHRHR	Gs	Growth hormone-releasing hormone
TRHR1/2	Gq	Thyrotropin-releasing hormone
CRFR1/2	Gs	Corticotropin-releasing hormone
Sst_1–Sst_5	Gi	Somatostatin
$V_{1A/1B}$	Gq/11	Oxytocin/vasopressin
V_2	Gs	Oxytocin/vasopressin
OT	Gq/11, Gi/o	Oxytocin/vasopressin
D_1	Gs/s,Gs/olf	Dopamine
$D_{2/3/4}$	Gi/o	Dopamine
D_5	Gs	Dopamine
Y_1–Y_6	Gi/o	Neuropeptide-Y, peptide-YY, pancreatic polypeptide-PP
$VPAC_{1/2}$, PAC_1	Gs	Vasoactive intestinal peptide, adenylyl cyclase-activating peptide, growth hormone-releasing factor
Pituitary hormones		
MC_1–MC_5	Gs	Melanocortin set (MSH, ACTH)
TSHR	Gq, G12, Gs, Gi	Thyroid-stimulating hormone
FSHR	Gs	Follicle-stimulating hormone
LHR	Gq/11, Gs, Gi	Luteinizing hormone
Pineal gland hormones		
MTR1–MTR3	Gi/o	Melatonin

Table 1.20. Examples of neuropeptide receptors and their main targeted G proteins (Source: [82]).

Type	Main transducer
Neuropeptide-B/W receptors	
NPBW1/2	Gi/o
Neurotensin receptors	
NTS1/2	Gq/11
Orexin receptors	
OX1/2	Gq/11

gastrointestinal, and immune systems. Urocortin-3 targets receptor CRHR2 that exists particularly in the nucleus tractus solitarius in rats. Injections of Ucn3 into the nucleus tractus solitarius decreases mean arterial pressure and heart rate, as well as greater splanchnic nerve activity [83]. Bilateral vagotomy abolishes Ucn3-induced bradycardia.

Urocortins are potent and long-lasting hypotensive agents that increase the coronary blood flow [84]. Urocortin primes its cardiovascular effect by binding on central and peripheral CRH receptors. Receptors CRHR2 are located in the myocardium as well as intramyocardial coronary vessels [85]. They are also detected in blood vessels, in particular the media of the internal mammary artery. Urocortin-1 to -3 produce a potent, sustained vasodilation that reverses the endothelin-1-induced constriction in endothelium-denuded internal mammary artery. Endothelium-independent vasodilator action of urocortin-2 and -3 can counterbalance centrally mediated pressor effect of CRH and urocortin-1.

1.4.2 Hypophysis

The pituitary gland, or hypophysis, is connected to the hypothalamus by the infundibulum. It is composed of 2 regions, the anterior lobe, or adenohypophysis, and the posterior lobe, or neurohypophysis. The *adenohypophysis* synthesizes: (1) adrenocorticotropic hormone, (2) gonadotropic hormones (LH, FSH), (3) growth hormone, (4) prolactin, and (5) thyroid-stimulating hormone. These hormones target: (1) the cortex of the adrenal glands, (2) gonads (ovaries and testes), (3) bones, muscles, and other organs, (4) mammary tissues, and (5) thyroid, respectively.

Growth hormone[67] is synthesized by somatotropes in the anterior pituitary. It has direct effects on protein, lipid, and glucid metabolisms and indirect effects via insulin-like growth factor-1 (IGF1) secreted by the liver and other tissues in response to growth hormone.[68] Growth hormone operates via its receptor GHR, the JaK2–STAT5 pathway, and Lyn-activated ERK1 and ERK2 kinases. Enzyme Lyn signals via phospholipase-Cγ to activate Ras. Conformational change in the extracellular domain of GHR homodimer affects the signaling choice between JaK2 and Lyn [86].

Hypothalamic growth hormone-releasing hormone (GHRH) stimulates the synthesis of the growth hormone.[69] It increases bone mineralization and muscle mass, promotes lipolysis and protein synthesis, reduces liver uptake of

[67] Growth hormone is also called somatropin or somatotropin.

[68] The growth hormone activates the proto-oncogene cFos, using the JaK–STAT (Vol. 3 – Chap. 11. Receptors of the Immune System) and Ras–MAPK pathways (Vol. 4 – Chap. 5. Mitogen-Activated Protein Kinase Modules).

[69] Pituitary growth hormone (GH) expression and secretion are mediated by 2 hypothalamic peptides: GH-releasing hormone (GHRH) and somatostatin, or somatotropin release inhibiting factor (SRIF), as well as ghrelin. Ghrelin acts on the GH secretagogue receptor (GHSR) in pituitary glands. Growth hormone

glucose, but favors liver gluconeogenesis, contributes to the maintenance of pancreatic islets, and stimulates the immune system.

Somatostatin produced by several body tissues inhibits growth hormone release (negative feedback loop), whereas gastric ghrelin stimulates growth hormone secretion. *Prolactin* promotes the growth of the mammary glands during pregnancy and lactation after birth. Melanocyte-stimulating hormones (MSH) stimulate the production of melanin.

The *neurohypophysis* (or posterior pituitary) secretes *vasopressin* (or antidiuretic hormone), *oxytocin*, and lipotropin, among other substances. Vasopressin derives from a preprohormone precursor that is synthesized in the hypothalamus and stored in vesicles at the posterior pituitary.[70] Stored vasopressin is then released into the blood stream in response to: (1) reduced blood volume mediated by baroreceptors in veins, carotid arteries, and atria; (2) augmented plasma osmolality via osmoreceptors in the hypothalamus; and (3) release of cholecystokinin by the small intestine. In addition, a certain fraction of vasopressin is also secreted directly into the central nervous system with a circadian rhythm by several populations of smaller neurons. Vasopressin is involved in aggression response, temperature regulation, as well as memory formation. In kidneys, it enhances water and salt reabsorption by the nephron, as it favors plasmalemmal insertion of aquaporin-2 water channels. It also provokes a moderate vasoconstriction.

Magnocellular neurosecretory cells that manufacture vasopressin are adjacent to magnocellular neurons that produce oxytocin. In the pituitary gland, oxytocin binds to its carrier neurophysin-1, like vasopressin to neurophysin-2. Oxytocin is released during orgasm, labor, and breastfeeding. Oxytocin causes contraction of smooth muscle cells in uterus walls and ducts of mammary glands. It is also a neurotransmitter in the brain. It can also slightly reduce the excretion of urine.

Secretin participates in the regulation of renal water reabsorption directly on nephrons and indirectly, as it stimulates vasopressin production in the hypothalamus and release from the posterior pituitary during chronic plasma hyperosmolality [87]. Secretin thus acts a neurosecretory hormone. Secretin that has been originally isolated from upper intestinal mucosa modulates water and electrolyte transport in pancreatic ductal cells, liver cholangiocytes, and epididymal epithelial cells.

---

receptor is also targeted by synthetic GH-releasing peptides (GHRP). Synthetic hexapeptide GHRP6 stimulates GH release, as it antagonizes somatostatin.

[70] Vasopressin is synthesized by hypothalamic magnocellular neurosecretory cells of the paraventricular and supraoptic nuclei and travels along axons through the infundibulum within neurosecretory granules, particularly within Herring bodies, to nerve terminals and the posterior pituitary gland.

1.4.3 Epiphysis

The *pineal gland*, or epiphysis cerebri, located between the lateral thalami, contains pinealocytes, which synthesize *melatonin*. Melatonin is an antioxidant, controlled by circadian rhythm. Melatonin is anti-gonadotropic, inhibiting the secretion of LH and FSH in the anterior pituitary. The production of melatonin is stimulated by darkness and inhibited by light. The precursor to melatonin is the neurotransmitter serotonin, itself derived from tryptophan. Melatonin receptors are G-protein-coupled receptors, mainly found in the suprachiasmatic nucleus of the hypothalamus, anterior pituitary, and retina.

1.4.4 Thyroid and Parathyroids

1.4.4.1 Thyroid Hormones

The thyroid gland in the neck produces the thyroid hormones, *thyroxine* (T_4; about 95% of active thyroid hormones), and *tri-iodothyronine* (T_3). The thyroid hormones increase the basal metabolic rate and sensitivity to catecholamines. They regulate protein, lipid, and glucid metabolism. Thyroid-stimulating hormone stimulates the secretion of T_3 and T_4 hormones. Concentrations of thyroid hormones are kept within physiological ranges by feedback loops on the hypothalamus–hypophysis–thyroid axis.

Thyroid hormones T_3 and T_4 are carried in the blood bound to plasma proteins: *thyronine-binding globulin*, a glycoprotein that has a higher affinity for T_4 than T_3, *transthyretin* that has a higher affinity for T_3 than for T_4, and albumin with low affinity, but high capacity. Thyroid hormones have specific transport proteins on plasma membranes.

Thyroid hormone tri-iodothyronine (T_3) quickly relaxes vascular smooth muscle cells, as it activates the PI3K–PKB (phosphatidylinositol 3-kinase–protein kinase-B) pathway to produce nitric oxide by neuronal (NOS1), endothelial (NOS3), and inducible (NOS2) nitric oxide synthase in these cells [88].[71] Therefore, thyroid hormone raises nitric oxide production in both vascular endothelial and smooth muscle cells to promote vasodilation by endothelium-dependent and -independent mechanisms. Tri-iodothyronine markedly lowers angiotensin-2-induced phosphorylation of myosin light chains. Thyroid hormone is thus able to decrease peripheral vascular resistance.

1.4.4.2 Calcitonin

Calcitonin, also secreted by the thyroid gland, reduces calcemia. Furthermore, calcitonin lowers phosphate levels in the plasma.

[71] Tri-iodothyronine significantly increases the expression of all the NOS isoforms in smooth muscle cells, but only NOS1 and NOS2 play a significant role on NO production by these cells.

Table 1.21. Receptors of calcitonin (CtR), amylin (AmyRi or Amy$_i$, $i = 1,2,3$), CGRP (CGRPR), and adrenomedullin (AMRi or AM$_i$, $i = 1,2$), with their main targeted G proteins. The gene CALCR codes for calcitonin receptor; the gene CALCRL codes for calcitonin receptor-like receptor. (Source: [82]).

Type	Main transducer	Ligand
Calcitonin receptor		
CtR (CalcR)	Gs, Gq	Calcitonin, calcium, NO
AmyR1 (CalcR + RAMP1)	Gs	Amylin, calcium, NO
AmyR2 (CalcR + RAMP2)	Gs	
AmyR3 (CalcR + RAMP3)	Gs	
Calcitonin receptor-like receptor		
CGRPR (CalcRL + RAMP1)	Gs, Gq	Calcitonin gene-related peptide
AMR1 (CalcRL + RAMP2)	Gs	Adrenomedullin
AMR2 (CalcRL + RAMP3)	Gs	Adrenomedullin, CGRP

The calcitonin family of peptides includes *calcitonin*, *amylin*, *adrenomedullin*, and *calcitonin gene related peptides* (CGRP; Table 1.21). Two forms of calcitonin gene-related peptide exists: CGRP1[72] and CGRP2.[73] Calcitonin gene-related peptide is a neuropeptide and potent vasodilator.

Calcitonin binds to *receptor activity-modifying protein* (RAMP). This Gs-coupled receptor component possesses a single transmembrane domain. It determines the functional specificity of calcitonin gene-related peptide receptors [89]. It interacts with G-protein-coupled receptors of class 2.[74] The calcitonin receptor does not require RAMP to translocate to the plasma membrane and to bind calcitonin.

1.4.4.3 Parathyroid Hormone

Parathyroid glands secrete *parathyroid hormone* (PTH) that regulates the blood calcium level, in opposition to calcitonin. Parathyroid hormone increases

[72] A.k.a. α-type CGRP and calcitonin-related polypeptide α (Calcα).

[73] A.k.a. β CGRP and Calcβ.

[74] Class 2 G protein-coupled receptors include receptors for calcitonin, parathyroid hormone, glucagon, and vasoactive intestinal peptide and pituitary adenylyl cyclase-activating polypeptide. Receptor of CGRP (CGRPR) is formed by RAMP1 and the calcitonin receptor-like receptor (CRLR). Adrenomedullin receptor is generated by the assembling of RAMP2 or RAMP3 with CRLR. The association between calcitonin receptor (CtR) and RAMPs yields amylin receptors (AmyR). Receptors RAMP1, RAMP2, and RAMP3 connected to CtR lead to AmyR1, AmyR2, and AmyR3, respectively.

calcemia by enhancing calcium release by bones, calcium reabsorption by nephrons, and calcium absorption by intestine.

Parathyroid hormone is able to increase intracellular calcium concentration in blood mononuclear cells and cardiomyocytes by activating voltage-dependent calcium channels.[75] Parathyroid hormone heightens diastolic and peak systolic calcium concentrations via the cAMP–PKA axis [90]. In addition, PTH attenuates the expression of calcium-sensing receptor.

It also decreases phosphate concentration in the blood. It stimulates osteoblasts to promote bone formation, whereas transforming growth factor-β1 that is locally produced hinders osteoblast differentiation [91]. Bone remodeling is, indeed, coordinately regulated by the opposing effects of PTH and TGFβ. PTH couples bone resorption and formation, as it simultaneously causes endocytosis of TβR2 and type-1 PTH receptor that suppresses both TGFβ and PTH signaling.

Parathyroid hormone receptor-1 (PTH_1) binds both parathyroid hormone and parathyroid hormone-related protein (PTHRP).[76] This ligand-bound G-protein-coupled receptor activates both adenylyl cyclase and phospholipase-C, and secondarily protein kinases PKA and PKC. The cAMP–PKA pathway is predominant. Parathyroid hormone receptor-2 coupled to adenylyl cyclase binds PTH. Microvascular endothelial cells express PTH-related protein (PTHRP). Both hormones PTHRP and PTH are ligands of cognate cardiac receptors.

1.4.5 Adrenal Glands

The adrenal glands, located over the kidneys, comprise an outer cortex and inner medulla. The adrenal medulla is regulated by the hypothalamus via a nervous electrochemical command (sympathetic nerves).

1.4.5.1 Adrenaline and Noradrenaline

The adrenal medulla secretes 2 hormones, adrenaline (or epinephrine) and noradrenaline (or norepinephrine). *Adrenaline* is implicated in the short-term stress reaction. In particular, it increases heart rate and stroke volume, constricts arterioles of the skin and gut, and dilates arterioles in leg muscles. Adrenaline acts on α1, β1, and β2 receptors (Vol. 3 – Chap. 7. G-Protein-Coupled Receptors). *Noradrenaline* acts mainly on α1 receptors.

[75] Hyperparathyroidism is commonly associated with cardiac hypertrophy. In patients with congestive heart failure, plasma PTH level is correlated to cardiac function. Whereas at high concentration PTH favors cardiac hypertrophy, picomolar, sustained (24 h) PTH concentration improves mechanical response to electrical stimulation.

[76] Parathyroid hormone-related protein controls the cell life (proliferation, differentiation, and apoptosis) and the development of several tissues.

Noradrenaline is a hormone secreted from the adrenal medulla, and also a neurotransmitter of noradrenergic neurons released during synaptic transmission. This stress hormone is involved in situations that need immediate attention and reaction. It activates the sympathetic nervous system, increasing heart rate and muscle readiness, releasing energy from lipid stores.

1.4.5.2 Adrenomedullin

The adrenal medulla, as well as the lungs, heart, kidneys, gastrointestinal tract, spleen, thymus, endocrine glands, brain, and vascular endothelial and smooth muscle cells produce adrenomedullin. *Adrenomedullin* is produced by cleavage of preproadrenomedullin. Adrenomedullin belongs to a regulatory peptide family that includes calcitonin gene-related peptide, amylin and calcitonin. The adrenomedullin (selective) receptors, derived from the calcitonin receptor-like receptor, increase cAMP level, according to the mode of interaction with receptor activity-modifying proteins.[77]

Adrenomedullin has paracrine effects in the vasculature, raising cAMP levels in vascular smooth muscle cells and nitric oxide production in endothelial cells. Adrenomedullin is a potent vasodilator, and thus a potent hypotensive peptide. Moreover, adrenomedullin protects endothelial cells from apoptosis. Its receptor is involved in angiogenesis, as adrenomedullin promotes angiogenesis. Adrenomedullin also acts on vascular permeability. Adrenomedullin enhances sodium excretion either by directly acting on nephron function or inhibiting aldosterone secretion. Adrenomedullin receptor expression is upregulated in cardiac hypertrophy secondary to hypertension.

Adrenomedullin acts via: (1) calcitonin gene-related peptide receptors (CGRPR) and (2) specific adrenomedullin receptors (AMR). The calcitonin receptor-like receptor (CRLR), which is homologous to calcitonin receptor, combines with the receptor activity-modifying protein (RAMP) to form a functional receptor.[78] The coexpression of CRLR with RAMP1 constitutes CGRP1 receptor, and CRLR with RAMP2 or RAMP3 yields adrenomedullin receptors (AMR) [92]. Two AMR subtypes exist, AMR1 (or AM_1), composed of CRLR and RAMP2, and AMR2 (or AM_2), made of CRLR and RAMP3 [93]. The cytosolic receptor component protein (RCP) binds to CGRPR and acts as a downstream regulator. Receptor RAMP1 exerts a dominant effect over RAMP2 to produce CGRP receptors rather than AMR in cells able to express both. The major signaling pathway is activated by AMR and CGRPR targets Gs protein and adenylyl cyclase. Activated CGRPR also increases intracellular calcium level.

---

[77] Coupled to CRLR, RAMP1 generates CGRP receptors, whereas RAMP2 and RAMP3 produce selective adrenomedullin receptors.

[78] Neither RAMPs nor CRLR function in the absence of the other protein.

1.4.5.3 Corticosteroids

The adrenal cortex produces *corticosteroids*, a class of steroid hormones with 3 subsets: (1) *glucocorticoids* (cortisol) that regulate glucid, lipid, and protid metabolism, in addition to their anti-inflammatory activity; (2) *mineralocorticoids* (corticosterone and aldosterone) that control the concentrations of sodium and potassium; and (3) *gonadocorticoids* in small amounts in both sexes, androgens (dehydroepiandosterone [DHEA] and androstenedione) and estrogens.

The inner regions of the adrenal cortex, the zona fasciculata and zona reticularis, synthesize glucocorticoids and adrenal androgens, glucocorticoid production being dominant, whereas the outer zona glomerulosa manufactures aldosterone.

Angiotensin-2 stimulates plasmalemmal receptor coupled to phospholipase C of zona glomerulosa cells. Aldosterone is synthesized in the zona glomerulosa by aldosterone synthase (CyP450-11b2). This synthase production is elicited by angiotensin-2 via protein phosphatase PP3 [94].[79]

Adrenocorticotropic hormone (ACTH) released by the hypothalamus regulates the cortisol production. Plasmalemmal ACTH receptors activate adenylyl cyclase. A feedback loop controls the circulating levels of corticotropin releasing hormone, ACTH, and cortisol.

Aldosterone is released by the adrenal gland either by volume loss mediated by angiotensin-2 or hyperkalemia. Aldosterone favors sodium reabsorption via the sodium channel in the mineralocorticoid-sensitive segments of distal nephron.

1.4.5.4 Catestatin

Catestatin is a small, cationic, hydrophobic peptide that results from cleavage of chromogranin-A.[80] Both catestatin and its precursor chromogranin are stored and released with catecholamines from the storage vesicles in chromaffin cells of adrenal medulla and adrenergic neurons [95].[81] Human variants of catestatin exist with different potency for inhibition of catecholamine secre-

[79] A.k.a. calcineurin and protein phosphatase PP2b.

[80] Several active products that result from chromogranin-A cleavage coexist in secretory granules: (1) vasodilator vasostatin that inhibits parathyroid hormone release; (2) pancreastatin that impedes glucose-stimulated insulin release from pancreatic islets; and (3) catestatin that hampers catecholamine release from adrenal medulla and neurons.

[81] In chromaffin cells and adrenergic neurons, catcholamines are stored not only with chromogranins and its cleavage products, but also adenosine triphosphate, acetylcholine, neuropeptide-Y, and calcium. Both catestatin and neuropeptide-Y prevent catecholamine secretion, but catestatin is much more potent than neuropeptide-Y.

tion.[82] A catestatin isoform has central action, as it acts on nicotinic (cholinergic) synapses of the cardiovascular–baroreceptor center of the nucleus tractus solitarius in the brainstem to increase baroreceptor sensitivity and cardiac parasympathetic activity and decrease cardiac sympathetic activity [95].

1.4.6 Gonads

Gonads not only produce sperm and ova, but also *steroid hormones*, androgens and estrogens. *Inhibins*, which prevent FSH synthesis, are counteracted by *activins*. Inhibins are synthesized in the ovary. Four activin genes exist. The activities of activins are regulated by inhibitors *follistatins*.[83] Follistatin can bind heparan sulfate proteoglycans, like many growth factors, which rely on heparan sulfate for their signaling. Activins participate in LH-stimulated androgen synthesis in the ovary and testis. Activin also enhances spermatogenesis. Activin is produced not only in the gonads, but also in other organs, such as the pituitary gland, placenta, etc. *Anti-Müllerian hormone* (AMH) is secreted in males during embryo- and fetogenesis to prevent the development of the Müllerian structures (vagina, uterus and cervix, and oviducts). It is produced in women after puberty, whereas it decays in men during puberty. Activins, inhibins, and AMH belong to the transforming growth factor-β family (Sect. 3.8.2).

Gonadal steroid hormones, testosterone, progesterone, and estradiol are produced by the testes and ovaries under the control of follicle-stimulating hormone and luteinizing hormone by the pituitary gland on the one hand and gonadotropin-releasing hormone by the hypothalamus on the other hand. Low levels of circulating sex hormones reduce feedback inhibition on GnRH synthesis (long negative feedback loop), leading to elevated FSH and LH. Both FSH and LH act via the cAMP–PKA pathway after binding to plasmalemmal receptors.

Estrogens (estradiol, estriol, and estrone) are steroid hormones present in both men and women, but at higher levels in women of reproductive age. They promote the development of female secondary sex characteristics, and regulate the menstrual cycle. *Progesterone* is a steroid hormone produced in the adrenal glands, gonads, and brain that is involved in the menstrual cycle, pregnancy, and embryogenesis.

Androgens are steroid hormones that control the development and maintenance of male characteristics. Adrenal androgens constitute a subset of androgens. Androgens are estrogen precursors. Androgens promote protein synthesis. *Testosterone* has virilizing (maturation of the sex organs) and anabolic (muscle growth and bone maturation) effects.

---

[82] The CHGA gene encodes chromogranin-A. It consists of 8 exons separated by 7 introns and belongs to chromosome 14q32 [95].

[83] Follistatins belong to a protein group with noggin, gremlin, and chordin.

Testosterone (plasma concentration in adult men 11–36 nmol/l) as well as other androgens such as 5α-dihydrotestosterone (plasma concentration 1.0–2.9 nmol/l) cause rapid vasorelaxation of vascular smooth muscle cells in large arteries (aorta and coronary and umbilical arteries) at micromolar concentrations and small resistance arteries (mesenteric, prostatic, subcutaneous, and pulmonary) at nanomolar (physiological) concentrations via endothelium-dependent and -independent mechanisms [96].

Endothelium-dependent mechanisms, particularly nitric oxide-dependent process, may be responsible for SMC relaxation caused by testosterone at physiological concentration (<100 nmol) [96]. Nitric oxide-dependent processes rely on NOS3 and NOS1 synthesized in vascular endothelial and smooth muscle cells, respectively. Yet, at supraphysiological (>100 nmol) and pharmacological concentrations (>10 μmol), testosterone and its metabolites induce endothelium-independent relaxation [96].

Vasodilation then results from the inactivation of voltage-gated Ca_V1 channels and/or activation of K_V and large-conductance Ca^{++}-activated K^+ channels (BK; Vol. 3 – Chap. 3. Main Classes of Ion Channels and Pumps). In smooth muscle cells, testosterone augments intracellular cGMP concentration via NOS1 and its soluble guanylyl cyclase receptor. Accumulation of cGMP stimulates BK channel activity at nano- and micromolar concentrations [96].

Testosterone also raises the intracellular cAMP concentration via the sex hormone-binding globulin–testosterone complex that binds to the plasmalemmal sex hormone-binding globulin receptors. Messenger cAMP then triggers the phosphorylation of protein kinase-A that subsequently phosphorylates either phospholamban, an inhibitor of sarcoplasmic reticulum Ca^{++} ATPase, phospholipase-C that produces IP_3, or inositol trisphosphate receptor.

Testosterone can be converted into 17β-estradiol by vascular cytochrome-P450 aromatase. 17β-Estradiol causes acute and long-term vasodilation. However, inhibition of CyP450 aromatase and estrogen receptor antagonism do not preclude testosterone-induced vasodilation [96]. Furthermore, dihydro- (e.g., 5α- and 5β-dihydrotestosterone)[84] and tetrahydro-androgens cannot be transformed into estrogens.

1.4.7 Other Endocrine Organs

1.4.7.1 Pancreas

The pancreas has exocrine regions, which secrete digestive enzymes, and endocrine zones, the pancreatic islets, which release glucagon and insulin.

[84] Testosterone metabolites 5α- and 5β-dihydrotestosterone act as signaling mediators in the androgen target tissues in which they are formed [96]. In addition, 5β-dihydrotestosterone, a potent vasodilator, has little or no affinity for the intracellular androgen receptor (Vol. 3 – Chap. 6. Receptors). On the other hand, 5α-dihydrotestosterone possesses a high affinity for the androgen receptor, hence, a high androgenic activity.

Insulin and glucagon have opposite effects on the hepatic control (glucose storage and delivery) of glycemia.

Insulin is synthesized by the pancreatic β cells. Insulin increases the synthesis of glycogen and fatty acids, as well as amino acid and potassium uptake. Furthermore, it decreases proteinolysis, lipolysis, and gluconeogenesis.

Glucagon is produced by the pancreatic α cells. It converts glycogen into glucose in the liver and releases the latter into the blood. In the pancreatic islets, α cells secrete glucagon in response to low glycemia, whereas β cells release insulin in response to high glycemia. The glucagon receptor mediates the actions of glucagon on carbohydrate metabolism in the liver and insulin release by the pancreatic β cells. Glucagon G-protein-coupled receptors, positively and negatively regulated by glucose and glucagon, respectively, increase the intracellular cAMP level.

The insulin receptor is composed of 2 extracellular insulin-binding α and 2 transmembrane β subunits. The β subunits have cytoplasmic ATP-binding and Tyr kinase domains. Insulin binding induces autophosphorylation of β subunits, activating the catalytic activity of the receptor. Several intracellular proteins serve as substrates for the insulin receptor, such as insulin receptor substrate-1. Binding of insulin to receptors on such cells causes fusion of cytoplasmic vesicles containing glucose transporters GluT4 with the plasma membrane and insertion of the glucose transporters into the membrane. In the absence of insulin, glucose transporters are removed into the cytoplasm. Certain substances can alter the conformation of the cytoplasmic kinase domain or bind to modulator binding sites of the insulin receptor.

Amylin is also produced by pancreatic β cells. It inhibits food intake and postprandial glucagon secretion [97]. It also opposes insulin activity in skeletal muscles, stimulating glycogen degradation.

1.4.7.2 Thymus and Placenta

Other organs, the heart, thymus, digestive tract, and placenta produce hormones. *Thymosin*, produced by the thymus gland, is involved in the development of the immune system. The placenta is a temporary endocrine gland. Human chorionic gonadotropin (hCG), produced during pregnancy, maintains progesterone production. Human placental lactogen (hPL; or chorionic somatomammotropin) increases the production of insulin and IGF1, as well as insulin resistance.

1.4.7.3 Heart

The heart manufactures *natriuretic peptides* (Sect. 1.4.9.5). The kidneys produce *renin* (Sect. 1.4.9.1), *erythropoietin* (Vol. 5 – Chap. 2. Hematopoiesis), and *calcitriol* (vitamin-D3). Calcitriol increases the calcium absorption in the gastrointestinal tract.

1.4.7.4 Digestive Apparatus

The digestive tract[85] with its serial organs (stomach, duodenum, jejunum, ileum, and colon [in fact, all conduits from the mouth to the anus]) secretes various hormones acting via receptors (Table 1.22).

Cholecystokinin (CCK; or pancreozymin) causes the release of digestive enzymes from the pancreas and bile from the gallbladder. *Enteroglucagon* is derived from preproglucagon. *Gastrin* stimulates secretion of hydrochloric acid and pepsin for food digestion. *Glucose-dependent insulinotropic peptide* (GIP) induces insulin secretion. *Motilin* increases gastrointestinal motility and stimulates the production of pepsin.

Secretin is produced in the S cells of the duodenum crypts of Lieberkühn. Secretin stimulates the secretion of bicarbonate from the liver, pancreas, and duodenal Brunner glands to buffer the acidic chyme. It also reduces acid secretion from the stomach by inhibiting gastrin release from G cells. It enhances the effects of cholecystokinin. It promotes growth and maintenance of the pancreas.

Vasoactive intestinal peptide (VIP) stimulates the secretion of water and electrolytes in the intestine, dilates intestinal smooth muscles, activates pancreatic bicarbonate secretion, and inhibits gastrin effects. It can act on the circadian rhythm. It also induces coronary vasodilation and has positive inotropic and chronotropic effects.

Neuromedin-B (brain) and *gastrin-releasing peptide* (GRP) of the bombesin family stimulate gastrin release from G cells. They form with cholecystokinin a negative feedback to stop eating. They activate 3 G-protein-coupled receptors (BBR1–BBR3). Activated bombesin receptors stimulate tissue growth, smooth muscle contraction, and secretion in particular. Neuromedin-B and *neuromedin-C* (digestive tract) elicit amylase and insulin release.

Neurotensin is a neuropeptide found in several organs, such as the brain and intestine. It regulates luteinizing hormone and prolactin release. Epidermal growth factor and transforming growth factor-α stimulate the synthesis in cultured rat hepatocytes, which is inhibited by transforming growth factor-β [98]. Insulin, vasopressin, or angiotensin-2 cooperate with low concentrations of epidermal growth factor, but this amplifying effect does not occur with neurotensin-related peptides, such as kinetensin, a nonapeptide and histamine releaser, and neuromedin-N. Neurotensin acts as a comitogen, like insulin, vasopressin, and angiotensin-2, which can regenerate the liver.

1.4.8 Granin Family

The granin (or chromogranin–secretogranin) family of neuroendocrine secretory proteins yields sensors for regulators of cell metabolism and functioning,

[85] The digestive apparatus includes the accessory organs of digestion, such as accessory digestive glands (salivary glands, liver and gallbladder, and exocrine pancreas) as well as digestive structures such as teeth.

Table 1.22. Receptors of digestive tract hormones and related substances and their main G-protein transducers. Glucagon-like peptide-1 (GLP1), of short life duration, binds to a G-protein-coupled receptor, and potentiates insulin secretion in response to food intake (Source: [82]).

Type	Main transducer	Ligand
Ghrelin R.	Gq/11	Ghrelin
Secretin R.	Gs	Secretin
CCKR1	Gq/11, Gs	Cholecystokinin, gastrin
CCKR2	Gs	
APJ	Gi/o	Apelin
BBR1–BBR3	Gq/11	Gastrin-releasing peptide, neuromedin-B/C
Glucagon R.	Gs	Glucagon
GLPR1/R2	Gs	Glucagon-like peptide-1/2
GIPR	Gs	Glucose-dependent insulinotropic polypeptide (gastric inhibitory polypeptide)

such as plasma oxygen, glucose, and calcium. Granins comprise acidic soluble proteins, such as 48-kDa chromogranin-A (CGa),[86] 76 kDa chromogranin-B (CGb),[87] and 67-kDa chromogranin-C (CGc).[88] They regulate storage and release under various stimuli in the extracellular space and blood circulation of unprocessed prohormone chromogranin-A and its processing products with hormones, such as catecholamines, insulin, and parathyroid hormone. In addition, antimicrobial peptides result from the natural processing of CGa and CGb [99].

Chromogranin-A resides in secretory vesicles of neurons and endocrine cells, such as chromaffin cells of the adrenal medulla, cells of the paraganglia, enterochromaffin-like cells, and β cells of the pancreas. This prohormone belongs to the family of acidic proteins that are costored and cosecreted with other hormones and peptides from the diffuse neuroendocrine system. Other cells such as cardiomyocytes can also produce and secrete chromogranin-A and its derived peptides. Chromogranin-A is a low-affinity, high-capacity Ca^{++}-binding protein. Chromogranin-A causes the activation of microglial cells (resident macrophages in the central nervous system).

Chromogranin-A is cleaved by a prohormone convertase[89] into several functional peptides that act locally (auto- or paracrine factors) and/or remotely. It is, indeed, the precursor to [626]: (1) β-*granin* (CGa_{1--128}) in pancreatic β cells; (2) *vasostatin*-1 (VS1; CGa_{1--78}) and (3) vasostatin-2

[86] A.k.a. parathyroid secretory protein-1.

[87] A.k.a. secretogranin-1.

[88] A.k.a. secretogranin-2.

[89] Cleavage is carried out by intragranular enzymes, not only prohormone convertase-1 to -3, but also neuroendocrine-specific carboxypeptidase E/H, and Lys- and Arg-aminopeptidases [99].

Table 1.23. Effects of chromogranin-A-derived peptides vasostatin-1 and catestatin (Source: [101]).

Vasostatin-1	Catestatin
Vasoconstriction inhibition	Catecholamine release suppression
Endothelial barrier integrity	Baroreceptor activation
Negative inotropic effect	Negative inotropic effect
Microglial activation	Histamine release activation
Inhibition of VEGF-induced cell proliferation and migration	Inhibition of nicotinic acetylcholine receptor desensitization
Ischemia preconditioning	
Cell adhesion activation	
PTH release inhibition	

(VS2; CGa_{1--114}) in neutrophils and sympathetic axons in the spleen that act as vasoinhibitors and antibacterial agents (against Gram+ bacteria, fungi, and yeasts); (4) *chromofungin* (CGa_{47--66}) that binds to calmodulin in the presence of Ca^{++} and inhibits calmodulin-dependent phosphatase PP3; (5) *pancreastatin* ($CGa_{250--301}$) that precludes basal and insulin-stimulated glucose transport, lactate production, lipogenesis, and exocrine pancreatic and gastric acid secretion, as well as insulin release, but stimulates glucagon secretion; (6) *catestatin* ($CGa_{344--364}$), a vasodilator and an inhibitor of catecholamine release from chromaffin cells and noradrenergic neurons; (7) *cateslytin*, the active domain of catestatin ($CGa_{344--358}$), an antimicrobial agent against bacteria, fungi, and yeasts; (8) *parastatin* ($CGa_{347--419}$) in the parathyroid gland, where it hampers the secretion of parathyroid hormone and chromogranin-A; (9) *chromostatin* that prevents chromaffin-cell secretion; (10) *chromacin* ($CGa_{173--194}$) that inhibits growth of both Gram+ and Gram− bacteria; as well as (11) WE14 ($CGa_{316--329}$) that can modulate histamine release from mastocytes and (12) GE25 (C-terminus of chromogranin-A) predominantly in the pituitary gland, intestine, and pancreas.

At least some of these peptides inhibit cell action. They impede auto- and paracrine functions of neuroendocrine cells as well as activity of cells of various tissues, including the cardiovascular system. Vasostatin-1 and catestatin can be considered as regulators of the cardiovascular system. Pancreastatin and vasostatin-1 may operate as regulators of glucose and calcium homeostasis, respectively [101] (Table 1.23).

Vasostatin-1 precludes the phosphorylation (activation) of P38MAPK (Vol. 4 – Chap. 5. Mitogen-Activated Protein Kinase Modules) via $G\alpha_{i/o}$ subunit of guanine nucleotide-binding protein (Vol. 4 – Chap. 8. Guanosine Triphosphatases and their Regulators) in pulmonary and coronary arterial endothelial cells. Furthermore, it modulates myocardial inotropy via nitric oxide liberation from endocardial and coronary endothelial cells and the NO–sGC–cGMP–PKG signaling axis (sGC: soluble guanylate cyclase; cGMP: cyclic guanosine monophosphate; PKG: protein kinase-G). Vasostatin-1 primes inward rectifying G-protein-coupled K^+ channels (GIRKs). Both vasostatin-1 and catestatin inhibit adrenoceptor activation of cardiomyocytes in a noncompetitive manner.

Catestatin prevents catecholamine release from chromaffin cells and central and peripheral neurons. In addition, it may modulate baroreceptor activity. Like vasostatin-1, it favors nitric oxide liberation from vascular endothelial cells. In cardiomyocytes, it may inhibit endothelin-1 receptor. In mastocytes, catestatin can trigger histamine release via $G\alpha_{i/o}$-protein-regulated signaling.[90]

Chromogranin-B is an heparin-binding protein. It is released from cells of pancreatic islets, where it can exert an autocrine inhibitory effect on islet amyloid polypeptide action and insulin secretion [626]. Secretolytin is a product of the proteolysis of chromogranin-B ($CGb_{614--626}$). It acts as an antibacterial agent at micromolar concentrations.

Chromogranin-C corresponds to gonadotrope polypeptide GP87 released from pituitary cells upon gonadotropin-releasing hormone stimulation and 84–86-kDa tyrosine-sulfated protein (TSP86-84) in endocrine cells [626]. Secretoneurin derives from chromogranin-C in sensory afferent C-fibers under the control of hypothalamic magnocellular neurons. Secretoneurin can trigger migration of endothelial cells, monocytes, and eosinophils. In addition, it can prime the release of dopamine. This angiogenic neuropeptide stimulates the proliferation of endothelial cells and prevents their apoptosis. Manserin also derives from chromogranin-C in endocrine cells of the pituitary anterior lobe and neurons of several hypothalamic nuclei. Protein EM66 results from post-translational processing of secretogranin-2 in gonadotrophs, lactotrophs, thyrotrophs, and melanotrophs (but not corticotrophs) as well as in chromaffin cells in the adrenal medulla.

Secretogranin-3 encoded by the human SCG3 gene is a component of intraneuronal vesicles in axons and vesicle-like structures in dendrites [626]. It functions as a sorting receptor. It binds to chromogranin-A and targets it to the secretory granules in pituitary corticotrope cells, chromaffin cells, and other endocrine cells. It also links to adrenomedullin in chromaffin cells.

Secretogranin-4 is produced in gut endocrine cells (somatostatin [D cells] and pancreatic polypeptide [PP cells] cells of the gastrointestinal mucosa), thyroid parafollicular cells (C cells), adrenal medullary and pancreatic islet cells, anterior pituitary cells, specific hypothalamic neuroendocrine cells, and large pyramidal cells of the cerebral cortex [626].

Secretogranin-5 encoded by the human SCG5 gene is a secretory protein.[91] It lodges in secretory vesicles of endocrine cells and neurons. It intervenes in

[90] Upon activation, heterotrimeric G proteins dissociate into $G\alpha^{GTP}$ and $G\beta\gamma$ dimer that can interact with signaling effectors.

[91] A.k.a. secretory granule neuroendocrine protein-1 and pituitary polypeptide P7B2.

the proteolysis (activation) of proprotein convertase-2 [626]. Yet, it can keep the enzyme transiently inactive.

55-kDa Neuroendocrine secretory protein NESP55 is synthesized in the adrenal medulla, anterior and posterior pituitary gland, and various regions of the brain. It undergoes proteolysis. Among derived products, the tetrapeptide Leu-Ser-Ala-Leu (LSAL) antagonizes $5HT_{1B}$ receptor [626].

1.4.9 Main Endocrine Regulators of Blood Circulation

Late-adaptive mechanism of the regulation of blood circulation is provided by kidneys, as they control the volemia through Na^+ and water reabsorption under action of the renin–angiotensin–aldosterone system (Vol. 6 – Chap. 3. Cardiovascular Physiology). The renin–angiotensin system participates in the regulation of hydroelectrolyte balance, arterial tone, and blood pressure. Furthermore, it modulates growth and differentiation of vascular smooth muscle cells and cardiomyocytes.

1.4.9.1 Renin–Angiotensin System

Renin is synthesized, stored, and released by granular cells of the juxtaglomerular apparatus of the afferent arteriole walls at the entrance of the glomerulus in kidneys.[92] Renin is the rate-limiting enzyme in the renin–angiotensin system that is responsible for the control of blood pressure and fluid homeostasis. Sympathetic stimulation via β1-adrenergic receptors, renal artery hypotension, and decreased Na^+ delivery to the distal tubules stimulate the release of renin by the kidney. Increased intracellular cAMP concentration provokes renin release from juxtaglomerular cells.[93]

Renin cleaves angiotensinogen into angiotensin-1 (ATn1). Angiotensin-converting enzyme ACE1 converts angiotensin-1 into angiotensin-2 (ATn2). Angiotensin-2 constricts the arterioles, thereby raising systemic vascular resistance and arterial pressure. Angiotensin-2 acts on the adrenal cortex to release aldosterone that increases Na^+ and water retention by kidneys.

---

[92] In embryos, renin precursor cells reside in the metanephric mesenchyme. They give rise to juxtaglomerular and arteriolar smooth muscle cells, even before vessel development. Later, in fetuses, renin-containing cells localize to large intrarenal arteries and in the glomerular interstitium. As maturation continues, the number of renin-containing cells decays. Nonetheless, when homeostasis is disturbed, as in the case of a low salt intake, the number of renin-expressing cells in preglomerular arterioles can rise.

[93] The cAMP-response element (CRE) region in the renin promoter is controlled by cAMP via CRE-binding protein (CREB) with coactivators CREB-binding protein (CREBBP or CBP) and P300 that facilitate action of transcription factor TF2b to initiate transcription. Histone acetyltransferases CBP and P300 are required for renin synthesis [102].

Kinase WNK4 regulates the balance between ion reabsorption and secretion by inhibiting activity of renal transporters and channels, such as Na^+–Cl^- cotransporter, epithelial Na^+ channel (ENaC), and K^+ channel ROMK ($K_{IR}1.1$), as well as modulating paracellular Cl^- conductance. Phosphorylation of WNK4 by aldosterone-activated kinase SGK alleviates inhibition of ENaC and ROMK, hence favoring K^+ secretion. On the other hand, angiotensin-2 abrogates inhibition of Na^+–Cl^- cotransporter by WNK4, without influencing inhibition of ROMK by WNK4 [105]. Angiotensin-2-mediated activation of Na^+–Cl^- cotransporter requires Ste20-type kinase SPAK that is also inhibited by WNK4 in normovolemia. Angiotensin-2 can then react to hypovolemia by causing Na^+ and Cl^- reabsorption without concomitantly increasing K^+ secretion, whereas aldosterone is released by hypovolemia or hyperkalemia to promote Na^+ and Cl^- reabsorption and restore intravascular volume or K^+ secretion.

Angiotensin-2 acts via angiotensin type 1 receptors (AT1R or AT_1; Vol. 3 – Chap. 7. G-Protein-Coupled Receptors). These receptors abound in most cell types, especially epithelial cells along the nephron, in the glomerulus, and renal blood vessels. Once activated, they promote sodium reabsorption by stimulating both sodium–proton antiporter and sodium–potassium ATPase on apical (luminal) and basolateral parts of the plasma membrane, respectively, in the proximal tubule of the nephron. These G-protein-coupled receptors stimulate epithelial sodium channels in the collecting ducts. Furthermore, activated vascular AT1Rs provoke a vasoconstriction that subsequently reduces renal blood flow and sodium excretory capacity. Angiotensin-2 receptors type 2 (AT2R or AT_2) are much less abundant and can counteract pressor and other effects of AT1Rs. Upon angiotensin-2 binding, AT1R couples to Gq/11 protein to prime the inositol triphosphate–calcium axis.

Angiotensin-2 not only serves as a vasoactive peptide, but also as a growth factor. It stimulates cytosolic and mitochondrial oxidant production. It generates focal adhesion changes and hence regulates the association of the cytoskeleton with the extracellular matrix as well as organization of both intra- and extracellular media.

The renin–angiotensin system not only contain an endocrine axis, but also local elements for paracrine and intracrine regulation. It indeed includes multiple processing enzymes (thimet oligopeptidase, neprilysin, prolyl oligopeptidase, and ACE homologs ACE2) as well as active angiotensin-2-derived fragments (angiotensin-3 and -4), heptapeptide angiotensin$_{(1--7)}$, and proangiotensin-12. Cleavage of angiotensinogen or angiotensin-1 into smaller active peptides and/or inactive fragments can be carried out not only by ACE1 and ACE2, but also carboxypeptidase and chymase.

In rats, proangiotensin-12 has been isolated from small intestine, where it reaches its highest level, but resides in many organs. It is also detected in blood circulation, thus behaving as a hormone. Chymotrypsin-like serine peptidase chymase that is synthesized by mast, endothelial, and mesenchymal cells that are widely expressed in tissues, but not ACE1, processes proangiotensin-12 into angiotensin-2 [103]. Chymase is the main enzyme for ATn2 formation in tissues, whereas ACE1 is the dominant enzyme in vascular endothelia. Vasoconstrictor proangiotensin-12 that reduces coronary flow operates via AT1R. On the other hand, vasodilator and antimitogen $ATn_{(1--7)}$ antagonizes cardiovascular effects of ATn2.

A local renin–angiotensin system exists independently of the circulating system. Propeptide $angiotensin_{(1--12)}$ cleaved from angiotensinogen can act as an alternate precursor for the local production of angiotensins (angiotensin-1 and -2, as well as $angiotensin_{(1--7)}$) involved in intra-, auto-, or paracrine signaling. $Angiotensin_{(1--12)}$ is predominantly produced by ventriculomyocytes and, to a lesser extent, by vascular endothelial and smooth muscle cells [104]. It is also synthesized in the proximal, distal, and collecting renal tubules. Increased cardiac $angiotensin_{(1--12)}$ concentrations in spontaneously hypertensive rats is associated with a small reduction in renal $angiotensin_{(1--12)}$ levels. $Angiotensin_{(1--12)}$ causes arterial wall constriction that is abolished by inhibition of angiotensin-converting enzyme or AT1Rs.

1.4.9.2 Aldosterone

Aldosterone regulates the concentrations of Na^+ and K^+, as it favors Na^+ reabsorption in kidneys upon binding mineralocorticoid receptors in epithelial cells of distal nephron. Aldosterone also promotes vascular inflammation. It is indeed able to trigger exocytosis (Vol. 1 – Chap. 9. Intracellular Transport) and secretion of Weibel-Palade bodies in endothelial cells via the mineralocorticoid receptor [106]. Consequently, P-selectin is externalized and von Willebrand factor released (Vol. 5 – Chap. 9. Endothelium),[94] thereby favoring leukocyte adherence to endothelial cells for extravasation. Aldosterone also stimulates endothelial expression of intercellular adhesion molecule-1. Moreover, it excites mineralocorticoid receptors in vascular smooth muscle cells to increase expression of inflammatory substances, such as interleukin-6, plasminogene activator inhibitor PAI1, and regulator of the immune system CTLA4 (cytotoxic T-lymphocyte antigen). Aldosterone can also cause oxidative stress, as it raises activity of NADPH oxidase subunits P22phox and GP91phox in the vascular wall (Vol. 4 – Chap. 9. Other Major Signaling Mediators) that lead to an increased production of reactive oxygen species. Furthermore, aldosterone attenuates endothelial expression of glucose-6-phosphate dehydrogenase, hence augmenting oxidative stress and reducing nitric oxide activity. Finally, aldosterone activates mitogen-activated protein kinase and protein kinase-A, -C, and -D in vascular smooth muscle cells. It is able to

[94] Weibel-Palade bodies contain von Willebrand factor and P-selectin, whereas other endothelial vesicles store tissue thromboplastin activator. Whereas thrombin induces secretion of both tissue thromboplastin activator and von Willebrand factor, aldosterone selectively causes release of von Willebrand factor, but not tissue thromboplastin activator.

provoke calcium influx. It also stimulates phosphatidylinositol 3-kinase in endothelial cells.

1.4.9.3 Vasopressin

Vasopressin (or antidiuretic hormone, ADH) is secreted by the posterior pituitary. It increases water retention by the kidneys. Angiotensin-2 stimulates the release of vasopressin.

1.4.9.4 Catecholamines

Angiotensin-2 favors noradrenaline release from sympathetic nerve endings and inhibits noradrenaline reuptake by nerve endings, hence enhancing the sympathetic function.

Acute increase in noradrenaline stabilizes perfusion of crucial organs by redirecting blood flow away from skin, muscles, kidney, and gut, and increasing heart frequency and contractibility (but also energy consumption). Noradrenaline operates via β-adrenergic receptors, stimulatory G proteins, adenylyl cyclase, second messenger cAMP, and cAMP-dependent protein kinase-A. Desensitization is due to G-protein-coupled receptor kinase GRK5 that phosphorylates noradrenaline-bound β adrenergic receptors and causes uncoupling from stimulatory G protein and receptor endocytosis.

Sustained activation of the sympathetic nervous system and elevated plasma concentration of noradrenaline leads to adverse effects by developing cardiomyocyte maladaptive hypertrophy and imposing peripheral vasoconstriction that increases heart load. Although the amount of β-adrenergic receptors and amplifier protein phosphatase inhibitor-1 decays, G-protein-coupled receptor kinase GRK2 and inhibitory G proteins become abundant in cardiomyocytes of failing heart. In the long term, cardiomyocytes that lose responsiveness to noradrenaline cannot adapt to exercise, but are protected against prolonged β-adrenergic overstimulation.

1.4.9.5 Natriuretic Peptides

Natriuretic peptides (or natriuretic factors) control the body fluid homeostasis and blood volume and pressure. The primary sites of synthesis of natriuretic peptides are heart and brain. Additional production sites include intestine and kidney [107].

Natriuretic Peptide Family

The natriuretic peptide family includes atrial- (ANP, or A-type; 28 amino acids), B-type (BNP; 32 amino acids), C-type (CNP; 22 amino acids), D-type natriuretic peptide (or dendroaspis natriuretic peptide;[95] DNP; 38 amino

[95] Dendroaspis natriuretic peptide has been isolated from the venom of green mamba snake Dendroaspis angusticeps.

Table 1.24. Natriuretic peptide genes (Source: [108]).

Gene	Name	Aliases
NPPA	Natriuretic peptide precursor-A	
NPPB	Natriuretic peptide precursor-B	
NPPC	Natriuretic peptide precursor-C	
NPR1	Natriuretic peptide receptor-1	NPRa, GCa
NPR2	Natriuretic peptide receptor-2	NPRb, GCb
NPR3	Natriuretic peptide receptor-3	NPRc
CORIN	Serine peptidase	ATC2, CRN, Lrp4, proANP convertase, heart specific serine peptidase
MME	Membrane metalloendopeptidase	Neprilysn, neutral endopeptidase

acids), and urodilatin (Uro; or ularitide; 32 amino acids). Urodilatin that was isolated from human urine is synthesized only in kidneys, mainly in the distal tubule of renal nephron. It is then secreted in the nephron lumen to exert a paracrine effect chiefly at the inner medullar collecting duct. Natriuretic peptides ANP, BNP, and CNP are produced from precursors that are synthesized by natriuretic peptide precursor genes NPPA, NPPB, and NPPC, respectively (Table 1.24). Natriuretic peptide precursors are cleaved into active peptides by the serine peptidase *corin*.[96]

Natriuretic peptides are endogenous hormones that are released by the heart in response to myocardial stretch. Atrial and brain natriuretic peptides are synthesized by cardiomyocytes as preprohormones, which are processed to yield prohormones and ultimately hormones. C-type natriuretic peptide is synthesized by endothelial and renal cells, among others.

Cardiac natriuretic peptides constitute a group of peptides that are synthesized from 3 different genes and then stored as 3 distinct prohormones: 126-amino acid atrial (ANP), 108-amino acid brain (BNP), and 103-amino acid C-type (CNP) natriuretic peptide prohormones. Whereas in the heart and other tissues ANP prohormone is cleaved between amino acids 98 and 99 to produce urodilatin and kaliuretic peptide, in kidneys, ANP prohormone is cleaved between amino acids 94 and 95. The ANP prohormone gives rise to 4 peptide hormones that regulate blood pressure and maintain plasma volume: (1) *long-acting natriuretic peptide* (LANP; amino acids 1–30 of ANP prohormone); (2) *vessel dilator* (amino acids 31–67 of ANP prohormone); (3) *kaliuretic peptide* (amino acids 79–98 of ANP prohormone and, in the kidney, amino acids 79–94); and (4) *urodilatin* (amino acids 99–126 of ANP prohormone and, in the kidney, amino acids 95–126). The BNP and CNP genes synthesize a single known peptide hormone. Peptides ANP and BNP

[96] A.k.a. proANP-converting enzyme (encoded by the gene CORIN).

Table 1.25. Affinity of natriuretic peptides for their receptors. Urodilatin binding to NPRs has similar affinity to that of ANP. LANP, vessel dilator, and kaliuretic peptide do not bind to NPR1, -2, or -3, as they have their own distinct receptors.

Type	Potency order
NPR1	ANP > BNP ≫ CNP
NPR2	ANP ∼ BNP ≪ CNP
NPR3	ANP > CNP > BNP

Table 1.26. Dissociation of complexes formed by natriuretic peptides and their receptors (Source: [110]).

	NPR1	NPR2	NPR3
ANP	low	large	low
BNP	moderate	very high	moderate
CNP	huge	moderate	moderate

are predominantly manufactured in atrial and ventricular myocardium, respectively.

Effects of Natriuretic Peptides

Natriuretic peptides are released into the circulation at a basal rate. Augmented secretion follows hemodynamical or neuroendocrine stimuli. Atrial natriuretic peptide exerts its activity not only as a circulating hormone, but also as an auto- and paracrine factor.

Natriuretic Peptide Receptors

Natriuretic peptides act by binding to natriuretic peptide receptors: guanylyl cyclase-coupled receptors NPR1 (NPRa or GCa) and NPR2 (NPRb or GCb) as well as clearance receptor (C-receptor; a.k.a. NPR3 and NPRc; Tables 1.25 and 1.26). Both NPR1 and NPR2 that are generated by genes Npr1 and Npr2 exist in many organs, such as blood vessel wall (endothelial and smooth muscle cells), heart, and kidney.

Non-catalytic receptor NPR3 lacks the guanylyl cyclase domain, but can influence cell functioning via Gi inhibitory subunit of guanine nucleotide binding protein [108]. In addition, NPR3 leads to natriuretic peptide endocytosis followed by recycling or lysosomal degradation (clearance; Table 1.27).

The second clearance mechanism is provided by enzymatic cleavage due to *membrane metalloendopeptidase* (MME). Both NPR3 and MME are expressed in multiple tissues (e.g., lung and kidney).

Table 1.27. Natriuretic peptides and their plasmalemmal targets (GC: guanylyl cyclase; MME: membrane metalloendopeptidase; NPR: natriuretic peptide receptor).

Target	Event
NPR1 (GCa)	GTP–cGMP–PKG or PDE
NPR2 (GCb)	GTP–cGMP–PKG or PDE
NPR3	Endocytosis and recycling or degradation
MME	Cleavage and clearance

Both ANP and BNP bind specifically to NPR1. C-type natriuretic peptide that acts as a local regulator is a selective ligand for NPR2. Natriuretic peptide receptor-2 is more abundant in veins than arteries. Activation of NPR2 by CNP in vascular smooth muscle cells induces vasorelaxation. Natriuretic peptides exert their activities via NPR1 and NPR2 by increasing intracellular concentration of cyclic guanosine monophosphate that targets protein kinase-G and phosphodiesterase.

Effects in Kidney

Natriuretic peptides are identified as diuretic–natriuretic mediators that control salt and water homeostasis and hormones that lower blood pressure. Natriuretic peptides have direct and indirect renal actions. In the adrenal cortex, natriuretic peptides preclude aldosterone synthesis and release. In addition, natriuretic peptides inhibit nuclear factor-κB and transforming growth factor-β [108].

Natriuretic peptide ANP increases renal blood flow. It inhibits renin release by kidneys, raises the glomerular filtration rate, and decreases the tubular sodium reabsorption (diuresis and natriuresis). Natriuretic peptides ANP and BNP can act synergistically, especially during cardiac stress.

Urodilatin exerts natriuretic and diuretic effects similar to those of ANP. Renal sodium excretion depends on auto-, para-, and endocrine regulation factors. In the inner medullar collecting duct, urodilatin inhibits Na^+ entry through epithelial Na^+ channels via NPR1. It also stimulates renal dopamine uptake in outer and juxtamedullar cortex as well renal medulla via organic cation transporters (OCT).[97] Dopamine then acts via its D_1 receptor. Dopamine heightens natriuresis by inhibiting sodium reabsorption through Na^+–K^+ ATPase in both proximal and distal tubules. Urodilatin and dopamine cooperatively further decrease Na^+–K^+ ATPase activity [111].

[97] Electrogenic OCTs and electroneutral OCTNs are controlled by protein kinases PKA, PKC, and PKG (Vol. 4 – Chap. 4. Cytosolic Protein Ser/Thr Kinases). Dopamine can be conveyed by OCT2 in nephron.

Effect in Heart

In addition to their vasorelaxant activity, natriuretic peptides have anti-fibrotic, anti-hypertrophic, and lusitropic effects. Natriuretic peptides inhibit not only hypertrophy, but also apoptosis of cardiomyocytes. They also repress the proliferation of cardiac fibroblasts.

In congestive heart failure, vessel dilator enhances both natriuresis and diuresis, whereas atrial natriuretic peptide produces neither diuresis nor natriuresis [109].

In addition to vasorelaxation, CNP reduces cardiac preload, inhibits vascular smooth muscle cell proliferation, hypertrophy of cardiomyocytes, and growth of fibroblasts, suppresses aldosterone release, attenuates myocardial ischemia–reperfusion injury, and prevents maladaptive remodeling after myocardial infarction [112]. However, it lacks significant natriuretic or diuretic effects.

Effect on Vascular Smooth Muscle Cells

Vessel dilator decreases systemic and pulmonary vascular resistance and increases the blood flow rate. Natriuretic peptides relax vascular smooth muscle cells. They interact with endothelins. Both ANP and BNP inhibit endothelin-1 production, whereas endothelin-1 stimulates natriuretic peptide synthesis [113]. Natriuretic peptides prevent not only endothelin synthesis, but also sympathetic nerve activity. They also regulate proliferation of vascular smooth muscle cells.

Endothelial C-type natriuretic peptide can regulate the local vascular tone and growth of vascular smooth muscle cells. The diuretic and natriuretic effects of CNP are much weaker than those of ANP and BNP. Peptide CNP is the most prominent neuropeptide of the natriuretic peptide family. With its receptor NPR2, it can operate in the nervous system to coordinate the central aspects of body fluid homeostasis.

In all 3 types of cultured human vascular smooth muscle cells, TGFβ and PDGFbb increase the expression of C-type natriuretic peptide. These growth factors signal via transforming growth factor-stimulated clone-22 domain-1 protein (TSC22D1), a transcription factor, to raise the production of C-type natriuretic peptide [114].

2

Cell Growth and Proliferation

Cells evolve among many possible fates that include cell survival, growth, division, motility (Chap. 6), and death (Chap. 4), as well as differentiation for precursor cells and self-renewal for stem cells (Table 2.1). Cell fate depends on availability of nutrients as well as environmental factors.

Table 2.1. Cell fate events and examples of parameters that are implicated in mathematical models.

Cell fate	Parameters
Quiescence	Density of quiescent cells
	Recruitment function
Nutrient uptake	Nutrient uptake coefficient
Matrix production	ECM production rate
Proliferation	Density of proliferating cells
	Fraction of proliferating cells (0.1–0.3)
	Proliferation rate
	Cytoplasmic growth duration
	Cell cycle duration
	Demobilization function
Differentiation	Differentiation rate
Migration	Migration rate
Death	Death rate
	Fraction of lost cells per day (0.05–0.40)

Cells must respond specifically to different environmental stimuli to survive. Multiple signal transduction pathways are triggered on sensing these stimuli (Vols. 3 and 4). Both simultaneous and temporally ordered cues achieve specificity using coordinated pathways that are connected and strictly controlled, avoiding useless crosstalk by inhibition at the appropriate pathway level.

Cell adhesion and migration are regulated by coordinated activities of the cell cytoskeleton and junctions with connecting cells and the extracellular matrix. Many regulators of cell adhesions control cell survival, growth, proliferation, and differentiation. Cell junctions are made of adhesion molecules (Vol. 1 – Chap. 7. Plasma Membrane) and associated proteins, such as scaffolds, small (monomeric) guanosine triphosphatase regulators (GTPases),[1] and enzymes (e.g., kinases, phosphatases, and peptidases). Structural proteins (e.g., vinculin, talin, and actinin) link the cell environment (extracellular matrix and/or apposed cells) to the cytoskeleton. Signaling molecules (e.g., focal adhesion kinase and paxillin) transduce the stimuli from the neighborhood to cellular effectors for organization of cell adhesions and cytoskeleton.

Molecular transport, especially along microtubules, is required for cell life and during cell division. Vesicles transport proteins, lipids, and other molecules between cell compartments owing to the nanomotors associated with the cytoskeleton. In particular, myosin and its binding partners link the cytoskeleton to coated and uncoated vesicles.

2.1 Energy and Nutrients

Many signaling pathways monitor nutrient availability. Three basic nutrients — amino acids, glucose, and fatty acids — are sensed for suitable cell response. Moreover, nutrients, cellular energy, hormones (e.g., insulin), and growth factors (Chap. 3) cooperatively trigger a signaling pathway mediated by (mammalian) target of rapamycin (TOR)[2] that controls cell growth (Sect. 2.3). Target of rapamycin protein kinase (Sect. 2.3.1 and Vols. 1 – Chap. 4. Cell Structure and Function and 4 – Chap. 4. Cytosolic Protein Ser/Thr Kinases) acts in nutrient signaling and regulates cell growth and proliferation.

Target of rapamycin senses nutrient availability and intracellular energy status and determines cell metabolism and growth according to received extracellular stimuli (hormones and growth factors). The TOR pathway integrates diverse nutritional and environmental signals for transduction. Target of rapamycin forms 2 functionally and structurally distinct target of rapamycin complexes (TORC) that act as protein Ser/Thr kinases: (1) TORC1 with G-protein β subunit-like protein and regulatory associated protein of

[1] Regulatory GTPases are active and inactive when bound to GTP ($GTPase^{GTP}$) and GDP ($GTPase^{GDP}$), respectively. The intrinsic enzymatic activity of small GTPases converts GTP to GDP. Specific GTPase-activating proteins (GAP) accelerate this intrinsic enzymatic activity, thus inhibiting small GTPases. Guanine nucleotide (GDP-to-GTP)-exchange factors mediate the dissociation of GDP from GTPases so that monomeric (small) GTPases can bind to a new GTP molecule and resume their activity, thereby stimulating small GTPases.

[2] Rapamycin is a TOR inhibitor. The ubiquitin ligase complex SCF (SKP1–Cullin–F-box) and its regulators are linked to the TOR pathway.

Table 2.2. Target of rapamycin complexes (TORC): raptor (regulatory associated protein of TOR) and rictor (rapamycin-insenstive companion of TOR). Kinase TORC1 regulates cell growth in response to growth factors as well as energy and amino acid availability. Small GTPase Rag acts as an amino acid-specific regulator of the TORC1 pathway. The ragulator complex causes TORC1 translocation to lysosomal membranes in a Rag-dependent manner in response to amino acid stimulation, where TORC1 can colocalize with its RHEB (Ras homolog enriched in brain) activator.

Type	Regulatory associated protein	Other regulators	Role
TORC1	Raptor	RHEB, Rag, Ragulator, insulin, amino acids, ATP	Cell growth, protein synthesis, ribosome formation, autophagy
TORC2	Rictor		Phosphorylation of PKB, phosphorylation of PKC, cytoskeletal organization, stress fiber stimulation (paxillin and Rho GTPases)

TOR (*Raptor*) and (2) TORC2 with G-protein β subunit-like protein and rapamycin-insensitive companion of TOR (*Rictor*; Table 2.2). The former regulates nutrient transport, ribosome formation, and protein synthesis, as well as autophagy (Sect. 4.3). The latter phosphorylates protein kinases: all conventional protein kinases-C[3] and PKCϵ [115] that are involved in cell proliferation, differentiation, and apoptosis, as well as protein kinase-B for full activation [116].[4]

Adipocytes are strongly involved in the regulation of energy and nutrient homeostasis. They secrete various adipokines (Sect. 1.3). Adipocytes store lipids for energy consumption. Adipocytes release fatty acids into the blood

[3] Members of the protein kinase-C family can be classified into: (1) conventional (α, β, and γ) that are activated by diacylglycerol and calcium; (2) novel (δ, ϵ, θ, and η) that are stimulated by diacylglycerol, but not calcium, and (3) atypical (ζ and ι) PKCs that are not activated by diacylglycerol. Enzymes PKC are phosphorylated within: (1) the kinase domain by 3-phosphoinositide dependent protein kinase PDK1 for partial activation and (2) C terminus motifs by TORC2 for full activation.

[4] Kinases of the AGC superfamily have turn and hydrophobic motifs in their C-terminus. Phosphorylation of the hydrophobic motif increases their catalytic activity. Phosphorylation of the hydrophobic motif of S6K and PKB in response to growth stimuli is mediated by TORC1 and TORC2, respectively. Phosphorylation of turn motif of both PKA and PKC by TORC2 is required for proper C-terminal folding and stabilizes the kinase site.

circulation that are used when glucose level is low. Fatty acids generated from triacylglycerols contain more energy per unit mass than glycogen.

Adipose depots, the so-called visceral fat, surround the heart and large vessels, among other loci. These depots are made of brown adipocytes in infancy. Brown adipocytes disappear in adults, scattered brown adipocytes remaining within white fat pads.[5] Fat pads are innervated by sympathetic nervous fibers. The sympathetic activation elicits lipolysis.

Fatty acid oxidation involves several transcription factors, such as peroxisome proliferator-activated receptors (PPAR or NR1c), ERRα (or NR3b1), FoxA2, and CCAAT/enhancer-binding proteins (C/EBP).

Fatty acids interact with fatty acid-binding proteins (FABP) to travel inside cells. The mammalian FABP family includes 9 detected cytoplasmic proteins. Fatty acid-binding proteins, such as FABP4 and FABP5 that are expressed by adipocytes and macrophages, respectively, participate in the regulation of the body's metabolism. Fatty acid-binding proteins can translocate to the nucleus, where they interact with nuclear receptors.

In the absence of fatty acid-binding proteins in macrophages, the production of cytokines and chemokines decays [117]. Fatty acid-binding proteins hence merge nutritional and inflammatory pathways, at least in adipocytes and macrophages. As cell metabolism and inflammation can be associated, cells must be protected against inflammation that might happen due to fluctuations in nutrient availability.

Members of the family of metalloreductase Six transmembrane epithelial antigens of the prostate (STEAP1–STEAP4) are ferric and cupric reductases, especially STEAP4, that can respond to nutritional, metabolic, and inflammatory signals to protect cells, particularly adipocytes.[6]

Energy-regulating leptin is a circulating hormone released by adipocytes. It regulates food intake and nutrient flux between organs and use. Leptin that favors the switch from lipogenesis to lipolysis works well in case of reduced energy stores, but not in the reverse situation. During starvation, leptin level falls. Energy store decline is signaled to the hypothalamus, which provokes endocrine responses (reduction in thyroid hormone level, suppression of the reproductive axis, etc.).[7] Direct effects of leptin in peripheral tissues only

[5] White adipose tissues also contain endothelial cells, fibroblasts, leukocytes, and macrophages.

[6] STEAP4-deficient mice develop a metabolic syndrome with decreased insulin sensitivity, inflammation (macrophage infiltration of adipose tissue), dyslipidemia, and fatty liver disease, even under standard diet and in the absence of obesity.

[7] Leptin targets multiple sites of the central nervous system, such as hypothalamus (arcuate, paraventricular, dorsomedial, ventromedial, and premammillary nuclei), hindbrain, and striatum. Leptin has opposite effects on arcuate neurons, as it inhibits cells that synthesize appetite-promoting peptides neuropeptide-Y and agouti-regulated peptide and stimulates cells that produce prepro-opionmelanocortin (POMC), the precursor of the appetite-suppressing peptide α-melanocyte-stimulating hormone.

slightly contributes to energy homeostasis.[8] Hypothalamic neurons assign the sympathetic nervous system for restraining the endocannabinoid[9] tone and hence regulate lipid metabolism in adipocytes [118].

Glucose homeostasis is due to concerted actions of several organs. Their coordination is determined by the central nervous system. Pancreatic β cells secrete insulin (Sect. 3.5) in response to elevated glucose concentration in the plasma.

Cell internal state is kept stable within tolerable limits owing to integration of different cues and multiple adjustments (homeostasis). MicroRNAs affect the cell fate, for example, by modulating Hedgehog signaling (Vol. 3 – Chap. 10. Morphogen Receptors).

The mitogen-activated protein kinase pathway (MAPK; Vol. 4 – Chap. 5. Mitogen-Activated Protein Kinase Modules), in particular, leads to formation of extracellular signal-regulated kinases ERK1 and ERK2. The latter transduce signals from the extracellular medium to control cell fate. During cell rest, unphosphorylated ERK1 and ERK2 interact with cytosolic anchors, thus remaining in the cytoplasm. Once stimulated, ERK1 and ERK2 relocalize in different cellular compartments, including the nucleus. Anchoring proteins can sequester ERK1 and ERK2 in these new sites. Retention of either unphosphorylated or phosphorylated ERK1 and ERK2 can then occur.

In addition, mitogen-activated protein kinases antagonistically control cell proliferation and senescence [119]. Cell response to environmental stress and developmental stimulus depends on the balance between the 2 stress-signaling MAP2K7–JNK and MAP2K3/6–P38MAPK pathways. Developmental programs in embryos and adults and environmental agents (osmolarity change, DNA damage, heat shock, ischemia, cytokines, UV waves, oxidative stress, etc.) prime mitogen-activated protein kinases, especially P38MAPK and Jun N-terminal kinase (JNK).

Whatever the circumstances, cells must maintain the concentration of ATP at suitable levels for its energy-requiring functioning. Adenosine triphosphate is mostly synthesized in mitochondria that adapt the oxidative phosphorylation to the demand. Mitochondria induce a retrograde signaling that acts on nuclear gene expression, as well as mitochondrial number and function.

Members of the nuclear receptor class relay signals to the genes, in association with certain inducible transcriptional coactivators to fit the energy demands. Members of the peroxisome proliferator-activated receptor-coactivator PPARγ family, PPARγ coactivators PGC1α and PGC1β, and PGC1-related

[8] Leptin receptor is a member of the interleukin-6 receptor family. It engages the signal transducer and activator of transcription STAT3 and phosphoinositide 3-kinase pathways, among others.

[9] Endocannabinoids are lipid mediators produced from membrane phospholipids in the central nervous system and in peripheral tissues. Endocannabinoids regulate food intake and lipogenesis in adipose tissue and liver by activating CB_1 cannabinoid receptor. Obesity and insulin resistance are asociated with endocannabinoid abundance in white adipose tissue.

coactivator (PRC)[10] serve in the regulation of mitochondrial activity and cellular energy metabolism [120].

A decay in ATP synthesis triggers an increased gene expression of oxidative phosphorylation and mitochondrial expression of PPARα (nuclear receptor NR1c1) to recover appropriate levels. The compensatory mechanism in response to reduced mitochondrial oxidative phosphorylation occurs owing to a burst of intracellular calcium followed by activation of AMP-activated protein kinase (AMPK) and increase in cAMP-response element-binding (CREB)-regulated transcription coactivator (CRTC)[11] that promotes the production of the transcriptional coactivator PGC1α [121].

PPARγ coactivator PGC1α participates in mitochondrial genesis when new mitochondria are needed by the cell and increases the energy substrate uptake.[12] Factors PGC1α and PGC1β protect the cell against oxidative damages [122].

The expression of genes targeted by the metabolic regulator PPARβ is regulated by Krüppel-like transcription factor KLF5 that is targeted by small ubiquitin-related modifier to control energy level and metabolic rate. Krüppel-like transcription factors regulate cell growth and differentiation. They also control adipogenesis, glucose homeostasis, and amino acid catabolism.

Sumoylation (that is reversible due to SUMo-specific SENP peptidases) of transcription factors usually represses transcription. Specific SUMo proteins modify distinct nuclear receptors: SUMo1 targets PPARγ and KLF5 and SUMo2 and SUMo3 liver X receptors. Sumoylated KLF5 interacts with corepressors, such as nuclear receptor corepressor (NCoR) and silencing mediator of retinoic acid and thyroid hormone receptor (SMRT). Under basal conditions, sumoylated KLF5 and PPARβ form a transcriptional repressor complex with NCoR and SMRT [123]. Upon ligand stimulation of PPARβ,

---

[10] A.k.a. peroxisome proliferator-activated receptor-γ coactivator-related factor PPRC1.

[11] A.k.a. transducer of regulated CREB (TORC).

[12] PPARγ coactivator-1 interacts with several nuclear receptors, such as peroxisome proliferator-activated receptors PPARα, PPARβ(or PPARδ), PPARγ, estrogen receptor-related receptor ERRα, nuclear respiratory factor NRF1 and NRF2, as well as with the thyroid hormone receptor, glucocorticoid receptor, estrogen receptor, retinoid receptors, farnesyl X receptor, pregnane X receptor, hepatic nuclear factor-4, liver X receptor, and non-nuclear receptors, such as myocyte enhancer factor MEF2, forkhead transcription factor-1, and Sry-related HMG box-9 [120]. Nuclear respiratory factors regulate the expression of mitochondrial transcription factor-A acting on mitochondrial DNA, as well as the activity of nuclear genes encoding respiratory chain subunits and proteins of the mitochondrial function. To activate the transcription, PGC1α forms complexes with transcription factors and transcription regulators, such as proteins with histone acetyltransferase activity (steroid receptor coactivator-1, CREB-binding protein, and P300), the chromatin remodeling increasing access of the transcription regulators, or such as activating complexes directly interacting with the transcription initiators (thyroid hormone receptor-associated protein/vitamin-D receptor-interacting protein coactivator complex).

SUMo peptidase SENP1[13] is quickly recruited by PPARβ to target KLF5 for desumoylation. KLF5 Desumoylation facilitates KLF5 interaction with PPARβ and exchange of coregulators (loss in corepressors and gain in coactivators such as CREB-binding protein) to prime metabolic gene expression (e.g., genes involved in lipid oxidation and energy uncoupling, such as genes encoding carnitine-palmitoyl transferase Cpt1b and uncoupling proteins Ucp2 and Ucp3).

2.2 Sirtuins

Sirtuins[14] (SIRT) constitute a subclass of NAD^+-dependent histone deacetylases. Sirtuins and related substances can coordinate metabolism with diet, especially during dietary restriction and excess. They may also be involved in stress management and aging. In fact, sirtuins contribute to: (1) cellular stress resistance via interaction with Forkhead box class-O transcription factors (FoxO; Vol. 4 – Chap. 9. Other Major Signaling Mediators); (2) DNA repair; (3) energy metabolism via peroxisome proliferator-activated receptor-γ coactivator PGC1α that regulates gluconeogenesis and fatty acid oxidation in liver, insulin secretion, and mitochondrial genesis; (4) circadian clock; and (5) tumorigenesis [124]. Sirtuins SIRT6 and SIRT7 are nuclear sirtuins, SIRT3, SIRT4, and SIRT5 mitochondrial proteins, and SIRT1 and SIRT2 reside both in the nucleus and cytoplasm. Enzyme SIRT7 is the single sirtuin located in nucleoli in association with RNA polymerase-1.

Sirtuin SIRT1 deacetylates certain transcription factors, such as P53, forkhead box group O proteins, and DNA-repair factor KU complex,[15] thereby increasing the stress resistance of cells [125]. In white adipose tissue, SIRT1 inhibits adipogenesis in precursor cells and reduces fat storage in differentiated cells. It inhibits nuclear receptor peroxisome proliferator-activated receptor-γ. Sirtuin SIRT1 stimulates insulin secretion. In pancreatic β cells, SIRT1 represses the uncoupling protein gene UCP2, thus increasing ATP synthesis and insulin secretion in response to glucose. Enzyme SIRT1 also protects β cells against oxidative stress by deacetylation and increased activity of FoxO1 protein. In the liver, sirtuin-1 deacetylases (activates) PPARγ coactivator PGC1α, hence increasing the gene expression for gluconeogenesis.

Amino acids can serve as carbon and energy sources in periods of energy deprivation. Sirtuins SIRT3 and SIRT4 are imported into the mitochondrial matrix. Enzyme SIRT3 deacetylates acetyl-coenzyme-A synthetase-2. Agent

---

[13] Six known members of the SENP subset of the SUMo peptidase family exist.

[14] A portmanteau for silent information regulator-2 (two).

[15] The regulatory subunits Ku70 and Ku80 of the multimeric Ser/Thr kinase DNA-dependent protein kinase (DNAPK; Vol. 4 – Chap. 4. Cytosolic Protein Ser/Thr Kinases) that form a Ku70–Ku80 heterodimer are able to recognize and bind to DNA double-strand breaks.

SIRT4 transfers ^{ADP}ribose to glutamate dehydrogenase that converts glutamate to α-ketoglutarate. Consequently, SIRT3 promotes the use of acetate, whereas SIRT4 represses the use of glutamate and glutamine in metabolism. In addition, mitochondrial deacetylase SIRT3 maintains basal ATP levels and regulates the mitochondrial electron transport [126].

Fat metabolism and thermogenesis are stimulated in brown fat cells by peroxisome proliferator-activated receptor-γ coactivators PGC1α and -β [125]. In slow twitch myocytes, which exclusively use oxidative metabolism for energy production, PGC1α stimulates the gene transcription for mitochondrial proteins by connecting to transcription factors, such as nuclear respiratory factor-1 and -2 and estrogen-related receptors. Factor PGC1α also activates fatty acid oxidation by binding to PPARα and PPARβ (nuclear receptors NR1c1 and NR1c2). In the liver, PGC1α activates both fatty acid oxidation and gluconeogenesis by linking to transcription factors FoxO1 and hepatocyte nuclear factor-4α (or NR2a1). In this organ, PGC1β, coactivator for the forkhead protein FoxA2, activates cholesterol synthesis and export to blood by tethering to the lipogenic transcription factors sterol regulatory element-binding protein (SREBP) and liver X receptor (LXR or NR1h).

2.3 Cell Growth

Cell growth commonly means either cell reproduction or increase in cell size. However, cell growth refers to size augmentation and cell proliferation to cell division and elevation in cell population. Nevertheless, both processes require protein synthesis (Vol. 1 – Chap. 5. Protein Synthesis) that needs material and energy. Cells must then ensure that sufficient resources are available for protein production.

Cells such as cardiomyocytes can increase in size because they are subjected to adaptation stimuli. Growth factors (Chap. 3) act via several effectors, such as phosphatidylinositol 3-kinase (PI3K), phosphoinositide-dependent kinase-1 (PDK1), and protein kinase-B (PKB). Insulin signaling in the cardiovascular system involves 2 pathways: (1) the phosphatidylinositol 3-kinase pathway and (2) growth factor-like pathway with mitogen-activated protein kinase (Vol. 4 – Chap. 5. Mitogen-Activated Protein Kinase Modules). The mitogen-activated protein kinase signaling network regulates cell fate by transducing multiple growth-factor signals.

Phosphatidylinositol 3-kinase (Sect. 2.3.1 and Vol. 4 – Chap. 1. Signaling Lipids) mediates increased uptake of glucose and amino acids required for cell development. The PI3K family includes 4 sets. The PI3K set 4 includes related enzymes, such as protein kinase target of rapamycin. The PI3K–PKB pathway activates TOR.

When the access to energy or amino acids becomes limited, protein generation is downregulated, as cells must use restricted resources to survive.

Mechanisms of translational control are indeed sensitive to nutrient and energy availability, as well as various other inhibitors and stimulators (stress, hormones, and growth factors).

The limiting step of protein synthesis is translation initiation, during which small ribosome subunit is recruited to mRNA end owing to eukaryotic translation initiation factor eIF4f. Complex eIF4f assembles from 3 components: eIF4a, -4e, and -4g, once inhibitory eIF4e-binding protein 4eBP1 has been removed upon phosphorylation.[16]

Translation initiation factor eIF4f then scans toward start codon, where the complete ribosome is subsequently assembled to begin polypeptide formation.[17] The rate-limiting step is targeted by multiple regulators.

2.3.1 The PI3K–PKB–TOR Pathway

Target of rapamycin regulates cell growth as well as cell proliferation and motility [127]. Protein TOR integrates multiple cues from hormones (e.g., insulin), growth factors (e.g., IGF1 and IGF2), and mitogens. Protein Ser/Thr TOR kinase also senses nutrient and energy levels (Fig. 2.1).

Kinase TOR exists in the cell in 2 different complexes — TORC1 and TORC2 — with distinct roles, regulation, and effectors (Table 2.3).

2.3.1.1 TOR Complex-1

Constituents

The nutrient–energy–redox sensor TOR complex-1 (TORC1), composed of TOR and regulatory associated protein of TOR (raptor), among other components, controls protein synthesis and targets S6 kinase S6K1 that regulates cell growth [132]. Raptor serves as a scaffold for TOR substrates. The complex TORC1 also associates with the activator mammalian lethal with Sec13 homolog LST8[18] and inhibitor 40-kDa Pro-rich Akt (PKB) substrate PRAS40.[19]

---

[16] In the absence of phosphorylation, eIF4e-binding proteins bind tightly to eIF4e that then cannot attach eIF4g. Once the eIF4e–eIF4g precomplex is formed, ribosomal protein S6 kinase phosphorylates eIF4b that is recruited to this complex and associates with eIF4a to enhance its activity.

[17] As some mRNA species contain inhibitory secondary structures in the 5′ untranslated region (UTR) that prevent efficient scanning of small ribosome subunit to the start codon, RNA helicases that process mRNA secondary structure are required for mRNA translation. Kinase S6K1 modulates the activity of RNA helicase eIF4a.

[18] A.k.a. G-protein β subunit-like protein (GβL).

[19] A.k.a. proline-rich Akt1 (v-akt murine thymoma viral oncogene homolog; i.e., PKBα or PKB1) substrate-1 (Akt1S1).

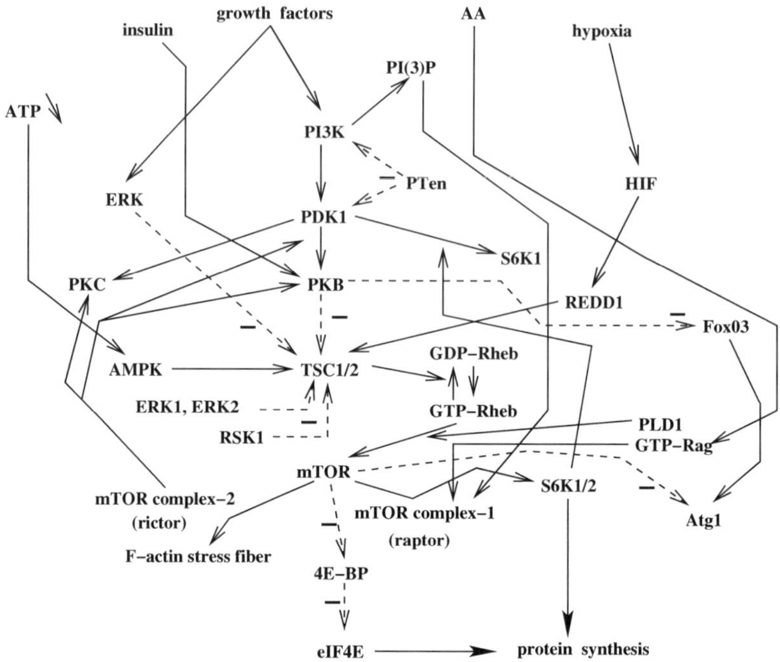

Figure 2.1. The PI3K–PKB pathway activates protein Ser/Thr kinase target of rapamycin (TOR; Sources: [115, 128–130]). It forms 2 complexes: TOR complex-1 (TORC1) with raptor and -2 (TORC2) with rictor. Tuberous sclerosis complexes TSC1 and TSC2 inhibit Ras homolog enriched in brain (RHEB) as RhebGAPs. Small GTPase RHEB activates TORC1 that, in turn, phosphorylates 4eBP1 and ribosomal S6 kinase S6K1 and stimulates the translation initiation factor eIF4e. Phospholipase PLD1, but not PLD2, is required for RHEB activation of TOR signaling to S6K1. Kinase TOR also impedes autophagy, as it phosphorylates autophagy-related gene kinase Atg1. In addition, protein kinase-B (PKB) phosphorylates (inhibits) transcription factor FoxO3 required in Atg synthesis. Insulin and energy inhibits TSC1 and TSC2, hence activating RHEB. Amino acids (AA) favor TORC1 interaction with RHEB^{GTP} via Rag GTPases that binds directly to raptor. TOR Complex-2 stimulates actin stress fibers. It phosphorylates PKB and facilitates PKB phosphorylation by phosphoinositide-dependent protein kinase PDK1. It also phosphorylates protein kinases-C, in addition to PDK1.

Regulation of Activity

The complex TORC1 is activated by multiple types of agents: (1) hormones such as insulin or growth factors; (2) glucose and amino acid availability; and (3) hypoxia. Target of rapamycin complex-1 is activated by insulin and glucose and amino acid availability.[20] Amino acids hence not only are substrates of cell

[20] Complex TORC1 is targeted by many pathways that involve the following mediators: AMP-activated protein kinase, extracellular signal-regulated kinase, glyco-

Table 2.3. Target of rapamycin complexes TORC1 and TORC2. (Source: [131]; GAP: GTPase-activating protein; HIF: hypoxia-inducible factor; S6K: P70 ribosomal S6 kinase; SREBP: sterol regulatory element-binding protein; TSC: tuberous sclerosis complex). Transcription of genes that encode enzymes of cellular metabolism depends on TORC1. Aberrant activation of TORC1 causes cancers and obesity.

TORC1	TORC2
Inhibition by rapamycin	No inhibition by rapamycin
Inhibition by TSC1–TSC2 (GAP)	Stimulation by TSC1–TSC2
Sensor of the metabolic state	
Promotion of protein synthesis	
Stimulation of glycolysis (HIF1α-dependent)	
Stimulation of the pentose phosphate pathway (SREBP1/2-dependent) (S6K1-mediated)	
Stimulation of lipid synthesis (SREBP1/2-dependent) (S6K1-mediated)	

metabolism but also regulate signaling pathways. Nutrients act with insulin or alone according to the cell type.

Many of these cues are integrated upstream of TORC1 by RhebGAPs[21] — the tuberous sclerosis complexes TSC1 and TSC2 — that inactivate Ras homolog enriched in brain (RHEB; Vol. 4 – Chap. 8. Guanosine Triphosphatases and Their Regulators). Small GTPase RHEB activates the kinase activity of TORC1. Insulin via protein kinase-B relieves inhibition by GTPase-activating proteic complexes TSC1 and -2 on RHEB that activates TOR, whereas amino acids directly stimulate TOR.

Class-3 phosphatidylinositol 3-kinase synthesizes phosphatidylinositol 3-phosphate (PI(3)P) that recruits FYVE-domain-containing proteins to endosomes used to build TORC1 signalosome and autophagosome.

In addition to LKB1- and AMPK-mediated TORC1 inhibition, hypoxia excites hypoxia-inducible factor HIF1α that stimulates expression of Regulated in development and DNA-damage response REDD1.[22] The latter competes with TSC2 for 14-3-3 binding and cytosolic sequestration, thereby inhibiting TORC1.

gen synthase kinase-3β, inhibitor of nuclear factor-κB kinase, phosphoinositide 3-kinase, ribosomal S6 kinase, and tumor-necrosis factor-α [133].

[21] Guanosine triphosphatase-activating protein for the GTPase Ras homolog enriched in brain (RHEB).

[22] A.k.a. DNA-damage-inducible transcript DDIT4.

Some cytokines such as tumor-necrosis factor-α can signal via inhibitor of NFκB kinase IKKβ and the Ras–Raf–ERK and PKB–IKKα axes to prime TORC1 activity.

2.3.1.2 TOR Complex-2

The TOR complex-2 (TORC2), made of TOR, rapamycin-insensitive companion of TOR (Rictor), among other components (e.g., stress-activated protein kinase-interacting protein SIP1 and, like in TORC1, LST8), stimulates actin stress fibers, paxillin, and Rho GTPases, hence regulating cytoskeletal organization.

Agent TORC2 phosphorylates protein kinase-B and facilitates PKB phosphorylation by phosphoinositide-dependent kinase PDK1 for full PKB activation [128]. The TORC2 complex also phosphorylates Rac1 and protein kinase-Cα [133]. Morphogen Wnt (Vol. 3 – Chap. 10. Morphogen Receptors) inhibits GSK3β and, in cooperation with AMPK, relieves GSK3β activation of TSC2.

2.3.1.3 Translation Initiation Factors

Target of rapamycin regulates protein translation that is initiated by at least 12 eukaryotic translation initiation factors (eIF; Vol. 1 – Chap. 5. Protein Synthesis). Eukaryotic translation initiation factor eIF4e-binding protein-1 (4eBP1) is another effector of TORC1 [127].[23] Protein TOR associates with adaptor raptor and LST8 to form TORC1 complex that phosphorylates translational modulators P70 ribosomal S6 kinases S6K1 and S6K2 and eIF4e-binding proteins. Phosphorylated translation inhibitor 4eBP1 is released from cap-binding protein eIF4e. Protein eIF4e can then bind mRNAs for its recruitment to the ribosomal initiation complex and initiation of translation with initiation factor eIF4b, leading to protein synthesis. Kinases S6K phosphorylate (activate) eIF4b (Table 2.4).

2.3.1.4 Synthesis of Ribosomal Components

Synthesis of ribosomal components requires: (1) RNA polymerase-1 to produce the large rRNAs; (2) RNA polymerase-2 to create the mRNAs that encode ribosomal proteins; and (3) RNA polymerase-3 to synthesize tRNA nd 5S rRNA. Kinase TOR stimulates transcription of rRNA genes by RNA polymerase-1, ribosomal protein genes by RNA polymerase-2, and tRNA genes by RNA polymerase-3.

[23] The eukaryotic translation initiation factor eIF4e forms a complex eIF4f with scaffold proteins eIF4g1 and eIF4g2 and RNA helicase eIF4a (eIF4a–eIF4e–eIF4g complex) and binds to mRNA.

Table 2.4. S6K1 targets (Source: [133]). 5′ Caps of newly synthesized mRNAs are associated with cap-binding protein CBP20 and CBP80. Protein CBP80 can be phosphorylated by S6K1 upon growth factor stimulation. Kinase S6K1 favors eIF4b recruitment to the translation initiation factor complex. Factor eIF4b enhances the activity of helicase eIF4a. Protein PDCD4 binds to and inhibits eIF4a. Kinase S6K1 phosphorylates PDCD4 and causes its ubiquitination and subsequent degradation. Scaffold SKAR that is connected to the exon–junction complex that links to newly generated mRNA after pre-mRNA splicing recruits hyperphosphorylated (activated) S6K1 to mRNAs (but not S6K2). Kinase S6K1 phosphorylates (inactivates) calcium–calmodulin-dependent kinase eEF2K that phosphorylates eukaryotic elongation factor eEF2, and thereby interferes with eEF2 binding to ribosome and peptide elongation rate. Kinase eEF2K is also phosphorylated by mitogen-activated protein kinase P38δ and the CcnB–CDK1 (cyclin–cyclin-dependent kinase) complex.

Parner	Effect
Cap-binding protein CBP80	Stimulation
Programmed cell death PDCD4	Inhibition
Ribosomal protein RPS6	Stimulation
Scaffold SKAR	Stimulation
Translation initiation factor eIF4b	Stimulation
Translation elongation factor-2 kinase eEF2K	Inhibition

The TOR pathway regulates the synthesis of tRNA and 5S rRNA by RNA polymerase-3, as TOR interacts with general transcription factor GTF3c that recognizes the promoters of tRNA and 5S rRNA genes [134].[24] Kinase TOR phosphorylates repressor of RNA polymerase-3 transcription Maf1 (Ser75). Protein Maf1, in fact, impedes transcription by all 3 nuclear RNA

[24] In Saccharomyces cerevisiae, the transcription of 5S rRNA and tRNAs relies on 3 factors: (1) GTF3a, a 5S gene-specific factor; (2) GTF3b, a 3-subunit initiation factor; and (3) GTF3c, a 6-subunit chromatin antirepression and activation complex. Complexes of TF3 factors assemble on the promoters of RNA polymerase-3 genes and direct the enzymatic activity of the 17-subunit RNA polymerase-3. Agent Maf1 represses the transcription of RNA polymerase-3 genes. (Maf1 homologs have been identified in various types of organisms, including humans.) Direct repressor of transcription by RNA polymerase-3 Maf1 homolog (Saccharomyces cerevisiae) interacts, in Homo sapiens, with casein kinase 2α polypeptide-1 and -2 (CK2α1 and CK2α2), and G1–S-specific cyclin-D3, among others [326]. Agent Maf1 exists in various phosphorylation states and interacts with RNA polymerase-3 in a dephosphorylated state. Its activity is modulated by a phosphorylation-dephosphorylation mechanism involving PP2 that controls its nuclear localization [135]. Nuclear accumulation of Maf1 is required for transcriptional repression. Inhibition of Maf1 by PKA results from the repression of its nuclear import [136]. Phosphorylation of Maf1 inside the nucleus directly reduces Maf1-mediated repression of RNA polymerase-3 and indirectly stimulates Msn5 binding and export of nuclear Maf1 to the cytoplasm [137].

polymerases. Protein TOR associates with TF3c, is recruited to RNA polymerase-3-transcribed genes, and relieves Maf1-induced repression.

2.3.1.5 Regulation of the TOR Pathway

Tumor suppressors hamartin TSC1 and tuberin TSC2[25] stimulate TOR via inhibition of RHEB GTPase.[26] Small GTPase RHEB binds TSC2 that acts as a GTPase-activating protein for RHEB and prevents RHEB activation of TOR. In response to growth factor stimulation and nutrient availability, $RHEB^{GTP}$ can interact with FKBP38, which binds to TOR and inhibits its kinase activity [138]. Protein RHEB hence prevents FKBP38 binding to TOR and frees TOR from inhibition. TSC2 is phosphorylated by PKB (inactivation) and AMPK (activation), as well as ERK and RSK (inhibition; Tables 2.5 and 2.6). Kinase AMPK thus impedes TOR activity by favoring the GTPase-activating protein activity of TSC2 toward RHEB and inhibiting Raptor. The TSC–RHEB module operates as a signal integrator of multiple cues, such as cellular energy level, growth factors, and hypoxia.

Another mechanism of RHEB stimulation of the amino acid-sensing TOR signaling pathway involves phospholipase-D isoform PLD1 [130]. Phosphatidic acid actually participates in the activation of TOR signaling associated with amino acid availability. This lipidic second messenger is produced in particular by phospholipase-D that hydrolyzes phosphatidylcholine in response to various stimuli.[27] Phospholipase-D1 is required for RHEB activation of TOR signaling via TORC1.

A member of the Rag family of 4 Ras-related small guanosine triphosphatases (RagA–RagD)[28] interacts with TORC1 in the presence of amino acids. Amino acids stimulate TOR by activating both the RagA(B)-RagC(D) heterodimer and class-3 PI3K. In the presence of amino acids, $RagC^{GTP}$ binds raptor, hence mediating amino acid signaling to TORC1. Rag GTPases promote TOR intracellular colocalization with its activator RHEB [129]. Both Rag and amino acids promote TOR translocation to the perinuclear region that contains RHEB and late endosomal Rab74. When bound to Rag, TORC1 is stimulated by RHEB for S6K1 and 4eBP1 phosphorylation.

Amino acids cause the transfer of TORC1 to lysosomal membranes, where Rag GTPases reside [139]. Endosomal adaptor mitogen-activated protein

---

[25] TSC: tuberous sclerosis complex.

[26] Tuberous sclerosis is a genetic disorder caused by mutations in one of 2 tumor suppressor genes: Tsc1 or Tsc2. The Tsc genes are, in particular, involved in the cell cycle regulation and cell size control. The activity of TSC1 and TSC2 depends on phosphoinositide 3-kinase and protein kinase-B.

[27] Phospholipase-D1 is activated by 3 types of enzymes: (1) protein kinase-C and (2) small GTPases of the (2.1) Rho class (CDC42, Rac1, and RhoA) and (2.2) of the Arf superfamily. Small GTPase CDC42 activates S6K1 via PLD1.

[28] Rag proteins act as heterodimers: RagA or RagB binds to RagC or RagD.

Table 2.5. Regulators of the target of rapamycin (TOR) pathway (**Part 1**; Sources: [130, 133]; AMPK; AMP-activated protein kinase; ERK; extracellular signal-regulated kinase; GSK: glycogen synthase kinase; IKK: inhibitor of nuclear factor-κB kinase; IRS: insulin receptor substrate; PDK: phosphoinositide-dependent kinase; PI3K: phosphoinositide 3-kinase; PKB: protein kinase-B; PP: protein phosphatase; PRAS40: 40-kDa Pro-rich AKT (PKB) substrate; Raptor: regulatory associated protein of TOR; RHEB: Ras homolog enriched in brain; RSK: P90 ribosomal S6 kinase; S6K: P70 ribosomal S6 kinase; TSC: tuberous sclerosis complex). TORC1 complex is stimulated by active RHEB and after phosphorylation of inhibitor PRAS40. The heterodimer RhebGAP TSC1–TSC2 inactivates RHEB. Growth factors and hormones signal via the PI3K–PKB and Ras–ERK pathways that inhibits the TSC1–TSC2 complex. The energy–nutrient–stress sensor AMPK for TORC1 is phosphorylated (activated) by LKB1.

Regulator	Target	Effect
		Kinase
AMPK	Raptor	Inhibition (14-3-3-mediated sequestration)
	TSC2	Stimulation
ERK	TSC2	Inhibition
GSK3β	TSC2	Stimulation
IKKα	Raptor	Stimulation
IKKβ	TSC1	Inhibition
PKB	TSC2	Inhibition
	PRAS40	Inhibition (dissociation from TORC1)
	IKKα	Stimulation
RSK	Raptor	Stimulation
	TSC2	Inhibition
TORC1	PRAS40	Inhibition
	S6K1/2	Protein synthesis
	4eBP1/2	Release to launch protein synthesis
	PP2a	Inhibition
	IRS1	Inhibition (feedback loop) via S6K1
TORC2	PRAS40	Inhibition
	PKB	Full activation (with PDK1)
	PKCα	
	Rac1	

kinase (MAPK) scaffold protein 1 adaptor protein (MAPKSP1AP)[29] associates with the cytoplasmic face of late endosomes. It binds to MAP2K1 and ERK1 kinase. It also interacts with MAPK scaffold protein MAP2K1IP1 (or MP1).

[29] A.k.a. Roadblock domain-containing protein RoblD3, P14, late endosomal–lysosomal MP1-interacting protein, and mitogen-activated protein-binding protein-interacting protein (MAPBPIP).

Table 2.6. Target of rapamycin (TOR) pathway regulators (**Part 2**; Sources: [130, 133]; PLD: phospholipase D; RHEB: Ras homolog enriched in brain; Raptor: regulatory associated protein of TOR). Small GTPase Rag that senses the concentration of amino acids interacts with the TOR complex TORC1 to control its subcellular localization in RHEB-containing regions.

Regulator	Target	Effect
Phospholipase		
PLD1	RHEB	Stimulation
Small (monomeric) GTPase		
$RHEB^{GTP}$	Raptor	Stimulation
Rag^{GTP}	TOR	Stimulation
CDC42	S6K1	Stimulation

Chromosome 11 open reading frame C11ORF59[30] is anchored to membrane rafts, predominantly of late endosomes. Adaptor C11ORF59 specifically binds to the MAPKSP1AP–MAP2K1IP1 complex in late endosomes. It also connects to MAP2K1 and ERK1 kinases, hence serving as a signaling component on late endosomes [140]. Adaptor C11ORF59 targets Rag GTPases to their subcellular location.

Ragulator is the complex formed by MAP2K1IP1 (or MP1), MAPKSP1AP (or P14), and C11ORF59 (or P18). The Ragulator–Rag complex serves as an amino acid-regulated docking site for TORC1 on lysosomal membranes, where TORC1 is activated [139]. On the lysosomal surface, the TORC1 activity is insensitive to amino acid and does not depend on Rag GTPase and Ragulator, but on RHEB GTPase.

In response to stimuli, such as nutrients and growth factors, multisubunit initiation factor complex eIF3 acts as a scaffold for both TORC1 and S6K1 that can then bind to mRNA. Inactive S6K1 attaches to the eIF3 complex. Activated TORC1 is recruited to eIF3 complex and phosphorylates S6K1 that can then dissociate from eIF3 and is subsequently phosphorylated (activated) by phosphoinositide-dependent kinase-1 [133]. Upon activation, the eIF3–TORC1 complex localizes to 5′ cap of mRNA.

2.3.1.6 Additional Effects of Target of Rapamycin

Splicing factor SF2/ASF[31] activates the TORC1 branch of the TOR pathway, bypassing the PI3K–PKB signaling [141]. It then increases eIF4e phosphorylation downstream of the Ras–MAPK pathway.

The PI3K–PKB–TOR pathway regulates cytokine production in myeloid dendritic cells. The PI3K–PKB–TOR–S6K signaling cascade also controls

[30] A.k.a. cyclin-dependent kinase inhibitor CKI1b (or $p27^{KIP1}$)-releasing factor from RhoA (p27RF-Rho), RhoA activator C11ORF59 homolog, and P18.

[31] Oncoprotein SF2/ASF and TOR signaling are upregulated in many cancers.

production of interferon-I in plasmacytoid dendritic cells[32] via 2 distinct, complementary mechanisms. Phosphorylation of early initiation factor eIF4e-binding protein mediated by TOR stimulates synthesis of interferon regulatory protein IRF7 upon stimulation by viral material of Toll-like receptors TLR7 and TLR9 (Vol. 3 – Chap. 11. Receptors of the Immune System) via TLR adaptor MyD88 and transcription factor IRF7 [143].[33]

The TOR pathway that is activated in monocytes, macrophages, and dendritic cells in response to bacterial infection or lipopolysaccharide exposure inhibits the innate inflammatory response. It indeed decreases the production of pro-inflammatory cytokines, such as interleukins IL12, IL23, and IL6, by inhibiting nuclear factor-κB (Vol. 4 – Chap. 9. Other Major Signaling Mediators), and increases the synthesis of anti-inflammatory cytokine IL10 by activating signal transducer and activator of transcription STAT3 [144]. In addition, TOR impedes helper T_{H1} and T_{H17}-cell responses.

2.3.2 The Glycogen Synthase Kinase Pathway

Glycogen synthase kinase-3 (GSK3) is a Ser/Thr kinase that controls various cellular functions, such as metabolism and protein synthesis, as well as cell growth and apoptosis. Kinase GSK3 has a unique feature among Ser/Thr kinases, as it is active in unstimulated cells.

Two isoforms exist: GSK3α and GSK3β. The activity of GSK3 is regulated by both phosphorylation-dependent and -independent mechanisms. Phosphorylation of GSK3 (Ser21 and Ser9 of GSK3α and GSK3β, respectively) by protein kinase-B blocks access of substrates to GSK3 catalytic domain, thereby inactivating GSK3. Isoform GSK3β, but not GSK3α, is also inhibited by phosphorylation by P38MAPK (Ser389). A phosphorylation-independent procedure of GSK3 regulation involves the Wnt pathway.

Glycogen synthase kinase-3 regulates cardiomyocyte growth and hypertrophy. Isoenzyme GSK3β phosphorylates activators of cardiac hypertrophy, such as eukaryotic translation initiation factor eIF2bϵ, nuclear factor of activated T cells, transcription factor GATA4, β-catenin, and myocardin. Phosphorylation by protein kinase-B of isoform GSK3α mediates maladaptive hypertrophy, whereas that of GSK3β impedes such an evolution, thereby serving as a compensatory agent [145].

[32] Plasmacytoid dendritic cells are main producers of interferon-I. Conventional dendritic cells recognize viral nucleic acids via cytoplasmic receptors RIG1 and Mda5, whereas plasmacytoid dendritic cells sense virus components via Toll-like receptors located in endosomes TLR7 and TLR9 [142].

[33] P70 Ribosomal S6 protein kinase regulates the formation of a proteic complex made of endosomal TLR9, TLR-activated adaptor MyD88, transcription factor IRF7, ubiquitin ligase TRAF6, Tyr kinase BTK, and kinases IRAK1 and IRAK4. The IRAK1–IRAK4 complex phosphorylates IRF7 to promote its translocation to the nucleus, thereby stimulating IRF7 transcriptional activity.

2.3.3 PIM Protein Kinases

Protein Ser/Thr kinase provirus insertion of Molony murine leukemia virus gene product (proto-oncogene PIM) fosters cell growth and survival. It impedes the phosphorylation (activation) of AMPK, hence promoting the activity of the target of rapamycin TORC1 complex. Among the 3 PIM kinases (PIM1–PIM3), PIM3 alone suffices to preclude AMPK activation and boost protein synthesis [146]. Kinase PIM3 raises the concentrations of transcription factors MyC and peroxisome proliferator-activated receptor-γ coactivator PGC1α as well as enzymes involved in glycolysis and mitochondrial biogenesis.

2.4 Cell Division

Most differentiated cells stop dividing and remain in a quiescent state for weeks, months, or even years. Nevertheless, cells can re-enter the cell division cycle, or cell cycle, and begin to proliferate again. Cell division is orchestrated by regulated signaling pathways that ensure the correct segregation of newly replicated chromosomes to the 2 daughter cells. The activity of numerous involved regulators is restricted to specific subcellular foci.

Growth factors (Chap. 3) cause cell proliferation. Two subclasses of growth factors exist: competence and progression factors. *Competence growth factors* such as platelet-derived growth factor provoke entry of quiescent cells into the cell cycle; then, *progression growth factors* such as insulin-like growth factor-1 promote cell cycle progression.[34] In addition, transcription repressor Hairy and enhancer of Split HES1 allows a quiescent cell to resume cell division and avoids inappropriate differentiation and irreversible senescence [163].

When the cell does not undergo division, unraveled chromosomes are located in a more or less outer nucleus region (chromosome sorting by radial arrangement; Vol. 1 – Chap. 4. Cell Structure and Function). Each chromosome is, indeed, confined to a given nuclear region. The spatial organization in chromosome territory is implicated in genome stability and gene regulation [148]. Genome sequences in various chromosome territories remain accessible to gene activators and inhibitors. The relative positioning of DNA sequences and their physical interaction can control the cell functions. The chromosome territory can move during cell differentiation. According to their activity, genome sequences can relocate from the nucleus periphery to the inner part.

Cell division depends on numerous proteic complexes. In particular mitotic spindle assembly and chromosome segregation rely on the γ-tubulin ring complex and anaphase-promoting complex, respectively [149].

---

[34] Insulin-like growth factor-1 triggers calcium entry that is required for cell cycle progression.

Both daughter cells contain the parent genetic code after DNA replication. The genetic code is duplicated owing to *replisome* that simultaneously separates double DNA strands and produces a copy of each strand. The 2 DNA strands are separated. Helicases unwind the DNA double helix into single strands. Then DNA complementary to each strand is created by DNA polymerase.

During cell division, the genetic material becomes visible and forms chromosomes. Each chromosome is constituted by 2 chromatids connected by a centromere with limbs of variable length terminated by a telomere. In chromosomes, chromatin compacts and organizes DNA. Centromeres in chromosomes are formed from specialized regions of chromatin that contain histone H3-variant centromere protein CenPa. Centromeres ensure equal chromosome distribution between daughter cells. Centromeres act not only for accurate segregation of the genome during cell division, but also as attachment sites for the mitotic spindle microtubules that separate duplicated chromosomes. Once the mitotic spindle forms, chromosomes line up perpendicular to the spindle's main filaments, at the midpoint between the spindle poles. Chromosome *kinetochores*[35] attach to microtubule tips during cell division, i.e., microtubules that extend out from the poles. Kinetochores are complexes that comprise more than 50 types of proteins. Accurate cell division and propagation of the genome by a proper partition of replicated DNA packaged into sister chromatids depend on adequate bi-oriented attachment of chromosomes to the microtubule-based mitotic spindle, more precisely kinetochores.

The kinetochore–microtubule attachment behaves like a catch bond similar to a seat belt that locks in place when pulled abruptly. A catch bond apparatus has both strongly and weakly bound states. Kinetochores are strongly tethered to growing microtubules and weakly linked to disassembling microtubules [150]. Applied force suppresses microtubule disassembly and therefore favors the strongly bound state. Tension also stabilizes bi-oriented kinetochore–microtubule attachments via inhibition of the destabilizing activity of Aurora-B kinase. Therefore, tension promotes proper kinetochore–microtubule attachments using a combination of mechanical stabilization and tension-dependent phosphoregulation [151].

Chromosomes then start to move toward the spindle poles. Formation of specialized CenPa-containing chromatin at centromeres that determines kinetochore assembly requires RNA interference-mediated formation of neighboring methylated heterochromatin [152]. Methylated heterochromatin near CenPa chromatin allows heterochromatin protein-1-related chromodomain proteins to bind kinetochore complex.[36]

[35] The kinetochore is the proteic structure on chromosomes where the mitotic spindle microtubules attach during cell division. Kinetochore assembles on the centromere.

[36] Methyltransferase Clr4 at $H3K_9$, ribonuclease Dicer, RNAi effector complex component Chp1 (RNA-induced initiation of transcriptional gene silencing), and Swi6 are also involved.

During most of the cell cycle, the outer membrane of the nuclear envelope remains contiguous with the endoplasmic reticulum membrane. At the beginning of mitosis, the nuclear envelope disassembles and, after the partitioning of chromosomes, reassembles to form nuclei that enclose segregated chromosomes in daughter cells. New nuclear envelopes are formed by the reorganization of the endoplasmic reticulum, as the chromatin recruits endoplasmic reticulum tubules. Transition of the endoplasmic reticulum from tubules to sheets afterward can limit the assembly of the nuclear envelope in daughter cells. When they are overexpressed, endoplasmic reticulum tubule-forming proteins, such as reticulon-3 and -4, slow down the nuclear reconstruction after mitosis [153].

The cell division axis determines the future positions of daughter cells. Cell anisotropy and polarity influence spindle orientation. In addition, the extracellular matrix connected to the actin cytoskeleton via transmembrane proteins contributes to the determination of the orientation of the division axis [154]. Both the architecture and adhesiveness of the cell microenvironment participate in the regulation of the mitotic spindle orientation. Activated cortical force generators interact with spindle microtubules [155].

2.4.1 Phases of the Cell Cycle

The cell cycle includes 4 phases (Fig. 2.2; Tables 2.7 and 2.8): gap 1 (`G1`), synthesis (`S`), gap 2 (`G2`), and mitosis (`M`).

2.4.1.1 Interphase

During interphase, i.e., from the beginning of G1 to the end of G2, the cell carries out the majority of tasks to prepare its division. The cell continues to take up indispensable nutrients and conducts its usual functions. Moreover, it grows and makes a copy of the genome. Most of the cell cycle duration is interphase.

Three phases constitute interphase. Each phase terminates with a checkpoint to control the accuracy of the stage's completion before proceeding to the next. Each phase is characterized by its progression speed, or duration, as well as the involved cell number in a population.

During gap 1, the cell size rises. The production of proteins and organelles heightens until the restriction point. Synthesis refers to DNA replication. During gap 2, the cell resumes its growth in preparation for mitosis.

The interphase is sometimes interrupted. The cell returns then to a quiescent state (G0 phase). After the completion of the G2 checkpoint, the cell proceeds to mitosis.

2.4.1.2 Mitosis

The M phase is subdivided into 2 main substages: (1) actual *mitosis* (chromosome copy distribution between daughter cells) and (2) *cytokinesis* (cytokinesis here stands for division of the plasma membrane and cytoplasmic content

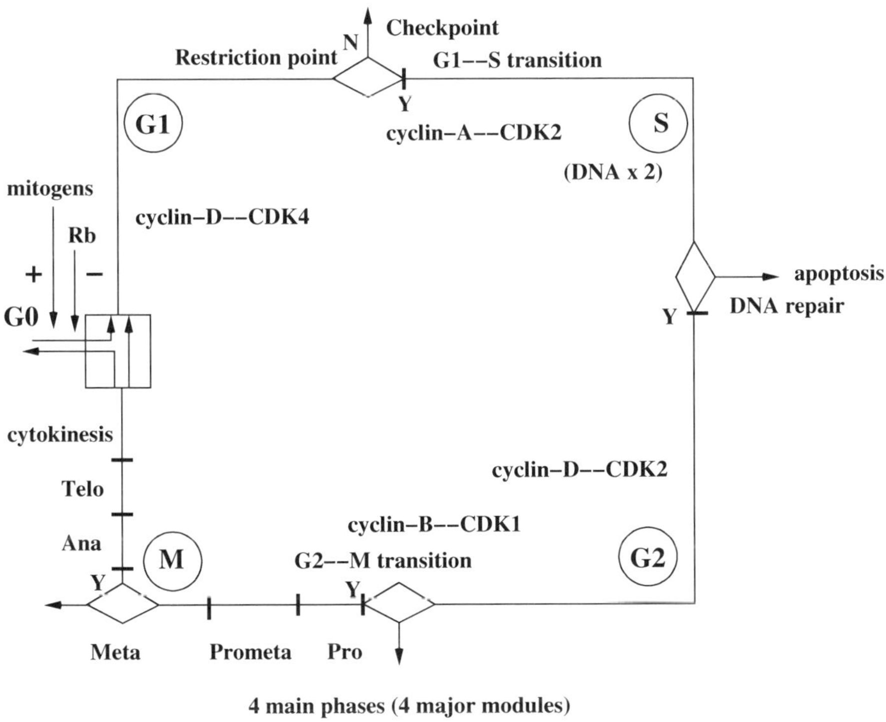

Figure 2.2. Cell cycle phases. The cell division cycle possesses 3 main features: (1) modularity with 4 phases; (2) fail-safe processes with checkpoints; and (3) regulation based on kinases and phosphatases for activation and inactivation with feedback loops, as well as ubiquitins for degradation. Cyclin (Ccn)–cyclin-dependent kinase (CDK) dimers are major regulators of the cell cycle. Different Ccn–CDK complexes promote distinct phases of the cell cycle. Kinases CDK4 and CDK6 complex with cyclins-D for G1-phase progression. The CcnE–CDK2 dimer commits a cell to enter S phase. The CcnA–CDK2 complex initiates and maintains S phase.

during separation of the 2 daughter cells and not cytoplasmic protrusions and retractions that occur during cell migration).

2.4.1.3 Mitosis Prophase and Prometaphase

The interphase includes the first 3 cell cycle phases (G1 to G2) before M phase. At the onset of the M phase (*prophase*),[37] chromatins condense to form replicated chromosomes with 2 twin chromatids bound at the centromere by the cohesion complex (Table 2.9).[38] The mitotic spindle is assembled to process the separation of chromosomes. During the *prometaphase*, the nuclear envelope ruptures and allows the mitotic spindle to contact the chromosomes.

[37] πρo: first, before.
[38] A chromatid corresponds to one of 2 DNA copies after chromosome replication.

Table 2.7. Classical model of the cell cycle. The cell division cycle is built upon the principle of an oscillator, the major components of which are cyclin-dependent kinases (CDK). Both external and internal signals control the activity of cyclins and cyclin-dependent kinases that govern DNA replication and chromosome segregation during the cell cycle. Three major transitions occur during the cell cycle: (1) beginning of S phase with the onset of DNA replication; (2) start of mitosis with breakdown of the nuclear envelope and chromosome condensation; and (3) metaphase–anaphase transition with segregation of the sister chromatids. Before (but not after) the restriction step, regression to G0 is possible. Cell cycle progression is controlled by a set of checkpoints that transiently halt cell division to avoid transmission of DNA mutations. The G1 phase of the cell cycle is particularly governed by cyclin D1 expression and assembly with CDK4 and CDK6. Cyclin-D1 links the cell environment to the cell cycle machinery. Other cyclins are periodically induced during the cell cycle progression. Complexes CcnD–CDK4 and -6 and CcnE–CDK2 control progression from G1 to S phase, whereas CcnB–CDK1 controls the G2–M transition.

G0	Quiescence	Antimitogens
G0–G1	Transition from quiescence to division	Mitogens
G1	Gap 1 (cell growth)	Ccn D–CDK4/6, Ccn E–CDK2
G1–S	Transition	Ccn D–CDK4, Ccn E–CDK2, Ccn A–CDK2
R	Restriction (no return point)	
S	DNA synthesis (chromosome duplication)	Ccns A/E–CDK2 Ccn A–CDK1/2
G2	Gap 2 (division preparation)	Ccns A/B–CDK1
G2–M	Transition (control of DNA replication)	Ccns A/B–CDK1
M	Mitosis (pro-, meta-, ana-, and telophase) Cytokinesis	Ccns B–CDK1
M–G1	Transition	Mitogens
or G0	Quiescence	Antimitogens

Table 2.8. Approximative duration of the cell cycle phases (total time 22–25 h).

Phase	Duration (%)
G1	0.40
S	0.50
G2	0.05
M	0.05

2.4.1.4 Mitosis Metaphase

During the metaphase,[39] the chromosomes gather in the equatorial region of the mitotic spindle. Chromosome segregation requires alignment of

[39] μετα: in the middle of.

Table 2.9. Phases of mitosis and their main features.

Phase	Features
Prophase	Chromatin condensation (early prophase) Twin chromatids bound at centromere (late prophase) Appearance of spindle fibers
Prometaphase	Degradation of the nuclear envelope Microtubule invasion inside the nucleoplasm Formation of the mitotic spindle to which chromosomes attach
Metaphase	Chromosome gathering and alignment in the equatorial plane Checkpoint
Anaphase	Division of centromere of twin chromatids Chromosome splitting Shortening of microtubules Chromosome motion in opposite direction
Telophase	Formation of nuclear envelopes Disappearance of spindle fibers Plasma membrane invagination Chromosome decondensation
Cytokinesis	Cleavage furrow Distribution of organelles Division of cytoplasm and plasma membrane

sister kinetochores on the metaphase plate. During chromosome alignment, kinetochores achieve stable bi-oriented attachments of chromosomes to the plus-end of spindle microtubules. Sister kinetochores must attach to and regulate the assembly of microtubules emanating from opposing spindle poles. This process, the so-called chromosome bi-orientation, requires the integrated activities of multiple kinetochore proteins.

Many interacting kinetochore proteins participate in kinetochore function. An error-correction mechanism is based on the protein kinase Aurora-B that detaches improper kinetochore–microtubule attachments, as it phosphorylates the KMN (KNL1–Mis12–Ndc80) kinetochore complex.[40]

[40] The KMN kinetochore complex is composed of: (1) kinetochore-null protein KNL1 (a.k.a. BUB-linking kinetochore protein and Spc105); (2) kinetochore complex during mitosis subunit Mis12 (a.k.a. kinetochore protein-2-associated protein Kntc2AP); and (3) Ndc80 (a.k.a. kinetochore-associated protein-2 [Kntc2] and highly expressed in cancer protein HEC1). Centromere-connecting protein homolog Mis12 localizes to the kinetochore with CenPa, CenPc, and Mis6. It determines metaphase spindle length and contributes to equal chromosome segregation. Loss of Mis12 causes misalignment of chromosomes in metaphase. The protein Mis12 forms a tetramer with KLN3 (a.k.a. kinetochore-associated protein DSN1), centromere protein-H (CenPh; a.k.a. interphase centromere complex protein ICen35, NNF1, and PMF1), and kinetochore-associated

The bundles of kinetochore-bound microtubules alternate between growth and shrinkage with a controlled rate of turnover that leads to regular oscillations along the spindle axis. This cycle results from coupled microtubule depolymerization at the leading kinetochore and polymerization at the trailing kinetochore.[41] In addition to the KMN complex, kinetochores contain another core complex, the CenPa nucleosome-associated (CenPaNAC) and distal (CAD) complex (CenPaNAC–CAD). The centromere protein CenPh, a subunit of the CenPaNAC–CAD complex, coordinates microtubule plus-end regular oscillations [156].[42]

2.4.1.5 Mitosis Anaphase

During the anaphase,[43] chromosomes are split apart and pulled to opposite sides of the cell. Chromosome compaction depends on kinase Aurora and microtubule dynamics. In anaphase, separating chromatin masses are compacted by DNA-binding plus-end-directed nanomotor kinesin-10[44] located to both chromosome arms and spindle microtubules. Nuclear envelope reformation is initiated by nuclear-pore complex assembly during late anaphase.

2.4.1.6 Mitosis Telophase

During the telophase,[45] the nuclear membrane reassembles around each set of separated, decondensing chromosomes to form 2 nuclei. Phosphorylated histones H3 during mitotic entry are dephosphorylated during mitotic exit, whereas chromatin dissociates and reassociates with heterochromatin protein-1 that contributes to nuclear envelope assembly.

---

protein NSL1 (a.k.a. Mis14, DC8, and DC31). Protein Ndc80 is a constituent of the outer plate of the kinetochore that stabilizes kinetochore–microtubule linkage. The tetramer Ndc80 contains 4 subunits; Spc24, Spc25, and Nuf2, in addition to Ndc80.

[41] Many (25–30) microtubule plus-ends bind each kinetochore to form microtubule bundles, the so-called kinetochore fibers. However, plus-ends of individual microtubules remain dynamic, with a rate of tubulin turnover 20-fold slower than that of plus-ends of free spindle microtubules [156]. Several microtubule-associated proteins accumulate onto end-on attached kinetochores and regulate dynamics of kinetochore microtubules, such as microtubule-depolymerizing member of the kinesin family KIF18a and KIF2c (a.k.a. kinesin-like protein KnsL6 and mitotic centromere-associated kinesin [MCAK]), and members of the cytoplasmic CAP-Gly domain-containing linker protein (CLIP)-associated protein (CLASP) and end-binding (EB) families that regulate microtubule plus-end dynamics.

[42] Centromere-specific, CenPa-containing nucleosomes (CenPaNAC–CAD) are constituted of 15 proteins (CenPh, -i, -k to -u, -w, and -x).

[43] ανα: culminating,

[44] Microtubule-associated kinesin-10 is also called chromokinesin Kid or KIF22. During prometaphase, it participates in chromosome movement toward the spindle equator.

[45] τελος: accomplishment.

2.4.1.7 Cytokinesis

After chromosomal segregation, cytokinesis corresponds to the final stage of the cell cycle. Mitosis usually refers to the process of chromosome duplication, whereas during the following M-phase substage of cytokinesis, the cell cytoplasm and plasma membrane divide and a single cell separates into 2 daughter cells. Specific proteins and lipids accumulate at the connecting bridge to ensure separation of daughter cells at the end of telophase and cytokinesis, i.e., abscission (ab: away from; scindere: to cut, divide, separate). The plasma membrane wrinkles in its central part, perpendicularly to the mitotic spindle, due to the contractile ring, thereby ensuring cell division. The other components of the cell (plasma membrane, cytoskeleton, organelles, and cytosolic content) are distributed to both daughter cells.

2.4.1.8 Cleavage Furrow

Cell splitting is achieved by formation and invagination of a cleavage furrow between chromosomes separated during anaphase. Cleavage furrow ingression is caused by assembly and contraction of the actomyosin contractile ring. A contractile ring of actin filaments and myosin nanomotors actually assembles via a stochastic, but reliable search-and-capture mechanism after a mesh of actin filaments appears. Membrane-bound nodes distributed around the cell equator that contain myosin capture actin filaments growing randomly [157]. Nanomotors pull on the attached node for a few seconds before releasing. These random motions condense the nodes into a continuous contractile ring around the cell equator for cell division-associated cytokinesis.

The ring of actomyosin contracts to form a cleavage furrow in the middle of the cell. This furrow ingresses until it reaches the midzone and creates an intracellular bridge containing a midbody that is finally eliminated to create 2 daughter cells.[46]

2.4.1.9 Abscission

At the end of mitosis, after separation of sister chromatids, the contractile actomyosin ring creates a narrow cytoplasmic bridge between the 2 reforming nuclei. The bridge consists of a thin tube of plasma membrane filled with 2 bundles of overlapping, anti-parallel microtubules with interdigitating plus ends at the midbody. Several kinesins are involved in microtubule bundling and anti-parallel gliding during central spindle formation and positioning specific midzone factors of cytokinesis, such as kinesin-4 (kinesin

[46] To initiate cleavage, the midbody 55-kDa centrosomal protein CeP55 recruits endosomal sorting complex required for transport ESCRT1 and ESCRT associated protein ALIX. During abscission, ALIX and ESCRT1 recruit ESCRT3 subunits. ALIX and ESCRT1 compete for the same binding domain of CeP55 [158].

family member KIF4, a chromokinesin), its partner the microtubule-bundling protein regulator of cytokinesis PRC1, KIF23 (kinesin-6), KIFC1 (kinesin-14A), among other kinesin types [159].[47] Protein KIF23 is a component of the centralspindlin complex. It recruits small GTPase Rho for constriction of the midzone, as well as Aurora-B and Polo-like kinase PLK1 to the central spindle. In addition, the transfer of endosomes toward the bridge participates in abscission.

Phosphatidylinositol(4,5)-bisphosphate abounds in the cleavage furrow and persists in the connecting bridge to recruit specific $PI(4,5)P_2$-binding partners [159]. The production and turnover of $PI(4,5)P_2$ in the connecting bridge allows the polarized recruitment of ezrin–radixin–moesin proteins as well as scaffolds and polymerizing GTPases (that lodge at the metaphase spindle) of the septin class that are required for completion of cytokinesis.[48] In addition, phosphatidylinositol 3-phosphate and centrosomal (during interphase), PI(3)P-binding, zinc finger, and FYVE domain-containing protein ZFYVE26[49] lodge in the bridge, where they contribute to abscission [159, 160].[50] Endosomal lipid PI(3)P recruits FYVE and PX domain-containing proteins, such as early endosomal antigen EEA1[51] to regulate endosome function and dynamics.

Class-3 phosphatidylinositol 3-kinase (PI3KC3)[52] and its partner beclin-1 locally generate PI3P on endosomes within the intercellular bridge to anchor ZFYVE26. Protein ZFYVE26 and its partner tetratricopeptide repeat (TPR) domain-containing protein TTC19 interact with the microtubule plus-end-directed kinesin class nanomotor KIF13a to move the ZFYVE26–TTC19 complex toward the midbody [160]. Protein TTC19 interacts with charged multivesicular body protein ChMP4b, a subunit of the endosomal sorting complex required for transport ESCRT3 involved in inward membrane budding as well as abscission [160].

2.4.1.10 Nuclear Envelope Disassembly and Reassembly during Mitosis

The nuclear envelope is made of inner and outer nuclear membranes separated by the perinuclear space and in close connection with the endoplasmic reticulum and nuclear lamina (Vol. 1 – Chap. 4. Cell Structure and Function). Inner and outer edges of the nuclear envelope are associated with the nucleo- and

---

[47] Kinesin family member KIF23 is also called mitotic kinesin-like protein MKLP1; KIFC1 kinesin-like protein KnsL2 and MKLP2.

[48] Septin-7 stabilizes the association of CenPe with the kinetochore.

[49] A.k.a. FYVECent, spastizin, and SPG15.

[50] During interphase, ZFYVE26 localizes to the centrosome. It relocalizes to the midbody during the stage of cytokinesis. Similarly, KIF13a and TTC19 reside in the centrosome and relocalize to the midbody during cytokinesis [160].

[51] A.k.a. zinc finger FYVE domain-containing protein ZFYVE2.

[52] A.k.a. vacuolar protein sorting VPS34.

cytoskeleton, respectively. The nuclear envelope is perforated by pores at sites of fusion between the inner and outer nuclear membranes that are filled with nuclear-pore complexes for molecular transport. During the last stage of the cell cycle (mitosis), the nuclear envelope is completely disassembled and then reassembled. Temporal and spatial control of nuclear disassembly and assembly is based on command of mitotic entry and exit as well as chromatin state identification. Chromatin identity is defined by small GTPase Ran (Vol. 4 – Chap. 8. Guanosine Triphosphatases and Their Regulators).[53]

During the prophase primed by cyclin-dependent kinase CDK1, chromatin starts to condense to fabricate chromosomes,[54] centrosome separates, and microtubule asters around centrosomes form and move apart along the nuclear envelope. Microtubules are attached to the nuclear envelope. Microtubule nanomotor dynein causes centrosome migration and nuclear envelope invaginations around centrosomes and holes on the opposite edge. The transition into prometaphase is actually characterized by a loss in nuclear envelope permeability. Whereas a cytoplasmic spindle is produced by microtubules[55] that emanate from duplicated cytoplasmic centrosomes, nuclear envelope breakdown is required for interaction between mitotic chromosomes and spindle microtubules. Nuclear envelope breakdown comprises [161]: (1) nuclear-pore complex disassembly; (2) nuclear lamina disintegration; and (3) nuclear envelope retraction into the endoplasmic reticulum. Soluble nucleoporins are dispersed in the cytosol. Nucleoporins participate in spindle assembly and sister chromatid separation.[56]

---

[53] In the nucleus during mitosis, a high concentration of Ran^{GTP} exists owing to chromatin-associated Ran guanine nucleotide-exchange factor (RanGEF) RCC1. In the cytoplasm, Ran GTPase-activating protein RanGAP1 converts Ran^{GTP} into Ran^{GDP}. The gradient of active Ran across the nuclear envelope yields the direction of nuclear transport in the interphase.

[54] Chromatin condensation that commences before nuclear envelope breakdown involves action of topoisomerase-2α and condensin-1 and -2 complexes, as well as phosphorylation of histone-H3. Topoisomerase-2α is recruited to centromeres after sumoylation that is promoted by 358-kDa nucleoporin NuP358, a SUMo ligase also termed Ran-binding protein RanBP2. It causes decatenation of sister chromatids at centromeres during early anaphase.

[55] During spindle assembly, microtubule minus-ends are focused into 2 poles at centrosomes, whereas plus-ends interact with chromosomes via kinetochores and align them on the metaphase plate.

[56] Nucleoporins operate in spindle assembly and kinetochore function. Nucleoporin RNA export RAE1 that forms a complex with NuP98 for RNA export during interphase binds to microtubules. Nucleoporin RAE1 actually interacts with microtubule-associated nuclear mitotic apparatus protein NUMA that promotes microtubule bundling at spindle poles for correct bipolar spindle formation (thus avoiding multipolar spindles). In addition, nucleoporin RAE1 shows homology with spindle assembly checkpoint protein BUB3. During prometaphase, the RAE1–NuP98 complex interacts with adaptor CDC20 homolog CDH1, an activator of anaphase promoter APC/C, and inhibits APC/C–CDH1-mediated

Lamina depolymerization begins in early prophase. Constituents of the inner nuclear membrane and nuclear lamina are phosphorylated by kinases, such as cyclin-dependent kinase CDK1, protein kinase-Cβ2, Aurora-A, and Polo-like kinase PLK1, thereby inducing lamina disassembly and retraction of membranes into the endoplasmic reticulum.[57]

Nuclear envelope reformation occurs around a compacted mass of segregated chromatin in each future daughter cell in late anaphase [161]. Nuclear reassembly is completed in telophase. In anaphase, nuclear envelope constituents are attracted back to chromatin. Nuclear-pore complex assembly is initiated by recruitment of nucleoporin NuP107–NuP160 complexes via factor Elys to chromatin and formation of chromatin-bound prepores. On the surface of mitotic chromatin, RanGTP generated by chromatin-bound RanGEF RCC1 helps to free nuclear-pore complex components from sequestration by importin-β. Meanwhile, chromatin decondensation is associated with structure and composition change. Dephosphorylation of nucleoporins, chromatin-associated factors, and nuclear envelope proteins requires the inactivation of mitotic kinases such as CDK1 and activation protein phosphatases such as PP1. Nuclear membrane is re-established by attraction of inner nuclear membrane proteins from the endoplasmic reticulum to the surface of decondensing chromatin that then flatten to form membrane sheets.[58] During telophase, retraction of membrane-bending proteins, the so-called *reticulons*, into the peripheral endoplasmic reticulum from the nuclear envelope in formation remodels tubules into sheets on the chromatin surface. A closed nuclear envelope is finally generated with nuclear-pore complexes and nuclear lamina.

---

ubiquitination of mitosis regulators (Sect. 2.4.2). Complex RAE1–NuP98 can prevent APC/C–CDH1-mediated ubiquitination of securin. In addition to inhibition of premature formation of the APC/C–CDH1 complex by CDK-dependent phosphorylation of CDH1 until anaphase onset, RAE1 also hinders APC/C–CDH1 activity during cell cycle exit from G1 phase [161].

[57] Remodeling of nuclear envelope and endoplasmic reticulum involves Rab5 and membrane-bending proteins. The NuP107–NuP160 complex is located in kinetochores during prometaphase for stable microtubule–chromosome attachment and recruitment of other nucleoporins (e.g., NuP358, or Ran-binding protein RanBP2, that forms a stable complex with sumoylated RanGAP1, and exportin CRM1) to kinetochores.

[58] During late anaphase, tips of endoplasmic reticulum tubules commence binding to chromatin surface. This event represents the first step in nuclear envelope membrane reformation. Tubule recruitment is followed by attraction of membrane material in sites located between tubule-chromatin contact sites that leads to fabrication of expanded, flattened patches of nuclear envelope on chromatin. Conversion of chromatin-associated tubules to sheets results from removal of endoplasmic reticulum proteins of the reticulon family from the forming nuclear envelope. During early telophase, different regions of chromatin attract distinct sets of nuclear envelope proteins. Whereas LBR, LAP2β, and lamin-B bind to peripheral chromatin regions, LAP2α, BAF, emerin, and A-type lamins localize regions that face spindle midzone and poles.

The major part of nuclear lamins reassembles into the nuclear lamina only after the nuclei have regained import capacity.

2.4.1.11 Transient Arrest and Re-Entry

A short time window exists in the cell cycle during which cells can respond to extracellular cues by withdrawing temporarily from the cell cycle. When these cells re-enter the cell cycle, several additional hours are necessary in the G1 phase before DNA replication [162]. Proliferating cells that experience exogenous or endogenous stress and damage adopt permanent cell cycle arrest (senescence).

2.4.1.12 Cell Division, Quiescence, and Senescence

In the transient or sustained absence of cell division, cells are in a quiescent state (more or less long-duration G0 phase). Senescent cells have permanently lost their ability to divide. Quiescent cells (but not senescent cells) are able to resume proliferation following prolonged (after weeks or months, even years) cell cycle arrest. Overexpression of cyclin-dependent kinase inhibitor CKI1a in fibroblasts causes cell cycle arrest. When CKI1a expression returns to normal levels, cells enter into senescence rather than cell division. However, transcription repressor Hairy and enhancer of split HES1 enables quiescent cells to resume proliferation following prolonged cell cycle arrest [163]. Moreover, HES1 inhibits transcription factor myogenic differentiation factor (MyoD) that leads to terminal differentiation of fibroblasts, i.e., to irreversible cell cycle arrest.

2.4.1.13 Checkpoints

The cell cycle proceeds through a defined sequence of phases. Each phase depends on completion of the previous phase to distribute completely and accurately genome replicas to daughter cells. Although process control is active throughout entire phases of the cell cycle, some instants yield an intensive care. Surveillance and quality-control checkpoints equip each stage of the cell cycle (Table 2.10). The main checkpoints that control the cell cycle actually include G1 restriction (or G1–S checkpoint), S, G2 (or G2–M), and mitosis (metaphase) checkpoints. Morphogenesis checkpoint searches for abnormality in cytoskeleton and arrests cell cycle at G2–M transition.

Cells must accurately copy their chromosomes and segregate them to daughter cells. When any dysfunction happens, cells activate DNA-damage checkpoint machinery that arrests the cell cycle. Checkpoint failure often causes gene mutations

Table 2.10. Checkpoints of the cell division cycle. Each cell cycle phase has a checkpoint that ensures cell division fidelity. According to the cell state, it either allows cell cycle progression when the cell cycle stage has been accurately completed or provokes cell cycle arrest until cell is ready to continue the process. However, in the absence of favorable treatment, apopotosis is triggered.

Phase	Controlled elements
G1 checkpoint	Adequate cell size? Nutrient availability? Presence of mitogens?
S checkpoint	DNA damage?
G2 checkpoint	Chromosome replication completed?
M checkpoint	Attachment of all chromosomes to mitotic spindle at equatorial plane

G1 Restriction Point

The restriction point occurs in late G1 phase. Cells that progress through this point are committed to enter S phase. When the cell is not ready or external conditions are not appropriate, then the cell can come back to G0 phase. Lack of mitogens (growth factors and hormones) and nutrients causes cells to arrest at the restriction point. The restriction point is mainly controlled by cyclin-dependent kinase inhibitor CKI2a that inhibits cyclin-dependent kinases CDK4 and CDK6 that, then, cannot interact with cyclin-D for the cell cycle progression. On the other hand, active CcnD–CDK4 and CcnD–CDK6 complexes phosphorylate retinoblastoma protein (RB) to relieve its inhibition on E2F transcription factor (Sect. 2.4.2). Factor E2F is then able to cause synthesis of cyclin-E that interacts with CDK2 to allow for G1–S transition.

S Checkpoint

In response to DNA damage, the cell cycle promptly arrests at specific stages (Fig. 2.3). Cell cycle arrest remains sustained in the absence of DNA-damage repair. After arrest, cells can re-enter the appropriate cell cycle phase. When DNA damages are too strong to be adequately processed, the cell primes apoptosis.

Ataxia telangiectasia mutated kinase (ATMK) activates the double-strand DNA break cascade. It, as well as other members of the PIKK kinase family (ATMK, ATRK, and DNAPK; Vol. 4 – Chap. 4. Cytosolic Protein Ser/Thr Kinases) phosphorylates histone H2ax and NBS1. Phosphorylated H2ax (Ser139) is called γ-H2ax. *Nibrin* (Nbn)[59] is a member of the MRe11–RAD50–

[59] A.k.a. Nijmegen breakage syndrom protein NBS1.

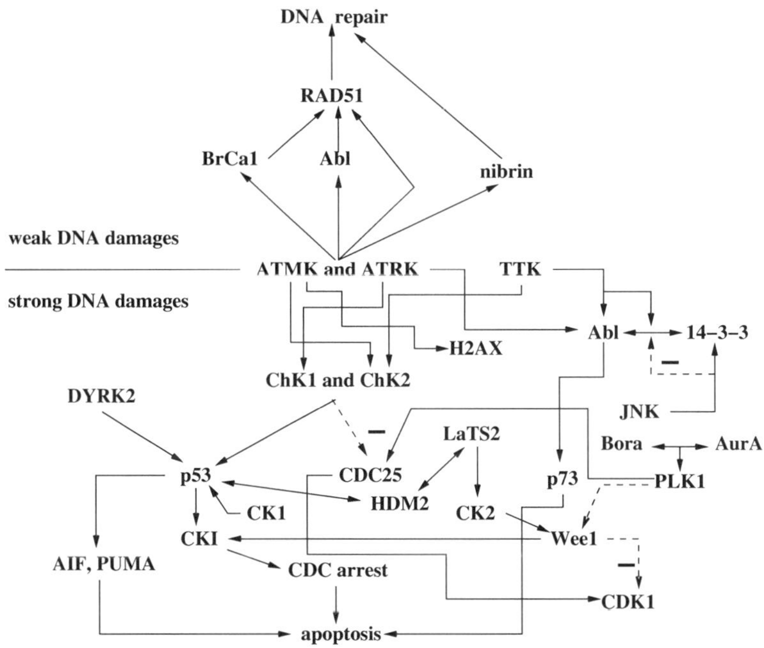

Figure 2.3. Synthesis checkpoint: DNA damage and repair or cell apoptosis (Abl: Abelson leukemia viral oncogene-related kinase; ATMK: ataxia telangiectasia mutated kinase; ATRK: ATMK and Rad3-related kinase; BrCa: breast cancer tumor suppressor; ChK: checkpoint kinase; CKI: cyclin-dependent kinase inhibitor; DYRK: dual-specificity Tyr (Y) phosphorylation-regulated kinase; H2AX: histone; RAD: recombination protein-A [RecA]-homolog DNA-repair protein).

NBS1 (MRN; Mre11: meiotic recombination protein-11; RAD50: recombination protein-A [RecA]-homolog DNA-repair protein, especially a radiation-induced DNA damage and cell cycle control factor)[60] double-strand DNA break repair complex. The RecA homolog and DNA-repair protein RAD51[61] operates in homologous recombination of DNA during ATP-dependent double-strand break repair.

G2 Checkpoint

The G2 checkpoint at the end of G2 phase triggers mitosis onset, when conditions are fulfilled. Mitosis-promoting factor (MPF), or maturation-promoting

---

[60] RAD50 is a member of the structural maintenance of chromosome protein family that participates in DNA double-strand break repair, cell cycle checkpoint activation, telomere maintenance, and meiosis.

[61] A.k.a. recombination protein-A (RecA), and BrCa1–BrCa2-containing complex subunit-5 (BrCC5).

factor, i.e., the CcnB–CDK1 complex, that is activated by CDC25 phosphatase, enables the entry in mitosis.

M or Spindle Checkpoint

During mitosis, the spindle checkpoint uses an anaphase waiting agent to delay the transition between meta- and anaphase until all chromosomes are correctly aligned and attached to microtubule spindles to ensure accurate chromosome segregation. When all the chromosomes are aligned at the mitotic plate, they experience tension created by the bipolar attachment that is sensed for anaphase entry.

The proteic complex cohesin links to duplicated chromosome pairs (sister chromatids) that attach to microtubules of the mitotic spindle via the proteic complex kinetochore. When all chromosome pairs are attached, cohesin is cleaved by separase, and sister chromatids move to opposite spindle poles. The spindle assembly checkpoint keeps separase inactive until all chromosomes are attached and properly oriented on the spindle. The sensor of chromosome-induced tension on the mitotic spindle then leads to degradation of cyclin-B. When all sister chromatids have achieved bi-orientation, degradation of cyclin-B relieves its inhibition on the anaphase-promoting complex that can thus target the separase inhibitor securin to liberate separase. The latter cleaves cohesin subunit Rad21 and then separates sister chromatids.

Components of the mitotic checkpoint regulator include: (1) budding uninhibited by benzimidazole proteins BUB1 (or BUB1α), BUB1β (BUBR1 or MAD3/BUB1-related protein kinase MAD3L), and BUB3; (2) mitotic arrest-deficient proteins MAD1 to MAD3; (3) microtubule-dependent kinetochore-associated mitotic kinesin nanomotor KIF10;[62] as well as (4) dual-specificity protein kinase TTK (Sect. 2.4.2.14).

Protein Ser/Thr kinase BUB1 is essential for the spindle assembly checkpoint. It phosphorylates MAD1 and mitotic checkpoint protein BUB3 as well as histone-2a [164]. Phosphorylated nucleosome protein H2a creates a binding site for shugoshin proteins (Sgo1–Sgo2),[63] hence supporting the recruitment at centromeres and correct function of Sgo1 and Sgo2 to control chromatid cohesion [165]. Shugoshin-1 protects chromatid cohesion at centromeres, as it prevents cohesin removal from the centromere. Shugoshin-2 operates in sensing of tension-less kinetochores, as it recruits Aurora-B that promotes proper chromosome attachment to the spindle and dismantles those that fail to produce tension. During mitosis, H2a phosphorylation is selectively eliminated from the bulk of chromatin, but retained at the centromere, where Sgo2 relocation from ends of chromosomes during interphase requires Bir1 subunit of a protein complex that contains Aurora-B [165].

[62] A.k.a. centromere-associated protein-E (CenPe).

[63] Japanese: guardian spirit, as these proteins protect the cohesin complex during prophase and afterward ensure correct chromosome segregation.

In addition, mitotic checkpoint protein BUB3 and activator CDC20 of Ub ligase APC/C are kinetochore proteins that interact with other kinetochore proteins of the mitotic spindle checkpoint complex, such as MAD2 and mitotic checkpoint protein Ser/Thr kinase BUB1β to yield a kinetochore protein complex that is capable of delaying anaphase by inhibiting ubiquitin ligation via the ligase anaphase-promoting complex/cyclosome. The concentration of BUB3 transcriptional repressor remains nearly constant throughout the cell cycle, whereas CDC20 accumulates during mitosis and is degraded prior to cytokinesis. Proteins BUB3 and CDC20 form a complex with histone deacetylases to repress transcription [166]. When the spindle checkpoint is activated at unattached kinetochores, mitotic checkpoint proteins BUB1β, BUB3, and MAD2 bind to and inhibit CDC20 to produce a diffusible waiting signal that delays the onset of anaphase.

The spindle-assembly checkpoint prevents cells from prematurely entering anaphase by inhibiting the ubiquitination (degradation) of cyclin-B1 and anaphase inhibitor securin by the APC/C Ub ligase. The spindle checkpoint complex actually blocks chromosome segregation until all twin chromatids are properly attached to the mitotic spindle by degrading CDC20 after ubiquitination by the Ub ligase APC/C [167].[64] Chromosomes are catalytically able to generate a diffusible CDC20 inhibitor or inhibit CDC20 already bound to the Ub ligase APC/C. The chromosome-produced inhibitor requires recruitment of mitotic arrest-deficient protein MAD2 by MAD1 and MAD2 becomes competent upon dimerization [169]. Spindle checkpoint protein BUB1β is then recruited. Whereas BUB1β prevents unscheduled degradation of specific APC/C substrates, CDC20 recognizes APC/C substrates and recruits them to APC/C for ubiquitination and degradation [170]. Upon maturation of the mitotic spindle assembly, inhibition of the cell cycle progression ceases,

[64] Adaptor cell division cycle protein CDC20 is also called Slp1 or Fizzy (Fzy). Engagement of Ub ligase APC/C to its substrates (e.g., cyclin-B1, securin, Polo kinase, and Polo-like kinase PLK1) relies on 2 adaptors: CDC20 and CDC20-related protein CDH1 (also termed Fizzy-related protein Fzr1 or Hct1). Its degradation starts when the nuclear envelope breaks. The activator of APC/C CDC20 interacts with spindle-assembly checkpoint component BUB1β using the checkpoint protein mitotic arrest-deficient MAD2. Protein BUB1β is an active kinase at unattached kinetochores that serves as a cytosolic inhibitor of the activity of the APC/C–CDC20 couple. Agent BUB1β also binds to CDC20 to inhibit APC/C activity in interphase, thereby allowing accumulation of cyclin-B in G2 phase prior to mitosis onset [168]. Ubiquitination of CDC20 promotes its degradation to maintain spindle-assembly checkpoint-associated cell cycle arrest. Protein MAD1 that is stably bound at unattached kinetochores binds MAD2 protein. The MAD1–MAD2 heterodimer recruits a second cytosolic MAD2 molecule to unattached kinetochores to bind CDC20 and promote BUB1β binding to the APC/C–CDC20 complex. Protein BUB1β, necessary for mitotic spindle checkpoint activation and inhibition of APC/C activity, forms a complex with MAD2 and CDC20. Spindle checkpoint inhibitor is generated by the sequential production of complexes MAD2–CDC20 and final BUB1β–CDC20 inhibitors [169].

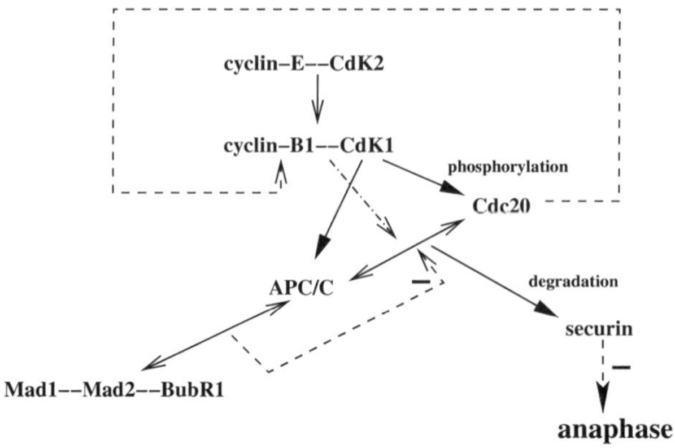

Figure 2.4. Spindle checkpoint. Members of the spindle assembly checkpoint include mitotic arrest-deficient proteins MAD1, MAD2, TTK (MPS1), BUB1 (BUB1α), BUBR1 (BUB1β), and BUB3. Activated MAD2 binds to and inhibits the CDC20–APC/C complex that then cannot degrade securin, an inhibitor of cohesin-cleaving separase, thereby impeding separase stimulation, as separase remains associated with securin. Cohesin holds sister chromatids together. Separase cleaves cohesin to initiate separation of sister chromatids in anaphase. The cyclin-E–CDK2 complex stimulates CDK1 at mitosis onset. The cyclin-B1–CDK1 complex can phosphorylate Cell division cycle CDC20, thereby lowering interaction between CDC20 and APC/C, as well as APC/C, thus heightening its affinity for CDC20.

MAD2 dissociates from CDC20 in the inhibitory checkpoint complex, and interaction between CDC20 and APC/C transiently increases (Fig. 2.4).

2.4.2 Control of the Cell Division Cycle

Execution of the cell cycle requires coordinated replication and separation of cellular material into 2 newly formed daughter cells. The cell cycle includes precisely regulated serial events: transcription and translation completion, chromosome condensation, nuclear envelope breakdown, nucleolus dissolution, and mitotic spindle formation. Each stage of the cell cycle ends before the next begins. Cell cycle regulation mainly relies on synthesis, phosphorylation, dephosphorylation, sequestration, and ubiquitination of actors (Fig. 2.5 and Tables 2.11, 2.12, and 2.13):

$$\begin{aligned}\frac{d}{dt}\text{agent} &= \text{synthesis rate} - \text{degradation rate} \\ &\quad + \text{activation rate} - \text{inactivation rate}. \end{aligned} \tag{2.1}$$

Progression through the cell cycle is controlled by cyclin-dependent kinases and successive expression and activation of transcription factors (Tables 2.14 and 2.15).

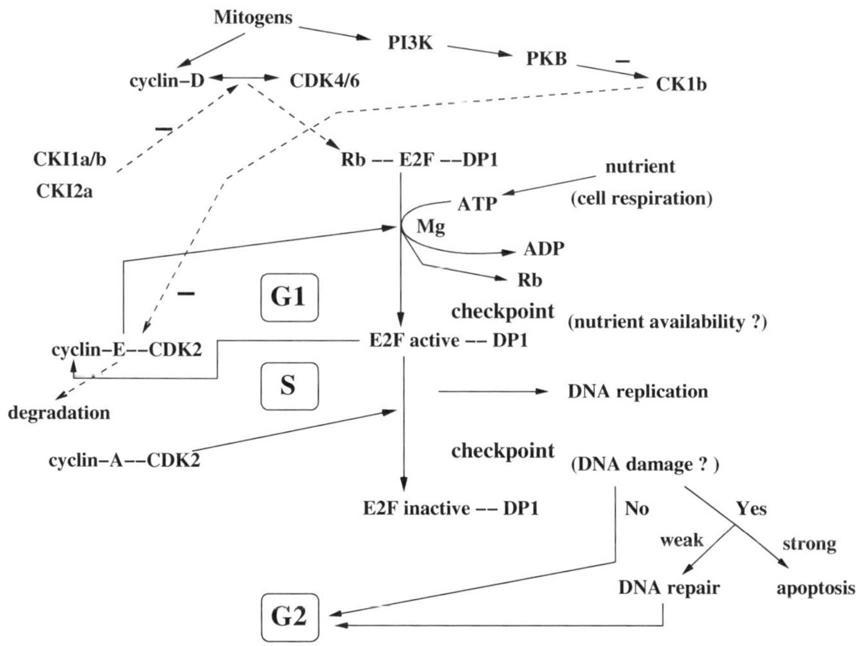

Figure 2.5. Main regulations during G1 and M phases of the cell division cycle. Each phase contains a checkpoint that allows cell cycle progression or causes cell cycle arrest according to the context.

Table 2.11. Cell cycle regulator fate.

Process	Influence factors
Synthesis	External signals, transcription factors
Activity	Activation (stimulatory kinases and phophatases) Inactivation (inhibitory kinases and phophatases, ubiquitin)
Elimination	Sequestration (reversible), degradation (irreversible)

Table 2.12. Proto-oncogene products and tumor suppressors.

Proto-oncogene products	Tumor suppressors
Ccn–CDK dimers	Retinoblastoma proteins
CDK activator kinases	P53
CDC25	CDK inhibitors (CKIs)
Abelson kinase	Breast cancer proteins
Monomeric GTPases	Ub ligase APC

2.4.2.1 Switch-like Behavior of Cell Cycle Controllers

Any received signal is transduced to produce a response. In general, an increase in the stimulus causes an augmentation in the output. When little

Table 2.13. Examples of proteins, the abundance of which rises in late G2 phase, during G2–M transition, or M phase (Source: [171]; CDCA: cell division cycle-associated protein; ChMP: chromatin-modifying or charged multivesicular body protein; FosL: Fos-like antigen; KIF: kinesin family member; NuSAP: nucleolar and spindle-associated protein; PTTG1: pituitary tumor-transforming gene product; SCFD1: Sec1 family domain-containing protein; UbE2S: ubiquitin-conjugating enzyme). M-phase-induced phosphorylation by many protein kinases involved in mitotic progression is required for mitotic functions. Cyclin-B1 and -B2 are highly upregulated during M phase, whereas cyclin-A2, -E1, and -E2 are produced in S phase. Amounts of PLK1 and Aurora kinases increase, but NEK2, Weel, and PTMYT1 kinases are reduced. Nucleolar and spindle-associated protein NuSAP1 is primarily nucleolar during interphase, then localizes mainly to central spindle microtubules during mitosis. Its concentration reaches a maximum at G2–M transition and abruptly declines after cell division. Microtubule-associated homolog TPX2 (TPX2: targeting protein for Xenopus laevis centrosomal kinesin-like protein XKLP2) is exclusively expressed in the nucleus of proliferating cells. It is strictly associated with the mitotic spindle during M phase, whereas during S and G2 stages, it distributes throughout the nucleus. Actin-binding anillin human homolog interacts with cleavage furrow proteins, actin and septins. Cysteine-rich angiogenic inducer CyR61 [a.k.a. insulin-like growth factor-binding protein IGFBP10 and connective tissue growth factor, Cys-rich protein, and nephroblastoma overexpressed gene product CCN1] is a secreted heparin-binding protein that promotes cell adhesion, migration, and growth factor-stimulated proliferation. TBC (Tre2/Bub2/Cdc16) domain-containing protein family member TBC1d15 operates as a RabGAP for Rab7 and, to a lesser extent, Rab11 GTPase. Completion of cell division during cytokinesis requires temporally and spatially regulated communications from the microtubule to actin cytoskeleton and plasma membrane.

Centromere–kinetochore complex components	Centromere protein CenPf (mitosin)
Cleavage-furrow components	Anillin
Chromosome condensation, chromatid separation	Topoisomerase-2α
Kinases	Aurora-A and -B, Polo-like kinase-1
Kinase partners	Cyclin-B1
Regulators	Cyr61, PTTG1 (securin), UbE2S
Spindle builders	NuSAP1, TPX2
Spindle stabilizers	CDCa8 (borealin)
Transcription factor complex components	FosL1
Transport	ChMP4a, KIF11, KIF20a, SCFD1, TBC1d15

signal processing occurs, a linear increase in the input induces initially a linear elevation in the response, until saturation. However, when the input rises above a certain amount, the response magnitude is lower for a given increase in input (hyperbolic signal–response curve) [172].

Table 2.14. Regulators of the cell cycle. (**Part 1**) Kinases, phosphatases, and cyclin-dependent kinase inhibitors.

Molecule	Effect
	Kinases and phosphatases
CDK-activating kinases	Phosphorylation of activating sites of CDKs
Checkpoint kinases	Control of CDK1
Wee1 kinase	Phosphorylation of inhibitory sites of CDKs
Polo-like kinases	Cohesin removal
	Activation of CDC25, inhibition of Wee1 and PKMYT1
	Activation of CDK1, inactivation of EMI1
	Cytokinesis
	Lamina depolymerization
Kinase Aurora	Chromosome compaction
	Lamina depolymerization
	Phosphorylation of Polo-like kinase-1 via Bora
	Phosphorylation of RacGAP1
Protein kinase-B	Phosphorylation of NFκB
	Inhibition of ChK1; SKP2 sequestration
	CKI1b activity
Protein kinase-C	Lamina depolymerization
	Dissolution of the mitotic ring
Phosphatases CDC25a/b/c	Dephosphorylation of inhibitory sites of CDKs
	Cyclin-dependent kinase inhibitors
CKI1a	Inhibition of G1–S- and S-associated CDKs in G1 upon DNA damage
	Blockade of phosphorylation of CDKs by CAK
CKI1b	Inhibition of G1–S- and S-associated CDKs in G1
CKI1c	Inhibition of CDK1
CKI2a	Inhibition of G1-associated CDKs in G1
	P53 stabilization (DM2 sequestration)

On the other hand, more complex signal processing by ultrasensitive systems is able to convert a high grade cue, the intensity of which is greater than a given threshold (the response is blunted at low messenger concentration), into an abrupt output (e.g., the response soars when the messenger concentration becomes greater than that of an inhibitor). Ultrasensitive systems respond to signals that range intermediate intensities, as, above a given value, the stimulus remains without significant effect (sigmoidal signal–response curve) [172].

A binary switch displays a very steep slope of the sigmoidal signal–response curve (all-or-none response). A positive feedback is able to induce a binary switch. Due to the steep slope, an intermediate activity between the extrema cannot remain stable. The system is bistable, as it has 2 stable states that

Table 2.15. Regulators of the cell cycle. (**Part 2**) Transcription factors, ubiquitin ligases, and others.

Molecule	Effect
	Transcription factors
P53	Activation of CKI1a
E2F	Gene transcrition for G1–S progression
NFκB	Expression of cyclin-D1, SCF
FoxM1	Upregulation of CDC25a
	Downregulation of CKI1a and CKI1b
	Ubiquitin ligases
SCF–SKP2	Ubiquitination of CDK1, CKI1a, CKI1b, EMI1
APC–CDH1	Ubiquitination of securin, shugosin, FoxM1
	Ubiquitination of M-associated cyclins
DM2	P53 ubiquitination and degradation;
	Inhibition of P53 transcriptional activation
	Miscellaneous
Retinoblastoma protein	Inhibition of E2F
Minichromosome maintenance	Regulation of genes expressed in G2–M and M–G1
proteins	Prereplicative complex
Small GTPases	Cytokinesis
Protein 14-3-3	Protein synthesis and sequestration

correspond to the 2 alternative states (nothing or maximal amplitude situations). The signaling can actually remain stable even when the stimulus source weakens.

In the case of a signaling based on kinases and phosphatases, the response deponds on the rates of stimulatory and inhibitory phosphorylations and inactivating and activating dephosphorylations. In a system endowed with a stimulatory kinase and an inhibitory phosphatase, when the rates of opposite reactions are balanced, the system remains at rest. Nevertheless, above a given threshold, the generation of mediators exceeds the capacity of inhibitory influences. Owing to the positive feedback that rapidly supplies activations even in the absence of further inputs, maximal kinase activity is quickly achieved.

2.4.2.2 Post-Translational Modification in Cell Cycle Control

Reciprocal regulation between kinases and phosphatases ensure adequate phosphorylation or dephosphorylation of a given mediator at a given time of the cell cycle (Table 2.16). Phosphorylation and dephosphorylation occur at appropriate instants to guarantee an orderly cell cycle. Among potential targets for phosphorylation and dephosphorylation, DNA primase is a RNA polymerase involved in the replication of DNA.[65] DNA Topoisomerases unwind and wind DNA to control protein synthesis and DNA replication. They cut DNA and, at the end of the process, reconnect it.

Kinases

Cyclin-dependent kinases phosphorylate a huge number of substrates. They themselves bear phosphorylation and dephosphorylation during mitosis by activators and inhibitors that can form feedback loops (Table 2.17). Activation of these enzymes requires binding to a cyclin. Moreover, cyclins can target CDK activity to particular substrates. At least, 10 cyclins (regulatory subunits) and 9 CDKs (catalytic subunits) exist. Once a cyclin is tethered to a CDK, the catalytic subunit undergoes either inhibitory (Thr14 and Tyr15) or stimulatory phosphorylation (Thr161) as well as activating dephosphorylation (Thr14 and Tyr15). In addition, cyclin–CDK dimers can be inhibited by several cyclin-dependent kinase inhibitors or experience proteolytic destruction of the cyclin component.

More than 1,000 proteins undergo change in phosphorylation level during the cell cycle [171]. More than 50 kinases are activated by phosphorylation. Phosphorylation level depends on protein types, some being much more heavily phosphorylated during mitosis than others. Phosphorylation influences protein conformation as well as its interactions, subcellular localization, and stability. Multiple phosphorylation serves for signal integration, creation of signaling thresholds, and ultrasensitivity. Dephosphorylation of numerous sites also generates conformational changes that temper localization. Phosphorylation as well as dephosphorylation either activates or inactivates substrates.

Major mitotic kinases comprise Aurora-A and -B, CDK1 and CDK2, PKMYT1, PLK1, and Wee1 that phosphorylate cell cycle regulators and are phosphorylated by timers. Moreover, their interacting proteins (e.g., abnormal spindle molecule ASpm1, protein Tyr^{P} phosphatase CDC25c, and microtubule-stabilizing translationally controlled tumor protein TPT1 for Polo-like kinase PLK1; and promyelocytic leukemia protein PML and targeting protein for Xklp2 [TPX2] for Aurora-A) are also regulated by phosphorylation.

---

[65] Primase catalyzes the synthesis of a short RNA segment, the so-called primer, complementary to a single-strand DNA template in the absence of DNA polymerase that can initiate the synthesis of a DNA strand without an initial RNA or DNA primer for temporary DNA elongation. Because DNA polymerase can continue, but not begin, a strand, primase must begin the process. DNA Polymerase forms a protein complex with 2 primase subunits to form the α-DNA polymerase primase complex.

Table 2.16. Substrates that can experience phosphorylation–dephosphorylation cycles during the cell cycle control (Source: [173]; APC: anaphase-promoting complex; BAD: BCL2 antagonist of cell death; BrCa: breast cancer tumor suppressor; CDC: cell cycle; (Cx): type-x cyclin; MyoD: myogenic differentiation; Pol: DNA polymerase; RBL1/2: retinoblastoma-like protein-1 [a.k.a. P107] and -2 [a.k.a. P130]; Top: topoisomerase; Ub: ubiquitin; WT1: Wilms tumor transcription factor). Condensins are large proteic complexes that intervene in chromosome assembly and segregation.

Protein	Phosphorylation effect	CDC phase	Kinase	Phosphatase
APC		M	CDK1	
BAD	Cell survival		PKB	
BrCa1	Dissociation from centrosome	M	CDK1	
Caspase-9	Cell survival		PKB	
CDC25	Activation, mitosis	G2/M	PLK	PP2a
CDK	Activation CDC progression	All phases	CAK	
CKI1a			CDK2(Ca)	
CKI1b	Recruitment by Ub	G1	CDK2(Ce)	
Condensin	Activation, chromosome condensation	M	CDK1	
DNA Polα Primase	Activation, DNA replication		CDK2 (Ca, Ce)	
DNA Top	Activation		PKC	
E2F	Inhibition of transcription	S–M	CDKs	
Lamin	Depolymerization of nuclear envelope	M	CDK1	PP1
MyoD	Degradation (no p21 stimulation)	G1	CDKs	
P53	Decreased inhibition by DM2 (Ser15)		CDKs	
RB	Inhibition of G1 arrest	Late G1–M	CDKs	PP1
RBL1	Inhibition of G1 arrest	Mid G1	CDKs	
RBL2	Inhibition of G1 arrest	Early-mid G1	CDKs	
WT1	Inhibition of G1 arrest	G1	CDKs	

Phosphatases

The cell cycle is regulated by periodic changes in activity of kinases, phosphatases, and ubiquitins. Phosphatases can intervene for the completion of the cell cycle stages. Protein phosphatase-2A is inactivated when cells enter mitosis and reactivated after proteolysis of mitotic cyclins [174].

Table 2.17. Examples of regulatory kinases and phosphatases. The progression of the cell cycle is regulated by cyclin-dependent kinases (CDK) that become active by linking to regulatory cyclins. The activity of cyclin-CDK complexes is stringently regulated via synthesis and degradation of cyclins and activation and inhibition (phosphorylation and dephosphorylation) of cyclin–CDK complexes. For example, G1 progression requires cyclin-D-dependent kinases CDK4 and/or CDK6 and the cyclin E-dependent kinase CDK2. Binding of CDK inhibitors CKI1a, CKI1b, and CKI1c to CcnD–CDK4, CcnD–CDK6, and CcnE–CDK2 inhibits them and provokes arrest. Inhibitors CKI2a to CKI2d compete with cyclin-D for CDK4 connection.

Regulator	Kinases	Phosphatases
Stimulator	CDK-activating kinases	CDC25
	Polo-like kinase-1	
Inhibitor	Wee1	CDC14

Table 2.18. Major cyclin–cyclin-dependent kinase (CDK) complexes that control the cell cycle.

Controller	Phases			
	G1	S	G2	M
cyclin	D	E	A	B
CDK	4/6	(1)/2	1/2	1

2.4.2.3 Cell Shape

Cell shape, i.e., mechanotransduction, intervenes in the control of cell cycle progression. In unspread cells in a medium filled with soluble growth factors, ERK1 and ERK2 kinases are activated. On the other hand, in spread cells, the cell cycle fails to progress through G1 and enter S phase [175]. Cyclin-D1 does not augment and cell cycle inhibitor CKI1b is not reduced. Cell proliferation is thus partly controlled by tension developed by the cytoskeleton.

2.4.2.4 Cyclins

At given stages of the cell cycle, cyclins are produced and then degraded, upon encountering ubiquitin ligase complexes such as anaphase-promoting complex or cyclosome (APC/C) (Tables 2.18 and 2.19). Two types of cyclin-A (A1–A2) and 3 types of cyclin-B (B1–B3) exist. Cyclin-A1 is expressed mainly in early zygotes and testis; cyclin-A2 is ubiquitous. At low concentration, cyclin detaches from cyclin-dependent kinases.

Transcription factors with their sequential waves act as an oscillator that regulates the sequential expression of cell cycle genes. In the absence of cyclins, a large proportion of cell cycle genes maintain cell cycle periodicity. A transcription factor-based oscillator can regulate the transcription of a significant

Table 2.19. Mediators of the cell cycle progression and blockage (Source: [176]).

Phase	Regulator(s)	Upstream regulator(s)
	Cell cycle progression	
G1 to S	CcnE–CDK2	SKP2 (inhibition of CKI1)
S to G2	CcnA–CDK2	CDC25a
G2 to M	CcnA–CDK1	CDK1, PLK1, CDCA3
	CcnB–CDK1	(inhibition of Wee1)
M to G1	APC/C–CDC20	CcnB–CDK1, UbE2C,
		PLK1 (inhibition of EMI1)
	APC/C–CDH1	UbE2C
	Cell cycle blockage	
G1 to S	CcnE–CDK2	CKI1a/b/c
S to G2	CcnA–CDK2	
G2 to M	CcnA–CDK1	Wee1
	CcnB–CDK1	Wee1
M to G1	APC/C–CDC20	EMI1
	APC/C–CDH1	

amount of cell cycle genes, although cyclins and cyclin-dependent kinases are required for successful cell cycle progression [177]. Cells lacking cyclins are indeed arrested at the G1–S border, as the cyclin–CDK oscillator governs major events of the cell cycle.

Cyclin-A

Cyclin-A accumulates from late G1 phase and is destroyed before metaphase [178]. The amount of cyclin-A that binds to CDK2 steadily rises from end G1 to G2 phase to decay more quickly during G2 and M phase.

Cyclin-A1 promoter is repressed during G1 phase and highly active during S phase. Cyclin-A1 then accumulates from early S phase and disappears during mitosis before cyclin-B. The single function of cyclin-A1 is related to spermatogenesis [179]. Cyclin-A1–CDK2 can interact with many partners.[66]

Cyclin-A2 starts to accumulate at early S phase, continues through S and G2 phase, and disappears during mitosis. The concentration of cyclin-A2 reaches its peak level in G2–M transition. It is about 30-fold less than that of CDK1 and 8-fold less than that of CDK2 [180]. Cyclin-A2 is quickly degraded during mitosis, as it is ubiquitinated for degradation by the APC Ub ligase.

[66] These interactors include G-protein-pathway suppressor GPS2, G protein, β-polypeptide-2-like GNB2L1, inhibitor of CDK interacting with cyclin-A1 (INCA1), kelch/ankyrin repeat-containing cyclin-A1-interacting protein KARCA1), Ku70, proline-rich cyclin-A1-interacting protein (PROCA1), and RNA-binding motif protein RBM4.

Both cyclin-A1 and -A2 form complexes with CDK1 and CDK2. Activation of CDK1 and CDK2 not only requires binding to a cyclin, but also phosphorylation (Thr161 and Thr160 in CDK1 and CDK2, respectively) by CDK-activating kinase. Protein phosphatase-2Cα and -β2 isoforms can dephosphorylate CDK2 (Thr160^P). CcnA1/A2–CDK1/2 complexes are inhibited by cyclin-dependent kinase inhibitors CKI1a to CKI1c.

Both cyclin-A1 and -A2 also binds to transcription factors, such as E2F-like protein E2F1 as well as retinoblastoma protein.[67] The CcnA2–CDK complex phosphorylates P73 (repression), P53 (stimulation), and estrogen receptor-α (enhanced activity). In addition, CcnA2–CDK complexes can be phosphorylated (inactivated; Thr14 and Tyr15 of CDK1 during the unperturbed cell cycle and both CDK1 and CDK2 after checkpoint activation) by kinases Wee1 and PKMYT1.

Cyclin-B

Cyclin-B is synthesized and degraded slightly later than cyclin-A. The amount of cyclin-B that binds to CDK1 to form the mitosis-promoting factor increases from the early stage of S phase to the beginning of M phase and then decreases abruptly due to degradation.

Cyclin-C

Cyclin-C participates in both positive and negative regulation of gene transcription in cooperation with cyclin-dependent kinases. The primary partner for cyclin-C is CDK8. Cyclin-C regulates both CDK8 substrate specificity and kinase activity. Cyclin-C is also able to associate with CDK3 to promote retinoblastoma protein-dependent exit from the quiescent state of the cell cycle [181]. Retinoblastoma protein phosphorylation indeed promotes transition from G0 to G1 phase.

Cyclin-C is stabilized by its interaction with CDK8, whereas the free form is unstable. The CcnC–CDK8 complex is phosphorylated, but not the free cyclin-C [181]. Unstable cyclin-C is degraded by the ubiquitin–proteasome axis.

Three alternatively spliced cyclin-C messenger-RNAs give rise to isoforms. The expression of the truncated isoform is regulated by the cell cycle, with maximal expression in G2–M transition [181]. Short isoform may compete with full-length isoform for CDK8 binding.

The CcnC–CDK8 complex is able to phosphorylate RNA polymerase-2. It regulates transcription by phosphorylating cyclin-H component of TF2H

[67] Retinoblastoma protein is a recruiter of several chromatin remodeling enzymes, such as methylases and acetylases. Hyperphosphorylated Retinoblastoma protein binds to E2F transcription factor. Protein RB represses transcription of CDK1, cyclin-A, Fos and Myc.

transcription factor. It attaches Mediator complex subunit-12 and -13 (Med12 and Med13)[68] to regulate transcription [181].

In addition, Mastermind homolog interacts with the CcnC–CDK8 complex and recruits the Notch intracellular domain (NotchICD). Kinase CDK8 hyperphosphorylates NotchICD and promotes its degradation [181].

Cyclin-D

The amount of cyclin-D that binds to CDK4 and CDK6 augments from early G1 to S phase, remains nearly constant until G2 phase to decline from G2 to M phase. The cyclin-D–kinase activity is maximal in early to mid-G1 phase. In early to mid-G1 phase, CcnD–CDK phosphorylates (inactivates) the cell cycle inhibitor retinoblastoma protein.

Three mammalian isoforms of cyclin-D (cyclin-D1–cyclin-D3) possess some functional redundancy, but different expression according to the cell type. Cyclin-D1, -D2, and -D3 bind to CDK4 and CDK6, the activity of which is required for G1–S transition.

Cyclin-E

The amount of cyclin-E that binds to CDK2 rapidly heightens during end of G1 phase to quickly fall during S phase. The CcnE–CDK2 complex phosphorylates CKI1b, an inhibitor of cyclin-D, that is then degraded.

Cyclin-H

Cyclin-H forms a complex with CDK7 to phosphorylate cyclin-dependent kinases. Cyclin-dependent kinase-7 is an important regulator of the cell cycle progression. It indeed forms a trimer with cyclin-H and CDK-activating kinase assembly factor Ménage à trois homolog MAT1. The CcnH–CDK7–MAT1 complex operates as a CDK-activating kinase (CAK).

2.4.2.5 Cyclin-Dependent Kinases

Cyclin-dependent kinases (CDK) are Ser/Thr protein kinases that are activated by association with cyclins (Table 2.20).[69] Cyclin-dependent kinases and corresponding cyclins form heterodimeric enzymes. The mitosis-promoting factor (MPF) corresponds to the CcnB–CDK1 complex. The Ccn–CDK complexes are regulated by kinases and phosphatases. In most cases,

[68] A.k.a. thyroid hormone receptor-associated protein complex 230-kDa and 240-kDa components (TRAP230 and TRAP240).

[69] Cyclin-dependent kinases are inactive without their cyclin partners. Orphan cyclins do not have identified CDK partners, but can be implicated in the cell cycle.

Table 2.20. Cyclins and cyclin-dependent kinases form activating complexes. Cyclin-dependent kinases constitute a family of protein kinases that regulate cell cycle progression (CDK1–CDK4 and CDK6), cell differentiation (CDK5), and transcription (CDK7–CDK10). The role played by CDK11 to CDK20 remains insufficiently known. Furthermore, in many cases, the regulatory cyclin subunits have not been identified. Kinases CDK14 (a.k.a. CDK-like PFTAIRE kinase PFTK1) and CDK15 (or PFTK2) bind to cyclin-Y and cyclin-Y-like-1 and target low-density lipoprotein receptor-related proteins LRP5 and LRP6 (Sect. 2.4.2.27).

Cyclin-dependent kinase	Cyclin partners and activators
CDK1	Cyclin-B (preferentially), -A, -D, and -E
CDK2	Cyclin-A and -E (strongly), also -B, and D
CDK3	
CDK4	Cyclin-D1, -D2, and -D3
CDK5	Neuronal-specific CDK5R1 (P35) regulator
CDK6	Cyclin-D1, -D2, and -D3
CDK7 (CAK)	Cyclin-H
CDK8	Cyclin-C
CDK9	Cyclin-T1, -T2a, -T2b, and -K
CDK10	
CDK11	Cyclin-L
CDK14	Cyclin-Y
CDK15	Cyclin-Y-like-1

the concentration of kinase and cyclin subunit is relatively constant and oscillates, respectively. Cyclin-dependent kinases, except CDK9, regulate the cell cycle, as well as transcription and mRNA processing. Active enzymes are inactivated by: (1) specific ubiquitin-mediated proteolysis of associated cyclins; (2) binding to an inhibitor; and (3) inhibitory phosphorylations.

The cyclin-dependent kinase-4(6)–cyclin-D complexes are expressed early in G1 phase. They stimulate the expression of genes required for G1 progression, as they phosphorylate retinoblastoma gene product (RB). Phosphorylation of RB causes dissociation of RB-transcription regulator complexes. Cyclin-dependent kinase-2 that first links cyclin-E and then cyclin-A afterward coordinates progression from G1 to S phase. Protein RB that is initially phosphorylated on multiple sites by CDK4 and CDK6 is additionally phosphorylated by CDK2.

New observations oppose the classical model of the cell cycle [174]. CDK1 controls the entire cell cycle in mouse embryonic fibroblasts. It can replace CDK2, as many different Ccn–CDK combinations (interchangeable binding partners) and redundancy exist. However, CDK2 is required for meiosis. Moreover, CDKs can have tissue-specific activity. CDK4 is mandatory in the pancreas and CDK6 in erythropoiesis. Cyclin-A1 is required in germ cells, but is absent in somatic cells.

The type of cyclins bound to a given CDK at different stages of the cell cycle may explain CDK control of diverse sets of reactions at different times. Moreover, cyclin species can influence kinase activity magnitude and substrate types according to subcellular location. Last but not least, the regulation of the cell cycle can depend not only on cell type, but also on organism type.

Cyclin-dependent kinase-8 phosphorylates the carboxy-terminal domain of the largest subunit of RNA polymerase-2, cyclin-H component of general transcription factor GTF2H1, and other transcription factors [186].

Cyclin-dependent kinase-11 has 2 isoforms encoded by distinct genes (CDC2L1 and CDC2L2). However, 3 major CDK11 isoforms exist: p110, p58, and p46. Isoform p110 that controls gene transcription interacts with cyclin-L1α and -L2α isoforms [187]. Isoform p58 is specifically translated at the G2–M transition. Isoform p46 is produced by cleavage of p110 and p58 by caspases after the initiation of apoptosis.

2.4.2.6 Cyclin-Dependent Kinase-Activating Kinases

Cell growth leads to cell division. Yet, cell size and cell division must be coordinated. A suitable cell size must indeed prime the cell division to avoid generating small or large daughter cells. Cells normally divide when they reach a certain size owing to cyclin-dependent kinase CDK1. Entry into mitosis depends on protein phosphatase CDC25 that relieves the inhibitory phosphorylation from CDK1 by Wee1 kinase.[70]

The activation of cyclin-dependent kinases depends at least on 2 steps: (1) binding to cyclin and (2) phosphorylation of CDK activation domain (Thr160) by CDK-activating kinase (CAK). The CDK-activating kinase corresponds to the trimer CcnH–CDK–MAT1.[71] It allows sorting of activation steps directed by Ccn–CDK complex specificity. Kinase CDK7 can only phosphorylate CDK1 in the presence of cyclin. On the other hand, CDK7 only phosphorylates the CDK2 monomer. Once phosphorylated, CDK2 is then able to bind cyclin [184].

---

[70] In fission yeast (Schizosaccharomyces pombe), a spatial gradient of the dual-specificity Tyr-phosphorylation-regulated kinase Pom1 that emanates from cell tips, where it lodges during interphase, serves as a length sensor of these rod-shaped cells and may regulate mitotic entry [182,183]. It induces cell division, as it removes action of CDK1 inhibitor Wee1. In yeasts, Wee1 is also inhibited by 2 related protein kinases of the GIN4–SAD family CDR1 (or AMPK-related non-inducible immunity kinase NIM1) and CDR2. Many proteins, such as the cell cycle regulators CDR1, CDR2, and Wee1, reside in interphase nodes that form a band in the medial cell cortex. In small cells, Pom1 phosphorylates (inactivates) CDR2, thereby allowing active Wee1 to inhibit CDK1 and preventing entry into mitosis. When cells elongate, Pom1 concentration lowers at the cell middle, hence CDR2 and CDR1 inhibit Wee1, preventing inhibitory phosphorylation of CDK1 and promoting mitotic entry.

[71] MAT1: Ménage à trois-1.

In humans, cyclin-A binds preferentially to CDK2 throughout G1 and S phase, despite greater abundance in CDK1. Yet, cyclin-A also assembles with CDK1, but efficient CDK1 binding does not begin until CcnA–CDK2 aggregation is nearly complete, i.e., only after CcnA–CDK2 complex level reaches a plateau during late S and G2 phases. Cyclin-B that normally binds nearly exclusively CDK1 can be diverted to CDK2 upon CDK7 inhibition [184]. Therefore, the activity and substrate preferences of CAK contribute to restrain pairing rules among cyclins and CDKs for complex formation.

Cyclin-dependent kinase CDK8 and cyclin-C are components of holoenzyme RNA polymerase-2, as CDKs are subunits of general transcription factor-2H, one of several general transcription factors that make up RNA polymerase-2 pre-initiation complex [185]. General transcription factor-2H is composed of 10 subunits, 7 of which form the core complex. Kinase CDK8 phosphorylates the largest subunit of the RNA polymerase-2 complex. It regulates transcription by targeting CcnH–CDK7 subunits of general transcription factor-2H.

The CDK-activating kinase subcomplex CcnH–CDK7–MAT1 is linked to the core of general transcription factor-2H. The GTF2h subunit CcnH–CDK7 can phosphorylate RNA polymerase-2 and possibly other proteins involved in the cell cycle.

2.4.2.7 Cyclin-Dependent Kinase Inhibitors

Cyclin-dependent kinase inhibitors encompass several protein types. Cyclin-dependent kinase inhibitor-1 belongs to the Cip–Kip family (Table 2.21). Inhibitor CKI1a accumulates transiently in late G1 phase and helps to prevent premature entry into S phase by inhibiting CcnE–CDK2. Afterward (during G1–S transition), it is targeted by the Ub ligase SCF for degradation. Inhibitor CKI1a again accumulates during G2 phase and contributes to preclude premature activation of mitotic Ccn–CDK1. It is then degraded in prometaphase following its ubiquitination by the Ub ligase APC [188].

Inhibitor CKI1b delays the transition from G1 to S phase until assembly of prereplication complexes at replication origins is complete. Inhibitor CKI1b binds CcnD–CDK4. However, once bound, it may or may not inhibit (bound non-inhibitor form) this complex depending on its phosphorylation status. Upon binding to CDK4, non-phosphorylated CKI1b also prevents its phosphorylation (activation) by CcnH–CDK7 [189]. Inhibitor CKI1c is expressed at the end of S phase and localizes to the nucleus, where it inhibits CDK1 activity and prevents entry into mitosis.

Structurally related cyclin-dependent kinase inhibitors, such as CKI1b and CKI1c share many characteristics, but also perform specific functions. Kinase CKI1c, but neither CKI1a nor CKI1b, mediates the effect of transforming growth factor-β in compensatory hypertrophy of nephron tubule [190]. Inhibitor CKI1a is less effective than CKI1b and CKI1c in hindering

Table 2.21. Cyclin-dependent kinase inhibitors (CKI) that control G1 progression and G1–S transition.

Type	Other name	Gene	Targets
	Cyclin-dependent kinase inhibitor-1		
CKI1a	p21, Cip1, Waf1	CDKN1A	CDK2, CDK4
CKI1b	p27, Kip1	CDKN1B	CDK1, CDK2, CDK4
CKI1c	p57, Kip2	CDKN1C	G1-associated Ccn–CDKs
	Cyclin-dependent kinase inhibitor-2		
CKI2a	p16, INK4a	CDKN2A	CDK4, CDK6
CKI2b	p15, INK4b	CDKN2B	CDK4, CDK6
CKI2c	p18, INK4c	CDKN2C	CDK4, CDK6
CKI2d	p19, INK4d	CDKN2D	CDK4, CDK6
	Cyclin-dependent kinase inhibitor-3		
CKI3	KAP, Cdi1	CDKN3	CDK2

CDK2-dependent phosphorylation of nuclear protein, ataxia telangiectasia gene product (NPAT) and transcriptional activation of histone-4 genes [191].

Cyclin-dependent kinase inhibitor-2 isoforms (CKI2a–CKI2d) of the INK4 family (INK4a–INK4d), unlike other universal CDK inhibitors, such as CKI1a and CKI1b, selectively inhibit activity of CDK4 and CDK6. Among the 3 alternatively spliced CKI2a variants, 2 isoforms inhibit CDK4. The remaining transcript contains an alternate open reading frame (ARF) that stabilizes protein P53. Both inhibitor isoforms and ARF product encoded by gene CDKN2A control G1-phase progression via regulators CDK4 and P53. Genes CDKN2B and CDKN2C encode proteins CKI2b and CKI2c that also control G1 progression, as they form a complex with CDK4 or CDK6 to impede their activation.

Cyclin-dependent kinase inhibitor-3 is a CDK2-associated dual-specificity phosphatase that dephosphorylates (inactivates) CDK2. Gene CDKN2A also generates a transcript CKI2a that also contains an alternate open reading frame. This product is a stabilizer of P53, as it interacts with and sequesters the Ub ligase DM2 that degrades P53.

Cell proliferation is controlled by both Ccn–CDK complexes and CDK inhibitors. In quiescent smooth muscle cells of embryonic rat thoracic aorta, both CKI1b and CKI1c are coexpressed. Upon proliferative stimuli, both CKI1b and CKI1c are repressed. The level of CKI1b is regulated by both transcriptional and post-transcriptional processes, whereas that of CKI1c is mainly controlled by a post-transcriptional mechanism [192].

Hypoxia significantly hampers CKI1b expression (98% decrease) and boosts CKI2c activity (81% increase), but does not significantly influence CKI1a, CKI1c, CKI2a, CKI2b, and CKI2d [193]. Heparin that precludes pulmonary artery smooth muscle cell proliferation by increasing levels of CKI1a

Table 2.22. Protein kinase Wee1 in S, G2, and M phases. Cell cycle transition from G2 to mitosis is regulated by the CcnB–CDK1 complex.

Phase	Event
S and G2 (Wee1)	Inhibitory phosphorylation of CcnB–CDK1
	Blockage of entry into mitosis
	Phosphorylation of CcnA/E–CDK2
	Translocation to the cytoplasm in late G2
	Phosphorylation and sequestration by 14-3-3 scaffold
	Activating dephosphorylation by CDC25 phosphatase of CDK1
	Inhibitory phosphorylation by CcnB–CDK1 and PLK1
	Additional phosphorylation by casein kinase-2
	Promotion of PLK1–Wee1 complex formation by CK2β
	Promotion of rapid CDK1 activity
M ($Wee1^P$)	Degradation by SCF1–β2-TRCP

and CKI1b restores CKI1b activity via kinases ERK1, ERK2, and P38MAPK, but not that of CKI2c.

2.4.2.8 Kinase NIM1

AMPK-related dual-specificity kinase and mitotic inducer Non-inducible immunity NIM1 (or CDR1) phosphorylates (inhibits) Wee1 kinase that then cannot phosphorylate CDK1 [194]. In humans, a Nim1-like kinase pathway acts at the G1–S transition.

2.4.2.9 Kinase Wee1

Protein kinase Wee1 (Vol. 4 – Chap. 6. Dual-Specificity Protein Kinases) hampers entry into mitosis. Its related kinase PKMYT1 also phosphorylates CDKs (Thr14 and Tyr15). Human mitotic inhibitor homologs Wee1-like protein kinase-1 and -2 that are encoded by genes WEE1 and WEE2 block cell division, as they phosphorylate (inhibit) CDK1. The latter allows the progression from G1 to S and G2 to M phase (Table 2.22).

Kinase Wee1 is expressed during S and G2 phases. Its degradation by the ubiquitin–proteasome system increases during M phase. Protein Wee1 is located in the nucleus during interphase of the cell cycle. Before mitosis, Wee1 moves into the cytoplasm.

Kinase Wee1 phosphorylates CDK1 and CDK2 (Tyr15; near the ATP-binding pocket), when targeted CDKs are associated with cyclins, but not CDK4 in vitro [195]. It phosphorylates (inactivates) CDK1 during S and G2 phases prior to mitosis, hence delaying the onset of mitosis, as well as

CcnA/E–CDK2 complexes. Its activity is regulated by protein phosphorylation and degradation. During S and G2 phases, Wee1 is phosphorylated (Ser642) to bind to scaffold 14-3-3β, -ζ, and -θ. In addition, human Wee1 can autophosphorylate (Tyr295 and Tyr362).

A member of the F-box protein class, β-transducin repeat-containing protein (βTRCP), binds to human Wee1. Ubiquitin ligase SCF causes Wee1 ubiquitination. Polo-like kinase-1 also binds phosphorylated $Wee1^P$ and phosphorylates $Wee1^P$ (Ser53) [195]. The CcnB–CDK1 complex phosphorylates Wee1 (Ser123) and thus favors Polo-like kinase-1 binding. Casein kinase CK2 phosphorylates (Ser121) phosphorylated (Ser123) $Wee1^P$ [195]. The casein kinase-2 regulatory β subunit participates in PLK1–Wee1 complex formation, thus independently of tetrameric CK2 kinase. Indeed, interaction of CK2β with Wee1 does not modify Wee1 activity. Instead, CK2β fosters CDK1 activity by preventing further inhibitory phosphorylation by Wee1 [196].

2.4.2.10 Aurora Kinase and the Chromosomal Passenger Complex

Aurora-B, in cooperation with inner centromere protein (InCenP), cell division cycle-associated protein-8 (CDCA8),[72] and Survivin, form the *chromosomal passenger complex* (CPC). Aurora-B corresponds to the CPC catalytic component, whereas InCenP, CDCA8, and Survivin are the CPC regulatory components. The chromosomal passenger complex operates during mitosis (and meiosis) by changing its localization [197]. At the beginning of M phase, this complex localizes to chromosomes, where it controls chromatin-dependent spindle assembly and processes at the centromere. During anaphase, the CPC complex dissociates from chromosomes and relocalizes to the spindle midzone to stimulate cytokinesis.

The chromosomal passenger complex can be separated into 2 functional modules: (1) the kinase module with Aurora-B and InCenP C-terminus and (2) the chromosome-localization module with CDCA8, Survivin, and InCenP N-terminus [197]. The chromosomal passenger complex is indeed recruited to chromosomes during mitosis, where it supervises kinetochore activity and cytokinesis [198]. Chromosomes attach to the microtubules of the mitotic spindle via the kinetochore–centromere assembly before their separation during mitosis. Phosphorylation of substrates at the centromere and kinetochore of sister chromatids by Aurora-B prevents premature stabilization of the kinetochore–spindle interface. Kinases Haspin and BUB1 as well as the CPC complex with Aurora-B are recruited to this interface.

At centromeres during mitosis, the mitosis-specific kinase Haspin interacts with Cohesin, a proteic complex required for sister chromatid cohesion. The resulting complex phosphorylates histone-H3 (Thr3) in the vicinity of these chromosomal regions [197–199]. Histone-H3 phosphorylation permits the recruitment of the chromosomal passenger complex to chromosomes, thereby

[72] A.k.a. Dasra-B and Borealin.

activating its Aurora-B kinase subunit. The chromosome-localization module permits the recruitment of CPC to histone-H3^P [197]. The Survivin subunit of the chromosomal passenger complex that binds to Aurora-B in this complex actually recognizes and tethers specifically to histone-H3^P at centromeres [197, 198].

Furthermore, the protein kinase budding uninhibited by benzimidazoles BUB1 phosphorylates histone-H2a (Thr120), thereby recruiting shugoshin (Sgo2 rather than meiosis-specific Sgo1), the centromeric adaptor for the chromosomal passenger complex. In particular, shugoshins may recruit CPC components Survivin or Borealin, once they have been phosphorylated by the CDK1 kinase. The CcnB–CDK1 dimer phosphorylates Survivin of the CPC complex to connect to centromeres as well as to support chromosome bi-orientation [200]. Survivin^P directly binds to shugoshin.

Therefore, the recruitment of the CPC complex depends on both histone-2A^P and -H3^P [199]. Histone-3^P promotes nucleosome binding of Survivin; histone-2A^P fosters the binding of shugoshin. In addition, BUB1, Sgo2, and Aurora-B are required for MCAK recruitment to the centromere and inner kinetochore.

Chromatin-induced Aurora-B activation is required for spindle assembly around chromosomes, as it suppresses microtubule-depolymerizing activities [197]. Aurora-B kinase acts as a quality factor in chromosome attachment to the mitotic spindle and ensures a proper checkpoint. The centromeric pool of Aurora-B corrects attachment errors and controls the normal segregation of chromosomes. For proper partitioning of chromosomes in mitosis, the chromosomal passenger complex must localize at the center of paired kinetochores, at the inner centromere.[73] Aurora-B targets proteins at the interface of the centromere and the inner kinetochore, such as centromere protein-A, mitotic centromere-associated kinesin (MCAK), NDC80 homolog, kinetochore complex component Ndc80, and Knl1. The ability of Aurora-B to connect to its kinetochore substrates is inversely proportional to the degree of tension exerted on kinetochores.

At the interface between the centromere and inner kinetochore, the extent of phosphorylation of Aurora-B substrates depends or not on local histone-H3^P. The recruitment of MCAK^P to centromeres and kinetochores depends on histone-H3^P, but not CenPa [198].

At the metaphase–anaphase transition, the CPC dissociates from chromosomes to allow subsequent chromosome decondensation and nuclear reformation. Chromosome decondensation also depends on the chromosomal passenger complex [197]. Histone-H3^P is dephosphorylated at the late stage of M phase

---

[73] The accurate segregation of chromosomes results from the capture of sister chromatids by microtubules from the opposite spindle poles. The trial and error process of kinetochore–microtubule attachment relies in particular on Aurora-B located at centromeres.

2.4.2.11 Polo-like Kinases

The Polo-like kinase family includes 4 detected members (PLK1–PLK4). Polo-like kinases are located to centromere and kinetochore regions of chromosomes owing to partners (e.g., Polo box-interacting protein PBIP1, inner centromere protein InCenP, mitotic checkpoint protein BUB1α), and spindle-checkpoint complex component BUB1β [201]. Polo-like kinase PLK1 is located at mitotic centrosomes, centromeres, kinetochores, spindle midzone in anaphase, and midbody in cytokinesis. Kinases PLK2 and PLK4 reside at or near centrioles throughout the cell cycle, whereas PLK3 localizes to the nucleolus.

Polo-like kinases participate in activation of cyclin-dependent kinase-1 that with cyclin-B controls the entry into M phase. Polo-like kinases phosphorylate: (1) CDC25 (activation) that stimulates CDK1 as well as (2) Weel and PKMYT1 (inactivation) that inhibit CDK1 [201].

They also operate in centriole and centrosome genesis and maturation. In addition, they control mitotic chromosomes. They promote the removal of cohesins in prophase and facilitate their cleavage by separase. They stimulate the degradation of early mitotic inhibitor EMI1, an anaphase-promoting complex (APC) inhibitor. They inhibit shugoshin that protects centromeric cohesion until anaphase. They favor kinetochore attachment to microtubules and/or spindle assembly checkpoint. Last but not least, they contribute to cytokinesis, as they interact with protein regulator of cytokinesis PRC1 and mitotic kinesin-like protein MKLP2.

Transition from G2 phase into mitosis is controlled by the synergistic action of Bora that has been transcribed during G2 phase and kinase Aurora-A (AurA) that interacts with Bora to phosphorylate Polo-like kinase-1, thereby leading to cyclin-dependent kinase-1 activation and mitotic entry [202]. Kinase PLK1 accumulates early in G2 phase, but PLK1 phosphorylation is only detectable late in G2 phase, as activator Bora allows access of PLK1 to AurA. During mitosis, Bora bound to PLK1 is phosphorylated by active PLK1 and degraded by an ubiquitin ligase in a PLK1-dependent manner.

In the centrosome, Polo-like kinase PLK1 interacts and colocalizes with minichromosome maintenance subunits and ORC2 component of the Origin recognition complex that is implicated in DNA replication initiation during S phase and localizes to kinetochores during mitosis.[74] Hence, PLK1 influences

[74] The DNA-binding Origin recognition complex (ORC) tethers in an ATP-dependent manner to origins of replication, or replication origins, i.e., particular genomic sequences at which replication is initiated. During the G1 phase, the Origin recognition complex recruits CDC6 and chromatin licensing and DNA replication factor CDT1 as well as the hexameric minichromosome maintenance MCM2-7 complex to form the pre-replicative complex and then license replication origins for replication. The MCM2-7 complex is recruited to chromatin by the origin recognition complex, CDC6, and CDT1. The MCM2-7 complex is activated at the G1–S transition by cell division control protein-45 homolog (CDC45L), an essential protein required for the initiation of DNA replication,

the activity of components of DNA replication [203]. Kinase PLK1 phosphorylates histone acetyltransferase binding to ORC1 of the pre-replication complex (HBO1)[75] as well as topoisomerase-2 [203].

Phosphorylation (Thr37) of cell division cycle protein CDC6, a DNA replication initiation factor, by PLK1 contributes to the regulation of chromosomal segregation [203].[76] During mitosis, the concentration of $CDC6^P$ soars in correlation with the PLK1 level. Kinase PLK1 is produced in the S, G2, and M phases with a peak activity in mitosis. Proteins CDC6 and PLK1 colocalize to the spindle pole and the central spindle in anaphase. Kinase PLK1 promotes the binding of $CDC6^P$ to cyclin-dependent kinase CDK1 that impedes CDK1 activity, activates separase, and subsequently causes chromosomal segregation, anaphase progression, and mitotic exit.

Kinase PLK4 is required for late mitotic progression. It localizes to the centrioles throughout the cell cycle and spindle midbody. It binds to and phosphorylates the mitotic RhoA guanine nucleotide-exchange factor RhoGEF31[77] that activates Rho GTPase during cytokinesis [204].

2.4.2.12 AGC Protein Kinases

Protein Kinase-B

Protein kinase-B (PKB) cooperates with IκB kinase to phosphorylate nuclear factor-κB during G1 and G2 phases, during which kinase ChK1 is phosphorylated (inhibited) by PKB [205]. NFκB and IκB kinases control cyclin-D1 expression. On the other hand, during S phase, PKB is inactivated and ChK1 phosphorylates: (1) NFκB subunit RelA to recruit HDAC1 to cyclin D1, Myc, and SKP2 promoters, and (2) IKKα to prevent NFκB2 processing.

NFκB also controls transcription of S-phase kinase-associated protein SKP2 that forms the Ub ligase SCF complex with components SKP1, Cullin-1,

---

and the protein kinases cell division cycle-7 homolog CDC7 and CDK2. CDC45L is a member of the complex that includes CDC6 (a.k.a. CDC18L), the minichromosome maintenance proteins, and DNA polymerase. A single CDC6 and a single CDT1 bind to each strand of DNA. Proteins CDC6 and CDT1 interact with each other and inhibit DNA helicase minichromosome maintenance. Minichromosome maintenance binds CDC6 and CDT1, once they are linked to the Origin recognition complex. Activated S-phase CDKs phosphorylate CDC6 for degradation. Agent CDT1 is inhibited by geminin. Once CDC6 and CDT1 are released, MCM can unwind the double-stranded DNA allowing the pre-initiation complex to bind. The Origin recognition complex is phosphorylated and DNA replication starts.

[75] A.k.a. histone acetyltransferase MYST2 and KAT7.

[76] CDC6 is also phosphorylated after the initiation of DNA replication. The phosphorylation of CDC6 by CcnA–CDK2 provokes the translocation of CDC6 from the nucleus to the cytoplasm to prevent DNA re-replication.

[77] A.k.a. epithelial cell-transforming sequence 2 oncogene product ECT2.

Really interesting new gene (RING) box-1 (RBx1),[78] and a member of the F-box domain-containing protein class that confers the specificity of the SCF complex.[79] Protein SKP2 promotes CDK1 degradation and CcnE–CDK2 activation and cell cycle progression. In G2 phase, NFκB hinders the expression of cyclin-D1, MYC, and Skp2 genes.

Inhibitor CKI1b belongs to the set of substrates of complex SCF–SKP2. Ubiquitin ligase SCF indeed targets CDK–Ccn–CKI1b. The PI3K–PKB pathway enhances destruction of cell cycle inhibitors CKI1b and CKI1a. Protein kinase-B controls CKI1b transcription, translation, localization, complex formation, and stability. It elicits SKP2 transcription [205]. On the other hand, PI3K antagonist lipid phosphatase PTen impedes SKP2 expression [206] and SCF–SKP2 omplex formation, as it prevents the Cullin-1 association with SKP1 or SKP2 [207].

Component SKP2 can be directly phosphorylated by PKB1 (Ser72 of CDH1-binding motif) [251, 252].[80] Protein kinase-B1, but not related kinases (e.g., PKB2, SGK, and S6K3), actually binds to SKP2. Phosphorylation of SKP2 (Ser72) primes its translocation to the cytoplasm and facilitates SKP2 linkage to 14-3-3 protein for sequestration.

Protein Kinase-C

Protein kinase-Cε needs to bear a set of phosphorylations to bind to 14-3-3 proteins and to localize to the mitotic cleavage furrow, which leads to daughter cell separation via its ingression driven by the constriction of an actomyosin ring.[81] Protein kinase-Cε activation and subsequent 14-3-3 tethering regulate the dissolution of the mitotic ring [210].

2.4.2.13 Kinases of the Cell Cycle Checkpoints

Checkpoints correspond to cell cycle stages during which, upon detection of DNA damage or defective DNA replication, cells respond by arresting the cell cycle to launch DNA repair or apoptosis. Cell cycle checkpoints are regulated by phosphoinositide 3-kinase-related enzyme ataxia telangiectasia mutated kinase (ATMK) that is activated by DNA damage. Kinase ATMK provokes P53-mediated neuronal apoptosis. In postmitotic neurons,

[78] A.k.a. regulator of cullins-1.

[79] The Ub ligase complex SCF–SKP2 aims at destructing numerous cell cycle regulators, such as CKI1b, FoxO1, and RBL2. Protein kinase-B promotes phosphorylation and subsequent cytoplasmic localization of pro-apoptotic FoxO1 and FoxO3a. Phosphorylation by PKB of CKI1a, CKI1b, and FoxO causes recruitment of protein 14-3-3.

[80] Ubiquitin ligase CDH1 promotes SKP2 destruction.

[81] PKCε, a member of the novel set of the PKC family, has 3 phosphorylation sites to control its binding with protein 14-3-3. These phosphorylations are executed by P38MAPK (Ser 350), GSK3 (Ser 346), and PKC itself (Ser 368).

DNA damage promotes phosphorylation by cyclin-dependent kinase CDK5 of ATMK that causes subsequent ATMK autophosphorylation to activate effectors and apoptosis [211]. Kinase CDK5 is controlled by activator p35. The ataxia telangectasia-mutated and Rad3-related kinase–checkpoint kinase-1 (ATRK–ChK1) pathway controls S-phase programs as well as CcnA2–CDK1 from middle to late S phase [212].

Activation of CDK1 by phosphatases CDC25a, -b, and -c causes entry into mitosis. Checkpoint kinase ChK1 is a Ser/Thr kinase that causes arrest of G2–M transition in response to DNA damage by phosphorylating (inhibiting) mitotic inducer CDC25c [213]. Phosphorylation (Ser216) indeed creates a binding site for protein 14-3-3 , hence leading to cytoplamic sequestration. In addition, when CDC25a is phosphorylated by ChK1 and ChK2, it is then rapidly degraded via ubiquitination. Kinase ChK1 also regulates Wee1. Moreover, ChK1 phosphorylates P53 (Ser15, -20, and -37). Following DNA damage, ChK1 is phosphorylated (Ser345) and then interacts with 14-3-3 protein.

2.4.2.14 TTK Kinase

Dual-specificity kinase TTK[82] has been initially detected in T cells. It is present at relatively high levels in testis and thymus that are characterized by a large number of proliferating cells [214]. Kinase TTK that can phosphorylate Ser, Thr, and Tyr of hydroxy-amino acids, is associated with cell division in several cell lineages. This mitotic checkpoint kinase localizes to the kinetochore. It interacts with kinetochore-associated nanomotor centromere protein-E (CenPe; i.e., KIF10) that accumulates in the G2 phase of the cell cycle. Two isoforms exist, one being a transmembrane molecule [215].

Kinase TTK also connects to components CDC16 and CDC27 of anaphase-promoting complex [326]. It phosphorylates ChK2 (Thr68). Moreover, TTK phosphorylates P53 (Thr18), hence disrupting the interaction with DM2 Ub ligase to abrogate DM2-mediated P53 ubiquitination [217].[83]

Abl Tyr kinase is a ubiquitously expressed proto-oncogene product. Nuclear accumulation of Abl kinase primes a pro-apoptotic response after DNA damage. DNA damage (as well as oxidative stress) induces phosphorylation of 14-3-3 proteins by Jun N-terminal kinase that disrupts the Abl–14-3-3 cytoplasmic complex, hence liberating Abl for translocation to the nucleus. On the other hand, phosphorylated Abl (Thr735) mainly remains in the cytoplasm. Protein Ser/Thr kinases STK3 and STK4[84] as well as dual-specificity

[82] Kinase TTK is the human homolog of yeast MPS1 that is also named phospho-Tyr (Y) picked Thr kinase (PYT), ESK, and monopolar spindle-1-like kinase MSP1L1.

[83] Protein P53 can also be phosphorylated by ATMK and ATRK (Ser15) as well as casein kinase-1 (Thr18) and ChK1 and ChK2 (Ser20). Ubiquitin ligase DM2 binds to P53 and promotes its degradation.

[84] Kinases STK3 STK4 and STK3 are also called mammalian Ste20-like kinase MSt2 and MSt1, respectively.

Table 2.23. The CcnB–CDK1 complex and its regulatory kinases and phosphatases ((+): stimulatory effect; (−): inhibitory action).

Enzyme	Substrate
CcnB–CDK1	MASTL (+), mitotic targets
CDC25	CcnB–CDK1^P (+)
PP2	CDC25^P (+), PKMYT1^P, Wee1^P (+)
PKMYT1, Wee1	CcnB–CDK1 (−)
MASTL	PP2^P (−)

protein kinases CDC-like kinase CLK1 and CLK2 (Vol. 4 – Chap. 6. Dual-Specificity Protein Kinases) are able to phosphorylate Abl (Thr735) [218]. In addition, upon exposure to oxidative stress, TTK phosphorylation (Thr735) also induces cytoplasmic sequestration of Abl kinase.

2.4.2.15 Microtubule-Associated Ser/Thr Kinase-like Protein

During the cell cycle, microtubule-associated Ser/Thr kinase-like protein (MASTL), a mitotic regulator, promotes mitotic entry. It actually inhibits protein phosphatase-2 that dephosphorylates the CcnB–CDK1 complex [219]. Inhibition of PP2 is mandatory for CcnB–CDK1 activation at mitotic entry. The CcnB–CDK1 activity is controlled during G2 phase by inhibitory phosphorylations (Thr14 and Tyr15) by kinases Wee1 and PKMYT1 membrane-associated Tyr–Thr protein kinase (Table 2.23). Upon mitotic entry, activating phosphatase CDC25 relieves this inhibition. Phosphatase PP2 also regulates CcnB–CDK1 activity, as kinases Wee1 and PKMYT1 and CDC25 phosphatase are PP2 substrates. At mitotic exit, CcnB–CDK1 is again inhibited by the ubiquitin-dependent degradation of cyclin-B. This inactivation results from PP2 reactivation. G2 Arrest results from dephosphorylation of CcnB–CDK1 as well as inhibition of CcnB–CDK1 by Wee1 and PKMYT1 in the absence of CDC25-induced dephosphorylation. The ability of cells to remain arrested in mitosis by the spindle assembly checkpoint is proportional to the MASTL amount. When MASTL is slightly reduced, cells arrest at prometaphase. More complete depletion in MASTL correlates with premature dephosphorylation of CcnB–CDK1, inactivation of the spindle assembly checkpoint, and subsequent exit from mitosis with severe cytokinesis defects. The balance between CcnB–CDK1 and PP2 is tightly regulated for correct mitotic entry and exit by MASTL kinase [219].

Table 2.24. Effects of WNK kinases (Source: [220]; HSP: heat shock protein; ERK: extracellular signal-regulated kinase; OSR: oxidative stress-responsive kinase; PKB: protein kinase-B; SGK: serum- and glucocorticoid-regulated kinase; SPAK: Ste20-related proline–alanine-rich protein kinase; STK: protein Ser/Thr kinase).

Effect	Mediators
Cell volume	OSR1, STK39 (SPAK), Cl^- cotransporters
Cell proliferation	PKB, ERK
Cell survival	HSP70, procaspase-3
Cell migration	ERK1/2/5
Cell differentiation	ERK1/2/5
Neurotransmission	Synaptotagmin-2
Paracellular permeability	Claudins
Transepithelial ion transport	Intersectin, clathrin, SGK1, OSR1, STK39 (SPAK)

2.4.2.16 WNK Protein Kinases

With no lysine (K) protein Ser/Thr kinases (WNK1–WNK4) impinge the functioning of ion carriers as well as the paracellular transfer of molecules in the nephron to achieve and maintain the fluid and electrolyte balance as well as in extrarenal tissues, in particular epithelia that are involved in chloride ion flux (Table 2.24; Vol. 4 – Chap. 4. Cytosolic Protein Ser/Thr Kinases). Like many other kinases, WNKs contain an auto-inhibitory domain that inhibits kinase activity and is located just beyond the kinase domain [220].

Subtype WNK1 is a substrate of protein kinase-B that limits insulin-stimulated proliferation of cells, in particular preadipocytes [221]. It operates in cell proliferation, migration, and differentiation of neural progenitor cells via the activation of the ERK1 and ERK2 and/or ERK5 axes [222] (Vol. 4 – Chap. 5. Mitogen-Activated Protein Kinase Modules). It phosphorylates WNK2 and WNK4 [223].

Among the members of the WNK family, WNK1 activates oxidative stress-responsive OSR1 protein kinase that regulates ion cotransporters. In the resting cell, the majority of OSR1 lodges in cytoplasmic puncta and the nucleus. Whereas WNK1 also localizes to cytoplasmic puncta in resting cells, during cell division, it is recruited to the mitotic spindle [224]. It participates in the chromosome segregation and abscission, independently of its effector OSR1 kinase.

On the other hand, WNK2 is a cell proliferation suppressor that inhibits mitogen-activated protein kinase kinase MAP2K1,[85] thereby controlling epidermal growth factor receptor signaling. Its depletion potentiates the cellular response to low epidermal growth factor concentrations. It impedes cell

[85] Kinase WNK2 fosters RhoA activation, but precludes that of Rac1 [225]. ($Rac1^{GTP}$ stimulates PAK1 that phosphorylates MAP2K1.)

proliferation (especially G1–S progression). Isoform WNK2 inhibits the extracellular signal-regulated kinases ERK1 and ERK2, but, unlike WNK1, has no effect on ERK5. Protein WNK3 can delay the onset of apoptosis via interaction of procaspase-3 with heat shock protein HSP70 [226].

2.4.2.17 PIM Protein Kinases

Protein Ser/Thr kinase provirus insertion of Molony murine leukemia virus gene product (proto-oncogene PIM) fosters cell growth and survival (Sect. 2.3.3) and impedes apoptosis. It promotes the cell cycle progression. All of the members of the PIM family (PIM1–PIM3) phosphorylate the cyclin-dependent kinase inhibitor CKI1b (Thr157 and Thr198), thereby leading to their cytosolic sequestration by 14-3-3 protein and then proteasome-dependent degradation [227]. In addition, PIM kinases phosphorylate (inactivate) transcription factors FoxO1a and FoxO3a, thus suppressing CKI1b transcription.

2.4.2.18 Cell Division Cycle Phosphatases

CDC14

Four CDC14 phosphatases (Vol. 4 – Chap. 7. Cytosolic Protein Phosphatases) exist: CDC14a, CDC14b, kinase-associated phosphatase (KAP), and an uncharacterized member. They are all related to type-3 CDC25 phosphatases that regulate the initiation of mitosis and DNA-damage checkpoint. Phosphatase KAP dephosphorylates CDK2, thereby impeding cell cycle progression. However, in tumors, an aberrant KAP splicing variant increases gene expression.

Phosphatases CDC14 contribute to centrosome maturation, spindle stability, and cytokinesis [228]. They are primarily regulated by sequestration in the nucleolus during interphase. Dual-specificity protein phosphatase CDC14a dephosphorylates CDC25a (Ser115 and Ser320) [229].

CDC25

Dual-specificity protein phosphatase CDC25 (Vol. 4 – Chap. 7. Cytosolic Protein Phosphatases) controls entry into and progression through various phases of the cell cycle, especially S and M phases. Phosphatases CDC25s activate cyclin-dependent kinases.

Three CDC25 phosphatase isotypes (CDC25a–CDC25c) have distinct subcellular compartmentation and abundance during the cell cycle, as well as, partly, functions. Phosphatase CDC25a is nuclear, whereas CDC25b and CDC25c shuttle between the cytosol and the nucleus throughout interphase depending on their interactions with 14-3-3 proteins [230].

Table 2.25. Interactions between antagonists CDK1 and PP2 during interphase and mitosis. Mitotic switch relies on CDK1 activation and inhibition of counteracting phosphatase PP2 ($\oplus \longrightarrow$: stimulation; $\ominus \longrightarrow$: inhibition).

Interphase	PP2 $\ominus \longrightarrow$ CDC25 $\oplus \longrightarrow$ CDK1, PP2 $\oplus \longrightarrow$ Weel, PKMYT1 $\ominus \longrightarrow$ CDK1
Mitosis	CDK1 $\oplus \longrightarrow$ MASTL $\oplus \longrightarrow$ Ensα, ARPP19 $\ominus \longrightarrow$ PP2

Phosphatase CDC25a is involved in cell cycle stage transitions and checkpoints. It can be active in all stages of the cell cycle. However, CDC25a concentration rises as the cell progresses from S phase to mitosis. Nonetheless, CDC25a can be sequestered by 14-3-3 proteins during interphase. Phosphatase CDC25a controls the G1–S transition. The Cdc25A gene is targeted by E2F factor. All CDC25 paralogs regulate the G2–M transition [230].

Phosphatase CDC25b dephosphorylates (activates) the CcnA–CDK2 and CcnB–CDK1 complexes. Phosphatase CDC25c is activated in mitosis, as it is phosphorylated by CcnB–CDK1 and Polo-like kinase PLK1 [230]. Phosphatases CDc25a and CDCc25c can be inhibited by checkpoint regulators. Checkpoint activation maintains CDC25c in a 14-3-3-bound form. Checkpoint kinase ChK1 phosphorylates CDC25a for proteolysis, especially to delay S–G2 transition after checkpoint activation.

2.4.2.19 Other Phosphatases

Protein Phosphatase-1

Protein Ser/Thr phosphatase PP1 contributes to the separation of daughter nuclei, anaphase completion, and chromosome segregation [173]. It dephosphorylates (inhibits) CDK1 (Thr161). In addition, PP1 dephosphorylates CDK1-substrate lamin, thereby allowing repolymerization to rebuild the nuclear envelope at the onset of cytokinesis. Phosphatase PP1 also targets RB tumor suppressor. Phosphatase PP1α is phosphorylated (Thr320) by CDK2 but not CDK4 during M phase as well as, in the presence of retinoblastoma protein, in late G1 and early S phase.

Protein Phosphatase-2

Entry into mitosis relies on in the balance between CDK1 and protein phosphatases. Protein phosphatases operate in 2 ways: (1) they inhibit activation of CDK1 by dephosphorylating protein kinases Weel and PKMYT1 and protein phosphatase CDC25; and (2) they antagonize phosphorylation by CDK1 of its substrates. The activity of PP2 is high in interphase and low in mitosis, in opposition to that of CDK1 (Table 2.25).

The heterotrimer PP2 is composed of catalytic ($PP2a_c$ subunit), scaffold (A subunit), and regulatory (B subunit) subunits (Vol. 4 – Chap. 7. Cytosolic Protein Phosphatases). Isoform PP2a(B55δ) is involved in the cell cycle.

Protein phosphatase-2 is able to inhibit entry into mitosis, as it inactivates CDC25 and activates Wee1, thereby preventing CDK1 activation [173]. Furthermore, PP2 can act downstream from CDK1, as it can dephosphorylate CcnB–CDK1 substrates.

Microtubule-associated Ser/Thr kinase-like protein (MASTL), the human ortholog of Greatwall, inhibits PP2a(B55δ) and hence facilitates mitotic entry. Kinase MASTL phosphorylates (stimulates) PP2 inhibitors α-endosulfine (Ensα) and 19-kDa cAMP-regulated phosphoprotein (ARPP19) [231, 232]. Both $Ens\alpha^P$ and $ARPP19^P$ tether to and inhibit PP2 enzyme. Kinase MASTL is phosphorylated (activated) by CDK1 (positive feedback loop).

2.4.2.20 Retinoblastoma Proteins

Retinoblastoma (RB) proteins preclude cell cycle progression in G1 phase and induce apoptosis in response to genotoxic stress. Retinoblastoma protein RB1, the first identified tumor suppressor gene product, limits the activity of transcription factor E2F and represses cell cycle genes, as it complexes with E2F1 [233].

Retinoblastoma family members also impede cell contact and formation of cellular structures [234]. In addition, expression of sterol regulatory element-binding proteins that regulate farnesyl diphosphate synthase and many prenyltransferases is hindered in an E2F-dependent manner to restrict nRas isoprenylation (activation) by retinoblastoma protein [235]. Elevated nRas activity induces DNA damage and retinoblastoma-like protein RBL2-dependent cellular senescence.

Retinoblastoma protein that blocks G1-phase progression can be phosphorylated at multiple Ser and Thr residues for S-phase entry by, at least, CcnD–CDK4 (Ser795), CcnE–CDK2, as well as CcnA–CDK2 [173].[86] Kinases that inactivate retinoblastoma protein are CDK4 and CDK6 coupled to cyclin-D1 and -D2 as well as, most likely, CcnE–CDK2 dimer in late G1 and CcnA–CDK2 during S phase. In addition, RB may contribute to the regulation of the passage through the restriction point.

Retinoblastoma protein is dephosphorylated by PP1 [173]. Therefore, PP1 is partly or entirely responsible for RB activation. Complete RB dephosphorylation requires another phosphatase that can act before or after PP1a. Cytoplasmic PP1 activity is the strongest during G0 and G1 phases. It is highly reduced in other phases of the cell cycle. On the other hand, nuclear, chromatin-associated PP1 activity is the highest in G0 and G1 phases as well

[86] In vitro, $Thr821^P$ and $Thr826^P$ repress binding of RB to LXCXE-containing proteins; $Ser807^P$ and $Ser811^P$ impede the connection with Abl kinase; and $Ser780^P$ suppresses the association with E2F transcription factor.

Table 2.26. Early mitotic inhibitor-1 in G2 and M phases.

Phase	Event
G2 (EMI1)	Inhibition of the Ub ligase complex APC/C, phosphorylation by CcnB–CDK1 and PLK1
M ($EMI1^{P}$)	Degradation by SCF1–β2-TRCP (prometaphase), activation of APC/C, degradation of CcnB–CDK1 and PLK1)

as in mitosis. Protein phosphatase-1 is phosphorylated (Thr320; inactivated) at the onset of mitosis.

2.4.2.21 Early Mitotic Inhibitor

Early mitotic inhibitor-1 (EMI1) inhibits the anaphase-promoting complex, or cyclosome (Table 2.26). The concentration of EMI1 oscillates during the cell cycle. It accumulates in the S phase and is rapidly degraded in prometaphase. Degradation of EMI1 in early mitosis is needed to activate APC/C in late mitosis. Inhibitor EMI1 is tagged for degradation by the Ub ligase complex SCF after phosphorylation by Polo-like kinase-1 [236].

2.4.2.22 Small Guanosine Triphosphatases

Cytokinesis ensures proper partitioning of genomic and cytoplasmic material between daughter cells. During cytokinesis, under control of the mitotic spindle, small guanosine triphosphatase RhoA orchestrates contractile ring assembly and constriction. During cytokinesis, active GTPase Rho is located to the narrow Rho activity zone. Activation of GTPase Rho precedes ^{F}actin accumulation.

Centralspindlin is a proteic complex made of kinesin-6 and GTPase-activating protein that inactivates small GTPase Rac, whereas small GTPase RhoA is transiently activated [237]. As Rac effector — the ARP2–ARP3 complex — promotes branching of actomyosin filaments, inactivation of Rac is necessary to impede formation of a branched actomyosin meshwork at the cleavage furrow. Centralspindlin moves to the plus-ends of equatorial and central spindle microtubules due to activity of its motor component. GTPase activating protein RacGAP1 tethers Rho to the contractile ring. Protein RacGAP1 is required during the entire duration of cytokinesis to regulate the spatiotemporal activation of Rho for the formation and maintenance of the Rho activity zone [238]. It transiently anchors Rho^{GTP}, thereby limiting the spreading of the Rho activity zone, and then promotes Rho inactivation. Flux of Rho then results from the GTPase cycle during cytokinesis that requires rapid actomyosin dynamics. Guanine nucleotide-exchange factor RhoGEF31[87] activates Rho that binds RacGAP1 and then its effectors, such as formins and RoCK kinase. Active Rho is inactivated by RacGAP1. Aurora-B phosphorylates RacGAP1 to stimulate its GAP activity.

2.4.2.23 Minichromosome Maintenance Proteins

Minichromosome maintenance protein MCM1 that is a combinatorial transcription factor regulates cell cycle genes expressed both in G2–M and M–G1 transitions with various regulators, thus controlling both entry into and exit from mitosis. Members (MCM2–MCM7) of the MCM family act as a hexamer at ARS sequence to initiate DNA replication.

The *prereplicative complex* that ensures a single genome replication per cell division forms during late M and early G1 phases to be activated at the G1–S border in particular by CDC7 and CDK2. Prereplicative complex assembly requires the association of minichromosome maintenance MCM helicase, a member of the AAA+ (ATPase-associated with various cellular activities) superclass of ATPases (hexamer MCM2–MCM7) with chromatin that depends on the Origin recognition complex (ORC), CDC6 homolog protein, an MCM loader, regulator of DNA replication, and member of the superclass of AAA+ ATPases, and chromatin licensing and DNA replication factor CDT1 (*DNA replication licensing system*) [239].[88] At the G1–S transition, CDC7 homolog kinase and CcnE/A–CDK2 recruit additional factors to the prereplicative complex to form the *pre-initiation complex*. During late S and M phase, CDKs induce the degradation of the prereplicative complex to impede DNA re-replication.

2.4.2.24 Forkhead Box Transcription Factors

Forkhead box factors regulate the cell cycle. Forkhead box factor FoxM1[89] is an activator of DNA replication and mitosis in various cell types. It controls progression into S phase, as it upregulates CDK2-activating phosphatase CDC25a and downregulates CDK inhibitors CKI1a and CKI1b. Cell cycle regulator cyclin-dependent kinase inhibitor CKI1b binds to and inhibits Ccn–CDK1/2.[90] Degradation of FoxM1 is stimulated by the anaphase-promoting

[87] A.k.a. ECT2.

[88] Cyclin-B and CDK1 phosphorylates MCM3 for its association with other MCM subunits and loading onto chromatin.

[89] A.k.a. FoxM1b, Trident, Forkhead-like protein FkHL16, hepatocyte nuclear factor-3 forkhead homolog HFH11, Winged-helix factor from insulinoma cell line Ins1 cells (WIn), and M-phase phosphoprotein MPP2.

[90] Inhibitor CKI1b content is high in G0 and early G1. It is required for G1 arrest induced by growth factor deprivation. Inhibitor CKI1b is inactivated due to nuclear export and sequestration by the CcnD–CDK4(6) complex that prevents CKI1b

complex (cyclosome) complex and adaptor cadherin-1. During embryo- and fetogenesis, FoxM1 is required for the development of the heart and vasculature as well as lung and liver, in particular. Ubiquitous FoxM1 in proliferating cells promotes G1–S as well as G2–M transition during the cell cycle, as it participates in the maintenance of genome stability and DNA-damage repair [240].[91] Expression of FoxM1 transcription factor is inhibited in quiescent and terminally differentiated cells.

In addition, FoxO transcription factor is required for the synthesis of cyclin-dependent kinase inhibitor CKI1 in response to starvation. FoxO promotes cell cycle arrest. Nutrient-sensitive insulin and IGF signaling hence control the cell division via FoxO [241].

2.4.2.25 E2F Transcription Factors

Factors E2Fs control the transcription of genes involved in DNA replication, cell cycle progression, and cell fate determination. The E2F family is composed of 6 distinct gene products that form heterodimers with partners of the DP family.

Three distinct subfamilies of E2Fs have been defined: subfamily-1 E2F1 to E2F3; subfamily-2 E2F4 and E2F5; and subfamily-3 E2F6. Members of the E2F subfamily-1 are tightly regulated during cell growth and the cell cycle, whereas those of other subfamilies are constitutively expressed. Members of the E2F subfamily-1 experience ubiquitination and degradation. They link to CcnA–CDK2 dimer to inhibit DNA binding. In addition, E2F1 to E2F3 act as activators of transcription. On the other hand, E2F4 and E2F5 are transcriptional repressors in cooperation with members of the retinoblastoma protein family. Retinoblastoma protein connects to E2F transcription factors to block cell division. Factor E2F6 also operates as a transcriptional repressor, but independently of retinoblastoma protein.

Factors E2Fs target genes that encode cell cycle regulators, such as transcription factor MyB-related protein MybL2, cyclin-A, cyclin-B, CDK1, retinoblastoma protein, and CDC6, a regulator of the DNA replication and loader of the minichromosome maintenance helicase.[92] Protein E2F1

interaction with Ccn–CDK1/2 and phosphorylates CKI1b for degradation by the ubiquitin–proteasome system. Kinases Src and Abl phosphorylate CKI1b that cannot subsequently inhibit CDK2. CDK2 then phosphorylates CKI1b for degradation.

[91] Phosphorylation of FoxM1 by the Raf–MAP2K–ERK module causes FoxM1 translocation from the cytoplasm to nucleus during late S phase, where it is further phosphorylated. Hyperphosphorylation during G2–M transition elevates its transcriptional activity.

[92] Cell division cycle protein CDC6 is a regulator of DNA replication that functions in the activation and maintenance of the checkpoints in the cell cycle. It is a constituent of the pre-replicative complex that is required for loading minichromosome maintenance proteins onto the DNA to initiate DNA synthesis. It is also a member of the superclass of AAA ATPases.

contributes to triggering of apoptosis in response to DNA damage in collaboration with Myc transcription factor. Protein E2F3 is particularly involved in cell proliferation.

E-box-binding factor TFE3 is an E2F3-specific partner. It cooperates with E2F3 to activate the P68-subunit gene of DNA polymerase-α [242].[93]

2.4.2.26 Merlin

Merlin,[94] a cytoskeletal protein encoded by the NF2 gene, is a member of the ezrin–radixin–moesin (ERM) family of proteins that belongs to the protein 4.1 superfamily of proteins. In humans, mutations of the NF2 gene cause multiple brain tumors, such as neurofibromatosis type-2, schwannomas, and meningiomas. Human merlin is predominantly found in cells of the nervous system, but resides also in multiple cell types. Merlin precludes mitogenic signaling at or near the plasma membrane. Ten known isoforms of human merlin exist.

Merlin is a scaffold protein that serves as both a membrane–cytoskeleton linker and a tumor suppressor. It mainly localizes to adherens junctions. Classical ERM proteins switch between a closed and open conformation that yield a controlled linkage between plasmalemmal proteins and the cortical cytoskeleton [243]. As it connects to plasmalemmal glycoproteins such as cell adhesion receptor integrins to the actin cytoskeleton, it impedes mitogenic signaling downstream from integrins [244]. It prevents the activation of extracellular signal-regulated kinase and small GTPase Rac1 [245].

Merlin undergoes a polyubiquitination by the Ub ligase complex (cullin-4A–RBx1–DDB1 complex).[95] Merlin polyubiquitination triggers its proteasomal degradation. Conversely, DCAF1 depletion stabilizes merlin, thereby repressing ERK and Rac activation.

Merlin also resides in the nucleus, where it can accumulate and inhibit the dimer formed by the ubiquitin ligase complex CRL4 and recruiter DCAF1 used for substrate recognition [246].[96] The latter can provoke an oncogenic

[93] Neither E2F1 nor E2F2 interact with the P68 promoter. Transcription factor TFE3 belongs to the class-E (or microphthalmia [MiTF–TFE] set) of basic helix–loop–helix leucine zipper transcription factors with TFEb (or bHLHe35), TFEc (or bHLHe34), and microphthalmia-associated transcription factor (MiTF or bHLHe32). Microphthalmia-associated transcription factor is involved in melanocyte and osteoclast development as well as the regulation of tissue-specific gene expression in mastocytes. They bind to DNA both as homo- or heterodimers.

[94] A portmanteau for moesin–ezrin–radixin-like protein; a.k.a. neurofibromin-2 and schwannomin.

[95] Merlin is recruited to the ubiquitin ligase complex cullin-4A–RBx1–DDB1 via WD40-containing adaptor Viral protein-R-binding protein VPRBP, also called damage-specific DNA-binding protein (DDB1)- and Cul4-associated factor DCAF1.

[96] Protein DCAF1 recruits specific targets to CRL4 for ubiquitination, as it binds both substrates and the DDB1 adaptor, a component of the CRL4 ubiquitin

program. The Merlin–CRL4–DCAF1 complex precludes ubiquitin ligase activity and suppresses cell proliferation. Conversely, NF2 mutations hampers merlin from hindering the activity of the CRL4–DCAF1 complex.

2.4.2.27 Wnt Signaling

Wnt morphogens (Vol. 3 – Chap. 10. Morphogen Receptors) are secreted glycoproteins that bind to Frizzled G-protein-coupled receptors. They then cause oligomerization of the coreceptors low-density lipoprotein receptor-related proteins LRP5 and LRP6 to activate β-catenin-dependent transcription. The LRP6 cytoplasmic domain is phosphorylated for signal transduction. Activation of LRP6 by Wnt results from phosphorylation by GSK3 glycogen synthase kinase and CK1 casein kinase. Phosphorylation og the LRP5–LRP6 complex results not only from the initial Wnt stimulus, but also an intracellular feedback loop, in which GSK3 and CK1 kinases, once recruited to the LRP5–LRP6 complex, also phosphorylate (activate) these coreceptors [248].

Moreover, intracellular Wnt effector Disheveled binds to type-1 phosphatidylinositol 4-phosphate 5-kinase PIP5K1 that produces $PI(4,5)P_2$. The latter promotes receptor aggregation, binding of the Axin–GSK3–CK1 complex, and, ultimately, LRP5–LRP6 phosphorylation. In summary, Wnt causes Frizzled LRP5–LRP6 interactions, Disheveled–PIP5K1–PIP_2-induced receptor aggregation, and Axin–GSK3–CK1 complex linkage to LRP5–LRP6 clusters.

Signaling from Wnt mediated by LRP6 is regulated during the cell cycle. It intervenes in the G1–S transition of the cell cycle by transcriptional activation of cyclin-D1 and Myc transcription factor. The G2–M transition that is associated with a peak in Wnt–β-catenin signaling primes Wnt signaling via the CcnY–CDK14[97] that also phosphorylates LRP6 (Ser1490 and Thr1479) [249].[98]

2.4.2.28 14-3-3 Proteins

During and immediately after metaphase, cap-dependent translation is suppressed, whereas cap-independent translation is activated. Non-capped mRNAs of cell cycle regulators indeed contain internal ribosome entry sites for translation.

During mitosis, 14-3-3σ binds to several translation initiation factors, thereby regulating protein synthesis.[99] Eukaryotic translation initiation factor

ligase complex [247]. The CRL4 DCAF1 dimer is more efficient than the CRL4 complex.

[97] A.k.a. CDK-like PFTAIRE kinase PFTK1.

[98] Residue Ser1490 is linked to phosphorylation sites (Thr1479, Thr1493, Ser1496, and Thr1499) targeted by CK1 casein kinase.

[99] Proteins 14-3-3 are encoded by 7 distinct genes (α/β, γ, δ/ζ, ϵ, η, θ/τ, and σ). They usually form homo- or heterodimers. They interact with phosphorylation sites of target proteins.

eIF4b that is sequestered by 14-3-3σ cannot bind to mRNA cap. Protein 14-3-3σ regulates mitotic translations by binding to translation initiation factors such as eIF4b.[100] In particular, protein 14-3-3σ is required for translation of protein Ser/Thr kinase isoform Cell division cycle protein-2-like CDC2L2[101] involved in cytokinesis [250].

2.4.2.29 Ubiquitins

Cell cycle progression is controlled by ubiquitin-mediated proteolysis of components of the cell cycle command. Ubiquitin–protein ligases are the protagonists of regulated protein destruction, hence have a major role in the progression of the cell cycle clock, which is comparable to that of other regulatory enzymes such as the cyclin-dependent kinases.

Two ubiquitin ligase complexes control the degradation of cell cycle regulators (Tables 2.27, 2.28, 2.29, and 2.30): (1) SKP1–Cul1–F-box protein (SCF1) with its substrate-recognizing subunit S-phase kinase-associated protein SKP2[102] and (2) anaphase promoting complex or cyclosome (APC/C) that is mainly controlled by 2 adaptors and coactivators, CDC20 and CDC20 homolog CDH1.[103]

The APC/C–CDC20 complex that acts mainly during the metaphase–anaphase transition and the APC/C–CDH1 complex that is activated during mitosis exit and G1 phase allow APC/C to target different substrates at distinct phases of the cell cycle. This switch from one APC/C complex to another

[100] Protein 14-3-3σ binds to eIF4b, eIF2α, and elongation factor EF1α. Other 14-3-3σ-associated proteins such as eIF4g interact indirectly with 14-3-3σ.

[101] A.k.a. galactosyltransferase-associated protein kinase P58GTA, CDK11^{P58}, and PITSLREb.

[102] Ubiquitin ligase SCF is active throughout the cell cycle. Its subunit SKP2 is phosphorylated at the G1–S transition of the cell cycle. Subunit SKP2 is a direct substrate of PKB1, but not PKB2 [251, 252]. Phosphorylation of SKP2 by PKB primes subsequent phosphorylation by casein kinase-1. Upon PKB-mediated phosphorylation, SKP2 accumulates in the cytoplasm and forms the SCF complex to ubiquitinate cyclin-dependent-kinase inhibitor CKI1b as well as FoxO1 and retinoblastoma-like protein RBL2.

[103] Ubiquitin ligase APC/C regulates metaphase-to-anaphase transition via degradation of *securin* and *shugoshin* and mitotic exit via mitotic cyclin degradation [254]. Coactivators CDC20 and CDH1 accumulate during G2 phase. Ub ligase APC/C–CDH1 complex remains inactive until late mitosis, as it is inhibited by EMI1 that is eliminated in prophase and phosphorylated by cyclin-dependent kinases that hinder CDH1 binding to APC/C until anaphase. During anaphase, CDH1 dephosphorylation leads to formation of active APC/C–CDH1 complex. Activity of the APC/C–CDH1 complex persists throughout G1 until CDH1 is inactivated at the G1–S transition by EMI1 binding. The APC/C–CDC20 complex is inactive during G2 phase. The APC-CDH1 complex also mediates SKP2 destruction [252]. Coactivator CDH1 regulates the G0 and G1 phases. Kinase PKB1 phosphorylates SKP2 to disrupt the interaction between CDH1 and SKP2 and cause cytoplasmic translocation of SKP2.

Table 2.27. Types and composition of ubiquitin ligase complexes (Source: [176]; APC: anaphase-promoting complex; BTB: broad complex–tramtrack–bric-à-brac; SOCS: suppressor of cytokine signaling). The SCF1, -2, and -3 complexes refer to their cullin subunit type (i.e., Cul1-based SCF1, Cul2-based SCF2, and Cul3-based SCF3). Cullin subunits foster Ub conjugase recruitment to SCF complex.

Type	Scaffold	RING subunit	Adaptor	Substrate adaptor
SCF1	Cullin-1	Rbx1	SKIP1	F-box-containing subunit
SCF2	Cullin-2	Rbx1	Elongin-B/C	SOCS-box-containing subunit
SCF3	Cullin-3	Rbx1		BTB-containing subunit
CRL4	Cullin-4	Rbx1	DDB2	DDB1
APC/C	APC2	APC11		

Table 2.28. Function of ubiquitin ligase complexes (Source: [176]; APC/C: anaphase-promoting complex or cyclosome; CDC: cell division cycle protein; CDH: CDC20 homolog; CKI: cyclin-dependent kinase inhibitor; CKS1: CDC28 protein kinase regulatory subunit-1 [cyclin-dependent kinase regulatory subunit-1]; CRL4: cullin-4A–RING ubiquitin ligase; DCAF1: DDB1- and Cul4-associated factor; DM: double minute; EMI: early mitotic inhibitor; SCF: SKP1–Cul–F-box Ub-ligase; SKP: S-phase kinase-associated protein; βTRCP: β-transducin repeat-containing homolog protein). The APC/C complex is mainly active in G1 phase and mitosis. It acts as the key antagonist of mitotic cyclin-dependent kinases (CDK). It primes the degradation of mitotic kinases (e.g., Aurora, and Polo-like and never in mitosis gene-A [NIMA]-related kinases), spindle regulators (e.g., kinesin-related proteic nanomotors for mitotic spindle assembly and chromosome segregation), and regulators of the DNA replication (e.g., CDC6).

Type	Substrate(s)	Phase and event(s)
APC/C	In cooperation with UbE2d and UbE2c	
APC/C–CDH1	SKP2, CKS1	G1-phase stabilization
APC/C–CDC20		Metaphase–anaphase transition
SCF	In cooperation with CDC34 and UbE2d	
	G1 cyclins	
SCF–SKP2–CSK1	CKI1a/b	S-phase entry
	CDK1, EMI1	
SCF–βTRCP	$EMI1^P$	Prophase
	$Wee1^P$	
	Securin	Anaphase (sister chromatid separation)
	IκBα, β-catenin	
CRL4–DCAF1	Merlin	Cell cycle inhibition (CRL4–DCAF1 inhibition)
DM2–KAT2		ChK2 degradation

Table 2.29. Components and regulators of ubiquitin ligase complexes APC/C and SCF (Source: [176]; CDCA: cell division cycle-associated protein; E2: ubiquitin-conjugase; E3: ubiquitin-ligase; EMI: early mitotic inhibitor; GRR: growth and reproductive region; RBx: really interesting new gene [RING]-box; ROC: regulator of cullins; TOME: trigger of mitotic entry protein; βTRCP: β-transducin repeat-containing homolog protein; UbCH: ubiquitin-conjugase homolog; UbE2: ubiquitin-conjugating enzyme E2).

Protein	Alias	Function	Regulation
CDC34	UbCH3 UbC3, UbE2r1	E2 for SCF	Autoubiquitination
UbE2r2	CDC34b		
UbE2d1/2/3	UbCH5a–c UbC4/5	E2 for RING finger E3s APC/C, SCF	
UbE2c	UbcH10	E2 for APC/C	Autoregulation
Cul1–Cul5		SCF subunit	Neddylation
APC2		APC/C subunit	Autoubiquitination
APC11		APC/C subunit	Autoubiquitination
CDC4		F-box subunit	Degradation by SCF
CDCA3	TOME1	F-box subunit, Weel inhibitor	Degradation by APC/C
EMI1		F-box subunit, APC/C inhibitor	Degradation by SCF
GRR1		F-box subunit	Degradation by SCF
RBx1	ROC1	SCF subunit	Autoubiquitination
SKP1			
SKP2		F-box subunit	Degradation by SCF and APC/C
βTrCP		F-box subunit	Degradation by SCF
CDC20		APC/C activator, recruiter	Degradation by APC/C–CDH1
CDH1		APC/C activator, recruiter	Degradation by APC/C, autoregulation
PLK1	CDC5	Mitosis entry, APC/C inhibitor	Degradation by APC/C–CDH1
CDK1		Mitotic regulator, APC/C activator	Degradation by APC/C

is controlled during the cell cycle by: (1) CDK-mediated phosphorylation of activating adaptors CDC20 and CDH1;[104] (2) their degradation; and (3) expression of APC/C inhibitors such as EMI1 Early mitotic inhibitor.

[104] Phosphorylation of CDH1 by CDKs impedes its ability to activate APC/C during S and G2 phases as well as mitosis.

Table 2.30. Substrates of the ubiquitin ligase complexes APC/C and SCF (Source: [176]).

Type	Substrates
APC/C	Cyclin-A/B, PLK1, CDC25a, SKP2, CDCA3, CDC20, CDH1, UbE2c
SCF1	Cyclin-E, CKI, Wee1, EMI1

The Ub ligase APC–CDH1 coordinates the degradation of cell cycle regulators. During exit from mitosis and G1 phase, the APC–CDH1 enzymatic complex degrades nuclear Jun N-terminal kinase [253]. However, initial inactivation of JNK during mitosis may start before its ubiquitination and degradation by the APC/C–CDH1 complex, due to a JNK-specific phosphatase. Derepression of the APC/C–CDH1 complex may result from CDH1 dephosphorylation by CDC14 phosphatase. Conversely, during G2 phase and early mitosis, activated JNK phosphorylates CDH1, thereby changing its subcellular localization and impeding its ability to activate APC/C Ub ligase complex during G2 and M stages. The JNK pathway may regulate histone H3, kinase Aurora-B, and CDC25c phosphatase.

The spindle checkpoint indeed prevents anaphase onset until completion of mitotic spindle assembly by restraining activation of the APC/C–CDC20 complex (Table 2.31). Several kinases that regulate microtubule-associated regulatory kinases, can activate the spindle checkpoint to avoid aberrant chromosomal segregation and aneuploidy (i.e., abnormal number of chromosomes that occurs during cell division when chromosomes do not separate properly between the 2 daughter cells).

Kinase ChK2 undergoes an ubiquitin-mediated proteasome degradation. Both Ub ligase DM2 that targets P53 as well as lysine (K) acetyltransferase-2 (KAT2; or P300 and CBP-associated factor PCAF2), a histone acetylase and Ub ligase, target ChK2 for ubiquitination and degradation. Damaged DNA stimulates phosphorylation of ChK2 (Ser456) that hinders DM2 linkage to ChK2 (stress-induced ChK2 stabilization).

Ubiquitin ligase KAT binds and ubiquitinates DM2 as well as ChK2. Furthermore, DM2 and KAT synergistically ubiquitinate ChK2. Ubiquitin ligases DM2 and KAT can form a large Ub ligase complex such as SCF or APC/C and serve mainly as a recognition factor and scaffold rather than an actual Ub ligase. Heterodimerization partner of DM2, DMx, could join this complex. Another potential partner is Cul4 that is also a binding partner of DM2 for P53 ubiquitination.

Factors P300 and CBP can yield a component, as P300 interacts with both DM2 and KAT owing to separate binding domains. Both P300 and KAT interact with DM2 using the same domain. Histone acetyltransferases P300 and CBP also act as non-histone acetylases that regulate multiple transcription factors, especially P53. Ubiquitin ligase P300 specifically targets P53

Table 2.31. Degradation of cell cycle proteins mediated by the ubiquitin ligase complex APC/C and regulation of APC/C activators in G1–S stage and early anaphase. Activity of the APC/C complex is regulated by activators Cell division cycle CDC20 and CDC20-related protein CDH1 and inhibitor mitotic arrest-deficient protein MAD2, as well as by its phosphorylation by Polo-like kinase and protein kinase-A and phosphorylation of its regulators CDC20 and CDH1 by the CcnB–CDK1 complex. Molecule APC/C that has a low level during S, G2, and early M phases, becomes activated in late metaphase and remains active at a high level until late G1 phase. At metaphase–anaphase transition, APC/C ubiquitinates regulatory proteins for initiation of sister chromatid separation. Later in anaphase and telophase (i.e., late mitosis), APC/C inactivates CDK1 by ubiquitinating cyclin-B for mitosis exit.

Stage	Event
1	Activation of degradation of mitotic proteins mediated by APC/C and the APC/C–CDC20 complex
2	Conversion from APC/C–CDC20 to APC/C–CDH1 complex in late anaphase and telophase
3	Degradation of CDC20 and other targets mediated by APC/C–CDH1 complex in late mitosis and early G1 phase
4	Autodegradation of multiubiquinated CDH1 by the APC/C–CDH1 complex

for polyubiquitination only when P53 is already oligoubiquitinated by DM2. Factor CBP shares a similar ubiquitination activity for P53.

2.4.3 Cell Division Cycle Arrest

In response to DNA damage, the cell cycle promptly arrests at specific stages (Fig. 2.3). Cell cycle arrest remains sustained in the absence of DNA-damage repair. After arrest, cells can re-enter the appropriate cell cycle phase. *Mathematical models* of the cell cycle and DNA-damage signaling aim at testing contributions of mechanisms devoted to cell cycle arrest and re-entry [255]. Cyclin inactivation independent of P53 confers immediate arrest, whereas P53-dependent cyclin downregulation allows sustained arrest. Inhibition of cyclin-dependent kinase activity by CKI1a prevents improper cell cycle re-entry.

2.4.4 Modeling of the Cell Division Cycle

The complexity and flexibility of the cell cycle control render difficult its full understanding, although the regulatory network that drives the cell division cycle has been investigated in a huge number of studies. Mathematical modeling of the cell cycle can then be aimed at checking assumptions in addition at predicting.

Mathematical models correspond to conceptual representations that capture the essential features of the investigated process, such as the cell cycle, and then omit details (i.e., elements that have negligible effects as well as elements that influence the explored behavior, but are assumed as secondary properties) to describe its mechanisms. Mathematical models cast a process in the form of equations of a particular type to predict the system behavior and possibly suggest complementary experiments for a better understanding.

Cell signaling that directs cell fate relies on a set of between-protein interactions that can lead to either activation or inactivation, especially using post-translational modifications such as the phosphorylation–dephosphorylation cycle. Signaling mediators are hierarchically organized according to time of mediation, number of connections, half-life, redundancy level, and role. Most effectors can be represented as nodes. Highly connected proteins that serve as hubs in the regulation networks are more important for cell growth and proliferation.

The cell cycle is mainly conducted by a complex network of cyclin-dependent kinases that tether to cyclins to be active and interact with substrates, which then undergo phosphorylation–dephosphorylation cycles. Cyclin-dependent protein kinases ensure the progression through the cell cycle. The activity of these major controllers relies on cyclins, the presence and action of which result from regulated synthesis, subcellular localization, and degradation, as well as nutrient availability and checkpoint activation. Combinatorial associations of CDKs with cyclins target distinct substrates that characterize cell cycle transitions. Other major players of the cell cycle control include kinases, such as Wee1 and PLK1, as well as phosphatases, such as PP2 and CDC25.

The control of the cell cycle is mainly based on protein phosphorylation and dephosphorylation. The mass action formalism can be employed using several types of possible interactions, based on the 2-variable Lotka-Voltera system for pedagogical purpose [256]:

$$\frac{d}{dt} v_i = f_i(v_i) - g_i(v_i, v_j), \; i, j = 1, 2, \; i \neq j; \qquad (2.2)$$

(1) if v_1 creates v_2, a term $+\alpha v_1$ is added to the v_2 equation; (2) if v_1 transforms v_2 from an inactive state into an active state with conservation of the total amount of v_2, a $+\alpha v_1(\beta - v_2)$ term is added to the v_2 equation, and vice versa for v_1 equation; (3) if v_1 and v_2 produce a new molecule v_2, or when v_1 destroys v_2, a $\pm\alpha v_1 v_2$ term is added to the v_2 equation; (4) if the creation of a new molecule is indirect, but the transformation of v_2 from an inactive into an active state or the decomposition of a complex occurs, a term $\pm\alpha v_1 v_2(1 - v_2)$ is added. More complex second-order interactions can also be used. For example, when a positive feedback loop in which v_1 creates itself and a new v_1 can be represented by a $+\kappa v_1^2$ term. The 2-variable models are schematic descriptions of feedback loops, but do not represent biological interactions.

Three-variable systems are expansions of 2-variable systems [256]. A negative feedback loop between v_1 and itself can result from ternary interactions: (1) v_1 activates v_2 that stimulates v_3 that inhibits v_1 or (2) v_1 inactivates v_2 that inhibits v_3 that inhibits v_1. Parallel positive and negative feedback loop are also represented by 3-variable systems. Every cell cycle controller, such as Ccn–CDK complexes and other major kinases, phosphatases, and transcription factors with their regulators, i.e., each step of the cell cycle, can be represented by 3-variable systems.

In general, models of the cell cycle focus on a part of the process and consider a subset of involved protein interactions to derive a set of governing kinetic equations. Models focus on cell cycle modules that are independent components of the regulatory network. These modules are supposed to be highly localized and transient, as well as weakly coupled to other modules. The module components undergo strong interactions among them.

One approach was based on regulatory modules with feedback loops based on the major cyclin types (cyclin-A, -B, -D, and -E) associated with their corresponding CDKs and their main regulators [257]. Among the set of possible descriptions of regulatory signals, 13 modules have been defined: (1) synthesis and degradation of cyclin-A, -B, and -E. Cyclin-A is active from S to early M phase, cyclin-B during mitosis, and cyclin-E primarily at the G1–S transition. (2) Regulation of Ub ligase anaphase-promoting complex that works in conjunction with its 2 coactivators CDC20 and CDH1 to ubiquitinate cyclin-B for proteasomal degradation. Adaptor CDH1 can be phosphorylated (inactivated) by cyclin-dependent kinases and dephosphorylated (activated) by CDC14 phosphatase. Modules correspond to irreversible processes, such as synthesis and degradation of cell cycle regulators, or reversible binding to activate or inactivate mediators.

A 2-variable model based on ordinary differential equations has been used to describe the kinetics of the set of chemical reactions that create, activate, and degrade the CcnB–CDK1 dimer, which triggers the G2–M transition [258]:

$$\frac{d}{dt}c_c = \kappa_1(c - c_c)(\kappa_2 + c_c^2) - \kappa_3 c_c, \tag{2.3}$$

$$\frac{d}{dt}c = \kappa_4 - \kappa_3 c_c, \tag{2.4}$$

where c_c is the ratio of the concentration of the active CcnB–CDK1 complex to the total concentration of CDK1, c the ratio of the sum of the concentrations of cyclin-B and active and inactive forms of the CcnB–CDK1 complex to the total concentration of CDK1, κ_1 the activation rate of the CcnB–CDK1 complex, when the entire amount of CDK1 is incorporated in the active CcnB–CDK1 complex (adjustable), κ_2 the ratio between the activation rate of the CcnB–CDK1 complex in the absence of already active CcnB–CDK1 complex and κ_1 (0.018/mn), κ_3 the breakdown rate of active CcnB–CDK1 complex (adjustable), and κ_4 the step rate constant (0.015/mn). Steady states are found for very low and high concentration of CcnB–CDK1 complex. A stable limit cycle (sustained oscillations) arises for intermediary values.

A similar model has been developed for the G1–S transition under the control of the transcription factor E2F activated by the release of its retinoblastoma protein (RB) inhibitor:

$$\frac{d}{dt}c_{ccnE} = 10^{-4} + 0.02\, c_{E2F} - 0.01\, c_{ccnE}, \tag{2.5}$$

$$\frac{d}{dt}c_{E2F} = \frac{1.5(c_K + c_{ccnE})(1 - c_{E2F})}{0.01 + 1 - c_{E2F}} - 4\frac{c_P c_{E2F}}{0.001 + c_{E2F}}, \tag{2.6}$$

where c_{ccnE} is the concentration of cyclin-E, c_{E2F} that of activated E2F transcription factor, c_K that (assumed constant) of a stimulatory kinase (bifurcation parameter), and c_P that of an inhibitory phosphatase. Activation is described by a zero-order sensitivity Goldbetter-Koshland switch. A bistable behavior can appear (Sect. 2.4.2.1).[105]

A minimal cascade model has been proposed for the activation at the G2–M transition of the mitotic CcnB–CDK1 oscillator [260]. This abrupt activation results from dephosphorylation by CDC25 stimulated by cyclin B. This activation is reversible due to phosphorylation by Weel. The kinetics model considers 3 players; cyclin-B, CDK1, and Ub ligase that degrades cyclin-B, which is produced at a constant rate. According to the values of involved parameters, the equation set gives rise to stationary and periodic (limit cycles) solutions.

In some species, the minimal core control network that defines independent cell cycle phases may rely on a single CDK oscillator. A single monomolecular CDK module endowed with 2 activity thresholds may indeed autonomously conduct the cell cycle in fission yeast (Schizosaccharomyces pombe) [261]. However, a single oscillating CDK module may explain a primitive regulation of the cell cycle in lower eukaryotes, but not in higher organisms.

Moreover, a single CDK module explains neither why several types of Ccn–CDK couples successively intervene to support the progression of the cell cycle in human cells, nor why they interact with a set of kinases and phosphatases. In addition, between-cell variation in size and cell cycle entry time as well as

[105] A zero-order sensitivity can arise when 2 antogonist enzymes, stimulatory and inhibitory, intervene in a given chemical reaction. Both active (P_a) and inactive (P_i) forms of the same protein P coexist. The activation kinetics is given by:

$$\frac{d}{dt}[P_a] = \frac{v_a[P_i]}{k_a + [P_i]} - \frac{v_i[P_a]}{k_i + [P_a]}.$$

Swith-like behavior (i.e., abrupt slope as soon as activation reaction speed (v_a) becomes greater than that of the inactivation reaction (v_i) for low-affinity values (k_a, $k_i \sim 10^{-3}$), i.e., the entire amount of the protein becomes active (all-or-nothing behavior associated with a threshold).

passage from quiescence to proliferation and conversely can influence the cell cycle of a cell population.

Therefore, models of the CDK network that enable the control of the progression along the successive phases of the cell cycle can contain 40 to 50 variables, especially when a DNA replication checkpoint is incorporated. Several approaches exist to reduce the complexity of cell cycle models.

> *"The objects that do not provoke the examination, "I said, "are all those that do not at the same time go over to the opposite sensation, but the ones that do summon the intellect, when the sensation does not reveal one thing any more than its opposite, regardless of whether the object strikes the senses from near or far off. "(Plato, The Republic, Book VII, 523c)*

Another strategy relies on cellular automata.[106] Cellular automata form a class of discrete models (discretized time and space as well as a finite number of discrete states and evolution rules) used in microstructure modeling that are defined by structural and functional features, i.e., topological and dynamical aspects of the cell network. Structural characteristics of cellular automata rely on a regular grid of cells that possess a finite number of states and have an allocated neighborhood function that yields cell connectivity for a given distance threshold. An initial state is selected and assigned to every cell. The functional properties of the cellular automaton are defined by a finite automaton (set of cell states, neutral state, local transition function). The state of every cell evolves at every iteration. A mathematical function that describes a physical law determines the evolution of every cell at every time, i.e., the new state of every cell with respect to the previous state of the cell and states of cells in its neighborhood.

An automaton model that progresses spontaneously through the 4 successive phases of the cell cycle has been carried out [263]. Each phase is characterized by a mean duration and a variability. Once the prescribed duration of a given phase has elapsed, the next phase of the cell cycle begins. The time at which the transition happens varies in a random manner according to a distribution of durations of the cell cycle phases. Upon completion of the mitosis, the 2 daughter cells immediately enter a new cycle. At each time step in any phase of the cell cycle, the cell has a given probability to exit the cell cycle,

[106] Cellular automaton is an oxymoron-like expression, as automaton focuses on artificial products, hence predictible behavior, and cellular refers to natural living structures, thus unpredictability. An oxymoron (οξυμωρος: inconsistency, ingenious alliance of contradictions; from οξυς: sharp and μωρος: dull; hence, sharp–dull, or [figurative] cunning–stupid, witty under a silly appearance) is a figure of speech that combines contradictory terms for emphasis (e.g., virtual reality).

> *"If you don't believe this lie is true, ask the blind man; he saw it too, through a knothole in a wooden brick wall." (Nathan Alterman).*

as the model incorporates a Markovian process. Cell population homeostasis is maintained by a balance between cell death and cell replication. Moreover, this stochastic automaton model incorporates the progressive desynchronization of cell populations as well as their entrainment by the circadian clock.

A 5-variable model with multiple oscillatory circuits for the CDK network can operate as the pillar of more detailed models of the cell cycle [262]. In the presence of sufficient amounts of growth factors, a transition occurs from a steady state to sustained oscillations of the various Ccn–CDK complexes, i.e., a switch from quiescence to cell proliferation. Sequential activation of the Ccn–CDK complexes allows the ordered progression along the G1, S, G2, and M phases with a restriction point in G1. The model contains the 4 main Ccn–CDK complexes, transcription factor E2F, and CDC20 phosphatase. The CcnD–CDK4(6) complex activated by growth factors acts in G1 phase. It stimulates E2F that then activates the CcnE–CDK2 complex at the G1–S transition as well as the CcnA–CDK2 complex during S phase. The CcnE–CDK2 and CcnA–CDK2 complexes stimulate E2F (positive feedback loop), thereby fostering the progression through the G1–S transition and the S phase, respectively. The CcnA–CDK2 complex stimulates the CcnB–CDK1 complex in G2 phase to trigger the G2–M transition. During mitosis, the CcnB–CDK1 complex phosphorylates (activates) CDC20. This phosphatase creates a negative feedback loop on both CcnA–CDK2 and CcnB–CDK1, as this interaction (dephosphorylation) leads to the degradation of these complexes, thereby enabling the completion of mitosis. This model exhibits an oscillatory behavior with self-sustained, quasi-periodic oscillations of the various Ccn–CDK complexes, which correspond to the evolution to a limit cycle, and deterministic chaos.

The last phase — mitosis — is characterized not only by a control that involves subcellular compartments, but also material transport from a disappearing nucleus to the mitotic spindle, then its splitting and reconstitution of 2 nuclei, as well as that of cytoplasmic constituents into 2 regions that are separated from a cleavage furrow. The growth of microtubules that form the mitotic spindle and actin fibers that generate the abscission ring are particular aspects of mass transfer. All of the events that occur during mitosis can be described by a set of transport–reaction equations:[107]

$$\frac{\partial \mathbf{c}}{\partial t} + \mathbf{v}\nabla\mathbf{c} = -\mathcal{D}\nabla^2\mathbf{c} + \mathcal{R}, \tag{2.7}$$

where $\mathbf{c}(\mathbf{x},t)$ represents the vector concentration of the mediator, $\mathcal{D}$ the diagonal matrix of diffusion coefficients, and $\mathcal{R}$ all local reactions.

---

[107] The following differential operators are used: (1) the gradient operator $\boldsymbol{\nabla} = (\partial/\partial x_1, \partial/\partial x_2, \partial/\partial x_3)$, (2) the divergence operator $\boldsymbol{\nabla}\cdot$, and (3) the Laplace operator $\boldsymbol{\nabla}^2 = \sum_{i=1}^{3} \partial^2/\partial x_i^2$. The gradient $\boldsymbol{\nabla} p$ of scalar p is a vector of component $(\nabla p)_i = \partial_i p$. The gradient $\boldsymbol{\nabla}\mathbf{v}$ of vector $\mathbf{v}$ is a second order tensor with component $(\nabla \mathbf{v})_{ij} = \partial_i v_j$.

Alternative approaches do not focus on intracellular dynamics of mediators, but on cell populations of various ages. Cell population dynamics are described by partial differential equations [264]. The cell population is divided into proliferative and quiescent compartments. The proliferative phase represents the complete cell cycle (G1–M) in a population committed to divide at its end. The model is structured by the time spent by a cell in the proliferative phase and amount (x) of CcnD–CDK4 (or CcnD–CDK6) complexes. Cells can transit from one compartment to the other, following transition rules which differ whether the tissue is healthy or tumoral. The system of ordinary differential equations is derived with respect to age (a) in the G1 phase. The numerous regulators of the production of active cyclin-D that intervenes in the G1 phase are represented by a single aggregation variable with an amount (y). The non-linear system of ODEs with respect to age a in the G1 phase for the 2 regulating variables can be written as:

$$\begin{aligned} \frac{d}{da}x &= \kappa_1 \frac{x}{1+x} y - \kappa_2 x, \\ \frac{d}{da}y &= \kappa_3 - \kappa_4 y, \end{aligned} \tag{2.8}$$

where the coefficients $\{\kappa_i\}_{i=1}^4$ are strictly positive (with the assumption $\kappa_1\kappa_3 > \kappa_2\kappa_4$), κ_1 and κ_2 are the production rate and linear degradation rate of cyclin-D and κ_3 and κ_4 the constant production rate and linear degradation rate of the lumped variable y. The possible saturation by the concentration of CcnD–CDK4(6) in its own production is the single non-linear part in this set of equations. The total weighted population $N(t)$ can be expressed with respect to the densities of proliferating ($p(t,a,x)$) and quiescent ($q(t,a,x)$) cells with age a and content x in CcnD–CDK4(6) complex at time t by:

$$N(t) = \int_0^\infty \int_0^\infty \Big(w_a(a,x)p(t,a,x) + w_q(a,x)q(t,a,x) \Big)\, da\, dx, \tag{2.9}$$

where the weights $w_a(a,x)$ and $w_q(a,x)$ represent mitogens and antimitogens that act on the populations of proliferating and quiescent cells, respectively. Cells exit from the quiescent compartment either by death at a rate d_q or entry in proliferation with a recruitment function R that depends on $N(t)$. Cells can leave the proliferative compartment either by death at a rate d_p or entry in quiescence according to demobilization function $\mathsf{D}(a,x)$.

The model is then described by the following set of equations with strictly postive coefficients:

$$\begin{aligned} \frac{\partial}{\partial t}p(t,a,x) &+ \frac{\partial}{\partial a}\Big(\mathsf{s}_a p(t,a,x)\Big) + \frac{\partial}{\partial x}\Big(\mathsf{s}_x(a,x)q(t,a,x)\Big) \\ &= -\Big(\mathsf{D}(a,x) + F(a,x) + \mathsf{d}_p\Big)p(t,a,x) + \mathsf{R}\big(N(t)\big)q(t,a,x)), \\ \frac{\partial}{\partial t}q(t,a,x) &= \mathsf{D}(a,x)p(t,a,x) - \Big(\mathsf{R}\big(N(t)\big) + \mathsf{d}_q\Big)q(t,a,x), \end{aligned} \tag{2.10}$$

where s_a (assumed constant) and s_x denote the evolution speed of age a and amount x of CcnD–CDK4(6) complexes with respect to time t, respectively, and $F(a, x)$ the fraction of dividing cells. This type of modeling enables the study of non-phase-specific cytotoxic drugs on transitions between quiescent and proliferative states.

Detailed models of cell proliferation include the control on progression speeds of each phase and transitions between phases (G1–S and G2–M) as well as switch from quiescent to proliferative phase (G0–G1) regulated by mitogens and antimitogenic factors, because cell cycle controllers have specific working time windows and can exert their action on phase progression (progression slowing and transient blockages) and transition (checkpoints). These models are needed to represent the action of anticancer drugs in mixed cell populations to easily test: (1) effects on cell cycle phases; (2) different targets on cell cycle control; and (3) action on both tumor and healthy cells to incorporate possible toxic side effects of anticancer drugs. Compartmental (quiescent vs. proliferating cell population) and time-structured (cell number in each cell cycle phase) models have been carried out, assuming that the cell number in each cell cycle phase follows asymptotically an exponential behavior tuned by a periodic function [265].

The simple set of partial differential equations that defined a time-structured model with passages between proliferation and quiescence is given by [265]:

$$\begin{aligned} &\frac{\partial}{\partial t} n_p(t,x) + \frac{\partial}{\partial x} n_p(t,x) + [\mathsf{D}(t) + \mathsf{d}_p(t)] n_p(t,x) = 0, \\ &\frac{\partial}{\partial t} n_q(t,x) + \frac{\partial}{\partial x} n_q(t,x) + [\mathsf{R}(t) + \mathsf{d}_q(t)] n_q(t,x) = 0, \end{aligned} \tag{2.11}$$

where n_p and n_q are the densities of proliferating and quiescent cells, d_p and d_q the death rates, D and R the exchange rates from cell proliferation to quiescence compartment and vice versa.

2.5 Tumorigenesis

Malignant growth in most types of cancers is characterized by cell proliferation, impaired apoptosis, angiogenesis (Vol. 5 – Chap. 10. Vasculature Growth), and metastasis (Sect. 6.6). These processes particularly involve receptor Tyr and Ser/Thr kinases (Vol. 3 – Chap. 8. Receptor Kinases).

2.5.1 Causal Genomic Alterations

Aneuploidy can cause cancer, especially an abnormal number of genes that encode components of the mitotic checkpoint machinery. Hemizygosity (i.e., absence of one copy of a gene) for mitotic arrest-deficient proteins MAD1 and

MAD2 as well as that of checkpoint protein and mitotic motor KIF10 (or CenPe) does not prevent anaphase proceeding. This phenomenon raises the rate of aneuploidy, chromosomal instability, and spontaneous tumorigenesis.

A wide range of genes are mutated in human cancers. Some cause hyperactivity of oncogenes (e.g., upregulation of catalytic activity of protein kinases). Others provoke loss of function of tumor suppressors. Acquisition of growth autonomy by tumor cells can result from mutation-induced activation of receptors, as well as oncogenic mutations of signaling components of signaling cascade, such as the PI3K–PKB (e.g., dysregulated lipid-derived signals), Ras–MAPK (Vol. 4 – Chap. 5. Mitogen-Activated Protein Kinase Modules), or SMAD pathway (Sect. 3.8 and Vol. 3 – Chap. 8. Receptor Kinases).[108] Loss of negative feedback loops also participates in tumorigenesis. Cancers result from the intracellular interaction of altered tumor suppressors and oncogenes.

2.5.2 Cancer Features

Seven main cell-autonomous alterations determine early oncogenesis. Cancer cells: (1) provide their own growth signals in order to proliferate via autocrine or paracrine mechanisms; (2) ignore growth-inhibitory cues; (3) resist apoptosis; (4) avoid immunosurveillance; (5) replicate without limits; (6) provoke angiogenesis; and (7) invade tissues.

Carcinogenesis can be divided into 3 phases: (1) initiation, with stable genomic alterations (DNA is mutated by chemical or physical carcinogens, with subsequent activation of oncogenes and/or inactivation of tumor-suppressor genes and proteins); (2) promotion, with proliferation of genetically altered cells; and (3) progression, with tumor spreading. During the progression stage, additional mutations can be acquired. When tumors progress, some newly acquired genetic alterations are required for tumor cell growth [268, 269]. Invasion occurs when malignant cells proliferate, spread, and are transported in other tissues. Tumor cell invasion is associated with a loss of structure and modification of production patterns.

Populations of malignant cells are either clustered as tumors or dispersed, as in leukemias. Cancer cells are characterized by uncontrolled multiplication and spreading. Although mutations occur in differentiated cells, cancers

[108] Small GTPases hRas, nRas, or kRas (Vol. 4 – Chap. 8. Guanosine Triphosphatases and Their Regulators) are mutated and remain in the active GTP-bound form, hence serving as oncogenes in many cancers. Small Ras GTPases activate tumor growth via the PI3K–PKB pathway. Blockade of the phosphorylation of endothelial nitric oxide synthase NOS3 (a PKB substrate; Vol. 4 – Chap. 9. Other Major Signaling Mediators) inhibits tumor initiation and growth [266]. Small GTPase RalB can couple the immune system to tumor cell survival. It interacts with SEC5 that activates TANK-binding kinase-1 (TBK1; TANK: tumor-necrosis factor receptor-associated factor [TRAF] family member-associated NFκB activator), an inhibitor of tumor cell apoptosis in response to oncogene activation [267]. Kinase TBK1 is also involved in the activation of the innate immune response.

contain stem cells. Cancers can thus be derived from stem cells. Stem cell alterations explain, at least partially as dedifferentiated cells mimic stem cell behavior, heterogeneity of most tumors, with different degrees of cell transformation. Cancer growth depends on the activity of these stem cells.

Tumors are made of malignant epithelial cells and of normal cells, particularly fibroblasts, endothelial cells, and smooth muscle cells of tumor blood vessels. The interactions between malignant cells and blood vessel cells influence tumor growth. Host stromal cells undergo genetic changes (amplifications and deletions) during cancer progression [270]. Myofibroblasts, activated fibroblasts, and perivascular mesenchymal cells such as pericytes intervene in cancer progression. Tumor cells can change into a different cell type and induce surrounding normal cells to do so. Epithelial-mesenchymal transition endows tumor (epithelial) cells with invasive ability. Mesenchymal cells can resemble activated fibroblasts that produce molecules required for metastasis. Transforming growth factor-β1 promotes epithelial–mesenchymal transition as well as endothelial–mesenchymal transition [271] (Vol. 1 – Chap. 1. Cells and Tissues).

2.5.3 Cancer and the Immune System

Many biological processes are involved in tumorigenesis. When carcinogenesis starts, an active antitumor immune response exists.[109] The immune system indeed recognizes cancer precursors and most often destroys these precursor cells. Cancer *immunosurveillance* is carried out by innate and adaptive immune cells that participate in the recognition and destruction of cancer cells.[110] Cancer cells avoid immunosurveillance by producing poorly or nonimmunogenic variants (*immunoselection*)[111] and suppressing the immune response (*immunosubversion*).[112]

---

[109] Adaptive immune responses require B lymphocytes that yield antibodies and T lymphocytes. Antigens are recognized by specific, plasmalemmal, dimeric receptors.

[110] CD8+ cytotoxic T lymphocytes and CD4+ T helper-1 lymphocytes secrete anti-angiogenic interferon-γ [272]. CD4+ T lymphocytes convert interleukin-10-producing M1 macrophages into interferon-γ-producing M2 macrophages. T helper-2 lymphocytes synthesize interleukin-4 and prevent angiogenesis via stromal fibroblasts. CD8+ cytotoxic T lymphocytes, interferon-producing killer dendritic cells (IKDC; another dendritic cell lineage distinct from conventional and plasmacytoid dendritic cells), natural killer T lymphocytes, and γδT lymphocytes with Vγ9Vδ2-containing T-cell receptors destroy tumor cells. In human tumors, Hsp70 is associated with cholesterol-rich nanodomains in the plasma membrane and, hence, can be targeted by the immune system that primes infiltrations of NK cells, macrophages, and granulocytes to prevent tumor growth [273].

[111] Reduction or loss of expression of major histocompatibility complex class-1 molecules impedes the immune response.

[112] Some tumoral factors can recruit myeloid suppressor cells and prevent their differentiation into mature dendritic cells via signal transducer and activator

Cancer results from accumulation of mutations within a genetically unstable heterogeneous population of cells, with the possible emergence of a malignant cell subgroup with the whole set of functions necessary for preventing the host defenses and invasion. Following a carcinogenic event, the initiated cell gives birth to several generations of daughter cells. Errors in DNA replication during cell division lead to genetically different subgroups of robust cells and tumor cell heterogeneity. These distinct tumor cells can cooperate to overcome certain host immune defenses [274]. Moreover, cancer cells interact with normal cells of the environment, such as endothelial, stromal, and nerve cells. Adjoining cells can protect each other from host immune defenses. Furthermore, during the early stage of development, cancer cells produce enzymes that neutralize glycosides of natural killer cells. The immune system is then unable to destroy tumor cells. Moreover, malignant cells produce soluble factors such as autocrine interleukins (Sect. 3.22) that can prevent inflammation and crosstalk between innate and adaptive immunity [272]. In addition, cancer cells that are insensitive to growth inhibitors proliferate in an environment rich in immunosuppressive transforming growth factor-β (Sect. 3.8). This growth factor reduces the amount of antigen presentation by dendritic cells, inhibits interferon-γ, represses T-lymphocyte proliferation, suppresses the cytotoxic activity of natural killer cells, and stimulates the proliferation of immunosuppressive regulatory T cells [272,275].[113] Tumor cells also develop resistance to apoptosis-inducing molecules of the innate and adaptive immune systems. Some molecules used by tumor cells to invade tissues affect their immunogenicity.[114] Moreover, tumors protect themselves against immune attack using extracellular adenosine generated in hypoxic tumor mass [276].[115]

---

of transcription STAT3. These myeloid suppressor cells inhibit tumor-resident T lymphocytes via overproduction of arginase-1 or nitric oxide synthase-2 [272]. Overexpression of pro-inflammatory cyclooxygenase-2 hinders antitumor immunity by impeding macrophage- and T-cell-mediated tumor degradation and favoring the balance of T helper cell responses toward T helper-2-cell stimulation that leads to imbalance between T helper-2 interleukin-10 and T helper-1 interleukin-12. Moreover, repetitive stimulation of naive T lymphocytes by immature dendritic cells causes T-cell anergy.

[113] Regulatory T cells, a subset of peripheral CD4+ T cells, mediate immunosuppression during tumor growth. Regulatory T cells are found in the tumor microenvironment. Regulatory, FoxP3+, IL10+ T cells suppress the activity of antigen-presenting cells and natural killer cells. CD8+ regulatory T cells can also be implicated.

[114] Nectin-like-2 that participates in the organization of epithelia and promotes the cytotoxicity of natural killer cells and production of interferon-γ by T lymphocytes is downregulated upon the inactivation of tumor suppressor in lung cancer gene product TSLC1 [272].

[115] During inflammation, activation of adenosine receptor (A2AR) on T cells stimulates adenylyl cyclase and increases intracellular cAMP level, thereby activating protein kinase-A.

Inflammation and innate immunity could exert protumorigenic effects, whereas adaptive immunity could have antitumorigenic effects. Inflammation acts via leukocytes (macrophages, tumor-associated macrophages, dendritic cells, neutrophils, mastocytes, and T lymphocytes), which are recruited to the tumor microenvironment. These leukocytes produce cytokines and growth and angiogenic factors (tumor-necrosis factor, interleukin-6, and vascular endothelial growth factor), as well as matrix metallopeptidases and their inhibitors.

Necrosis of cancer cells in the hypoxic tumor core leads to the activation of tumor-associated macrophages. Nitric oxide mediates iron and glutathione release by activated macrophages through transporter multidrug resistance-associated protein-1 to inhibit of tumor cell proliferation [277] (Vol. 3 – Chap. 4. Membrane Compound Carriers). Interferons (Sect. 3.23), interleukins IL4, IL12, and IL13, and granulocyte–macrophage colony-stimulating factor (Sect. 3.19) may increase tumor rejection, inducing maturation of dendritic cells and promoting initiation of an adaptive immune response. On the other hand, colony-stimulating factor-1 (mCSF) and interleukin-6, produced by tumor-associated macrophages, can prevent maturation of dendritic cells. Together with IL1 and tumor-necrosis factor, CSF1 and IL6 can promote tumor progression.

Viral and bacterial infections,[116] subsequent inflammation, and cancer can be connected. Inflammation can induce tumor growth by the activation of nuclear factor-κB [279] (Vol. 4 – Chap. 9. Other Major Signaling Mediators).

Nevertheless, major histocompatibility complex-disparate T-cell lineage-committed lymphoid precursor cells can be used for immunotherapy, as allogeneic T-cell precursors can develop into host-MHC-restricted T cells characterized by dual tolerance and selection of a functional receptor repertoire [280]. In addition, calreticulin[117] exposure on tumor cell surface by drugs, such as anthracyclines, triggers the immune responses leading to cell apoptosis, avoiding release of cytoplasmic content and damages to normal cells [281, 282].

2.5.4 Oncogenes

Tumor cells are characterized by limitless replicative potential, as tumor cells result from genetic alteration of signaling pathways that promote cell growth and survival. In small-cell lung cancers,[118] fibroblast growth factor-2 increases expression of anti-apoptotic proteins XIAP and BCLxL (Sect. 4.6), via the formation of a specific multiprotein complex including GTPase B-Raf, protein kinase-Cε, and ribosomal S6 kinase S6K2 [283]. Factor CKI1b is dysregulated

[116] Papillomaviruses, polyomavirus [278], hepatitis-B and -C viruses, and Epstein-Barr virus are risk factors for malignancies.

[117] Calreticulin is a calcium-binding protein predominantly located in the endoplasmic reticulum, where it participates in proper folding of proteins.

[118] Small-cell lung cancer represents about 20% of lung tumors.

during tumorigenesis [284]. The CDK inhibitor CKI1b controls the cell division in G1 phase, regulating CcnE– and CcnA–CDK2 complexes.[119]

Although receptor Tyr kinases typically activate oncogenic pathways, the ephrin receptor Tyr kinase EPHb4 inhibits breast tumor growth in mice [285]. However, the expression of the EPHb4 ligand ephrin-B2 (Vol. 3 – Chap. 8. Receptor Kinases) is lost.[120]

2.5.5 Cancer and the Circadian Clock

The circadian clock that is composed of a central pacemaker located in the hypothalamus and local oscillators in peripheral tissues (Chap. 5) is involved in cell division and tissue homeostasis. The core clock oscillator in both the central pacemaker and peripheral tissues consists of 2 transcriptional factors, CLOCK and BMAL1, and their targets, Periods (Per1–Per3) and Cryptochromes (Cry1 and Cry2). Both Period and Cryptochrome inhibit CLOCK–BMAL1 activity (autoregulatory feedback loop). The circadian regulation also includes various clock-controlled genes such as cell cycle regulator genes. Gene PER1 that regulates the circadian clock can function as tumor suppressor, acting on checkpoint pathways [286].[121]

2.5.6 Tumoral Angiogenesis

Tumor cell expansion relies on nutrient supply, thus on angiogenesis orchestrated by the transcriptional factor hypoxia-inducible factor (Vol. 4 – Chap. 9. Other Major Signaling Mediators). Tumors are indeed exposed to hypoxia, especially quickly growing tumors, because of high oxygen consumption and inadequate blood supply. The transcription factor hypoxia-inducible factor is activated for angiogenic signaling and glycolysis. A decay in tumoral hypoxia-inducible factor activity can then reduce tumor growth. In response to hypoxia, tumor cells secrete angiogenic factors such as vascular endothelial growth factor. These molecules diffuse in the extracellular matrix and reach neighboring blood vessels. Endothelial cells at the interface between flowing blood and vessel wall respond to these growth factors. They degrade their adhesion with matrix and adjoining cells, then move toward high concentrations of chemotactic substances. Once in place, endothelial cells proliferate and self-organize to create capillaries.

[119] Phosphorylation of CKI1b leads to degradation of CKI1b by the binding of CKI1b to SKP2 and subsequent polyubiquitination by the SKP1-Cullin-F-box Ub ligase. Besides, ERK1 and ERK2 activation in most of the lung tumors is correlated with the loss of CKI1b proteins.

[120] Normally, the negative effect of EPHb4 on tumor growth via ligand binding results from phosphorylation by CRK kinase that binds CAS scaffold protein.

[121] In order to cope with possible DNA damages, cells have developed a network of checkpoint pathways that initiate either cell cycle arrest and DNA repair or apoptosis.

Cancer cells use the Ras–ERK and PI3K–PKB cascades that are activated by various growth factors, hormones, and extracellular matrix proteins. These 2 pathways also control the expression of vascular endothelial growth factor-A. Transcription factor HIF controls the expression of angiogenic factors VEGFa and angiopoietin-2. It controls not only neovascularization, but also energy metabolism, cell survival, intracellular pH, and cell migration. Two types of oxygen-sensitive compounds control stability and activity of HIF1α subunit, which is regulated by oxygen (but not HIFβ): prolyl and asparaginyl hydroxylases [287]. 2-Oxoglutarate-dependent and iron-dependent dioxygenases are oxygen sensors that control the hypoxic response. Prolyl hydroxylases have a much lower affinity for oxygen and 2-oxoglutarate than asparaginyl hydroxylase. Decreases in oxygen level according to the distance with respect to the closer capillary are associated with the increase in extracellular acidity due to accumulation of lactate and CO_2. Hypoxia induces the expression of carbonic anhydrase-9 that favors a relatively neutral intracellular pH. Hypoxic stress also leads to decay in protein synthesis owing to inhibition of the TOR pathway.

2.5.7 Tumoral Invasion

Tumor progression and invasion require that the cells promote extracellular matrix remodeling, increased migration, and digestion of basement membrane. Hypoxia-inducible factor activates peptidases (cathepsin-D, urokinase-type plasminogen-activator receptor, matrix metallopeptidase-2) and migration factors (cytokine phosphoglucose isomerase, or autocrine-motility factor, transforming growth factor-α, and hepatocyte growth factor receptor). Hypoxia also increases free NADH level that causes binding of transcriptional corepressor CtBP, which senses free nuclear NADH levels, to E-cadherin transcriptional repressors, thus reducing E-cadherin expression concomitantly with dedifferentiation and invasion of epithelial cells during tumorigenesis [288]. Hypoxia-inducible factor inhibits E-cadherin expression.

Tumors are characterized by abnormal and dysfunctional blood vessels. Tumor cells can then easily migrate via lymphatic and blood vessels to reach healthy tissues, inducing metastasis. On the other hand, most of an injected drug either does not reach tumor cells or the small amount that does reach them is unevenly distributed. Restoration of a normal architecture of a tumor vasculature allows cancer therapies to penetrate the tumor and to act more effectively.

Endothelial cell G proteins, particularly Gα_q and Gα_{11} (Vols. 3 – Chap. 7. G-Protein-Coupled Receptors and 4 – Chap. 8. Guanosine Triphosphatases and Their Regulators), interact with and mediate intracellular signaling stimulated by vascular endothelial growth factor receptor-2. Regulators of G-protein signaling shorten the duration of active $G\alpha_i$ and $G\alpha_q$ and act on mitogen-activated protein kinase activity and cell migration and proliferation.[122]

2.5.7.1 Metastasis

Most human cancer deaths are caused by metastasis. Metastatic cells must successfully survive in blood flow and invade a foreign tissue. Metastasis can begin as an early rather than a late event in cancer progression, as mammary cells with delayed oncogene induction travel to, reside, and survive in lungs during 16 weeks at most when these cells remain untransformed and do not form ectopic tumors [289]. Therefore, mammary cells that do not bear oncogenic transformation have a capacity for long-term residence in the lungs. Upon oncogene activation, genetic changes increase the survival duration in remote tissues. MicroRNAs (Vol. 1 – Chap. 5. Protein Synthesis) serve as biomarkers to identify cancer origin when the primary site remains unknown [290].

抗癌

Figure 2.6. In chinese, kang ai means against cancer.

Cancer cells can express metastasis suppressors[123] that prevent the growth of neoplastic cells at metastatic sites but not cell proliferation at primary sites. Tetraspanin TSpan27[124] binds the Duffy blood group glycoproteic chemokine receptor[125] (DARC) that is expressed on the endothelium [291]. The interaction between Kai1 and DARC suppresses the growth of cancer cells and leads to senescence of tumor cells by modulating TBX2 and CKI1a expression.

Tumor cells metastasize with the help of host cells that modify their microenvironment to support the survival and growth of newly arriving tumor cells. Tumor cells can secrete granulocyte colony-stimulating factor that expands and mobilizes bone-marrow, lymphocyte antigen Ly6G+, and Ly6C+ granulocytes and facilitates their homing into the lung before the arrival of tumor cells [292]. These granulocytes then secrete enhancers of the invasiveness

[122] RGS4 inhibits mitogen-activated protein kinases and vascular endothelial growth factor receptor-2 expression. Other stimulators of G-protein-coupled receptors (thrombin, angiotensin-2, endothelin-1, prokineticin-1 and -2, etc.) are also implicated.

[123] Several metastasis suppressors exist, such as MAP2K4 enzyme and RhoGDI2.

[124] A.k.a. CD82 and Kangai-1 [Kai1] metastasis suppressor or suppressor of tumorigenicity. The ubiquitous transmembrane protein Kai1 (chinese kang ai: against cancer [Fig. 2.6]) acts as a metastasis suppressor. Kai1 interacts with manifold proteins, such as integrins, PKCα, B-cell receptors, other tetraspanins, and receptor Tyr kinases.

[125] A.k.a. CD234 and glycoprotein GPd or GPfy.

of tumor cells, such as matrix metallopeptidases and cell migration stimulator prokineticin-2 that targets PKR1 prokineticin receptor. Furthermore, the co-travel of circulating tumor cells with primary tumor-associated stromal cells such as fibroblasts improve their viability [293].

2.5.7.2 Tumor Growth Modeling

Tumor growth modeling deals with a multiscale problem, as it involves genetic and metabolic level with gene mutations, cell level with cell–cell and cell–matrix interactions, tissue level with cell invasion, and body level with metastasis. Modeling of tumor cell behavior incorporates 2 major processes, cell proliferation and migration, knowing that certain cell types divide, but do not move, and vice versa. Two types of models have been developed. Discrete models (e.g., cellular automata and agent-based models) describe evolution of individual cells. They are hence suitable for early stages of tumor growth. Contінous phenomenological models rely on partial differential equations of the reaction–diffusion type, using a set of parameters associated with cell birth and death to describe the cell-density evolution as well as cell-density and nutrient-availability thresholds, in addition to cell flux across the boundary of the domain.

Tumoral invasion can be represented by the *reaction-diffusion equation*. an initial value problem (IVP) with unknown u. This non-linear parabolic equation that degenerates at points $u(x,t) = 0$ is given in a two-dimensional space (i.e., a single space variable x and time t) by:

$$u_t(x,t) = f(u)_{xx} + g(u), \quad t > 0, x \in \mathbb{R}, \tag{2.12}$$

with the initial condition $u(x,0) = u_0(x)$, where $u_t \equiv \partial u/\partial t$ and $f(u)_{xx} \equiv \partial f/\partial x^2$, $f(u)$ and $g(u)$ are non-negative continuous functions on the normalized [0,1] range, $f'(u) > 0$, $f(0), f'(0) = 0$, and $g(0) = g(1) = 0$ [294] or with a bistable, unbalanced function $f(0) = f(1) = 0$, $f'(0), f'(1) < 0$, $f < 0 \,\text{on}\, (0,\theta)$, $f > 0 \,\text{on}\, (\theta,1)$, and $\int_0^1 f\,dx > 0$ [295].

When $g(u) = 0$, equation 2.12 corresponds to a non-linear diffusion equation for mass transfer with anisotropic diffusivity or heat equation with a temperature-dependent heat conductivity.

When $f(u) = u$, equation 2.12 corresponds to the Kolmogorov-Petrovsky-Piskounov (KPP) equation that describes the propagation of advantageous genes in populations and can be applied to a propagating front of tumor cells. The asymptotic behavior of solutions of the KPP equation with an initial condition given by a non-negative function bounded by 1 whose support is restricted to the negative half of the real line was exhaustibly studied [296]. Some particular analytical solutions of the KPP equation can be obtained using the Painlevé expansion and bilinear method, among others.

When $f(u) = u$ and $g(u) = u(1-u)$, i.e., for reaction–diffusion equations with a non-linear, but relatively simple reaction term, a family of traveling wave solutions in the 2-dimensional space (x,t) exists:

$$u(x,t) = U(x - ct),$$

with wave speed c.

The simplest formulation of the Kolmogorov-Petrovsky-Piskounov-Fischer can be written as a diffusion–reaction equation of the population density n [297]:

$$n_t(x,t) = \mathcal{D} n_{xx} + \mathsf{r} n (1 - n/\mathsf{c}), \tag{2.13}$$

where r and c in the reaction term are the linear reproduction rate and carrying capacity of the environment ($n_{xx} \equiv \nabla^2 n$). The first and second right-hand side terms of the equation 2.13 represent the diffusion and reaction components, the latter modeling the medium modification due to the progressive invasion of cancerous cells to the detriment of healthy cells. This equation possesses constant-speed traveling wave solutions. A reaction–diffusion KPP-type equation in periodic media has pulsating wave solutions.

The generalization of the equation 2.13 leads to the following diffusion–reaction equation:

$$\mathbf{u}_t(x,t) = \nabla \cdot (\mathcal{D} \boldsymbol{\nabla} \mathbf{u}) + \mathbf{S}, \tag{2.14}$$

where $\mathbf{S}$ is the vector source term ($\boldsymbol{\nabla}\mathbf{u}$: tensor, $\nabla \cdot (\mathcal{D}\boldsymbol{\nabla}\mathbf{u})$: vector).

2.5.8 Pharmacokinetics and Pharmacodynamics of Antimitogens

Pharmacokinetics (PK)[126] is the study of the fate of substances (nutrients, metabolites, hormones and growth factors, as well as toxins, and mainly drugs) administered externally to a living organism. Investigations focus on mechanisms of absorption and distribution of an administered drug, the delay of action and duration of effect, chemical changes experienced by the substance in the body, and effects and routes of excretion of drug metabolites.

Phenomenological compartment-based models of pharmacokinetics that rely on kinetic models aimed at predicting the substance concentration at any time provide initial guesses and insights into the absorption, distribution, metabolism, and excretion of drugs. Standard PK models describe the relationship between the drug dose and the dynamics of the drug concentration at the target site as the resultant of 4 processes: absorption, distribution, metabolism, and excretion. These processes are assumed to be linearly connected and depend linearly on drug concentrations in the relevant compartments.

Pharmacodynamics (PD)[127] is the study of the biochemical and physiological effects of drugs on the body, mechanisms of drug action (chemical

---

[126] φαρμακον: drug and κινεω: to move, set in motion, disturb, or stir; i.e., "what the body does to the drug".

[127] "What the drug does to the body".

reactions, interactions with receptors, carriers, ion channels and pumps, structural proteins, and enzymes), and the relationship between drug concentration and effect.

Phenomenological models of pharmacodynamics rely, in particular, on the relationship between the drug concentration at the target and the effect on the functioning of the target (dose–response models).

More mechanistic PK–PD modelings aim at associating metabolic and regulatory networks relevant to the target that have been explored in the framework of systems biology. Non-linear and feedback-driven signal-transduction pathway models are indeed needed to improve PK–PD models.

Modern chemotherapy is governed by several basic principles, among which (1) limitation of toxicity to cancerous lesions; (2) limitation of acquisition of drug resistance; (3) optimization of delivery time during the circadian rhythm; and (4) optimization of drug synergy.

Genomic identity of patients enables a personalized treatment based on predictions on patient-specific drug sensitivity and degradation rate. Yet, most drugs distribute to the entire body. A general toxicity and poor acceptance of treatments by patients then results. The targeted delivery of chemotherapeutics to cancer and stromal cells in tumors at a proper time of the day is one of the main challenges in therapeutic strategies to minimize the cytotoxicity on healthy cells. Cancerous cells express molecules (receptors, adhesion molecules, peptidases, etc.) that are different or differently expressed with respect to normal cells. Optimal drug nanovectors contain hydrophilic polymers to be furtive to the immune system and homing heads that recognize cancer cell markers. Nanoaerosols are used for drug delivery on bronchopulmonary cancers using a cannula or catheter that can be introduced in the tracheobronchial tree under visual guidance assisted by virtual fibroscopy (Vol. 6 – Chap. 5. Images, Signals, and Measurements).

Conceiving, designing, implementing, and operating efficient procedures of chronotherapy partly rely on modeling with optimal control. Mathematical modeling and numerical simulations of the effect of anticancerous drugs require a model for the cell cycle with its different phases that includes exchanges between populations of quiescent and proliferating cells as well as control by the circadian rhythm (Sect. 2.4.4).

A modulation of the death rate by a periodic function that represents a circadian entrainment accelerates tumor growth with respect to a constant death rate with the same average. On the other hand, a modulation of transition rates from one cell cycle phase to the next $\mathsf{T}_{i \rightarrow i+1}$ yields results in agreement with observation [298]. In this model, the population of cells is represented by a partial differential equation for the density $n_i(t, x)$ of cells with age a in the cell cycle phase ($i = 1, , \ldots, 4$) at time t:

$$\frac{\partial}{\partial t} n_i(t, x) + \frac{\partial}{\partial a} n_i(t, x) + [\mathsf{d}_i(t, x) + \mathsf{T}_{i \rightarrow i+1}(t, x)] n_i(t, x) = 0, \tag{2.15}$$

associated with boundary and initial conditions, where d_i is the apoptosis rate. Therefore, the pharmacological control is preferentially exerted during checkpoints of the cell cycle.

Chronotherapy takes into account the influence of circadian rhythm of cell behavior to maximize the treatment efficiency on tumor cells and minimize its toxicity on healthy cells. A mathematical model with pharmacokinetics and pharmacodynamics parts has been carried out for a set of 6 coupled differential equations that govern the time evolution of both tumor (n_t) and healthy (n) cell population, the latter to be shielded from unwanted side effects during the treatment [299]. This model was aimed at addressing neither drug resistance nor cell cycle phase specificity, but balance between effects on tumor and healthy cells by cytotoxic drugs, considering the drug concentrations in the plasma c_p, target tissue c_t, and cancer c_c:

$$\begin{aligned}
\frac{d}{dt}c_p &= -\kappa_1 c_p + \frac{\mathsf{i}(t)}{V_{dist}}\Phi(t), \\
\frac{d}{dt}c_t &= -\kappa_2 c_t + \kappa_3 c_p, \\
\frac{d}{dt}c_c &= -\kappa_4 c_c + \kappa_5 c_p, \\
\frac{d}{dt}n &= \mathsf{r} - \kappa_6, \\
\frac{d}{dt}\mathsf{r} &= -[\kappa_7 + \mathsf{m}(c_t)]\mathsf{r} - \kappa_8 n + \kappa_9, \\
\frac{d}{dt}n_t &= -\kappa_7 n_t \ln\{n_t/n_{t_{max}}\} - \mathsf{m}(c_c)n_t,
\end{aligned} \tag{2.16}$$

where r is the cell renewal rate, $\mathsf{i}(t)$ the drug perfusion (input) flow rate, V_{dist} the distribution volume (i.e., the volume of the plasmatic compartment in which the active drug is infused), $\Phi(t)$ a function equal to 1 during infusion period and 0 otherwise, and $\mathsf{m}(c_t)$ and $\mathsf{m}(c_c)$ circadian modulation functions. This model yields the optimal drug perfusion rate for a given fraction of healthy cells that must be preserved.

In addition, efficient models of the cell cycle can be coupled to models of drug signaling from the cell surface and catabolism inside the cell. Deterministic pharmacokinetics–pharmacodynamics (PK–PD) models focus on the drug fate and effects within the body. Besides, the optimal administration time of chemotherapeutics coincides with the time of least toxicity for healthy cells. Possible drug resistance that adds a constraint in therapy optimization can be represented by a stochastic extension of deterministic models of drug fate.

Validated, detailed biochemical models of the cell cycle, reductionist versions (that contain a reduced number of variables) of these models, and automaton models that are coupled to models of population dynamics enable the study of the response of a population of proliferating cells to the circadian rhythm-determined administration of anticancerous drugs that act specifically or not on a particular phase of the cell cycle.

2.6 Cell Metabolism and Fate

Numerous metabolites participate in cell energy status, such as glucose,[128] lipids, amino acids, and nucleic acids. Cell metabolism, in particular the tricarboxylic acid cycle (Vol. 1 – Chap. 4. Cell Structure and Function),[129] glycolysis,[130] pentose phosphate pathway,[131] and fatty acid synthesis,[132] is an important determinant of a cell's decision between proliferation and death.

Signaling molecules integrate metabolic inputs and regulate the activities of cyclin-D, cyclin-dependent kinases, ubiquitin ligase anaphase-promoting complex, P53 transcription factor, caspase-2,[133] and anti-apoptotic protein B-cell lymphoma BCL2, among others [300] (Sect. 2.4 and Chap. 4 as well as

[128] Glucose enters the cell using glucose transporters. Inside the cell, it is phosphorylated by hexokinases into glucose 6-phosphate. The latter is used in glycolysis to produce ATP, coenzyme NADH, and pyruvate, or in the pentose phosphate pathway to synthesize ribose 5-phosphate and NADPH. Agent NADPH is an anti-oxidant that maintains glutathione in a reduced state to prevent oxidative damage. It is also a cofactor in the synthesis of amino and fatty acids as well as nucleotides. Agent NADH is used in the mitochondrial oxidative phosphorylation for ATP production.

[129] Pyruvate produced by glycolysis is converted to acetylCoA that enters the tricarboxylic acid cycle, in which 2 ATP and 6 NADH molecules are synthesized per glucose. In addition to glucose, amino acids can also participate in the tricarboxylic acid cycle. Tricarboxylic acid cycle intermediates comprise α-ketoglutarate, succinylCoA, fumarate, malate, and oxaloacetate. In mitochondria, acetylCoA combines with oxaloacetate to form citrate. Citrate is exported and degraded in the cytosol into acetylCoA by ATP citrate lyase. In the cytoplasm and nucleus, acetylCoA can be used as a precursor for post-translational acetylation of proteins, including histones.

[130] Two major enzymes of glycolysis are phosphofructokinase-1 (PFK1) and pyruvate kinase. Phosphofructokinase-1 catalyzes the conversion of fructose 6-phosphate to fructose (1,6)-bisphosphate. Pyruvate kinase produces pyruvate and ATP in the final stage of glycolysis.

[131] The entry of glucose 6-phosphate in the pentose phosphate pathway relies on glucose 6-phosphate dehydrogenase. In the pentose phosphate pathway, glucose 6-phosphate is converted to ribose 5-phosphate and produces 2 NADPH.

[132] AcetylCoA produced from the tricarboxylic acid cycle and glycolysis is a precursor of fatty acid synthesis. AcetylCoA can be converted to malonylCoA by acetylCoA carboxylase. AcetylCoA and malonylCoA are both used by fatty acid synthase to produce and elongate fatty acid chains.

[133] Caspase-2 is inhibited by the glucose flux and resulting production of NADPH by the pentose phosphate pathway. Caspase-2 is phosphorylated (inhibited) by Ca^{++}–calmodulin-dependent protein kinase CamK2, itself activated by NADPH. It is then sequestered by 14-3-3 protein. Caspase-2 activation depends on the metabolism-regulated release of the phosphorylated 14-3-3 protein. When the

Table 2.32. Crosstalk between cell metabolism and proliferation (Source: [300]; ↑: increase; ↓: decrease; ACC: acetylCoA carboxylase; ACL: ATP citrate lyase; APC: anaphase-promoting complex [or cyclosome]; CDC25: cell division cycle phosphatase; CDH: CDC20 homolog; CDK: cyclin-dependent kinase; CKI: cyclin-dependent kinase inhibitor; FAS: fatty acid synthase; GluT: glucose transporter; PFKFB3: 6-phosphofructo 2-kinase–fructose (2,6)-bisphosphatase isoenzyme-3; PPP: pentose phosphate pathway).

Metabolic agent	Cell cycle regulators	Outcome
ACC	Inhibition by cyclin-D1	Fatty acid synthesis
ACL	Histones	Histone acetylation
		S-phase progression
FAS	Inhibition by cyclin-D1	Fatty acid synthesis
Hexokinase-2	Cyclin-D1 ↓	Pentose phosphate pathway
		Glycolysis
	Inhibition by cyclin-D1	Glycolysis ↓
PFKFB3	Cyclin-D3 ↑	G1-phase progression
	CDC25c ↑	Entry in mitosis
	CKI1b ↓	CDK1 activity ↑)
		Entry in mitosis
	Inhibition by APC–CDH1	Glycolysis ↓
		G1 repression
Pyruvate kinase	Inhibition by cyclin-D1	Glycolysis
GluT	Cyclin-D2 ↑	G1–S transition

Table 2.32). In addition, a reduction in mitochondrial ATP production can lead to cyclin-E ubiquitination and degradation, hence G1–S arrest.

Bifunctional enzyme 6-phosphofructo 2-kinase and fructose (2,6)-bisphosphatase isoenzyme-3 (PFKFB3) interconverts fructose-6-phosphate and fructose (2,6)-bisphosphate.[134] Enzyme PFKFB3 thus promotes the production of cyclin-D3 and CDC25c as well as impedes that of the cyclin-dependent kinase inhibitor CKI1b [300]. Reciprocally, cyclin-D1 overexpression impedes the synthesis of hexokinase-2, pyruvate kinase, fatty acid synthase, and acetyl-CoA carboxylase.

The complex made of the anaphase-promoting complex and CDC20 homolog CDH1 is active during G1 phase and from late mitosis. It ubiquitinates PFKFB3 for degradation. The APC–CDH1 complex thus decreases glycolysis. A concomitant increase in glucose flux via the pentose phosphate pathway supports cell protection from oxidative stress-induced apoptosis.

NADPH level decays, 14-3-3 is released from caspase-2 that is dephosphorylated (activated) by PP1 protein phosphatase.

134 Fructose (2,6)-bisphosphate is a coactivator of the rate-limiting glycolytic enzyme phosphofructokinase PFK1. Among PFKFB enzymes, cytoplasmic PFKFB3 is an isoform with a high kinase to phosphatase activity ratio (740/1).

Table 2.33. Crosstalk between metabolism and apoptotic regulators (Source: [300]; MOMP: mitochondrial outer membrane permeabilization; BAD: BCL2 antagonist of cell death; BAK: BCL2-antagonist–killer; BAX: BCL2-associated X protein; BCL: B-cell lymphoma protein; GSK: glycogen synthase kinase; MCL1: BCL2-related myeloid cell leukemia sequence protein-1; maPKA: mitochondrion-associated protein kinase-A; PKB: protein kinase B; PP2: protein phosphatase-2; PUMA: P53-upregulated modulator of apoptosis; VDAC: voltage-dependent anion channel).

Metabolic agent	Effect on BCL2 family member
Ceramide	PP2-mediated BCL2 dephosphorylation (pro-apoptotic)
	Cooperation with BAK to promote MOMP
GSK3	MCL1 ubiquitination and degradation (pro-apoptotic)
Hexokinase-2	VDAC binding to limit BAX-induced MOMP (anti-apoptotic)
P53	PUMA and Noxa expression (pro-apoptotic)
	BAX expression and activation
maPKA	BAD phosphorylation and inhibition (anti-apoptotic)
PKB	Inhibition of GSK3 activity (anti-apoptotic)
	BAD phosphorylation (relieve BCL2 inhibition)
	BAX inactivation

Glucose metabolism raises citrate production. Enzyme ATP–citrate lyase links glucose metabolism to replication and transcription. It indeed generates acetylCoA from citrate. AcetylCoA is a substrate for histone acetylation that fosters DNA replication and gene transcription, especially those involved in metabolism, e.g., those that encode phosphofructokinase-1, hexokinase-2, ^{L}lactate dehydrogenase-A chain, and glucose transporter-4 [300].

When intracellular glucose concentration is low, P53 transcription factor is phosphorylated by AMP-activated protein kinase to promote cell cycle arrest (G1–S period; Sect. 2.4). Factor P53 intervenes in cell metabolism, as it controls the synthesis of guanidinoacetate ^{N}methyltransferase,[135] synthesis of cytochrome-C oxidase SCO2, glutaminase-2, (T)P53-induced glycolysis and apoptosis regulator (TIGAR), and phosphoglycerate mutase [300]. Agent P53 also influences apoptosis, as it regulates the production of BCL2 homology BH3-only proteins P53-upregulated modulator of apoptosis (PUMA) and Noxa as well as pro-apoptotic proteins BCL2-associated X protein (BAX) and P53-induced protein with a death domain (PIDD; Chap. 4 and Table 2.33).

135 Factor P53 modulates creatine synthesis during both genotoxic and nutrient stress, as it upregulates guanidinoacetate ^{N}methyltransferase (GAMT). The latter converts the glycine metabolite guanidoacetate to creatine to yield an alternative source of ATP via increased fatty acid oxidation and promote genotoxic- and glucose starvation-induced apoptosis.

Life is real! Life is earnest!
And the grave is not its goal;
Dust thou art, to dust returnest,
Was not spoken of the soul.

(A Psalm of Life, Henry Wadsworth Longfellow [1807-1882])

3

Growth Factors

... Of vista — Suppose some sight in arriere, through the formative chaos, presuming the growth, fulness, life, now attain'd on the journey;
(But I see the road continued, and the journey ever continued;)
— Of what was once lacking on earth, and in due time has become supplied ...

(Thoughts, 1855, Walt Whitman [1819–1892])

Any cell needs extracellular signals for its growth, proliferation, and survival, among other events. Chemical, mechanical, electrical, and other physical interactions of cells with the extracellular matrix yield functional and structural signals for normal cellular activity as well as formation and maintenance of three-dimensional tissues. Environmental chemical signals are transmitted by hormones (Sect. 1.4) and growth factors (Table 3.1), in addition to nervous cues (Sect. 1.1).

In the sustained absence of growth factors, cells can undergo apoptosis (Chap. 4). Growth factor withdrawal causes loss in plasmalemmal transporters for nutrients and receptors for extracellular molecules, such as glucose, amino acids, low-density lipoprotein, iron, etc. [305]. Yet, transiently, cells have a defense system against stress for survival.

Many growth factors (GF) are molecules, most often proteins, but also lipids, that bind to receptors on the cell surface (Fig. 3.1; Vol. 3, Chap. 6. Receptors), mainly to activate cellular proliferation and/or differentiation. The terminology currently used comes from the initial discovery context.

Many growth factors have a pleiotropic nature. They thus stimulate almost every cell type, whereas others are specific to particular cell types. The growth factors have functional redundancy, but they can have distinct effects. Their production and activation depend on the balance between inhibitory and stimulatory factors as well as their transport and storage capacities. In

Table 3.1. Examples of growth factors that regulate development and homeostasis of blood cells and vessels (Sources: [301–304]; gCSF: granulocyte colony-stimulating factors [CSF3]; gmCSF: granulocyte–macrophage colony-stimulating factors [CSF2]; mCSF: macrophage colony-stimulating factor [CSF1]; EGF: epidermal growth factor; Epo: erythropoietin; FGF: fibroblast growth factor; IGF: insulin-like growth factor; PDGF: platelet-derived growth factor; TGF: transforming growth factor; TNF: tumor-necrosis factor; VEGF: vascular endothelial growth factor; Bφ: basophil; EC: endothelial cell; Eφ: eosinophil; FB: fibroblast; Lφ: lymphocyte; Mφ: macrophage; Mo: monocyte; Nφ: neutrophil; RBC: red blood cell [erythrocyte]; SMC: smooth muscle cell; TC: thrombocyte [platelet]). Concentration gradients of growth factors are required for cell chemotaxis.

Factor	Sources	Targets	Function
gCSF	Mφ, EC	Nφ	Hematopoiesis
gmCSF	Mo, Mφ, T Lφ, EC, FB	Mφ, Nφ, FB	Immunity
mCSF	T Lφ, Mφ	Mo, Mφ	Hematopoiesis
EGF	TC, Mo, Mφ,	FB, EC	Cell division, chemotaxis, angiogenesis
Epo	Kidney	RBC	Erythropoiesis
FGF1,2	Mo, EC, FB, Mφ	FB, EC, SMC	Cell division, chemotaxis, angiogenesis
IGF	Mφ, FB, TC	FB, EC, SMC	Cell division, collagen synthesis
PDGF	EC, SMC, Nφ, TC, Mφ, FB	Nφ, Mo, Mφ, FB, EC, SMC	Cell division, chemotaxis, angiogenesis, collagen synthesis,
TGFα,β	Mφ, T Lφ, EC, TC, FB	FB, Mφ, Mo, Lφ, EC, SMC	Chemotaxis, angiogenesis, matrix production, healing, inhibition of inflammation
TNF	Mφ, T Lφ, Nφ	Mφ, FB	Angiogenesis, inflammation
VEGF	Mφ, FB, SMC	EC	Angiogenesis

particular, cell growth is controlled by a balance between growth-promoting and -inhibiting factors.

Growth factors promote not only cell division, maturation, and functioning, but also tissue growth and remodeling. Growth factors act on [307]: (1) growth factor-producing cells (*intra- and autocrine* effects),[1] (2) neigh-

[1] Autocrine growth factor is involved in autostimulatory growth control, in which a cell secretes a factor that binds to its receptor, which is also expressed by the

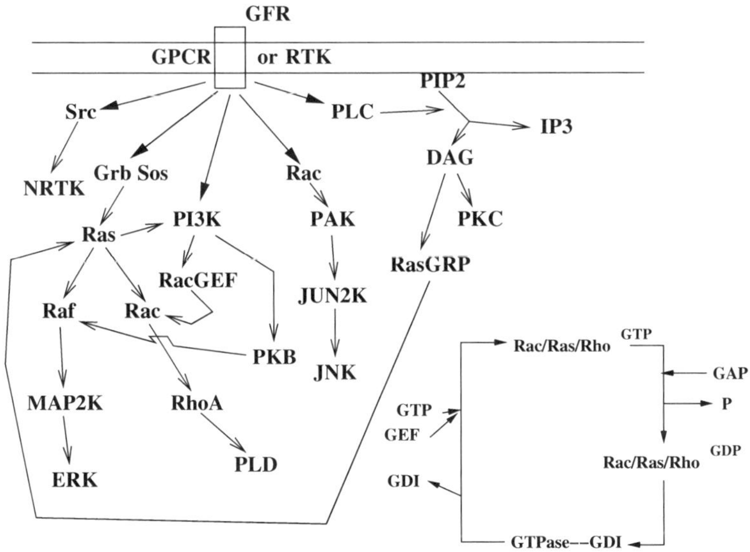

Figure 3.1. The growth factor receptor and biochemical cascades (Source: [306]). Growth factor-bound receptor (GFR) activates the adaptor GRB coupled to the guanine nucleotide (GDP-to-GTP)-exchange factor (GEF) SOS (GRB–SOS complex), and subsequently Ras GTPase, Raf kinase, and mitogen-activated protein kinase kinase (MAP2K) successively, and then extracellular regulated kinase (ERK) on the one hand (MAPK module), and small GTPases Rac and RhoA, and phospholipase-D (PLD) on the other hand. Ligand-bound receptors also activate among other pathways: (1) phosphatidylinositol 3-kinase (PI3K) and protein kinase-B (PKB), (2) Src kinase, (3) Rac GTPase and Jun N-terminal kinase (JNK), and (4) phospholipase-C (PLC). (**Lower right corner**) Activation–inactivation cycle of small, monomeric guanosine triphosphatases (GTPase; e.g., Rac, Ras, and Rho; $\text{GTPase}^{\text{GTP}}$, $\text{GTPase}^{\text{GDP}}$: active and inactive forms) by GEF and GTPase-activating protein (GAP) and sequestration by guanine nucleotide-dissociation inhibitors (GDI).

boring cells (*paracrine effect*),[2] (3) connecting cells (*juxtacrine effect*), and (4) distant cells after blood transport (*endocrine effect*).

The set of growth factors includes *cytokines* and *chemokines* (chemotactic cytokines). Cytokines,[3] or immunocytokines, are growth factors that modulate

---

producing cell. This autogenous loop creates a substance that acts back on the cells that produce it.

[2] Paracrine growth factor influences the growth and functional activities of surrounding cells expressing the corresponding receptor. Paracrine growth factors modulate the microenvironment. In particular, they are involved in angiogenesis, stroma formation, modulation of immune response, activation of proteolytic enzymes, etc.

[3] κυτος: hollow object [organ] and also [like Latin cutis] envelope [by extension cell]; κινεω: to move.

activities of immunocytes. These immunomodulators, or immunotransmitters, are secreted primarily by leukocytes, either by lymphocytes (*lymphokines*) or monocytes and macrophages (*monokines*). Hence, cytokines are separated from other regulatory peptides that modulate the proliferation and activities of non-immunocytes. However, some cytokines are produced by almost the entire cell spectrum.[4] Cytokines stimulate both the humoral (antibody) and cellular immune responses.

Among lymphokines, *interleukins* (IL) are growth factors targeted to hematopoietic cells. *Interferons* (Ifn) are glycoproteic cytokines produced by cells of the immune system in response to foreign agents (viruses, bacteria, parasites, tumor cells). immunocytes have reciprocal activations.[5]

Organ development is controlled by a set of growth factors and their regulators (e.g., Notch). Vascular endothelial growth factor-A is the most important agent required for vasculogenesis (migration and differentiation of angioblasts and formation of primitive vascular structures), as well as angiogenesis before and after birth and vessel remodeling (Vol. 5 – Chap. 10. Vasculature Growth). Growth factors yield cell response using several regulated mechanisms, the so-called signaling pathways (Vol. 3 – Chap. 1. Signal Transduction) that lead up to substance release as well as gene expression and production of alternatively spliced variants (Vol. 1 – Chap. 5. Protein Synthesis).

3.1 Connective Tissue Growth Factor

Connective tissue growth factor (CTGF)[6] is a member of the CCN gene family that is activated by several factors such as transforming growth factor-β.

The *CCN family* comprises 6 known members [308]: cysteine-rich protein-61 (CyR61 or CCN1), connective tissue growth factor (CTGF or CCN2), nephroblastoma overexpressed protein (NOv or CCN3), and Wnt-induced secreted protein-1 (WISP1 or CCN4), -2 (WISP2 or CCN5), and -3 (WISP3 or CCN6; Table 3.2).[7] These proteins regulate cell survival, differentiation, proliferation, adhesion, and migration, as well as extracellular matrix production. Therefore, members of the CCN family are involved in implantation,

---

[4] Heart expresses cytokines, particularly tumor-necrosis factors TNFα and -β, interleukin-1α and -1β, -2 to -6, and -10, as well as interferon-γ.

[5] T lymphocytes synthesize lymphokines (gmCSF, IL3, and Ifn) that stimulate macrophages. Afterward, macrophages produce monokines (IL1) that activate T lymphocytes.

[6] A.k.a. hypertrophic chondrocyte-specific protein HCS24, CCN family member CCN2, and insulin-like growth factor-binding protein IBP8 and IGFBP8.

[7] Alias CCN comes from the first 3 members of the family CyR61, CTGF, and NOv. These multimodular proteins contain an insulin-like growth factor (IGF)-binding (module 1), a von Willebrand type-C (module 2), a thrombospondin-1 domain (module 3), and a C-terminus (module 4) [308].

Table 3.2. Members of the CCN family (Source: [309]; CEF: chicken embryo fibroblast; CTGF: connective tissue growth factor; CTGFL: connective tissue growth factor-like protein; CyR: cysteine-rich protein; ELM: expressed low in metastasis; FISP: fibroblast-inducible secreted protein; HCS: hypertrophic chondrocyte-specific gene product; HICP: heparin-inducible CCN-like protein; IGFBP: insulin-like growth factor-binding protein; IGFBPRP: IGFBP-related protein; NOv: nephroblastoma overexpressed protein: WISP: Wnt-inducible signaling pathway protein).

CCN member	Other aliases
CCN1	CyR61, CEF10, IGFBP10, IGFBPRP4, CTGF2, angiopro
CCN2	CTGF, FISP12, IGFBP8, IGFBPRP2, HCS24, ecogenin
CCN3	NOv, IGFBP9, IGFBPRP3
CCN4	WISP1, ELM1
CCN5	WISP2, CTGFL, CTGF3, HICP, COP1
CCN6	WISP3

placentation, embryo- and fetogenesis, as well as wound healing and fibrosis. They bind to and activate integrins. At least in fibroblasts or endothelial cells, CCN family members CCN1 to CCN3 activate focal adhesion kinase and mitogen-activated protein kinase [308]. Both estrogen-inducible CCN1 and CCN2 regulate endothelial cell proliferation and angiogenesis [308].

The CTGF transcript is detected in fibroblasts, epi- and endothelial cells, vascular smooth muscle cells, and chondrocytes. Transforming growth factor-β, bone morphogenic protein-2, platelet-derived (PDGF), epidermal (EGF), and fibroblast (FGF) growth factor stimulate CTGF synthesis [310]. In addition, CTGF interacts synergistically with EGF, PDGF, IGF1, and FGF2.

Connective tissue growth factor promotes fibroblast proliferation, adhesion, and migration, as well as extracellular matrix formation. It intervenes in extracellular matrix remodeling during embryo- and fetogenesis, wound healing, etc. It is involved in diverse auto- and paracrine actions in several cell types, such as vascular endothelial and smooth muscle cells, as well as epithelial and neuronal cells and cells of supportive skeletal tissues [311]. In some circumstances, CTGF has antimitotic and apoptotic effects.

Connective tissue growth factor not only regulates cell proliferation and apoptosis, as well as angiogenesis and tissue fibrosis, but also tumor growth and metastasis. Sphingosine 1 phosphate isoform S1P2 (Sect. 3.17), but not S1P1, upregulates CTGF in a concentration- and time-dependent manner via small GTPase RhoA and kinases RoCK and Jun N-terminal kinase [312].

Connective tissue growth factor binds to transforming growth factor-β. It is able to suppress transcription of related SMAD7 factor (Sect. 3.8).

3.2 Superfamily of Epidermal Growth Factors

3.2.1 Epidermal Growth Factor

Epidermal growth factor (EGF) binds to the HER family of receptors (Vol. 3 – Chap. 8. Receptor Kinases) of responsive cells and primes phosphorylation of EGF receptor and other proteins. The kinase domain of EGF receptor is auto-inhibited at rest. Activation of EGFR needs formation of an asymmetrical dimer [313]. Epidermal growth factor promotes receptor dimerization and activates the intracellular Tyr kinase domain. Activated receptors phosphorylate each other on various interaction sites and liberate recruitment motifs for enzymes or adaptors. Epidermal growth factor fosters proliferation of cells, especially fibroblasts. Mutations of the Egfr gene can thus lead to cancer.

Epidermal growth factor generates an initial rapid (20-mn) wave of transcription of a small number of *immediate-early genes* (IEG), such as components of Activator protein-1, Fos and Jun, that encode transcription factors for signaling responses. Certain proteins, such as bimodal regulator of receptor Tyr kinase Sprouty2 and ErbB (HER) receptor feedback inhibitor ErRFI1[8] with a peak expression 60 to 120 mn after stimulated EGF receptor degradation, generate a refractory period by inhibiting EGF receptors to avoid repetitive stimulation. In addition to transcription factors encoded by immediate early genes, epidermal growth factor initiates a coordinated transcriptional program of microRNAs that are attenuators of growth factor signaling [314].

The coordinate expression of *delayed-early genes* (DEG) impedes the action of immediate early genes (Fig. 3.2). The delayed-early genes have a peak expression 40 to 240 mn after growth factor stimulation. Epidermal growth factor signaling encompasses genes that are coexpressed in feedbacks for signaling attenuation at specific nodes of the chemical reaction cascade. A node of the mitogen-activated protein kinase module of the EGF pathway is inhibited by dual-specificity MAPK phosphatases [315]. Other inhibitors also determine the activation duration. These inhibitory feedback loops are triggered by the signaling pathway itself to limit its activity duration. These inhibitors comprise transcription regulators and RNA-binding attenuators.[9]

3.2.2 EGF Superfamily Members

Epidermal growth factor (or β-urogastrone) is the founding member of the EGF superfamily. Members of this superfamily share similar structural

[8] A.k.a. mitogen-inducible gene 6 protein (MIG6).

[9] Transcription attenuator most often forms a complex with and attenuates the action of a transcription activator that is activated earlier. Krüppel-like factor KLF6 and avian musculoaponeurotic fibrosarcoma oncogene MAFF are activated by EGF [315].

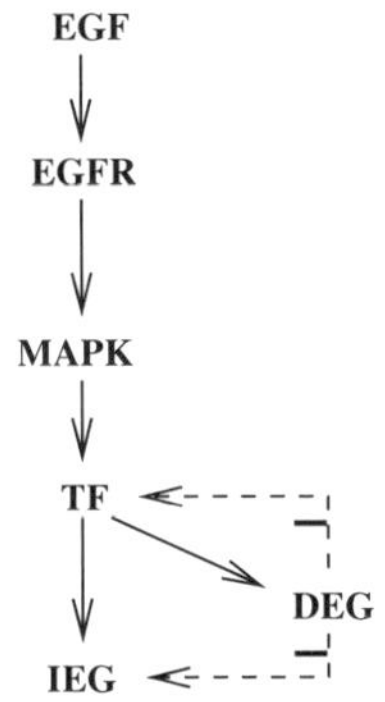

Figure 3.2. Negative feedback for signaling attenuation (Source: [315]). Signaling pathway triggered by EGF activates immediate-early genes (IEG; first post-stimulation wave of transcription) via mitogen-activated protein kinases (MAPK) and transcription factors (TF). Activated delayed-early genes (DEG; second post-stimulation wave of transcription) impedes the action of immediate-early genes.

and functional characteristics. The EGF superfamily also includes: heparin-binding EGF-like growth factor, transforming growth factor-α, amphiregulin, epiregulin, epigen, βcellulin, and neuregulin-1 to -4.

3.2.2.1 Heparin-Binding EGF-like Growth Factor

Heparin-binding EGF-like growth factor (HBEGF) intervenes in cardiogenesis and functioning, cardiac hypertrophy and wound healing. Heparin-binding EGF-like growth factor is synthesized as a membrane-anchored precursor (proHBEGF) that is cleaved to release a soluble HBEGF by specific metallopeptidases. In fact, proteolytic cleavage of proHBEGF yields N- and C-terminal fragments (HBEGF and HBEGFc) that both function as signaling molecules [316].

3.2.2.2 Transforming Growth Factor-α

Transforming growth factor-α (TGFα) is produced in macrophages, brain cells, and keratinocytes. It promotes epithelial development.

3.2.2.3 Amphiregulin

Amphiregulin (AReg) is an autocrine growth factor as well as a mitogen for astrocytes, Schwann cells, and fibroblasts.

3.2.2.4 Epiregulin

Epiregulin (EReg) is another autocrine growth factor for human keratinocytes.

Table 3.3. Examples of signaling pathways of epidermal growth factor (EGF) that suppress cell apoptosis and foster cell mobility. Neuraminidase-3 (Neu3), or membrane sialidase-3 (Sial3) that, in particular, localizes to the extracellular leaflet of the plasma membrane removes sialic acid from sialoglycoproteins and -lipids (with high specificity for gangliosides, i.e., sialoglycosphingolipid [composed of ceramide, oligosaccharide, and sialic acids]). It can intervene in EGF signaling, as it enhances Tyr phosphorylation of epidermal growth factor receptor (EGFR) in response to EGF.

Pathway	Effect
EGF–EGFR–Ras–ERK	Cell motility
EGF–EGFR–Ras–PI3K–PKB	Cell survival

3.2.2.5 Epigen

Epigen (Epgn)[10] is a widely expressed transmembrane glycoprotein that undergoes cleavage to release a soluble EGF-like domain-containing fragment. It can promote the proliferation of epithelial cells, hence favoring wound healing.

3.2.2.6 βCellulin

βCellulin (BtC) primes the phosphorylation of all the 3 EGFRs on endothelial cells (HER2–HER4) [317]. It provokes the phosphorylation of components of mitogen-activated protein kinase (MAPK) modules, such as extracellular signal-regulated kinases ERK1 and ERK2 and P38MAPK, and triggers the PI3K–PKB (phosphatidylinositol 3-kinase–protein inase-B) pathway (Table 3.3).

3.2.2.7 Neuregulins

Neuregulins (NRg1–NRg4) are encoded by 4 genes. Neuregulins are characterized by numerous alternatively spliced variants that explain structural and functional diversity. Neuregulins mediate intercellular interactions in the nervous system, heart, breast, and other organ systems. They contribute to the regulation of cell growth, proliferation, adhesion, migration, differentiation, and apoptosis.

Neuregulin signaling between apposed cells is bidirectional. Forward signaling from a NRg-producing cell to a NRg-responsive cell relies on receptor Tyr kinases of the HER family. Reverse signaling (backward or back signaling) occurs from HER-expressing cells to NRg-producing cells. Neuregulin then serves as receptor and HER as agonist. This bidirectional communication is

[10] A.k.a. epithelial mitogen homolog.

Table 3.4. Neuregulin-1 functions in the cardiovascular and ventilatory systems (Source: [319]).

Heart	Development of ventricular wall trabeculae, atrioventricukar septum, cardiac valves, and nodal tissue, growth and survival of cardiomyocytes
Blood vessels	Angiogenesis
Lung	Development of respiratory epithelium

illustrated by motor neurons that produce NRgs and inform HER-synthesizing Schwann cells and skeletal myocytes.

Neuregulins are ligands for HER3 and HER4 receptors. The HER2 receptor[11] is activated on neuregulin binding to HER3 and HER4 receptors. The HER2 receptor is required for neuregulin signaling [318]. Therefore, neuregulins bind to HER3, HER4, or both HER3 and HER4 that then form homo- or heterodimers often including HER2 (e.g., HER2–HER4 and HER3–HER4), with which neuregulins interact only after linking to HER3 or HER4 [319].

Neuregulin-1

Neuregulin-1 operates in the development of the nervous system and heart (Table 3.4). Like EGF, TGFα, and other ligands of receptors of the HER family, NRg1 is synthesized as proprotein. Neuregulin-1 acts in both juxta- and paracrine signaling. Most NRg1s are produced as transmembrane proteins. Nevertheless, after proteolytic cleavage and shedding or secretion (e.g., NRg1-2β3 isoform), NRg1s serve as diffusible messengers for short-distance intercellular communication. Adamlysins ADAM17[12] and ADAM19[13] can cleave plasmalemmal NRgs [319]. Soluble NRg1 fragment produced by shedding allows communication between endocardium and myocardium. As a paracrine factor released by microvascular endothelial cells, NRg1 ensures cardioprotection. Shedding of NRg1 can also produce an autocrine signal. Plasmatic NRg1β can serve as a biomarker in patients with heart failure, especially ischemic cardiomyopathy [320].

Neuregulin-1 is encoded by a single gene. However, 15 known isoforms arise due to differential splicing (NRg1-1–NRg1-6). Type-1, -2-, and -3 NRg1 splice variants are further subdivided in α and β subtypes. Type-1 and -2 NRg1s are sometimes referred to as immunoglobulin-like domain-containing

[11] A.k.a. metastatic lymph node gene 19 protein [MLN19], ErbB2, CD340, and Neu.

[12] A.k.a. tumor-necrosis factor-α-converting enzyme (TACE).

[13] A.k.a. meltrin-β and metallopeptidase and disintegrin dendritic antigen marker (MADDAM).

NRgs (IgNRg) and type-3 NRg1 as cysteine-rich domain-containing NRgs (CrdNRg).

Heregulin (HRg) is the type-1 NRg1;[14] *glial growth factor* (GGF) type-2 NRg1; and *sensory and motor neuron-derived factor* (SMDF) type-3 NRg1. The major NRg1 isoforms that act as glial growth factors are type-3 NRg1s (GGF1–GGF2), not type-2 NRg1, originally called glial growth factor [319]. Isoforms of NRg1 that differ in their N-terminus or EGF-like domain have distinct functions. Heregulins participate in synapse development.

Some NRg1 splice forms are produced as transmembrane precursors that are processed and released, whereas others are soluble. The sequences of NRg1-1β1a and NRg1-3β1a differ only in their N-terminus. Nonetheless, NRg1-1β1a is a transmembrane protein that can be liberated once cleaved, whereas cleavage of NRg1-3β1a creates a transmembrane N-terminal fragment [319]. Unlike NRg1-1β3, transmembrane type-1 NRgs (i.e., NRg1-1β1a, NRg1-1β2a, and NRg1-1β4a) are released.

Type-1 and -2 NRgs serve as paracrine factors, type-3 NRgs as juxtacrine messengers [319]. The retention of type-3 NRg1s in the plasma membrane confines the range of signaling. In addition, type-1 and -2 NRgs bind to heparin and other glycosaminoglycans (carbohydrate side chains of proteoglycans) in the extracellular matrix and on cell surfaces, but not type-3 NRg1s.

Paracrine signals transmitted by NRg1s from endocardium contribute to myocardial differentiation. During cardiogenesis, type-1 and low concentrations of type-3 NRgs are produced by the endocardium. On the other hand, type-2 NRgs are not expressed by the embryonic myocardium [319]. Neuregulin-1 functions in the heart beyond embryo- and fetogenesis. In adults, the endothelium of the cardiac microvasculature may be a source of paracrine NRg1 signals.

Intracellular or membrane processing of tissue-specific proheregulin isoforms (e.g., α- and βHRg) generate corresponding soluble forms. γ-Heregulin that also targets HER3 and HER4 receptors can function as an autocrine growth factor [318].

NRg1β-induced migration relies on ERK1 and ERK2, JNK, PI3K, PKB, SRC family kinases, and RoCK1 and RoCK2 [321]. On the other hand, NRg1α does not exert any effect on the migration of certain types of malignant peripheral nerve sheath tumor cells, but can inhibited the migration of other cell lines.

Neuregulin-2

Neuregulin-2 elicits growth and differentiation of epithelial, neuronal, and glial cells, among others. At the neuromuscular junction, NRg1 and NRg2 cause acetylcholine receptor transcription [322]. Like neuregulin-1, NRg2 as well as HER2 and HER4 receptors are required for cardiogenesis.

---

[14] A.k.a. Neu differentiation factor (NDF) and acetylcholine receptor-inducing activity protein (ARIA).

Neuregulin-2 activation of HER4 homodimers can elicit different patterns of receptor phosphorylation and signaling than those resulting from activation of HER4 homodimers in the same cell type by NRg1 [319]. Like NRg1, alternative transcripts encode distinct isoforms (NRg2α–NRg2β).

Neuregulin-3

Neuregulin-1, -2, and -3 are expressed in the mammalian nervous system. Neuregulin-3 binds to and activates HER4, but not HER2 and HER3 receptors. It acts in epidermal morphogenesis. An NRg3 isoform (human fetal brain neuregulin-3 [HFBNRg3]) is specific to human embryonic central nervous system. This type-1 glycosylated plasmalemmal protein is shed into the extracellular space, from which it activates HER4 [323].

Neuregulin-4

Neuregulin-4 activates HER1 receptor (EGFR). Various NRg4 splice variants exist (NRg4a1–NRg4a2, NRg4b1–NRg4b3). Type-A variant (NRg4a1) localizes to the membrane, whereas type-B variant (NRg4b1) lacks the transmembrane domain and resides in the cytosol [324]. NRg4 is involved in the differentiation of the somatostatin-expressing δ cells of pancreatic islets [319].

3.3 Superfamily of Fibroblast Growth Factors

Fibroblast growth factors (FGF) that are also characterized by Tyr kinase activity are particularly involved in embryo- and angiogenesis as well as wound healing. There are several kinds of fibroblast growth factors.[15] Twenty-two fibroblast growth factors encoded by 22 genes (FGF1–FGF14 and FGF16–FGF23, as FGF19 is a human ortholog of mouse FGF15) have been identified in humans. Many of the Fgf gene products also exist in multiple isoforms generated by alternative splicing of messenger RNA.

Fibroblast growth factors signal via their cognate receptor Tyr kinases encoded by 4 Fgfr genes (FGFR1–FGFR4; Vol. 3 – Chap. 8. Receptor Kinases). Except FGF11 to FGF15, FGF19, FGF21, and FGF23, all canonical FGFs activate FGFRs with different degrees of specificity [325]. Once secreted, canonical FGFs bind to FGFRs and induce their dimerization and phosphorylation of specific cytoplasmic Tyr residues.

Two factors — FGF1 and FGF2 — were originally isolated as growth factors for fibroblasts. Fibroblast growth factor FGF1[16] binds to FGFR1 to

[15] For example, FGF2, FGF4, FGF7 to FGF10, FGF17, and FGF18 are expressed in the limb bud and developing skeleton.

[16] A.k.a. acidic fibroblast growth factor (aFGF), FGFα, endothelium cell growth factor ECGFa, ECGFb, β-ECGF, and ECGFβ, as well as heparin-binding growth factor HBGF1.

FGFR4 receptors. Fibroblast growth factor FGF2[17] promotes endothelial cell proliferation and formation of tubular, endothelial structures. It binds to the receptor FGFR1 as well as some heparan sulfate proteoglycans, such as glypicans Gpc3 and Gpc4 and syndecans Sdc2 to Sdc4 [326].

The EGF superfamily encompasses other proteins that are similar to FGF1 or -2, such as heparin-binding EGF-like growth factor (HBEGF). The latter is a member of the EGF superfamily that is synthesized as a membrane-anchored precursor (proHBEGF). It is cleaved to by specific metallopeptidases that thus release a soluble HBEGF protein (N-terminal region [$HBEGF_N$]) and a C-terminal fragment ($HBEGF_C$). Both fragments function as signaling molecules.

The mammalian FGF superfamily can be decomposed into 7 families [325] (Table 3.5): (1) *canonical* FGF families (cFGF) that encompass the FGF1 (with FGF1, -2, and -5), FGF3 (with FGF3, -4, and -6), FGF7 (with FGF7, -10, and -22),[18] FGF8 (with FGF8, -17, and -18), and FGF9 (with FGF9, -16, and -20) subfamilies; (2) *intracellular* FGF11 family (iFGF; with FGF11, -12, -13, and -14);[19] and (3) *hormone-like* FGF19 family (hFGF; with FGF19, -21, and -23).[20]

3.3.1 FGF Signaling

Seven FGFR isoforms (FGFR1b, FGFR1c, FGFR2b, FGFR2c, FGFR3b, FGFR3c, and FGFR4) result from 4 Fgfr genes. The isoforms transcribed from the genes Fgfr1, Fgfr2, and Fgfr3 derive from alternative mRNA splicing that specifies the C-terminus structure, as it utilizes one of 2 unique exons in 2 different versions (FGFR*i*b and FGFR*i*c, $i = 1, 2, 3$) of Ig-like domain III. The FGFR*i*a splice form encodes a secreted (extracellular) FGFR form (sFGFR) without known signaling capability. Yet, sFGFR1 oligomerizes upon ligand binding. Secreted sFGFR1 preferentially binds FGF2 over FGF1 [328].

The FGFR*i*b and FGFR*i*c splice forms are used according to cell lineage. The b isoform is restricted to epithelial lineages and c isoform is preferentially expressed in mesenchymal lineages [329]. The FGF7 subfamily is expressed in mesenchymal tissues (connective tissue, bone, cartilage, and lymphatic and blood circulatory networks). Its members have the greatest affinity for

[17] A.k.a. basic fibroblast growth factor (bFGF), FGFβ, and HBGF2.

[18] FGF7 and FGF10 are also called *keratinocyte growth factors* KGF1 and KGF2.

[19] Originally the family of FGF homologous factors (FHF1–FHF4).

[20] Another decomposition of the FGF superfamily exists at least in mice [327] as well as Fig.1 (Evolutionary relationships within the mouse Fgf gene family) of Ref. [325], with the FGF1 family (FGF1 and FGF2); FGF4 family (FGF4, FGF5, and FGF6); FGF7 family (FGF3, FGF7, FGF10, and FGF22); FGF8 family (FGF8, FGF17, and FGF18); FGF9 family (FGF9, FGF16, and FGF20); FGF11 family (FGF11, FGF12, FGF13, and FGF14); and FGF19 family (FGF15, FGF21, and FGF23).

Table 3.5. Families of FGFs and their functions (Source: [325]; FGF*i*F [i: integer]: FGF*i* family). Redundant FGF functions exist within and among FGF families of the FGF superfamily. The FGF1 subfamily of the canonical FGF family contains the prototype FGFs (FGF1 and FGF2). Fibroblast growth factors participate in cell growth and differentiation, especially in organ development, tissue maintenance, and wound repair. Effective activation of extracellular FGF signaling (except the FGF11 family of intracellular FGFs) requires FGF binding to FGF receptor in association with extracellular matrix components such as heparan sulfate proteoglycans. After endocytosis, some FGF–FGFR complexes translocate to the nucleus where they influence gene expression. In addition, some intracellular FGF isoforms may function as nuclear signaling factors.

Family (subfamily)	Members	Function
		Conventional FGFs
FGF1F	FGF1/2/5	Cardiovascular, skeletal, and neuronal development
FGF3F	FGF3/4/6	Development of the central nervous system
FGF7F	FGF7/10/22	Limb development, muscle regeneration
		Branching morphogenesis (lung, pancreas, etc.)
FGF8F	FGF8/17/18	Central nervous system morphogenesis
		Cerebellar, skeletal, and pulmonary development
FGF9F	FGF9/16/20	Development of heart, vasculature, and lung
		Intracellular FGFs
FGF11F	FGF11/12/13/14	Neurological and neuromuscular function
		Hormone-like FGFs
FGF19F	FGF19/21/23	Cardiogenesis
		Phosphate, vitamin-D, and bile-acid metabolism
		Energy homeostasis

FGFR2b (Table 3.6). The FGF8 subfamily is observed in epithelia. Its members preferentially activate FGFRc splice forms.[21] On the other hand, FGF1 prototype of the FGF superfamily can activate all the FGFR receptors.

All the members (FGF9, -16, and -20) of the FGF9 subfamily activate FGFR4, FGFR3b, and all the FGFRc isoforms with higher affinity than that for FGFR3b [329]. The FGF20 factor has a slight affinity for FGFR2b and no activity for FGFR1b. Isoforms FGF9 and FGF16 do not show activity toward FGFR1b or -2b.

Members (FGF19, -21, and -23) of the FGF19 family possess a low affinity compared to that of classical members of the FGF superfamily. With

[21] All the members (FGF7, -10, and -22) of the FGF7 subfamily strongly activate FGFR2b. Isoforms FGF10 and FGF22 show weak affinity for FGFR1b. All the members (FGF8, -17, and -18) of the FGF8 subfamily specifically activate FGFR1c, -2c, and -3c, in addition to FGFR4 [329].

Table 3.6. Members of the FGF superfamily and receptor specificity (Source: [329]; NA: not applicable).

Family	Members	Target FGFR
FGF1F	FGF1	All FGFRs
	FGF2	FGFR1c/3c > FGFR1b/2c
FGF4F	FGF4/5/6	FGFR1c/2c > FGFR3c
FGF7F	FGF3/7/10/22	FGFR2b > FGFR1b
FGF8F	FGF8/17/18	FGFR3c > FGFR2c > FGFR1c ≫ FGFR3b
FGF9F	FGF9/16/20	FGFR3c > FGFR2c > FGFR1c/3b
FGF11F	FGF11/12/13/14	NA
FGF19F	FGF19/21/23	FGFR1c/2c/3c (weak)

concentrations high enough (concentration 3–800 nm, heparin concentration >10 μg/ml), they target FGFR1c, -2c, -3c, and -4 (but not FGFR1b, -2b, and -3b) [329].

Extracellular FGF monomers have a reduced heparin-binding affinity. Homodimerization of FGF regulates receptor binding, at least for members of the FGF9 subfamily (FGF9 and FGF20), and concentration gradients in the extracellular matrix due to heparan sulfate-dependent diffusion [330]. Dimerization of FGF thus operates as an autoregulatory mechanism for growth factor activity.

Intracellular FGFs interact with intracellular domains of voltage-gated sodium channels and mitogen-activated protein kinase scaffold, the MAPK8-interacting protein MAPK8IP2 [325].[22]

Members of the canonical and hormone-like families are released from cells. FGF3 to FGF8, FGF10, FGF17 to FGF19, and FGF21 to FGF23 are secreted proteins with cleavable N-terminal signal peptides. On the other hand, FGF9, FGF16, and FGF20 contain uncleavable bipartite signal sequences [325]. Isoforms FGF1 and FGF2 might be released by exocytosis.

All canonical FGFs possess binding sites for acidic glycosaminoglycans, such as heparin and heparan sulfate. In the presence of heparan sulfate, FGF binding to FGFR is stabilized, once (2:2:2) FGF–FGFR–heparan sulfate hexamer is formed [331]. Acidic glycosaminoglycans in the form of heparan sulfate proteoglycans retain secreted FGFs in the vicinity of FGF-producing cells. They can then act as paracrine regulators. On the other hand, hFGFs have low affinity for heparin and operate as endocrine regulators.

[22] A.k.a. Jun N-terminal kinase-interacting protein JIP2 and islet brain IB2 (Vol. 4 – Chap. 5. Mitogen-Activated Protein Kinase Modules).

3.3.2 Hormone-like FGF19 Family

The fibroblast growth factor-19 family that encompasses FGF19 (human ortholog of murine FGF15 isotype), FGF21, and FGF23 regulates glucose metabolism, bile acid synthesis, mineral ion homeostasis, and phosphate and vitamin-D metabolisms.

Factors FGF19 and FGF21 act as endocrine hormones. Whereas FGF19 has both metabolic and proliferative effects, FGF21 has only metabolic effects because of structural differences that determine distinct receptor interactions [332]. The FGF19 factor reduces the plasma glucose concentration and heightens insulin sensitivity. The FGF21 factor is predominantly synthesized in liver, where it promotes the production of GluT1 glucose transporter. Factors FGF19 and FGF23 are secreted from ileal enterocytes and bone cells, respectively, to circulate in the blood stream.

Fibroblast growth factor-21 acts as a liver-derived endocrine factor that stimulates glucose uptake in adipocytes after binding the complex formed by FGF receptors and β-Klotho[23] and activating the mitogen-activated protein kinase cascade [333]. Klotho confers tissue-specific activity to FGF factor. In addition, FGF21 adapts cells to starvation, as it stimulates fatty acid release from adipocytes and promotes their conversion in hepatocytes to ketones that can be used as an energy source when glucids are scarce. However, unlike in vitro observations, β-Klotho is not essential for FGF21 signaling in adipocytes in vivo, because FGF21 cues are transduced in the absence of β-Klotho, but relies on a cofactor [334].

On the other hand, the mineral homeostasis is regulated by FGF23 and the feedback control of α-Klotho and calcitriol[24] Similarly, the bile acid–cholesterol metabolism depends on FGF19 and feedback from β-Klotho and bile acids [334]. Tissue-specific signaling from FGF19 family members relies on interactions between FGF19, FGF21, and FGF23 with β-Klotho cofactor and α-Klotho, respectively.[25]

3.3.3 FGF and Control of Cilium Length

Cilia contribute to intercellular signaling. Specialized groups of cilia are involved in embryogenesis. Monocilia (organelles of almost all cell types) are

[23] The β-Klotho cofactor is predominantly expressed in liver, pancreas, and adipose tissue.

[24] A.k.a. (1,25)-dihydroxycholecalciferol and (1,25)-dihydroxyvitamin-D3 (alias $(1{,}25)(OH)_2D$).

[25] In kidneys, α-Klotho interacts with FGF23 to maintain plasma calcitriol levels, as it regulates 25-hydroxyvitamin-D 1α-hydroxylase (CyP27b1) and 24-hydroxylase (CyP24) [334]. In distal convoluted tubules and choroid plexi as well as parathyroid glands during secretion of parathyroid hormones, α-Klotho binds to Na^+–K^+ ATPase and FGF23 for transepithelial Ca^{++} transport [334]. Therefore, the Na^+–K^+ ATPase–FGF23–α-Klotho complex contributes to the regulation of calcium and phosphate concentrations.

sites of signal reception that can modulate Hedgehog morphogen, platelet-derived growth factor, and Wnt signaling. Conversely, ciliogenesis that entails ciliogenic transcription factors FoxJ1 and Rfx2 is a regulated process. Fibroblast growth factors FGF8 and FGF24 and their FGFR1 receptors control cilium length [337].

3.3.4 FGF and Synaptogenesis

Fibroblast growth factors FGF7 and FGF22 are expressed by CA3 pyramidal neurons in the hippocampus. Factors FGF22 and FGF7 promote the formation of glutamate-mediated excitatory and GABA-mediated inhibitory synapses on dendrites of CA3 pyramidal neurons via FGFR1 and FGFR2 receptor, respectively [335]. These postsynaptic neuron-derived presynaptic organizers participate in the local differentiation of axons into functional presynaptic terminals. This presynaptic differentiation includes the clustering of synaptic vesicles, formation of active zones, cytoskeletal restructuring and assembly of vesicle recycling machinery.

Hence, fibroblast growth factors FGF7 and FGF22 work in synergy with other synaptogenic molecules such as neuroligins, ephrins, brain-derived neurotrophic factor, Wnts, cell adhesion molecule-1 (CAdM1),[26] netrin-G ligands, thrombospondins, signal regulatory proteins (SiRP),[27] leucine-rich repeat transmembrane neuronal proteins LRRTM1, and neuronal PAS domain protein NPAS4[28] as well as glial cells to coordinate the formation of appropriate synapses.

3.4 Hepatocyte Growth Factors

Hepatocyte growth factor (HGF)[29] contributes to tissue morphogenesis, angiogenesis, wound repair, and tissue regeneration. It belongs to the family of "serine peptidase" growth factors, but does not have peptidase activity as key catalytic residues are missing. It is secreted by mesenchymal cells. It targets primarily epithelial and endothelial cells, but also acts on hematopoietic progenitor cells.

---

[26] A.k.a. synaptic cell adhesion molecule SynCAM1, nectin-like protein NecL2, and immunoglobulin superfamily member IGSF4.

[27] The family of signal regulatory proteins comprises at least 15 members [336]. Signal regulatory protein-α is also called Tyr-protein phosphatase non-receptor type substrate PTPNS1. The transmembrane polypeptide SiRPα1 is a substrate of activated RTKs that represses cell responses caused by growth factors and insulin [336].

[28] A.k.a. PAS domain-containing protein PASD10. It is selectively expressed in the nervous system. It activates the production of the dendritic-cytoskeleton modulator at synapses: developmentally regulated brain protein (drebrin).

[29] A.k.a. hepapoietin-A and scattering factor.

Protein HGF possesses an α chain that contains its N-terminus and 4 kringle domains and a β chain with a serine peptidase domain [338]. It is synthesized as a single inactive polypeptide proHGF, although this precursor can bind hepatocyte growth factor receptor (HGFR). ProHGF cleavage (Arg494) by serine peptidases into a 69-kDa α chain and 34-kDa β chain leads to its active form. A disulfide bond between the α and β chains produces the active heterodimer. Mastocyte chymase, neutrophil elastase, and plasma kallikrein also cleave HGF (Cys487) and generate a free α chain, a competitive inhibitor of HGF.

It has 2 natural splice variants — NK1 and NK2 — that contain the N-terminus and the first kringle (K1) or the first 2 kringle domains of HGF [338]. The splice variant NK1, an agonist of HGFR, forms an NK1 homodimer. Heparan sulfate is necessary for its full activity. On the other hand, NK2, a monomer and HGFR antagonist that possesses the N-terminus and first 2 kringle domains, impedes HGF activity.

The HGFR receptor[30] dimerizes upon HGF binding and transmits signals via the MAPK modules and PI3K–PKB pathways to prime cell survival, proliferation, and motility.

Lymphatic endothelial cells express higher levels of hepatocyte growth factor receptor than blood vascular endothelial cells [339]. The HGF factor promotes the proliferation of lymphatic endothelial cells, their migration (which is partially mediated via α_9 integrin), and the formation of lymphatic vessels.

3.5 Insulin

Insulin is a major anabolic hormone. Pancreatic β cells secrete insulin in response to elevated glucose level in the plasma. Insulin promotes glucose uptake by all cell types, particularly adipocytes and myocytes, and prevents glycogenolysis and gluconeogenesis in the liver (Table 3.7). Insulin regulates glucose homeostasis by activating phosphoinositide 3-kinase and increasing glucose uptake rate, especially into myocytes and adipocytes, using the GluT4 transporter. Insulin also stimulates the CAP–CBL axis via the initiator SH2B2 that recruits both CBL and CBL-associated protein (CAP) to the insulin receptor (IR). Subsequent CBL phosphorylation dissociates the CAP–CBL complex from IR. This complex then migrates to flotillin.[31] The resulting recruitment of the CRK–RapGEF1 complex leads to the activation of the small GTPase RhoQ. Activated RhoQ causes actin remodeling and enables GluT4 to dock to the plasma membrane. Conversely, a low glucose concentration leads to a low insulin level combined with an elevated level in antagonist hormones

[30] A.k.a. mesenchymal–epithelial transition factor (MET or cMET; Vol. 3 – Chap. 8. Receptor Kinases).

[31] Flotillins are membrane nanodomain (raft and caveolae)-associated, integral membrane proteins.

Table 3.7. Insulin signaling, glucose input, and glucose transporter GluT4 mobilization (Sources: [340, 341]; CAP: CBL-associated protein; CBL: Casitas B-lineage lymphoma adaptor and Ub ligase; CRK: v-Crk sarcoma virus CT10 regulator of kinase homolog (adaptor and proto-oncogene product); G6P: glucose 6-phosphate; RapGEF: Rap guanine nucleotide-exchange factor; SH2B2: SH2B adaptor protein-2 [a.k.a. adaptor protein with pleckstrin homology (PH) and Src homology (SH2) domains (APS)]; SNAP: soluble ^{N}ethylmaleimide-sensitive factor attachment protein; SNARE: SNAP receptor; Stx: syntaxin (SNARE^{Q}); StxBP: syntaxin-binding protein; VAMP2: vesicle-associated membrane protein-2 [vSNARE; a.k.a. synaptobrevin-2]). Nutrient influx and resulting insulin secretion prevents hepatic neoglucogenesis and promotes the utilization and storage of glucose (glycogeno- and lipogenesis). The PI3K and RhoQ pathways are triggered by insulin to translocate Glut4 to the plasma membrane owing to actin–myosin-1C filaments, microtubules, and intermediate filaments. In addition, plasmalemmal atypical protein kinase-Cζ, a component of the PI3K pathway, phosphorylates RhoQ. The transcription factor FoxO1 contributes to the metabolic adaptation to fasting. It integrates and controls metabolic, survival, and mitogenic pathways. In the hepatocyte, it coordinates all the metabolic pathways that maintain a constant glycemia during fasting. It favors neoglucogenesis and prevents glycolysis and lipogenesis. It activates some members of the mitogen-activated protein kinase family, such as ERK1, ERK2, JNK1, JNK2, and P38MAPK. The last-mentioned fosters the formation of the TORC2 complex that phosphorylates PKB.

Glucose metabolism and nutrient input	
Feeding	Fasting
Glucose–G6P (phosphorylation)	G6P–glucose (dephosphorylation)
G6P–glycogen (glycogenogenesis)	Pyruvate–G6P (neoglucogenesis)
G6P–pyruvate (glycolysis)	
Pyruvate–lipid (lipogenesis)	
Signaling axes	
Insulin–IR–IRS1/2/3/4–PI3K–PDK1/2–PKB–Rab8/10/14	
Insulin–IR–IRS1/2/3/4–PI3K–PDK1/2–PKB–Fox0	
Insulin–IR–IRS1/2/3/4–PI3K–PDK1/2–PKC–RhoQ	
SH2B2–CAP–CBL–flotillin–CRK–RapGEF1–RhoQ	
Glut4 transfer	
SNAREs: Stx4–SNAP23–StxBP1–StxBP4–VAMP2	

(glucagon, adrenaline, and corticosteroids) and favors glucose production in the liver.

The pancreas has pancreatic islets that release glucagon and insulin. Insulin and glucagon produced by the pancreatic α cells and secreted in response to low glycemia have opposite effects on hepatic control (glucose storage and delivery) of glycemia. Insulin increases the synthesis of glycogen and fatty acids, as well as amino acid and potassium uptake, and decreases proteinolysis, lipolysis, and gluconeogenesis. Glucagon GPCR transmits signal destined for carbohydrate metabolism in the liver and insulin release from the pancreatic β cells.

Insulin receptor (Vol. 3 – Chap. 8. Receptor Kinases) is composed of 2 extracellular insulin-binding α and 2 transmembrane β subunits. β Subunits have cytoplasmic ATP-binding and Tyr kinase domains. Insulin binding induces autophosphorylation of β subunits, activating the receptor catalytic activity.

Insulin signaling targets many mediators, such as glucose transporter-4 (Vol. 3 – Chap. 4. Membrane Compound Carriers), insulin-receptor substrate proteins, CCAAT-enhancer-binding protein-α, and peroxisome proliferator-activated receptor-γ (NR1c3; Vol. 3 – Chap. 6. Receptors), as well as long-chain fatty acid acylCoA synthase.

In adipocytes, insulin is the main regulator of hormone sensitive lipase that is the rate-controlling enzyme for triglyceride hydrolysis. In response to insulin, adipocytes also secrete several small soluble proteins, such as adiponectin, adipsin, and leptin.

In any cell, binding of insulin to receptors causes fusion of cytoplasmic vesicles that sequester glucose transporters GluT4 with the plasma membrane and insertion of glucose transporters into the membrane. In the absence of insulin, glucose transporters are internalized. Certain substances can alter conformation of the cytoplasmic kinase domain or bind to modulator-binding sites of the insulin receptor.

Insulin binds to and stimulates insulin receptor Tyr kinase that activates phosphatidylinositol 3-kinase and protein kinase-B to mediate most effects. Upon stimulation by insulin, insulin receptor recruits and phosphorylates *insulin-receptor substrate proteins* that excite the PI3K–PKB pathway. Activated PKB phosphorylates effector kinases and transcription factors.

A second, PI3K-independent insulin pathway exists that involves a complex made of insulin receptor, scaffold protein β-arrestin-2 (Vol. 1 – Chap. 9. Intracellular Transport),[32] Src kinase, and protein kinase-B [342]. Insulin favors interaction between β-arrestin-2, protein kinase-B, and Src kinase with activated insulin receptor. The absence of β-arrestin-2 indeed attenuates glucose metabolism and insulin sensitivity on the one hand and augments hepatic glucose production, although PI3K-dependent insulin signaling remains

[32] β-Arrestins regulate the activity of G-protein-coupled receptors (Vol. 3 – Chap. 7. G-Protein-Coupled Receptors), as they link activated receptors with accessory and effector proteins to determine the specificity and efficiency of insulin signals.

efficient. β-Arrestin-2 loss reduces insulin-stimulated Src binding and Tyr phosphorylation of protein kinase-B.

Type-1 diabetes is caused by loss in insulin-producing β cells of the pancreas by autoimmune destruction. Chronic hyperglycemia causes endothelial dysfunction and vascular damages. On the other hand, insulin resistance (Vol. 6 – Chap. 7. Vascular Diseases) that precedes the onset of and characterizes type-2 diabetes is a defect of insulin stimulation of its receptor. Insulin signaling is disturbed because of protein modifications such as phosphorylation of insulin-receptor substrate proteins. tumor-necrosis factor-α, interleukin-6, and free fatty acids that are secreted at high quantities by enlarged adipocytes intervene in the development of insulin resistance. These substance cause Ser phosphorylation of insulin-receptor substrate-1 and -2, which reduces their capacity to be phosphorylated by insulin receptor.

Moreover, in obesity, activated Jun N-terminal kinases (Vol. 4 – Chap. 5. Mitogen-Activated Protein Kinase Modules) mediate Ser phosphorylation of insulin-receptor substrate protein IRS1. β-Arrestin-2 level is reduced in a mouse model of type-2 diabetes [342].

Caveolin-1 stabilizes the insulin receptor. MicroRNA-103 and -107 target the CAV1 gene and contribute to insulin resistance [343].[33]

3.6 Insulin-like Growth Factors

Insulin-like growth factor-1[34] (IGF1) can be produced in response to growth hormone. It acts as a growth and differentiation factor. The cellular response to IGF1 depends on the differentiation state, plasmalemmal receptor density, and cell environment. Insulin-like growth factor-1 is synthesized in numerous cell types. Its production varies according to the cell type.

The cardiac synthesis of IGF1 is higher in athletes than sedentary control subjects, whereas the production of endothelin-1 and angiotensin-2 does not change [344]. Heterotetramer IGFR1 is a receptor Tyr kinase. Insulin-like growth factor-1 acts exclusively via this class-2 receptor Tyr kinase (Vol. 3 – Chap. 8. Receptor Kinases).

On the other hand, IGF2 is almost exclusively expressed in embryonic and neonatal tissues. Whereas IGF1 operates during both pre- and postnatal development, IGF2 intervenes only during embryo/fetogenesis [345]. The expression of components of the insulin-like growth factor signaling (IGF1 and IGF2, IGF1R and IGF2R receptors, and soluble IGF-binding proteins [IGFBP1–IGFBP7] and IGF-binding protein-like molecule IGFBPL1) is ubiquitous

[33] MicroRNA-103 and -107 are upregulated in obese mice. Repression of miR103 and miR107 improves insulin-stimulated glucose uptake. Conversely, upregulation of miR103 and miR107 impairs glucose homeostasis.

[34] A.k.a. somatomedin-C.

during intra-uterine and postnatal development.[35] Insulin-like growth factor-2 at relatively low concentrations mainly binds to type-2 IGF receptor [346]. Type-2 IGFR is coupled to calcium gating. However, IGF2 can also operate via type-1 IGF receptor.

Mitogenic IGF2 targets somatic cells in various tissues. In the brain, it is highly expressed in the hippocampus. Memory consolidation involves synthesis of proteins, especially the transcription factors of the families of cAMP response element-binding proteins and CCAAT enhancer-binding proteins.[36] In rats, IGF2 enhances memory during a specific time window after learning (sensitive period of memory consolidation) [347].[37]

Insulin-like growth factor-binding proteins (IGFBP) bind to and modulate the activity of insulin-like growth factors. Insulin-like growth factor binders include 6 members (IGFBP1–IGFBP6; Table 3.8).). In addition, IGFBPs also have IGF-independent activities. Proteins IGFBP1, IGFBP2, and IGFBP6 are modest Wnt inhibitors, but IGFBP4 is the most potent repressor of the canonical Wnt pathway (Vol. 3 – Chap. 10. Morphogen Receptors). Proteins IGFBP3 and IGFBP5 do not intervene in Wnt–β-catenin signaling. Protein IGFBP4 binds and inhibits components of the canonical Wnt signaling pathway to promote cardiomyocyte differentiation, independently of IGF1 and IGF2.[38]

---

[35] Insulin-like growth factor receptor IGFR2 is also called mannose 6-phosphate receptor (M6PR). It binds with high affinity to IGF2, but not IGF1. It lacks the major intracellular domains used in signal transduction. Its primary function is the inhibition of the action of IGF2 in utero. It controls the interstitial IGF2 concentration, as it fosters its endocytosis and lysosomal degradation [345].

[36] The transcription factor C/EBPβ is upregulated for more than 28 h after training [347].

[37] The concentrations of IGF2 messenger RNA and protein increase between 20 and 36 h after training, but neither immediately, nor 3 d after training. Enhancement of IGF2-dependent memory requires cytoskeletal-associated proteins and glycogen-synthase kinase GSK3 [347]. Moreover, IGF2 increases the expression of the α-amino 3-hydroxy 5-methyl 4-isoxasolepropionic acid (AMPA)-type ionotropic glutamate receptor GluR1 subunit. In the hippocampus, IGF2 promotes a long-term potentiation after weak synaptic stimulation. The effect of the memory enhancer IGF2 is selectively mediated by IGF2R (mannose 6-phosphate receptor), but not IGFR1 receptor.

[38] Protein IGFBP4 binds to the extracellular domain of Wnt receptor Frizzled Fz8 and coreceptor low-density lipoprotein receptor-related protein LRP6 [348]. It is expressed in cells adjacent to cardiomyocyte progenitors (paracrine regulation). It competes with Wnt3a to bind to Frizzled-8 and impedes the Fz8–LRP6–Wnt3a–β-catenin signaling. Proteins IGFBP1, -2, and -6 are also able to bind LRP6 and Frizzled-8, but they are only slightly expressed in heart.

Table 3.8. Insulin-like growth factor-binding proteins (CCN: cysteine-rich CYR61, connective tissue growth factor (CTGF), and nephroblastoma overexpressed (NOv) family; PP: placental protein).

Member	Other aliases
IGFBP1	AFBP, IBP1, IGFBP25, PP12
IGFBP2	IBP2, IGFBP53
IGFBP3	BP53, IBP3
IGFBP4	Band-3A, BP4, Deb2, HT29-IGFBP, IBP4
IGFBP5	IBP5
IGFBP6	IBP6
IGFBP7	AGM, FSTL2, IBP7, IGFBPrP1 (rat), MAC25
IGFBP8	CCN2, CTGF, HCS24, IBP8, NOv2
IGFBP9	CCN3, IBP9, NOv
IGFBP10	CCN1, CYR61, CTGF2, GIG1, IBP10

Table 3.9. Members of the family of platelet-derived growth factors (PDGF) and their receptors (PDGFR). Protomers of PDGF ligands and PDGFR receptors constitute homodimers and some can form heterodimers.

Subunit	Dimers
	PDGF
PDGFa	PDGFaa, PDGFab
PDGFb	PDGFbb, PDGFab
PDGFc	PDGFcc
PDGFd	PDGFdd
	PDGFR
PDGFRα	PDGFRαα, PDGFRαβ
PDGFRβ	PDGFRββ, PDGFRαβ

3.7 Platelet-Derived Growth Factors

The family of platelet-derived growth factors (PDGF) comprises 4 protomers (PDGFa–PDGFd) that homo- (mainly) and heterodimerize (Table 3.9). Platelet-derived growth factors are potent activators of the PI3K pathway. They bind to 2 plasmalemmal receptor Tyr kinases, PDGFRα and PDGFRβ receptors that also homo- and heterodimerize upon PDGF binding.

Platelet-derived growth factor consists of different combinations of 2 subunits among 4 protomers (encoded by the PDGFA to PDGFD genes) to form homodimers (PDGFaa, PDGFbb, PDGFcc, and PDGFdd) as well as a single heterodimer (PDGFab). The PDGFaa factor specifically interacts with PDGFRαα homodimer, whereas PDGFbb links to both homo- and heterodimers formed by PDGFRα and PDGFRβ. The isoform PDGFcc associates to PDGFRαα and PDGFRαβ, whereas PDGFdd activates PDGFRαβ and PDGFRββ. Subunits PDGFRα and PDGFRβ are predominantly expressed on vascular endothelial and smooth muscle cells, respectively.[39] Binding of PDGF to its cognate receptors primes signaling cascades that initiate proliferation, migration, and differentiation of various cell types, such as fibroblasts and smooth muscle cells.

A combination of FGF2 and vascular stabilizer PDGFbb that target mainly endothelial cells and vascular mural cells (pericytes and smooth muscle cells), respectively, promotes collateral growth and stabilization [349].[40] Both PDGFaa and PDGFbb are able to synergistically stimulate angiogenesis with FGF2, but only PDGFbb can stabilize in the presence of FGF2. The PDGFbb receptor actually needs to cooperatively work with FGF2 for vessel stability. PDGFRα is mainly involved in angiogenesis, whereas PDGFRβ mediates vessel stability.

Angiogenesis requires both migration and proliferation of endothelial cells and coverage of vascular sprouts by vascular smooth muscle cells or pericytes for vessel stabilization via vascular endothelial and platelet-derived growth factors, respectively. Growth factor PDGF also induces neovascularization by stimulating vascular smooth muscle cells and pericytes to release pro-angiogenic mediators.

3.8 Transforming Growth Factors and Related Proteins

Transforming growth factors comprise 2 classes of polypeptidic growth factors, TGFα and TGFβ, that act via different receptors. The former induces epithelial development, whereas the latter operates in cell growth and differentiation, embryogenesis, tissue regeneration, and regulation of the immune system.

3.8.1 Transforming Growth Factor-α

Transforming growth factor-α (TGFα), secreted by activated macrophages and platelets, as well as brain cells and keratinocytes, binds to EGF receptor (EGFR) and its own receptor.

3.8.2 Transforming Growth Factor-β Superfamily

In addition to TGFβ isotypes, the TGFβ superfamily includes (Tables 3.10 and 3.11): (1) dimeric *activins*, (2) dimeric *inhibins*, (3) *Nodal*, (4) *anti-*

---

[39] PDGFRβ and -α are also expressed in endothelial cells and smooth muscle cells, respectively.

[40] Combination of PDGFbb and FGF2, but not PDGFbb and VEGF or FGF2 and VEGF, is able to induce angiogenesis and arteriogenesis.

Table 3.10. Members of the transforming growth factor-β (TGFβ) superfamily (AMH: anti-Müllerian hormone; BMP: bone morphogenetic protein; GDF: growth and differentiation factor). Members of the TGFβ superfamily specify the anterior–posterior and dorsal–ventral axes, endoderm, mesoderm and ectoderm, left–right asymmetry and organ development. Subtypes TGFβ1, TGFβ2, and TGFβ3 participate not only in the organogenesis, but also wound healing and immune response, as well as in the regulation of tumor cell fate.

Class	Components
TGFβ	TGFβ1–TGFβ3 and their isoforms
AMH	(dimeric glycoprotein encoded by the Amh gene)
Activin	Activin-A (β_A–β_A dimer)
	Activin-AB (β_A–β_B dimer)
	Activin-AC (β_A–β_C dimer)
	Activin-B (β_B–β_B dimer)
	Activin-BC (β_B–β_C dimer)
	Activin-C (β_C–β_C dimer)
	Activin-E (β_E–β_E dimer)
Inhibin	Inhibin-A (α–β_A dimer)
	Inhibin-B (α–β_B dimer)
BMP	BMP2–BMP7, BMP8a–BMP8b, BMP10, BMP15
GDF	GDF1–GDF3, GDF5–GDF11, GDF15
Nodal	Nodal, Nodal–BMP4, and Nodal–BMP7 dimers

Müllerian hormone (AMH),[41] (5) members of the family of *bone morphogenetic proteins* (BMP),[42] and (6) components of the family of *growth and differentiation factors* (GDF).

The transforming growth factor-β superfamily comprises: (1) 3 TGFβ isoforms (TGFβ1–TGFβ3); (2) 4 activin β chains (β_A–β_C and β_E; activin-β_A and -β_B are identical to the 2 β subunits of inhibins), that form 2 inhibins (inhibin-A and -B, i.e., α–β_A and α–β_B dimers, respectively) and 7 activins (activin-A, -AB, -AC, -B, -BC, -C, and -E, i.e., β_A–β_A, β_A–β_B, β_A–β_C, β_B–β_B, β_B–β_C, β_C–β_C, and β_E–β_E); (3) 10 bone morphogenetic proteins (BMP2–BMP7, BMP8a and -8b, BMP10, and BMP15); and (4) 11 growth and differentiation factors (GDF1–GDF3, GDF5–GDF11, and GDF15). These

[41] A.k.a. Müllerian-inhibiting factor (MIF), hormone (MIH), and substance (MIS).

[42] Metallopeptidase BMP1 that acts on procollagen-1, -2, and -3, does not belong to the TGFβ superfamily. Bone morphogenetic proteins with several other signals, such as Wnt and FGF, are involved in heart development. Noggin, a bone morphogenetic protein antagonist, is transiently and strongly expressed in the heart-forming region during gastrulation [350]. Cardiomyocytes can be obtained from mouse embryonic stem cells by inhibition of BMP signaling.

Table 3.11. Common members of the bone morphogenetic protein (BMP) and growth differentiation factor (GDF) families. Factor GDF2 is one of the most potent BMPs in bone formation. Factor GDF5 (a.k.a. cartilage-derived morphogenetic protein CDMP1) participates in development of the central nervous system as well as skeleton and joints. It also promotes survival of dopamine-sensitive neurons. It targets AcvR2a and AcvR2b associated with AcvR1 receptors. Factor GDF6 contributes to the control of eye development. It binds to BMPR2 linked to BMPR1a or BMPR1b receptors. Factor GDF7 induces the formation of sensory neurons in the dorsal spinal cord. Agents GDF10 and BMP3 are considered as a separate subgroup within the TGFβ superfamily. Factor GDF11 controls anterior–posterior patterning.

BMP Member	Corresponding GDF member
BMP3b	GDF10
BMP9	GDF2
BMP11	GDF11
BMP12	GDF7
BMP13	GDF6
BMP14	GDF5

molecules generally form homodimers, but heterodimers exist, such as Nodal–BMP4 and Nodal–BMP7 complexes, in addition to inhibins and activin-AB, -AC, and -BC.

3.8.3 Synthesis, Secretion, and Storage in the Extracellular Matrix

Transforming growth factor-β isoforms as well as other TGFβ superfamily members are synthesized as dimeric preproproteins. A large N-terminal prodomain is required for the proper folding and dimerization of the C-terminal growth-factor domain.[43] It also confers latency, at least, on some superfamily members and enables storage in the extracellular matrix, once complexed particularly with latent TGF-binding proteins (LTBP) or fibrillins.[44]

Precursors of the TGFβ superfamily members are processed and secreted by cells in an inactive form. In particular, transforming growth factor-β is stored in the extracellular matrix as a latent TGFβ. Activation of members of the TGFβ superfamily thus follows their liberation from latency.

[43] The prodomain of Nodal, which links to Cripto, is cleaved by peptidases secreted by neighboring cells [351]. Anti-Müllerian hormone is secreted mostly uncleaved; its prodomain potentiates its activity. Protein Lefty is cleaved to enable access of the growth-factor domain to its receptors.

[44] Many members of the TGFβ superfamily remain associated with their prodomains after secretion, such as BMP4, BMP7, BMP10, GDF2, GDF5 and GDF8. Many of these prodomains bind to fibrillin-1 and -2. Binding to LTBPs or fibrillins strengthens the latent prodomain–growth factor complex [351].

Precursors of TGFβ are cleaved by peptidases of the subtilisin-like proprotein convertase family such as furin.[45] Mature dimeric growth factors can also be secreted. TGFβ isoforms, GDF8,[46] and GDF11 are secreted as propeptides. They undergo an additional cleavage by specific enzymes of the bone morphogenetic protein-1–Tolloid-related family of metallopeptidases to release the active form, like endorepellin (angiostatic C-terminal fragment of perlecan). However, Nodal precursor binds to receptors and activates signaling without being processed [353].

Components of the extracellular matrix, such as fibrillins, emilin, and auxiliary receptor endoglin, control the availability of active extracellular TGFβ (extracellular regulation of TGFβ signaling) [354]. Fibrillins have a dual role, because they concentrate TGFβ at sites of function (positive activity), but sequester TGFβ and inhibit its activation (negative activity).

Emilin-1 protects TGFβ from proteolysis by furin convertase, and then inhibits TGFβ signaling. Emilin-1 particularly impedes excessive TGFβ signaling characterized by a reduction in arterial lumen with a resultant increase in vascular resistance and hypertension.

Sequestered TGFβ can be rapidly released by peptidases (plasmin, mastocyte chymase, thrombin, and matrix metallopeptidases MMP2 and MMP9). Activated TGFβ can then act on neighboring cells. Factor TGFβ can also be activated by thrombospondin-1 as well as β_6 and β_8 integrins.

All of the 33 members of the TGFβ superfamily (nodal, activins, inhibins, bone morphogenetic proteins, and growth differentiation factors) undergo a folding of their prodomain that shields the growth factor from recognition by receptors [351]. Activation of TGFβ requires the binding of $\alpha_V\beta_6$ integrin to an RGD sequence in the N-terminal prodomain and exertion of force on this domain by latent TGFβ-binding proteins in the extracellular matrix.

3.8.3.1 TGFβ Isoforms

Among the 3 structurally and functionally distinct TGFβ isoforms, TGFβ1 is the prevalent isoform, whereas TGFβ2 and TGFβ3 are expressed in a limited number of tissues. Isotype TGFβ1 is particularly synthesized by endothelial cells, vascular smooth muscle cells, myofibroblasts, macrophages, lymphocytes, and hematopoietic cells, and TGFβ2 by keratinocytes among others.

[45] Furin is a portmanteau for Fes upstream region protein. It is also called paired basic amino acid cleaving enzyme (PACE). In addition to transforming growth factor-β precursor, its substrates include proparathyroid hormone, proalbumin, proβsecretase, membrane type-1 matrix metallopeptidase, β subunit of pro-nerve growth factor, and von Willebrand factor. Furin and furin-like proprotein convertases target the iron regulator hemojuvelin (or repulsive guidance molecule RGMc), a glycosyl-phosphatidylinositol-anchored protein and soluble glycoprotein of the liver and striated muscles [352].

[46] A.k.a. myostatin (Mstn). It interacts with secreted glycoproteic inhibitor follistatin-like FSTL3 and AcvR2b receptor.

Factor TGFβ is a growth inhibitor for endothelial cells, fibroblasts, and other cell types.

Several factors regulate TGFβ synthesis. In vascular smooth muscle cells, angiotensin-2 stimulates TGFβ expression and promotes its conversion to active form. Matrix metallopeptidase MMP2 enhances active TGFβ1 in vascular smooth muscle cells. In addition, angiotensin-2 and endothelin-1 stimulate thrombospondin-1 that increases separation of active TGFβ from inactive latent complex.

The 3 TGFβ isoforms act as auto-, para-, and sometimes endocrine factors. They regulate via their cognate receptor Ser/Thr kinases and SMAD effectors (homologs of Caenorhabditis elegans SMA and Drosophila MAD; Vol. 3 – Chap. 8. Receptor Kinases) the fate (cell growth, adhesion, migration, differentiation, and apoptosis) of many cell types, especially endothelial and vascular smooth muscle cells, both during embryogenesis and after birth, during childhood as well as in adults. Effectors SMADs transmit the signal down to DNA. Signaling magnitude and duration determine cell response features in a given cell type and signaling context [353]. Endocytosis kinetics of TGFβ receptors could regulate the duration and magnitude of signaling. The SMAD signaling is generally slow and sustained.

Transforming growth factor-β initiates a signaling cascade not only by binding and then activating its cognate receptors, but also by triggering the assembly of an active heterotetrameric receptor. TGFβ receptor-1 (TβR1) acts downstream from high-affinity TGFβ receptor-2 (TβR2)[47] and determines the signaling specificity.

3.8.3.2 Signaling by Members of the TGFβ Superfamily

Members of the TGFβ superfamily target 2 main classes of receptors: the TGFβ receptor-1 (TGFBR1 class) and -2 (TGFBR2 class) sets (Tables 3.12, 3.13, and 3.14). The TGFBR1 class is composed of 2 subsets: (1) TGFBR1-class subset 1 with activin receptor-like kinases ALK4, ALK5, and ALK7 and (2) TGFBR1-class subset 2 with ALK1, ALK2, ALK3, and ALK6. The TGFBR2 class includes TβR2, activin receptor-2 (AcvR2), BMP receptor-2 (BMPR2), and anti-Müllerian hormone receptor-2 (AMHR2). Each TβR2 associates with a TβR1 of a given subset. For example, bone morphogenetic proteins signal via BMPR2 and a receptor of the TβR1 subset 2.

Regulation of Signaling by TGFβ Superfamily Members

Two main sets of proteins control ligand–receptor interactions. (1) Accessory TGFβ receptors (TGFβ receptor-3) comprise *βglycan*, *endoglin*, and *Cryptic* family members. The expression of plasmalemmal coreceptors is needed for receptor activation by given cell types, thus allowing these cells

[47] Ligand-bound TβR2 recruits low-affinity TβR1.

Table 3.12. Signaling by members of the TGFβ superfamily (Sources: [353, 354]). The TGFβ superfamily comprises: (1) 3 TGFβ isoforms (TGFβ1–TGFβ3); (2) 4 activin-βs (activin-βA–activin-βC and activin-βE) that build 2 inhibins (inhibin-A and -B) as well as 7 activins (activin-A, -AB, -AC, -B, -BC, -C, and -E); (3) Nodal; (4) 10 bone morphogenetic proteins (BMP2–BMP7, BMP8A–8B, BMP10, BMP15), (5) 11 growth and differentiation factors (GDF1–GDF3, GDF5–GDF11, GDF15), and (6) anti-Müllerian hormone (AMH). members of the TGFβ superfamily have 2 types of receptors: TβR1 and TβR2 receptor Ser/Thr protein kinases that are encoded by the genes TGFBR1 (transforming growth factor-β receptor-1; a.k.a. 53-kDa activin-A receptor type 2-like kinase [ALK2]) and TGFBR2 ([70–80-kDa] TGFβ receptor-2). Transforming growth factor-β receptor-3 (> 300-kDa) encoded by the TGFBR3 gene is a cell-surface chondroitin sulfate–heparan sulfate proteoglycan. Signaling by members of the TGFβ superfamily uses a limited number of receptors of both types, but multiple kinds of interactions occur (1) between ligands and receptors on the one hand and (2) between receptor types on the other. Receptor AcvR2b binds various TβR1: ALK4 in response to activin, ALK4 or ALK7 to Nodal, ALK3 to BMP2, ALK2 to BMP7, ALK4 or ALK5 to GDF8, and ALK4, ALK5 or ALK7 to GDF11. Coreceptors modulate ligand binding to receptors.

Receptor class	Types or ligands	Effector
TGFBR1	ALK4, ALK5, ALK7	SMAD2, SMAD3
	ALK1, ALK2, ALK3, ALK6	SMAD1, SMAD5, SMAD9 (SMAD8)
TGFBR2	TGFβ, Acv, Nodal	TβR1 subset 1
	BMP, GDF, AMH	TβR1 subset 2
TGFBR3	β-Glycan, endoglin, Cryptic	Coreceptor

to respond to corresponding ligands. Elevated levels of endoglin in endothelial cells of developing blood vessels cause endothelial dysfunction, reducing the activity of nitric oxide synthase NOS3 [354]. (2) Extracellular ligand-sequestering proteins include *chordin*, *noggin*, and *Twisted gastrulation*; members of the *Cerberus* family; and *sclerostin*. These diffusible inhibitors modulate the signaling amplitude and duration.

Latent TGFβ-Binding Proteins and Latent TGFβ Complex

Natural TGFβ antagonists include numerous binding partners of members of the TGFβ superfamily (Table 3.15). Latent TGFβ-binding proteins (LTBP1–LTBP4) control TGFβ availability.

Regulator TGFβ is synthesized as a large, homodimeric, inactive precursor (*preproTGFβ*). The dimeric precursor is cleaved intracellularly to produce the *small latent TGFβ complex* (SLTC), formed by the C-terminal region that will generate the mature TGFβ and latency-associated peptide (LAP or

Table 3.13. Receptors of the TGFβ superfamily (ALK: activin receptor-like kinase).

Subclass	Type(s)	Alias
	Class-1 receptors	
TβR1		ALK5
AcvR1	AcvR1a	ALK2
	AcvR1b	ALK4
	AcvR1c	ALK7
	AcvRL1	ALK1
BMPR1	BMPR1a	ALK3
	BMPR1b	ALK6
ALK class 1: ALK4, ALK5, and ALK7		
ALK class 2: ALK1, ALK2, ALK3, and ALK6		
	Class-2 receptors	
TβR2		
AcvR2		
BMPR2		
AMHR2		

Table 3.14. Signaling from TGFβ superfamily members (ALKC1[2]: class-1[2] activin receptor-like kinase). Latent TGFβ-binding proteins (LTBP1–LTBP4) control TGFβ availability. Once liberated from the small and large latent complexes, TGFβ family members can be intercepted by diffusible proteins that bind these ligands and inhibit their access to cognate receptors.

Agonists	Type-2 receptor	Type-1 receptor	Inhibitors
TGFβ1–3	TβR2	TβR1 (ALKC1)	LTBP1, thrombospondin-1, decorin, gremlin
AMH	AMHR2	ALKC2	
Activins	AcvR2	AcvR1 (ALKC1)	Follistatin
BMP	BMPR2	BMPR1a/b (ALKC2)	Noggin (BMP2/4–7/13), chordin, Cer1
GDF	AcvR2	ALKC2	Noggin (GDF5/6)
Nodal	AcvR2b	AcvR1b/c (ALKC1)	Lefty

TGFβ propeptide) [354].[48] The dimer SLTC is then secreted, but cannot bind to the TGFβ receptors. It binds to latent TGFβ-binding protein (LTBP1–LTBP4). In fact, the SLTC dimer usually binds to LTBP to form a trimer, the *large latent TGFβ complex* (LLTC). The latter binds (via LTBP) to collagen, fibronectin, and fibrillin-1, hence it is anchored to the extracellular matrix.

[48] Unlike most propeptides that have a low affinity for mature proteins, latency-associated protein (LAP) strongly binds to the TGFβ region to build the dimeric small latent TGFβ complex.

Activation of TGFβ then requires dissociation from the latent TGFβ complex. Once liberated from these latent aggregates, TGFβ superfamily members can be intercepted by diffusible proteins that bind these ligands and inhibit their access to cognate receptors.

Receptors that activate LAP such as the scavenger receptor ScaRb3,[49] thrombospondin Tsp1, and $\alpha_V\beta_6$ integrin, are expressed on monocytes, endothelial cells, and dendritic cells, but not on T lymphocytes. Regulatory T lymphocytes that are activated by TGFβ then required their interaction with antigen-presenting cells.

DAN Family

The DAN family of TGFβ endogenous, glycoproteic antagonists comprises multiple members that restrict the activity of bone morphogenetic proteins and/or Wnt morphogens. The founding member of this family, Differential screening-selected gene aberrative in neuroblastoma (DAN),[50] localizes to axons. Seven BMP antagonists of the DAN family have been identified in rodents, such as tumor suppressor DAN, head-inducing factor Cerberus-related protein Cer1, gremlin, protein related to DAN and Cerberus (PRDC). Some members of the Cerberus–DAN family are implicated in left–right patterning owing to their asymmetrical distribution. Unlike Lefty proteins, DAN family members bind directly to extracellular Nodal and prevent signaling. Cerberus-related protein Cer1[51] antagonizes GDF5, GDF6, and GDF7 factors (Table 3.15).

Chordin and Chordin-like Proteins

Chordin is involved in the body's patterning during embryogenesis. Chordin-like protein-1 (ChrdL1)[52] is upregulated by hypoxia, especially in retinal pericytes [355]. Chordin-like protein-2 (ChrdL2)[53] is expressed preferentially in chondrocytes of developing cartilage and degenerated (osteoarthritic) joint cartilage [356]. It precludes generation, maturation, and regeneration of articular chondrocytes.

Decorin and Follistatin

Decorin is a proteoglycan that interacts with fibronectin, thrombospondin, and TGFβ, among others. It regulates TGFβ activity. Follistatin is a follicle-stimulating hormone (FSH)-suppressing, activin-binding protein. This ubiquitous glycoprotein is an autocrine regulator that serves as a safeguard against uncontrolled cellular proliferation.

[49] A.k.a. CD36, thrombospondin receptor, platelet collagen receptor, fatty acid translocase, and glycoproteins GP3a and GP4.

[50] A.k.a. DAN domain family member-1 (DAND1), neuroblastoma candidate region, suppression of tumorigenicity-1 (NBL1) and NO3.

[51] A.k.a. DAN domain-containing protein DAND4.

[52] A.k.a. neuralin-1 (Nrln1), neurogenesin-1, and ventroptin (Vopt).

[53] A.k.a. breast tumor novel factor BNF11.

Table 3.15. Inhibitors of the signal transduction triggered by members of the TGFβ superfamily (DAN: differential screening-selected gene aberrative in neuroblastoma; FLRG: follistatin-like related gene product; FSRP: follistatin-related protein; GASP: GDF-associated serum protein).

Inhibitor	Binding partners of the TGFβ superfamily
Cerberus	Activin, BMP2/4/7, Nodal
Chordin	BMP2/4/7
ChrdL1	BMP4/5/6
ChrdL2	BMP
DAN	BMP2/4/7, GDF5/6/7
Decorin	TGFβ1/2
Follistatin	Activin, BMP2/4/6/7, GDF8/11, TGFβ1
FLRG	Activin, GDF8/11, TGFβ1
FSRP	Activin, BMP2/6/7
GASP1	GDF8/11
Gremlin	BMP2/4/7
α2 macroglobulin	Activin, inhibin, TGFβ1/2
Noggin	BMP2/4/5/6/7, GDF5/6
Sclerostin	BMP5/6

Lefty Proteins

Proteins Lefty-1 and -2 serve as left–right asymmetry determination factors during embryogenesis that antagonize Nodal action.[54] Noggin inactivates members of the TGFβ superfamily such as BMP4. It can create morphogenic gradients. Noggin is an important regulator of cartilage and bone patterning. Sclerosteosis, or sclerostin, belongs to the DAN family of glycoproteins. It is specifically expressed by osteocytes and inhibits osteoblast differentiation [357].[55] Thrombospondin-1 binds and antagonizes fibronectin and TGFβ, among others.

Heat Shock Protein

Heat shock protein HSP90 interacts with TβR1 and TβR2 to regulate TGFβ signaling.[56] Chaperone HSP90 that stabilizes its target impedes TβR2

[54] Lefty represses the phosphorylation of receptor-regulated SMAD (rSMAD) after TGFβ and BMP stimulation. Lefty-1 restricts the expression of Lefty2 and Nodal to the left side. Lefty-2 and Nodal are actually involved in the left-side patterning during embryogenesis.

[55] In fact, sclerostin mainly restricts the canonical Wnt signaling by binding to Wnt coreceptors, the low-density lipoprotein receptor-related proteins LRP5 and LRP6.

[56] Regulatory domain of HSP90 contains an ATP-binding site. Upon ATP binding and hydrolysis, the HSP90–protein complex associates with cochaperones such as cell division cycle CDC37 homolog, an HSP90 cochaperone, for protein

ubiquitination by SMURF2 ubiquitin ligase and degradation, thus promoting activation of SMAD2 and -3 [358].

Connective Tissue Growth Factor

Factor TGFβ binds to connective tissue growth factor for intense, prolonged TGFβ activity. TGFβ also causes CTGF synthesis that leads to SMAD7 transcriptional suppression, as it primes transcription factor TGFβ-inducible early gene product TIEG1, thereby relieving SMAD7-mediated negative feedback loop. In vascular smooth muscle cells, CTGF is a signaling mediator of angiotensin-2 and TGFβ via SMAD and the RhoA–RoCK–NADPH oxidase–ROS–MAPK pathway (Vol. 3 – Chap. 7. G-Protein-Coupled Receptors).

Angiotensin-2

Angiotensin-2 regulates TGFβ expression as well as its activation and secretion. Angiotensin-2 can activate the SMAD pathway independently of TGFβ via AT_1 receptors [359]. In vascular smooth muscle cells, angiotensin-2 provokes rapid SMAD2 phosphorylation via P38MAPK mitogen-activated protein kinase and nuclear translocation of phosphorylated SMAD2 as well as SMAD4.

Arkadia Ubiquitin Ligase (RNF111)

Ubiquitin ligase Arkadia[57] is a potentiator of transforming growth factor-β, as it provokes ubiquitin-dependent degradation of several inhibitors of TGFβ, such as inhibitory SMAD7 as well as members of the SKI family of corepressors of TGFβ signaling such as v-Ski sarcoma viral oncogene homolog (Ski) and Ski-like protein SkiL that interact with SMAD proteins [360]. In particular, Arkadia interacts with Axin that sequesters SMAD7 in the cytoplasm and forms a complex with Axin and SMAD7 to support SMAD7 polyubiquitination and degradation. In addition, Arkadia ubiquitinates μ2 subunit of the clathrin adaptor complex AP2, hence limiting EGFR endocytosis by AP2 complex [361].[58]

---

folding and stabilization. On the other hand, $HSP90^{ADP}$ associates with different cochaperones such as HSP70 for ubiquitin-mediated degradation mediated by C-terminus heat shock cognate-70-interacting protein (CHIP) of target proteins, such as HER2 plasmalemmal receptor and protein kinase-B.

[57] A.k.a. RING finger protein RNF111.

[58] Heterotetramer AP2 consists of large adaptins α and β2, a medium subunit μ2, and small τ2 component. Adaptin α recruits endocytotic accessory proteins Epsin, Eps15, and clathrin coat-associated protein AP180 (91-kDa synaptosomal-associated protein SNAP91). Adaptin β2 triggers clathrin assembly.

Signal Transduction Primed by TGFβ Superfamily Members

Distinct modes of graded and switch-like assembly of receptor Ser/Thr kinases exist for signal transduction [362]. Dimers TβR3 form cooperative heterohexamers with pairs of TβR1 and TβR2. Bone morphogenetic proteins form distinct heterotetramers of receptor pairs.

Factor TGFβ causes a rapid activation of the mitogen-activated protein kinase modules that are involved in the Ras–ERK, MAP3K7—MAP2K4–JNK, MAP3K7—MAP2K3/6–P38MAPK, Rho/Rac/CDC42–MAPK, and PI3K–PKB pathways.

Following ligand binding, TGFβ receptor-2 transphosphorylates TGFβR1, which stimulates additional trans- and autophosphorylation before phosphorylating receptor-associated SMADs (rSMAD), thereby promoting their nuclear translocation and gene regulation. In addition, SMAD-independent pathways exist.

Moreover, active TGFβR1 is sumoylated to enhance SMAD3 recruitment, binding, and phosphorylation [363]. Sumoylation requires kinase activities of both TβR1 and TβR2. The TβR1–TβR2 complex afterward undergoes ubiquitination and degradation.

Canonical signaling triggered by TGFβ superfamily members includes 2 main intracellular pathways according to SMAD mediators. Receptor TβR2 phosphorylates (activates) TβR1 that is then able to recruit receptor-regulated SMADs. Receptors of the TGFBR1 subset activate only a subset of rSMADs: (1) receptors of ALK class 1 that specifically phosphorylate SMAD2 and SMAD3 and (2) receptors of ALK class 2 that are specific for SMAD1, SMAD5, and SMAD8 (or SMAD9).

Consequently, SMAD2 and SMAD3 are phosphorylated by TGFβ, activins, and Nodal, whereas SMAD1, SMAD5, and SMAD8 are activated by bone morphogenetic proteins, growth and differentiation factors, and anti-Müllerian hormone. However, in endothelial cells, TGFβ can recruit both ALK5 (TGFBR1 subset 1) and ALK1 (TGFBR1 subset 2) to form a single receptor complex, and thus activate both pathways [353].

In addition, the Nodal–SMAD2 and Nodal–SMAD3 pathways recruit lysine (K)-specific demethylase KDM6b to counteract repression by Polycomb-group proteins on Nodal target genes [364]. Polycomb-group proteins remodel chromatin that is composed of DNA wrapped around histones, hence reducing DNA accessibility and causing epigenetic silencing of genes.[59] Therefore, intercellular signaling via Nodal balances epigenetic regulation mediated by Polycomb-group proteins, especially during the embryogenesis.

---

[59] Histones undergo numerous post-translational modifications, such as acetylation and methylation. The Polycomb repressive complex PRC2 trimethylates histone-H3 (Lys47) to yield a binding site for the Polycomb repressive complex PRC1 that represses gene expression [364]. This process is reversible, as histone demethylases target methylated histone-$H3^{M}$.

Once TβR1 has phosphorylated specific intracellular SMAD effectors, the latter form homomers that link common mediator SMAD4. The SMAD complexes impart the signal into the nucleus and regulate — positively or negatively — target gene transcription.

Signaling from TGFβ is modulated by other signaling pathways and post-translational modifications, as SMADs are controlled by phosphorylation, acetylation, ubiquitination, and sumoylation (Vol. 1 – Chap. 5. Protein Synthesis).

Phosphorylated receptor-regulated SMADs ($rSMAD^P$) form heteromers with their partner SMAD4 that translocate and accumulate in the nucleus and bind to DNA. The DNA-binding domain can be phosphorylated by kinases, such as mitogen-activated protein kinases, glycogen synthase kinase-3β, and cyclin-dependent kinases. Mediators SMADs can hence interact with other signaling pathways. Another binding domain is involved in SMAD interactions with receptors, other SMADs, transcription factors, coactivators, and corepressors.

Nuclear accumulation of active SMAD complexes is correlated to the degree of receptor activation. The SMAD complexes preferentially reside in the nucleus, whereas rSMAD monomers localize to the cytoplasm. Both rSMAD phosphorylation and SMAD nuclear accumulation are maintained when receptors are active. Nuclear phosphatases dephosphorylate SMADs that then return to the cytosol. Nuclear accumulation of SMADs during active signaling then overcomes SMAD dephosphorylation. In the absence of signal, SMADs travel continuously between the cytoplasm and the nucleus. However, SMADs are exported from the nucleus faster than they are imported.

Inhibitory SMADs (iSMAD) are activated by TGFβ superfamily members. They prevent TβR1 activation and dephosphorylate (inactivate) or degrade active receptors. Isoform SMAD7 can bind to SMAD-responsive elements and inhibit SMAD-dependent promoter activation. In particular, the pathways activated by TβR2 linked with ALK5 and coreceptor β-glycan on the one hand, and BMPR2 associated with a member of TGFBR1 subset 2 (ALK1–ALK3 and ALK6) and endoglin on the other hand, with their corresponding rSMAD effector, SMAD2 and -3 as well as SMAD1, -5, and -8, are inhibited by SMAD7 and SMAD6, respectively [354].

Receptors of TGFβ can relay cues via non-SMAD pathways. The receptor TβR2 phosphorylates the polarity protein Par6 and TβR1 scaffold protein SH2 domain-containing transforming protein SHC1 [353]. In addition, TGFβ stimulates G-protein-coupled receptor kinase-2 (GRK2) that desensitizes G-protein-coupled receptors and inhibits TGFβ signaling (negative feedback loop) [365].

Factor TGFβ also activates MAP3K7 mitogen-activated protein kinase. Enzyme MAP3K7, or TGFβ-activated kinase TAK1, phosphorylates (activates) P38MAPK and Jun N-terminal kinases, which leads to apoptosis. Factor TGFβ causes autoubiquitination and activation of tumor-necrosis factor receptor-associated factor TRAF6 that binds to TGFβR1 [366]. Ubiquitin ligase TRAF6 then ubiquitinates (activates) MAP3K7 kinase.

Factor TGFβ promotes production and association of transcription factor human immunodeficiency virus type-1 enhancer binding protein HIVEP2 (Schnurri-2 homolog) and chloride intracellular channel ClIC4, as well as their nuclear accumulation.[60] In the nucleus, ClIC4 mediates TGFβ transcriptional response by binding to $SMAD2^P$ and $SMAD3^P$, hence precluding their dephosphorylation by magnesium-dependent PPM1a protein phosphatase [367].

Growth factor TGFβ activates protein kinase-B[61] in glomerular mesangial cells via the microRNAs miR216a and miR217 that target phosphatase and tensin homolog, an inhibitor of PKB activation [368]. It intervenes in hypertrophy and survival of glomerular mesangial cells.

Signaling from TGFβ is attenuated and terminated by the activation of the TMEPAI gene, as its product transmembrane prostate androgen-induced protein (TMePAI)[62] impedes TGFβ signaling. Protein TMePAI reduces TβR1-mediated phosphorylation of SMAD2 and SMAD3 effectors, as it competes with and sequesters adaptor SMAD anchor for receptor activation (SARA)[63] away from TβR1 [369].[64] Therefore, TGFβ-induced TMePAI reduces the production of proteins from TGFβ-activated genes, such as JunB, Myc, plasminogen activator inhibitor PAI1 (or Serpin-E1), and cyclin-dependent kinase inhibitor CDKI1a.

TGFβ in Inflammation

Among the 3 isoforms, TGFβ1 is predominantly expressed in immunocytes. Transforming growth factor-β possesses both positive and negative roles in inflammation (Tables 3.16 and 3.17).[65] In the presence of interleukin-2, immunoregulator TGFβ stimulates FoxP3+ regulatory T cells [370]. In the presence of interleukin-6, it activates T_{H17} cells.[66] In addition, pleiotropic cytokines TGFβ and interleukin-10 suppress the immune response.

Transforming growth factor-β [370]: (1) suppresses pro-inflammatory effector T_H cell differentation (Table 3.18); (2) converts naive T cells into

---

60 Cellular stress such as DNA damage causes ClIC4 nuclear translocation. Transcription factor HIVEP2 also acts in bone morphogenetic protein signaling.

61 Protein kinase-B is particularly activated by transforming growth factor-β1 in diabetic kidneys.

62 A.k.a. prostate transmembrane protein, androgen-induced PMePAI.

63 A.k.a. mother against decapentaplegic homolog-interacting protein (MADHIP), SMADIP, and zinc finger, FYVE domain-containing protein ZFYVE9.

64 Protein SARA presents inactive SMADs to the TβR1–TβR2 complex to facilitate SMAD phosphorylation and transduce the TGFβ signal.

65 Among immunocytes, T helper cells are major regulators of immune responses. After activation by antigenic stimulation, naive T_H cells differentiate into effector or regulatory T cells that are responsible for positive and negative regulation of

Table 3.16. Effect of TGFβ in immunocytes (Source: [370]; IL: interleukin; GATA: GATA-binding protein [globin transcription factor]; SMAD: small mothers against decapentaplegic homolog; STAT: signal transducer and activator of transduction; TBx21: T-cell-specific T-box transcription factor). Transforming growth factor-β: (1) inhibits T_{H1} and T_{H2} cell differentiation; (2) causes differentiation into T_{H17} and iT_{Reg} cells; and (3) impedes maturation of other immunocyte types, such as CD8+ CTL-, NK-, and dendritic cells and macrophages. Nuclear receptor NR1f3-2 (or RORγ2) is a master transcription factor for T_{H17} cells. SMAD-Independent differentiation of naive T cells into T_{H17} cells relies on STAT3 factor. Transcription factor FoxP3 is involved in the differentiation of naive T cells into induced regulatory T cells with SMAD2 and SMAD3 as well as STAT5. Factors STAT3 and STAT5 inhibit FoxP3 and NR1f3-2, respectively.

Cell type	TGFβ Effects
T cell	Inhibition of IL2
T_{H1}	Inhibition of STAT1/4 and TBx21
T_{H2}	Inhibition of STAT6 and GATA3
T_{H17}	Activation of NR1f3-2
iT_{Reg}	Activation of FoxP3

anti-inflammatory regulatory T cells;[67] (3) hinders the proliferation of T and B lymphocytes; (4) prevents effector cytokine production, such as interleukin-2

immunity, respectively. Autoimmune and inflammatory diseases can be caused by excess immune reactions and decreased immune suppression.

[66] Interleukin-6 may suppress FoxP3 activity to strenghten the differentiation into T_{H17} cell with respect to induced regulatory T cells. Interleukin-6 also maintains high levels of NR1f3-2 (a.k.a. RORγ2), as the transcription factor FoxP3 hampers the transcriptional activity of NR1f3-2. Factor NR1f3-2 is activated by TGFβ and interleukin-6 to promote T_{H17}-cell differentiation. Signal transducer and activator of transduction STAT3, which is also involved in T_{H17}-cell differentiation, also represses FoxP3 function. In addition, interferon regulatory factor IRF4 and transcription factor MAF (V-Maf musculoaponeurotic fibrosarcoma oncogene homolog) that are upregulated by STAT3 support NR1f3-2 expression [370].

[67] After emigrating from the bone marrow, thymocyte progenitors enter the thymus. After positive selection, CD4+ or CD8+ single-positive cells migrate as naive T cells. Naturally occurring CD4+, CD25+, FoxP3+ regulatory T cells (nT_{Reg}) also develop in the thymus from immature CD4+ T cells. Dendritic cells in the presence of TGFβ and CD8+, CD205+ dendritic cells promote expansion of nT_{Reg}s and selectively repress that of effector T cells. Naive T cells are activated by antigen-presenting cells and differentiate into effector or memory T cells.

Table 3.17. Pro- and anti-inflammatory factors (BMP: bone morphogenetic protein; CCL: chemokine CC-motif ligand; CNTF: ciliary neurotrophic factor; CSF: colony-stimulating factor; CT1: cardiotrophin-1; IL: interleukin; IL1RA: IL1-receptor antagonist; LIF: leukemia-inhibitory factor; OSM: oncostatin-M; PKA: protein kinase-A; PlGF: placental growth factor; S1P: sphingosine 1-phosphate; TGF: transforming growth factor; TNF: tumor-necrosis factor). Cytokines are growth factors involved in immunity and hematopoiesis. They allow communications between immunocytes as well as between immune and other cell types. Chemokines are inducible chemotactic cytokines that mobilize nearby responsive cells according to a chemotactic direction. C1q is the recognition protein of the classical complement pathway.

Pro-inflammatory agents	Anti-inflammatory agents
TGFβ	TGFβ
BMP2/4	
TNFα	
C1q	
Adipocytokines (adiponectin, leptin, resistin, visfatin)	Adiponectin
PlGF	
Midkine, pleiotrophin	
CSF	Soluble receptors
S1P	(sTNFR, sIL1R)
IL8, CCL2, CCL5	IL1RA, IL1R2
IL1, IL5, IL6,	IL4, IL6
IL12, IL17, IL22, IL23	IL10, IL13, IL32, IL35
CNTF, CT1, LIF, OSM	
Interferons	

and -4 and interferon-γ;[68] and (5) inhibits macrophages, granulocytes, antigen-presenting dendritic and natural killer cells, and mastocytes.

3.8.3.3 Bone Morphogenetic Proteins

Transforming growth factor-β and bone morphogenetic proteins are involved in lung formation and cardiomyo- and vasculogenesis as well as embryonic and adult angiogenesis in normal conditions and diseases. Bone morphogenetic protein BMP4 provokes capillary sprouting of endothelial cells. In

[68] Transforming growth factor-β prevents TBx21 activity, hence Ifnγ production, on the one hand, and STAT6 function, thus IL4, on the other. In addition, upon TGFβ stimulus, the SMAD2–SMAD3–SMAD4 complex excites protein inhibitor of activated STAT1 (PIAS1) that then hinders Ifnγ transcriptional effect. Conversely, Ifnγ receptor (IfnGR1) activates STAT1. The STAT1 target genes such as SMAD7 preclude TGFβ signaling. Moreover, Ifnγ hampers TGFβ1 responses via sequestration of the nuclear coactivator P300 by STAT1, hence preventing SMAD–P300 linkage and SMAD transcriptional activity.

Table 3.18. Differentiation of T_H cells from naive T cells (Source: [370]). The T_{H1} cell aims at eradicating intracellular pathogens (bacteria, viruses, and some protozoa). It produces interferon-γ (Ifnγ). Polarization into T_{H1} is primarily due to interleukin-12 (IL12) and interferon-γ. Repression by TGFβ of T_{H1} cell differentiation is associated with reduced expression of IL12 receptor subunit IL12Rβ2 and T-cell-specific T-box TBx21 transcription factor. The T_{H2} cell is involved in the elimination of parasitic infections. It synthesizes interleukins IL4, IL5, IL10, and IL13 that stimulate the synthesis in B lymphocytes of immunoglobulin IgE, recruits eosinophils, and promotes mucus production and hypermotility, respectively. Maturation of T_{H2} cells is primarily directed by IL4. Differentiation of T_{H17} cells is initiated by IL6 and TGFβ and mediated by STAT3 agent. Factor STAT3 fosters the production of nuclear receptor NR1f3-2 (or RORγ2) that supports T_{H17}-cell differentiation by stimulating IL23 receptor. The T_{H17} cell secretes IL17a, IL17f, IL21, and IL22. Agent IL23 contributes to the maturation and maintenance of T_{H17} cells. Once activated by TGFβ, FoxP3 antagonizes NR1f3-2, which is also stimulated by TGFβ, to inhibit T_{H17}-cell differentiation (in the absence of signaling from IL6 and STAT3). Differentiation into induced FoxP3+, regulatory T cells (iT_{Reg}) involves TGFβ and relies on FoxP3 and SMAD factors. These transcription factors are a major source of active TGFβ in iT_{Reg} cells.

Cell	Activators	Mediators	Products	Target cells	Target pathogens
T_{H1}	IL12/18, Ifnγ	STAT1, TBx21	Ifnγ	Macrophage	Intracellular bacteria, viruses, protozoa
T_{H2}	IL4	STAT6, GATA3	IL3/4/5/13	Eosinophil, Mastocyte	Helminths
T_{H17}	IL6 TGFβ IL23	STAT3, NR1f3-2	IL17a/17f, IL21/22	Neutrophil	Extracellular bacteria, fungi
iT_{Reg}	IL2 TGFβ STAT5	SMAD, FoxP3	TGFβ, IL10	All types	Immune tolerance

addition, BMP2 and BMP4 exert pro-inflammatory effects on the endothelium.

Both TGFβ and BMP stimulate several microRNAs such as miR21 that favors expression of SMC contractile genes, such as smooth-muscle α-actin, calponin-1, and transgelin (or 22-kDa actin-binding smooth muscle protein SM22α) [371].

The BMP pathway also activates the transcription of SMC-specific contractile genes, as it promotes the nuclear translocation of 2 transcription activators myocardin-related transcription factor MRTFa[69] and MRTFb (or MKL2) that interact with the transcription factor myocardin.

[69] A.k.a. myocardin-like protein MKL1, RNA-binding motif protein-15, and megakaryoblastic leukemia-1 fusion protein.

The activity of bone morphogenetic proteins is regulated by: (1) extracellular modulators, (2) cell-surface receptors, and (3) intracellular mediators. Extracellular BMP-binding *chordin* that is encoded by the gene CHRD assists in transporting BMPs. It actually protects BMPs from degradation. However, chordin is a BMP antagonist, as it interferes with interactions between BMPs and their receptors.

Several BMP inhibitors exist in addition to chordin, such as the ubiquitous, activin-binding, glycoproteic, autocrine regulator *follistatin*[70] (encoded by the FST gene), *noggin* (encoded by the gene NOG), and *sclerostin* (produced by osteoclasts from the SOST gene).

BMP endothelial cell precursor-derived regulator (BMPER) is an extracellular BMP modulator that controls BMP4 activity in endothelial cells, such as sprouting and migration. Regulator BMPER is not strongly diffusible, thereby accumulating BMP activity where it localizes. Synthesis of BMPER is activated by the transcription factor Krüppel-like factor KLF15 [372]. Endothelin-1 that operates via its receptor ET_B is a potent inhibitor of KLF15 and, hence, BMPER production in endothelial cells.

Antagonism between PDGF and TGFβ Superfamily Members

In response to vascular injury, vascular smooth muscle cell experiences a phenotype change from a quiescent contractile to a proliferative phenotype. Both bone morphogenetic proteins and transforming growth factor-β promote the contractile phenotype, whereas platelet-derived growth factor-BB favors a switch to the proliferative phenotype. Factor PDGFbb stimulates miR24 that impedes activity of Tribbles-like protein Trb3 that is associated with reduced BMP and TGFβ signaling to prevent resulting inhibition of proliferation and migration of vascular smooth muscle cells [371]. In addition, PDGFbb suppresses the activity of myocardin, a smooth muscle and myocardium-specific transcriptional coactivator of serum response factor. Furthermore, PDGFbb induces the expression of miR221 that impedes those of SCFR receptor and cyclin-dependent kinase inhibitor CKI1b. Downregulation of SCFR lowers myocardin activity, whereas that of cyclin-dependent kinase inhibitor CKI1b promotes proliferation.

[70] A.k.a. follicle-stimulating hormone-suppressing protein.

Table 3.19. Members of the VEGF family and their receptors (PlGF: placental growth factor). Two neuropilins — Nrp1 and Nrp2 — are glycoproteic coreceptors for class-3 semaphorins and VEGF. Neuropilins generally work as dimers. Different combinations have different affinities for ligands.

Isoforms	Variants	Receptors and coreceptors
VEGFa	$VEGFa_{121/165/189/206}$	VEGFR1, VEGFR2, VEGFR3
VEGFb	$VEGFb_{167/186}$	VEGFR1, Nrp1
VEGFc		VEGFR2, VEGFR3
VEGFd		VEGFR2, VEGFR3
PlGF	PlGF1–PlGF4	VEGFR1, Nrp1
	Exogenous	
VEGFe		VEGFR2

3.9 Vascular Endothelial Growth Factors

Vascular endothelial growth factor (VEGF)[71] regulates the development of vascular endothelia and endocardium. It increases nitric oxide activity that stimulates endothelial proliferation via the protein kinase-G pathway. It permanently acts on endothelial cells to maintain the vasculature in a suitable state. Inhibition of vascular endothelial growth factor indeed causes capillary regression associated with endothelial fenestrations [373].

The VEGF family consists of 5 members (Table 3.19) that include VEGF isoforms (VEGFa–VEGFd encoded by the genes VegfA–VegfD)[72] and placental growth factor (PlGF). Vascular endothelial growth factor VEGFa (or VEGF) is considered as the master regulator of angiogenesis and vascular permeability. Isoform VEGFa is a mitogen for endothelial cells involved in vasculo- and angiogenesis, whereas other VEGF isoforms contribute to lymphangiogenesis.

3.9.1 VEGFa Isoform

The balance between different alternatively spliced VEGFa variants regulates vessel growth and patterning. In humans, the alternative splicing of a single precursor mRNA gives rise to 4 VEGFa products: $VEGF_{121}$ and

[71] A.k.a. vascular permeability factor. Vessel wall fenestrations induced by VEGFa (distinguished from other isoforms [VEGFb–VEGFd]) allow leakage of small molecules. Large substances are transported via caveolae, vesiculovacuolar organelles, and transendothelial pores. Permeability created by VEGFa depends on nitric oxide.

[72] The gene VegfD is also termed Fos-induced growth factor gene (Figf). Proteins related to VEGFs comprise those encoded by viruses (VEGFe) and those found in the venom of some snakes (VEGFf).

$VEGF_{165}$ isoforms are secreted as homodimeric glycoproteins; $VEGF_{189}$ and $VEGF_{206}$ isoforms are bound to extracellular matrix constituents.

The concentration of vascular endothelial growth factor VEGFa evolves in a limited range during development.[73] The transcriptional coactivator and histone acetyltransferase P300 stimulates the production of VEGFa.

Concentrations of circulating vascular endothelial growth factor-A and hepatocyte growth factor are higher in women and smokers [374]. They are also correlated with mean flow velocity in brachial artery and body mass index. On the other hand, plasmatic level of soluble receptor VEGFR1 is lower in women and smokers.

Isoform VEGFa interacts with the Notch pathway (Vol. 3 – Chap. 10. Morphogen Receptors). Delta-like (Notch) ligand DLL4 is expressed in response to VEGFa. On the other hand, Notch activated by DLL4 suppresses the expression of VEGFR2 [375]. Ligand DLL4 is thus expressed in VEGF-activated endothelial cells that are selected for sprouting, whereas neighboring Notch-activated cells remain quiescent.

Furthermore, VEGFa reduces DNA methylation in the promoter regions of the genes of octamer-binding transcription factor Oct4[74] and Reduced expression protein REx1[75] in endothelial progenitor cells [376]. In addition, VEGF hinders the production of miR101, thereby counteracting the repression by miR101 of the histone methyltransferase enhancer of zeste homolog EZH2 of the Polycomb group family.[76]

3.9.2 VEGFb Isoform

Vascular endothelial growth factor VEGFb is expressed in the heart, skeletal muscle, and adipose tissue, as well as smooth muscle cells in adults. Two different VEGFb subtypes exist: heparin-binding $VEGFb_{167}$ and diffusible $VEGFb_{186}$.

Unlike other VEGF family members, in most conditions, vascular endothelial growth factor-B does not markedly influence angiogenesis and blood vessel permeability. It can even be anti-angiogenic. Isoform VEGFb is rather involved in the maintenance of newly formed blood vessels. Although VEGFb is dispensable for growth of blood vessels, it is critical for their survival [377]. Moreover, survival effect of VEGFb is exerted not only on vascular endothelial cells, but also on pericytes, smooth muscle cells, and vascular stem and progenitor cells. Therefore, VEGFb targets vascular cells of the blood–wall interface as well as coverage cells of blood vessels of various bore (pericytes and smooth muscle cells).

[73] A single VEGFa allele as well as a 2-fold increase impede a normal development.
[74] A.k.a. Oct3 and POU domain, class-5, transcription factor POU5F1.
[75] A.k.a. zinc finger protein ZFP42.
[76] A.k.a. lysine N-methyltransferase KMT6.

Factor VEGFb, indeed, regulates the expression of many vascular prosurvival genes via both receptors VEGFR1 and neuropilin Nrp1, a receptor for semaphorins (or collapsins).[77]

Isoform VEGFb is poorly angiogenic in most tissues. On the other hand, dietary lipids in the blood circulation must be transported through the vascular endothelium to be metabolized by cells. Isotype VEGFb abounds particularly in tissues enriched in mitochondria that use fatty acids as a chemical energy source, such as the heart, skeletal muscles, and brown adipose tissue [378]. In endothelial cells, VEGFb regulates the fatty acid transport. It indeed augments the abundance of fatty acid transport proteins (FATP) via VEGFR1 and neuropilin-1 [378]. Overexpression of FATP3 or FATP4 raises uptake of long-chain fatty acids.

3.9.3 VEGFc Isoform

Vascular endothelial growth factor-C participates in angio- and lymphangiogenesis. It also contributes to endothelial cell growth and survival. Moreover, it can influence the permeability of blood vessels. Secreted VEGFc undergoes a proteolytic maturation and generates multiple processed forms which bind and activate VEGFR3 receptors. Only the fully processed form can bind and activate VEGFR2 receptors.

The 2 isoforms VEGFc and VEGFd constitute a subfamily of VEGF proteins. Both VEGFc and VEGFd are secreted as propeptides. The Ser peptidase plasmin cleaves both propeptides [379]. Enzymatic cleavage gives rise to their mature forms. Dimers of the central VEGF homology domain (VHD that spans about one-third of the VEGFc precursor) bind receptors with much greater affinity than the full-length forms. They can signal via VEGFR2 and VEGFR3 receptors that launch programs for angio- and lymphangiogenesis, respectively.

Lymphangiogenic factors VEGFc and VEGFd are synthesized by activated tumor-associated macrophages to promote peritumoral lymphangiogenesis [380]. These cells also produce VEGFc- and VEGFd-specific VEGFR3 receptor. These cells derive from circulating monocytes that do not express VEGFc and VEGFd isoforms. Once stimulated by various factors, such as tumor-necrosis factor-α and VEGFd, these monocytes start to synthesize VEGFc.

3.9.4 VEGFd Isoform

Vascular endothelial growth factor-D is secreted and undergoes a proteolytic maturation. The primary gene product has long N- and C-termini.

[77] Semaphorins constitute a large family of secreted, transmembrane, and membrane-associated glycosyl-phosphatidylinositol (GPI)-anchored proteins that convey information in the immune system and during organogenesis, particularly neuro-, vasculo-, and angiogenesis.

Proteolytic processing releases the central VEGF homology domain (VHD) that binds and activates VEGFRs. In fact, multiple resulting forms bind and activate VEGFR2 and VEGFR3 receptors. In adult humans, VEGFd reaches its highest levels in heart, lung, skeletal muscle, and intestinal tract. It serves as a mitogen for endothelial cells.

Lymphangiogenic factors VEGFc and VEGFd cause the migration of microvascular endothelial cells that produce $\alpha_9\beta_1$ integrins and VEGFR3 [381].

3.9.5 Placental growth Factor

Placental growth factor was initially described in placenta, but it is produced in other organs, such as lungs and heart.[78] In humans, the PLGF gene leads to 4 isoforms (PlGF1–PlGF4) by alternative splicing. These PlGF isoforms differ in binding affinities and secretion modalities [382]. Factor PlGF can form heterodimers with VEGF, hence increasing VEGF-induced chemotaxis of endothelial cells. It is a pleiotropic cytokine with pro-inflammatory and -angiogenic activities. It promotes infiltration of macrophages and T lymphocytes in tissues.

Factor PlGF binds to VEGF receptor-1 (VEGFR1) and to one of its semaphorin-related coreceptor neuropilins Nrp1, besides Nrp2 and heparan sulfate proteoglycans that can favor retention of lipoproteins [382]. Isoform PlGF2 augments the expression of vascular cell adhesion molecule VCAM1 at the luminal side of endothelial cells to promote monocyte adhesion and rolling [383].

3.9.6 VEGF Receptors and Signaling

Receptor Tyr kinases of VEGF (Vol. 3 – Chap. 8. Receptor Kinases) are encoded by the Fms-like Tyr kinase (Flt) gene family.[79] The Flt1 gene encodes vascular endothelial growth factor receptor-1 (VEGFR1).[80] The Flt4 gene encodes VEGFR3 for VEGFc and VEGFd that are involved in lymphangiogenesis and maintenance of the lymphatic endothelium. Vascular endothelial growth factor receptor-2[81] is encoded by the Kdr gene.

[78] In normal conditions, circulating PlGF level is undetectable. Concentration of PlGF increases in atherosclerosis, cutaneous wounds, bone fractures, cancers, etc.

[79] The composition of the Flt gene resembles that of the Fms gene (FMS: feline McDonough sarcoma viral oncogene homolog).

[80] The gene Fms-like Tyr kinase-2 (Flt2) encodes fibroblast growth factor receptor 1 (FGFR1) or cluster of differentiation CD331. The gene Flt3 encodes a cytokine receptor FLT3, a.k.a. fetal liver kinase-2 (FLK2) and CD135, that is expressed on the surface of hematopoietic progenitor cells. The FMS-like Tyr kinase-3 ligand (FLT3L) acts as a growth factor that increases the number of immunocytes, as it activates the hematopoietic progenitors

[81] A.k.a. fetal liver kinase FLK1, kinase insert domain receptor (KDR), and cluster of differentiation CD309.

Subtype VEGFa is a ligand for both VEGFR1 and VEGFR2; VEGFb for VEGFR1 and neuropilin-1.[82] As VEGFa binds to both VEGFR1 and VEGFR2, it stimulates endothelial cell proliferation, migration, and survival for angiogenesis as well as vascular permeability.

Receptors VEGFRs have distinct functions in the growth regulation of blood and lymph vessels. Some VEGF receptor subtypes that are mainly observed on lymphatic endothelial cells are activated neither by VEGFa nor VEGFb, but VEGFc [384]. Isoform VEGFc provokes the proliferation of the lymphatic vessels, but not blood vessels. Subtype VEGFd is also able to trigger growth of lymphatic vessels [385].

Protein VEGF leads to phosphorylation of protein kinase-B and endothelial nitric oxide synthase (NOS3) in arterioles. Phosphorylation by VEGF of PKB, but not NOS3, is significantly reduced in coronary arterioles of old rats with respect to that of young rats. Decay in flow-induced vasodilation in coronary arterioles with aging involves the VEGFR2–phosphatidylinositol 3-kinase–PKB pathway [386].

Vascular endothelial cells continuously perceive mechanical and chemical stimuli and coordinate the activity of the corresponding signaling pathways. Once stimulated by chemicals, VEGF quickly binds VEGFR2, recruiting in particular adaptor NCKβ to trigger the mitogen-activated protein kinase (MAPK; Vol. 3 – Chap. 5. Mitogen-Activated Protein Kinase Modules) pathway.[83] Mechanical stimuli, such as wall shear stress, activate similar plasmalemmal targets (e.g., integrins and VEGFR2) and effectors (e.g., extracellular signal-regulated protein kinases and Jun N-terminal kinases). These convergent modules of the pathway triggered by mechanical and chemical stimuli then require insulation by different molecular complexes to keep their specificity, i.e., to possess divergent modules responsible for the specific response to different stimuli. VEGF, but not shear stress, induces the formation of a VEGFR2–NCKβ complex (Fig. 3.3) [387]. The wall shear stress activates extracellular signal-regulated protein kinase via VEGFR2 and recruitment of CBL adaptor, but not NCKβ.[84] It bypasses VEGFR2 phosphorylation for JNK activation, but uses small GTPases Rho, and Src, PI3K, and RoCK kinases.

[82] Orf-virus derived protein VEGFe is a specific ligand for VEGFR2 and Nrp1 receptors.

[83] Activated VEGFR2 interacts with various proteins, such as GRB2, NCK, PI3K, SHC, and SOS mediators. Adaptor NCKβ binds various receptor Tyr kinases, such as EGFR and PDGFR. It also associates with other proteins, such as P21-activated kinase, thereby serving as a docking protein to expose the protein to its activators (CDC42 and Rac1 for PAK). Adaptor NCKβ can then lead to the stimulation of effectors ERK and JNK kinases of the MAPK cascade.

[84] Wall shear stress can stimulate ERK via manifold proteins, such as integrins, PECAM1, FAK, Src, and PKCϵ. Both integrins or PECAM1 can interact with VEGFR2.

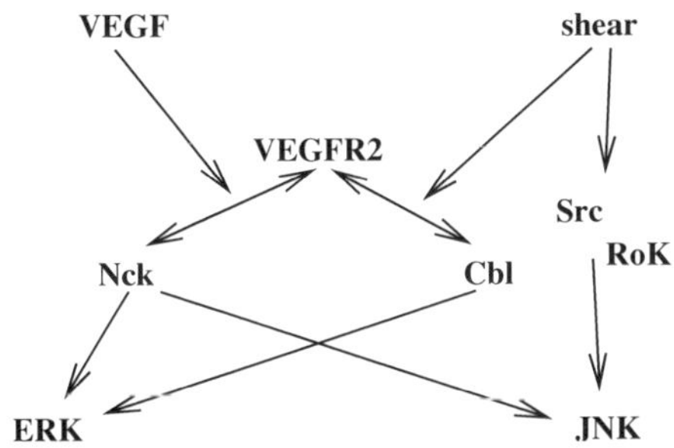

Figure 3.3. VEGF and insulation (Source: [387]). Both VEGF and wall shear stress stimulate VEGFR2. Because of the presence of this convergent node, insulation by different molecular complexes is required for specificity. Whereas VEGF induces the formation of a complex between VEGFR2 and NCKβ to activate its effectors, such as ERK and JNK, wall shear stress leads to the recruitment of CBL to stimulate ERK. Wall shear stress activates JNK owing to Src, PI3K, and RoCK kinases.

Angiogenesis is achieved by the concerted migration and proliferation of endothelial and mural cells (pericytes and vascular smooth muscle cells). Endothelial cells form new vessels, whereas mural cells support and stabilize nascent vessels. However, upon stimulation of mural cells by PDGF to initiate angiogenesis by releasing pro-angiogenic mediators, in the presence of synergistic stimulation of VEGF and PDGF, VEGF is able to impede neovessel maturation, as it can ablate pericyte coverage of nascent vascular sprouts and hence destabilize these nascent vessels [388]. During PDGF–VEGF costimulation, VEGF activates VEGFR2 that suppresses PDGFRβ signaling in vascular smooth muscle cells, as VEGF induces the formation of VEGFR2–PDGFRβ complex. Dual expression of PDGFRβ and VEGFR2 is limited to α-smooth muscle actin+ perivascular cells. Inhibition of VEGFR2 prevents assembly of this receptor complex. In the case of joint exposure to both VEGF and PDGF, an antagonistic competition occurs between these 2 factors, as they both combine with FGF2. Tumors express high VEGF levels and develop tortuous, leaky, immature blood vessels with minimal pericyte coverage due to coexpression of VEGFR2 and PDGFRβ on perivascular cells. Anti-VEGF therapy normalizes tumor vasculature, hence accelerating tumor growth, but allows efficient delivery of cytotoxic agents.

Certain organs such as kidneys (especially glomerular endothelial cells) constitutively express VEGF in adulthood. Glomerular VEGF is increased in response to hypertension and activation of the renin–angiotensin system [389]. VEGF upregulation can correspond to adaptation to hypertension, as VEGF increases expression of endothelial nitric oxide synthase.

Table 3.20. VEGF signaling in cardiomyocytes (Source: [391]). Copper heigthens the VEGFR1-to-VEGFR2 ratio (DAG: diacylglycerol; ERK: extracellular-regulated kinase; GRB: growth factor receptor-bound protein; IP3: inositol trisphosphate; MAP2K: mitogen-activated protein kinase kinase; PKC: protein kinase-C; PKG: cGMP-dependent protein kinase; PLC: phospholipase-C; SOS: Son of sevenless; SHC: Src homology and collagen-like protein).

Effect	Pathway
Proliferation	VEGFR2–SHC–GRB2–SOS–Ras–Raf–MAP2K1/2–ERK1/2
CMC growth	VEGFR2–PLCγ–DAG/IP3–PKC–ERK1/2
Regression of hypertrophy	VEGFR1–PKG1

3.9.7 VEGF Signaling in the Heart

Endothelial growth factor receptors are differentially expressed in distinct types of vessels in the human heart. Endothelial receptor Tyr kinases include 3 members of the vascular endothelial growth factor receptor family and 2 members of the angiopoietin receptor (TIE) family. In addition, $VEGF_{165}$ isoform binds to semaphorin receptor neuropilin-1. In human fetal hearts, the endocardium contains VEGFR1, VEGFR2, Nrp1, TIE1, and TIE2, but not VEGFR3 [390]. Superficial coronary vessels possess VEGFR1, Nrp1, Tie1, and Tie2, but neither VEGFR2 nor VEGFR3. Myocardial capillaries and epicardial blood vessels possess VEGFR1, VEGFR2, Nrp1, and Tie1 (weakly Tie2). Epicardial lymphatic vessels are labeled by VEGFR2 and VEGFR3 (weakly Tie1 and Tie2), but neither VEGFR1 nor Nrp1 receptor. Besides, endothelial and smooth muscle cells as well as cardiomyocytes develop from a common VEGFR2+ cardiovascular progenitor.

Vascular endothelial growth factor binds to its receptors to participate in cardiovascular development, as it favors stem cell differentiation into cardiomyocytes as well as stem cell migration and survival [391]. In adults, because it activates mitogen-activated protein kinases, it can elicit re-entry of cardiomyocytes into the cell division cycle, thereby promoting cardiac adaptive (due to repeated exercises) and maladaptive (in response to pressure overload) hypertrophy.

In mouse myocardium, VEGFR1-specific ligand $VEGFb_{186}$ and VEGFR2-specific agonist VEGFe are equally potent in inducing angiogenesis, but less efficient than $VEGFa_{165}$ that stimulates both VEGFR1 and VEGFR2 [392].

In cardiomyocytes, VEGF causes cardiac hypertrophy or regression according to the prevalent binding to VEGFR2 or VEGFR1, respectively [391]. Copper increases the ratio of VEGFR1 to VEGFR2, hence switching VEGF signaling from cell growth to reversal of cardiomyocyte hypertrophy via cGMP-dependent protein kinase PKG1 [391] (Table 3.20).

3.10 Midkine and Pleiotrophin

Midkine (mid-gestation and kidney protein [MdK]), or neurite growth-promoting factor-2 (NeGF2),[85] and pleiotrophin (Ptn or NeGF1)[86] constitute the NEGF family. Midkine is strongly produced during mid-gestation. Midkine and pleiotrophin expression is restricted in adults.

Midkine and pleiotrophin are pleiotropic, as they act in cell proliferation, survival, and migration, especially during angiogenesis and neurogenesis, as well as inflammation and fibrinolysis in vascular endothelia [393]. They also operate in epithelial-mesenchymal interactions during organogenesis.

Both midkine and pleiotrophin bind to transmembrane receptor Tyr kinase anaplastic lymphoma kinase [394] (ALK; Vol. 3 – Chap. 8. Receptor Kinases). Midkine targets a complex formed by chondroitin sulfate proteoglycan, receptor protein Tyr phosphatase PTPRb (PTPζ), low-density lipoprotein receptor-related protein (LRP), anaplastic leukemia kinase, and syndecans [393]. Signaling mediators of both midkine and pleiotrophin include phosphatidylinositol 3-kinase and mitogen-activated protein kinase.

In endothelial cells associated with smooth muscle cells, midkine not only increases their proliferation rate, but also synthesis of glycosaminoglycan [395]. Interaction between endothelial and smooth muscle cells is required.

Upon exposure to midkine, synthesis of interleukin-8 by smooth muscle cells rises. Once secreted, IL8 can act on endothelial cells (paracrine regulation). In addition, midkine and pleiotrophin are upregulated during ischemia. Midkine is also involved in intimal hyperplasia.

3.11 Semaphorins

Semaphorins form a family of membrane-bound and secreted proteins that are short-range repulsive cues for endothelial cells during angiogenesis and for neurons during neurogenesis. They prevent axons to develop toward inappropriate regions. In particular, semaphorin-3 controls both axon guidance and angiogenesis. Semaphorins also intervene in immune defense, cardiogenesis, and osteogenesis.

There are 8 classes of semaphorins. Class-1 and -2 semaphorins exist in invertebrates only, class-5 semaphorins are specific to virus and observed in both vertebrates and invertebrates, and class-3, -4, -6, and -7 semaphorins in vertebrates only. In humans, they are encoded by the genes SEMA3A to SEMA3G,

---

[85] A.k.a. retinoic acid-inducible heparin-binding protein (RIHB) and amphiregulin-associated protein (ARAP). This alias is also used for ArfGAP with RhoGAP, ankyrin repeat, PH domains.

[86] A.k.a. heparin-binding brain mitogen [HBBM], heparin-binding growth associated molecule [HBGAM], heparin-binding growth factor HBGF8, and osteoblast-specific factor OSF1.

SEMA4A to SEMA4G, SEMA5A and SEMA5B, SEMA6A to SEMA6D, and SEMA7A.

Semaphorins signal via the plasmalemmal receptors *plexins* and *neuropilins.* In fact, many semaphorins interact with receptor complexes that are formed by neuropilins and plexins. However, semaphorin-3E can signal independently of neuropilin via plexin-D1 that is expressed in endothelial cells. Loss of plexin-D1 causes aberrant sprouting. Plexin-A1 is the principal receptor component for class-3 and -4 semaphorins. Independently of neuropilin, plexin-A1 operates in neurogenesis and cardiogenesis as a receptor for class-6 semaphorins (Sema6c and Sema6d).

Neuropilin-1 (Nrp1) is a coreceptor for semaphorin-3A and vascular endothelial growth factor-A that acts with VEGFR2, both in vascular endothelial and smooth muscle cells.[87] Neuropilin-1 increases VEGF binding in both vascular endothelial and smooth muscle cells, in which it regulates VEGFR2 expression. Neuropilin-1 enhances VEGF signaling in endothelial cells, whereas it hinders VEGF activity in smooth muscle cells. Neuropilin-1 is also targeted by heparan sulfate or chondroitin sulfate [396].

Several semaphorins are costimulators that favor the activation of B and T lymphocytes, macrophages, and dendritic cells. In the immune system, plexin-A1 is expressed specifically in dendritic cells. Dendritic cells are antigen-presenting cells that reside in tissues as sentinels and move to ensure their immune role at the right site at the appropriate instant. After antigen exposure, they enter into lymphatics and migrate to lymphoid organs, where they activate T lymphocytes. Plexin-A1 is required for the entry of dendritic cells into lymphatics [397]. Semaphorin-3A produced by lymphatic cells, but neither Sema6c nor Sema6d, is involved in the transmigration of dendritic cells through the lymphatic endothelium. Semaphorin-3A promotes phosphorylation of the myosin light chain, hence actomyosin contraction at the trailing edge of migrating dendritic cells.

3.12 Angioneurins

Angioneurins constitute a group of substances that influence both neural and vascular cell fate, especially survival and growth, as they are both neurotrophic and angiogenic factors [398]. Neurovascular crosstalk uses common signaling pathways, such as Notch cues. Angioneurins are secreted by endothelial cells, neural stem cells, and astrocytes.

Angioneurins not only support neuroregeneration, but also prevent neurodegeneration.[88] Angioneurins protect neurons from ischemia and control

[87] The composition of Nrp1 glycosaminoglycan chains differs between endothelial cells and smooth muscle cells. These cell types then differentially respond to VEGF and neuropilin.

[88] Several neurodegenerative disorders (Alzheimer's disease, Parkinson's disease, amyotrophic lateral sclerosis, etc.) are characterized by vascular defects.

Table 3.21. Angioneurin types classified by discovered effects (Sources: [398]).

Neurotrophic and axon-guidance factors	Nerve growth factor, brain-derived neurotrophic factor, neurotrophin NT3 and NT4, semaphorins and neuropilin receptors, netrins, slits, ephrins
Angiogenic factors	Vascular endothelial growth factor-A, -B, and -C, fibroblast growth factor-2, platelet-derived growth factor, transforming growth factor-β1, insulin-like growth factor-1, hepatocyte growth factor, epidermal growth factor, angiogenins, angiopoietin-1, progranulin

the blood–brain barrier integrity. They also promote neurogenesis, because they influence synaptic transmission, stimulate axon sprouting, neurite extension and branching, and elicit myelination or remyelination via migration and proliferation of Schwann cells.

Many molecules act on the development of both the nervous and vascular systems. Neurogenic and angiogenic factors guide blood vessels and nerves along predestined tracks to their final destination during genesis. Vascular endothelial growth factor-A controls both angiogenesis and neurogenesis. The VEGFa isoform enhances vascular perfusion and promotes the survival of multiple types of neurons, as well as astrocytes and microglial cells. Brain-derived neurotrophic factor (BDNF) acting via its receptor NTRK2 on endothelial cells stimulates angiogenesis in the heart, skeletal muscle, and skin, and maintains the stability of cardiac vessel walls during development. Neurotrophin NT4 binds to NTRK2 and has similar functions. Neurotrophin NT3 that binds to NTRK3 and leukemia-inhibitory factor (LIF) of the interleukin-6 family to inhibit the growth of some endothelial cells. Nerve growth factor (NGF) increases VEGF expression and enhances vascular cell growth. Ciliary neurotrophic factor (CNF) and glial-cell-line-derived neurotrophic factor (GDNF) do not act on angiogenesis.

Fibroblast growth factor FGF2, transforming growth factor TGFβ1, and hepatocyte growth factor ensure the integrity of the blood–brain barrier. FGF2 acts on astrocyte architecture and HGF and FGF2 upregulate tight-junction proteins. Angiopoietin-1, stromal-derived factor-1α, and vascular endothelial growth factor are synthesized by endothelial cells to release neurogenic factors, such as neurotrophins and BDNF. Astrocytes secrete FGF2, IGF1, and VEGF that target neural stem cells. Erythropoietin is also involved.

3.13 Adrenomedullin

Adrenomedullin is a peptide discovered in 1993 in pheochromocytoma in adrenal medulla [399]. Mature active adrenomedullin shares moderate structural similarity to the calcitonin family of regulatory peptides (amylin, calcitonin, and calcitonin gene-related protein).

The main sources of plasma adrenomedullin are vascular endothelial and smooth muscle cells. Adrenomedullin is also expressed in adventitial fibroblasts. Adrenomedullin circulates in blood in high concentrations. Circulating adrenomedullin (plasma half-life 22 mn) includes mature, amidated and predominant, inactive, glycated form.

Adrenal cells secrete adrenomedullin as well as proadrenomedullin peptide (PAMP). The PAMP peptide causes a dose-dependent elevated release of all steroids (aldosterone, cortisol, and dehydroepiandrosterone). On the other hand, adrenomedullin only increases aldosterone and cortisol secretion, but not that of dehydroepiandrosterone [400].

Both adrenomedullin and PAMP can act as auto- and paracrine regulators of adrenal steroid secretion. Moreover, PAMP, but not adrenomedullin, is an inhibitor of adrenal catecholamine release in vivo [401].

The human adrenomedullin gene (AM)[89] encodes a 185-amino acid precursor peptide that can be differentially cleaved to form 52-amino acid adrenomedullin and various peptides: adrenomedullin fragments AM_{22--52} and $AM_{95--146}$ and adrenotensin.

Oxidative stress increases adrenomedullin production. The blood concentration of adrenomedullin increases in cardiovascular diseases, such as atheroma, heart failure, hypertension, and septic shock [402]. Adrenomedullin protects against progression of vascular damage and maladaptive remodeling.

Adrenomedullin binds to heterodimers formed by calitonin receptor-like receptor (CalcLR) and receptor activity-modifying protein (RAMP),[90] as well as calcitonin gene-related protein receptors (Vol. 3 – Chap. 7. G-Protein-Coupled Receptors).

Adrenomedullin stimulation produces cyclic adenosine monophosphate and nitric oxide. It indeed activates adenylyl cyclase and nitric oxide synthase [403]. In fact, adrenomedullin stimulates inducible nitric oxide synthase NOS2 in interleukin-1β-stimulated vascular smooth muscle cells via, at least partly, a cAMP-dependent pathway.

Ventriculomyocytes and fibroblasts also produce adrenomedullin [404]. Cytokines, such as IL1β and TNFα, regulate its synthesis and secretion in the cardiac ventricles.

Adrenomedullin is a potent and long-lasting vasodilator that is able to initiate hypotension, as it reduces peripheral resistances. Reduction in arterial

[89] The AM gene is located on chromosome 11 with 4 exons and 3 introns. It is expressed in most organs.

[90] Receptor activity-modifying protein RAMP2 has a higher affinity than RAMP3 protein.

pressure is associated with a rise in cardiac output, as adrenomedullin has positive inotropic effect.

Adrenomedullin attenuates the effect of angiotensin-2, but not that of noradrenaline. In addition, it increases plasma renin activity, but impedes an angiotensin-2-induced rise in plasma aldosterone level [405].

Adrenomedullin lowers oxidative stress and prevents endothelial cell apoptosis. It also regulates endothelial permeability and stabilizes the endothelial barrier [406]. It diminishes thrombin- and H_2O_2-induced myosin light-chain phosphorylation as well as stress fiber formation and gap formation in cultured human umbilical vein endothelial cells.

Adrenomedullin contributes to the differentiation of bone marrow-derived mononuclear cells into endothelial progenitors [407]. It favors survival, adhesion, and differentiation of transplanted mononuclear cells.

Adrenomedullin hinders the expression of hepatocyte growth factor and cyclooxygenase-2 in gastric mucosa via CGRP receptors [408].

Adrenomedullin works via CalcLR–RAMP2 complex, cAMP, and phosphorylated extracellular signal-regulated kinases during vasculogenesis and lymphangiogenesis (Vol. 5 – Chap. 10. Vasculature Growth). This lymphatic growth factor promotes also lymphangiogenesis after injury [409].

Prospero-related homeobox transcription factor Prox1 that is specific to venous and lymphatic endothelial cells upregulates CalcRL receptor. In venous endothelium, Prox1 controls the transition from venous to lymphatic endothelial cells and lymphatic sprouting from veins. Adrenomedullin elicits proliferation, migration, and lymph vessel formation of cultured human lymphatic endothelial cells. In injury sites, adrenomedullin increases the number of both lymphatic and blood vessels, at least partly, via the cAMP–ERK pathway.

3.14 Progranulin

Progranulin (PGrn)[91] is a pleiotropic, cysteine-rich growth factor[92] that is glycosylated [411]. Its half-life equals about 40 h [410]. Proteolytic cleavage of this precursor by extracellular peptidases, such as neutrophil elastase, serine proteinase-3 (Prtn3),[93] MMP14, and ADAMTS7, generates 6-kDa, pro-inflammatory granulins (Grn) [410, 412].[94]

[91] A.k.a. acrogranin, acrogranulin (from acrosome of the sperm head), granulin–epithelin precursor (GEP), proepithelin (PEpi), prostate cancer (PC) cell-derived growth factor (PCDGF), and epithelial transforming growth factor (TGFe).

[92] Progranulin contains seven and a half repeats of a cysteine-rich motif (CX5-6CX5CCX8CCX6CCXDX2HCCPX4CX5-6C; X, any amino acid) in the order P-G-F-B-A-C-D-E, where A-G are full repeats and P is the half-motif [410].

[93] A.k.a. leukocyte proteinase-3, neutrophil proteinase-4 (NP4), myeloblastin (MBn), and Wegener granulomatosis auto-antigen.

[94] A.k.a. epithelins.

In the central nervous system, this neurotrophic factor is synthesized by microglial cells and neurons.[95] It is also produced by epithelial cells, especially those of rapidly cycling epithelia (skin and gastrointestinal and reproductive tracts), and much less strongly in weakly proliferating epithelia (lung alveoli and nephron) [413]. It is expressed by mesenchymal cells such as fibroblasts[96] and chondrocytes, as well as leukocytes, particularly immunocytes [413].[97]

This multifunctional, secreted protein mediates cell cycle progression and cell motility. It is indeed manufactured during tissue development and remodeling. In particular, this autocrine growth factor is a potent mitogen for fibroblasts. It increases the expression of cyclin-D and -B. Progranulin binds the matrix proteins perlecan and chondrocyte oligomeric matrix protein (COMP) [414]. Perlecan decreases the proliferative activity of progranulin, whereas COMP fosters it.

Progranulin is implicated not only in embryo- and fetogenesis, but also in tissue repair and inflammation. Overproduction provokes tumorigenesis. Progranulin is involved in inflammation with secretory leukocyte peptidase inhibitor (SLPI)[98] produced by macrophages and neutrophils that binds to and protects it from proteolysis [411]. It interacts also with elastase released in large quantities by neutrophils that cleaves it into smaller peptides. Agent SLPI indeed binds directly to both progranulin and elastase. Hence, anti-inflammatory effect of progranulin is protected by its binding partners, not only secretory leukocyte peptidase inhibitor, but also apolipoprotein-A1. Yet, during inflammation, neutrophils and macrophages release peptidases that digest progranulin into granulins. Granulins can neutralize the anti-inflammatory effects of progranulin.

Progranulin is a potent inhibitor of the inflammatory cytokine tumor-necrosis factor-α. It actually binds directly to the tumor-necrosis factor receptor and antagonizes the TNFα–TNFR interaction, thereby preventing TNFα-initiated signaling [410].

[95] Progranulin is highly expressed in specific subsets of neurons, such as cortical neurons, Purkinje cells of the cerebellum, and granule cells of the hippocampus [413]. Mutations in the PGRN gene cause frontotemporal lobar degeneration [411].

[96] Progranulin is not or weakly detected in most mesenchymal tissues, such as connective and adipose tissues, cardiac, skeletal, and smooth muscle, and vascular conduits [413].

[97] Progranulin is highly expressed in lymphoid tissues of the lung, gut, and spleen. In the spleen, the progranulin transcript is confined mainly to marginal cells of the periarteriolar lymphoid sheath (PALS).

[98] A.k.a. antileukoproteinase (ALP), mucus peptidase inhibitor (MPI), seminal peptidase inhibitor, and WAP four-disulfide core domain-containing protein WAP4 and WFDC4.

Progranulin contributes to wound healing, during which it promotes angiogenesis [412]. This paracrine factor stimulates proliferation and migration of fibroblasts and endothelial cells.[99]

Progranulin activates the extracellular-regulated kinase modules and the PI3K–PKB and PI3K–S6K cascades [414]. It promotes phosphorylation of focal adhesion kinase, hence cell motility.

3.15 Prokineticins

Prokineticins are small, secreted chemokine-like peptides that constitute a family of 2 known members: prokineticin-1 and -2. They signal via 2 ubiquitous Gq-coupled prokineticin receptors PKR1 (PK_1) and PKR2 (PK_2) [415]. Prokineticin-1 and -2 as well as PKR1 are synthesized in the human myocardium by endothelial cells and cardiomyocytes [416].

Prokineticins and their receptors are also highly expressed in bone marrow cells, circulating leukocytes, and peripheral immunocytes. They act as cytokines [417]. Prokineticin-2 is a more potent ligand at both PKR1 and PKR2 than prokineticin-1.

Prokineticins promote proliferation and differentiation of granulocytic and monocytic cell lineages in bone marrow. They also stimulate gut smooth muscle contraction, especially in myenteric and submucosal enteric plexuses of proximal colon.

Prokineticin-1 elicits angiogenesis in various endocrine glands [418].[100] Signaling via Gq and G13 indeed contributes to vessel formation.

The receptor PKR1 favors cardiomyocyte survival and reduces hypoxia-induced apoptosis. Signaling via $G\alpha_{q/11}$ regulates the cardiac development and hypertrophy. Other Gq-coupled receptors are involved in cardiomyocyte protection.

Overexpression of PKR1 by cardiomyocytes upregulates prokineticin-2 that has paracrine activity and provokes the proliferation and differentiation of epicardial-derived progenitor cells [419]. The receptor PKR1 also enhances the proliferation of endothelial cells in the myocardium. Moreover, PKR1 in coronary endothelial cells favors the postnatal coronary angiogenesis.

On the other hand, PKR2 overexpression by cardiomyocytes causes: (1) cardiac hypertrophy with increased sarcomere numbers and without dysfunction (autocrine activity) and (2) abnormal endothelial cell shape with altered distribution of tight junction protein zona occludens-1 (Vol. 1 – Chap. 7. Plasma Membrane) that leads to endothelial fenestrations and vascular leakage (paracrine activity) without angiogenesis [420].

---

[99] Although quiescent fibroblasts and endothelial cells synthesize low amounts of progranulin, these cells can rapidly raise their production when they deploy a tissue remodeling program with elevated cell proliferation and migration.

[100] Hence its original name endocrine gland-derived vascular endothelial growth factor (egVEGF).

Prokineticin-2 expression depends on light entrainment in the suprachiasmatic nucleus of hypothalamus. The gene PK2 is controlled by the circadian clock (Chap. 5) and its transcription is modulated by light [421]. It transmits circadian rhythm with a positive feeback on its expression and operates as an output for the circadian locomotor rhythm. In the central nervous system, prokineticin-2 is also involved in olfactory bulb morphogenesis that persists in adult brain [422].

3.16 Melatonin

Melatonin is synthesized and released into the circulation by the epiphysis (epiphysis cerebri), or pineal gland, under the control of the circadian clock. Exogenous melatonin does not modify the cardiac frequency and mean arterial pressure, as well as the cerebral blood flow. However, it influences the blood flow distribution. The blood flow renal blood flow velocity lowers after the ingestion of melatonin with respect to administration of sucrose (40.5 ± 2.9 vs. 45.4 ± 1.5 cm/s) [423]. On the other hand, the forearm blood flow rises upon melatonin stimulation (2.4 ± 0.2 vs. 1.9 ± 0.1 ml×100 ml/mn). Renal vasoconstriction results from a melatonin-mediated increase of the sympathetic input to the renal arterial bed.

3.17 Sphingosine 1-Phosphate

Sphingosine 1-phosphate (S1P) is an active metabolite of sphingolipids at the plasma membrane that acts as a lipid growth factor. Sphingosine 1-phosphate has a short life, as it is degraded by plasmalemmal S1P lyase or dephosphorylated by plasmalemmal S1P phosphatase.

Sphingolipids in cholesterol-enriched membrane rafts are rapidly metabolized upon stimulation of plasmalemmal receptors that convert sphingomyelin and glycosphingolipids to ceramide by sphingomyelinases and subsequently to sphingosine (Sph) by ceramidases. Sphingosine kinases SphK1 and SphK2 phosphorylate sphingosine to generate lysosphingolipid.

Sphingosine 1-phosphate has a cell-intrinsic (e.g., acts on calcium flux) and -extrinsic activity via cognate G-protein-coupled receptors (S1PR1–S1PR5 or $S1P_1$–$S1P_5$; Vol. 3 – Chap. 7. G-Protein-Coupled Receptors). It is exported from producing cells by ATP-binding cassette or other transporters.

Among its intrinsic effects, sphingosine 1-phosphate participates in TNFα signaling and NFκB activation in anti-apoptotic, inflammatory, and immune processes. Ubiquitin ligase tumor-necrosis factor receptor-associated factor TRAF2 is involved in polyubiquitination of TNFRSF receptor-interacting kinase RIPK1 that then recruits and stimulates IκB kinase to activate the transcription factor NFκB. It binds to sphingosine kinase-1 as well as sphingosine 1-phosphate [424]. The latter, which stimulates the ligase activity of TRAF2, is required for Lys63-linked polyubiquitination of RIPK1, thereby preventing RIPK1 switching from a prosurvival to pro-apoptotic adaptor.

Blood concentration of S1P (in the low-micromolar range) is mainly determined by the secretion from erythocytes and endothelial cells. Free and albumin-bound S1P are more easily degraded than high-density lipoprotein-bound S1P.

Tissular concentration of S1P is low compared with that in lymph (in the hundred-nanomolar range) and blood. Sphingosine 1-phosphate is also secreted by platelets and mastocytes activated by thrombin or IgE-bound antigen, respectively.

3.17.1 Sphingosine 1-Phosphate in Platelets

Sphingosine 1-phosphate produced from phosphorylation of sphingosine of platelet membrane by sphingosine kinase, activates endothelial cells by phospholipase-D, independently of protein kinase-C and Ca^{++} [425]. S1P is produced and is then secreted [426].[101]

In addition, S1P in synergy with thrombin causes tissue factor release from endothelial cells to promote blood coagulation. Secretion of S1P by platelets increases adhesion of leukocytes on the endothelium.

3.17.2 Sphingosine 1-Phosphate in Endothelial Cells

Endothelial cells have an intracellular reserve of functional $S1P_1$ receptors in caveolae. Sphingosine 1-phosphate is a ligand for some types of the family of G-protein-coupled lysosphingolipid receptors.[102]

Sphingosine 1-phosphate mediates endothelial cell maturation, migration, and angiogenesis (Fig. 3.4). It controls vascular permeability (barrier-enhancing effect), as it regulates Rho GTPases and cortical actin polymerization.[103]

Moreover, S1P enhances assembling of adherens junctions and focal adhesions [429, 430], using the Rho and Rac pathways. The S1P–$S1P_1$ complex

[101] Lysophospholipids sphingosine 1-phosphate and lysophosphatidic acid regulate various cell functions [426]. Lysophosphatidic acid is most often synthesized extracellularly in microvesicles shed from cell membranes. Similar to most small lipids in extracellular spaces, the newly generated and secreted lysophospholipids are bound by plasma proteins.

[102] A.k.a. endothelial differentiation gene receptors (EDGR).

[103] Monomeric GTPases CDC42, Rac1, and RhoA are rapidly activated by wall shear stress. Small GTPase RhoA can be activated within 5 mn after shear stress stimulation, leading to cell rounding via Rho kinase. Endothelial cells then elongate as RhoA activity returns to baseline and Rac1 and CDC42 that are required for cell elongation reach peak activation [427].

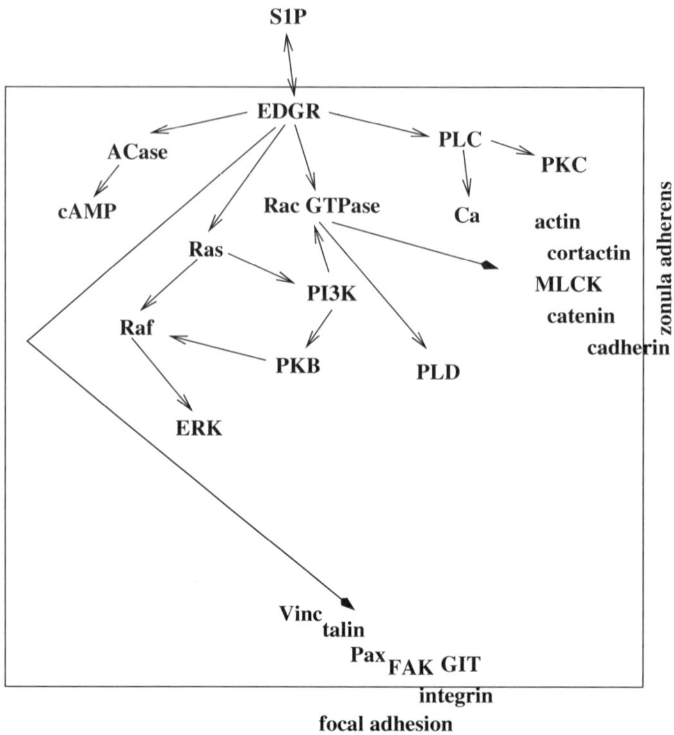

Figure 3.4. Sphingosine 1-phosphate (S1P) effects on the endothelial cell (Source: [428]). Sphingosine 1-phosphate acts from G-protein-coupled receptors on adenylyl cyclase (ACase), small GTPase Ras, phosphatidylinositol 3-kinase (PI3K), and phospholipase-C (PLC). S1P is strongly involved in cell adhesion.

also tightens adherens junctions between endothelial cells. Lipid S1P stimulates the translocation of actin-binding proteins, such as cortactin that enhances actin polymerization (whereas actin-severing proteins such as cofilin are inactivated) and myosin light-chain kinase for adhesion stabilization. It redistributes cadherin and catenin to the cell cortex, as these molecules are involved in adherens junctions.

3.17.3 Sphingosine 1-Phosphate in Immunocytes

3.17.3.1 Sphingosine 1-Phosphate Receptors in Immunocytes

Receptor $S1P_1$ is expressed by most immunocytes, whereas other receptors have a more limited distribution in the immune system. Receptors $S1P_1$ and $S1P_4$ are synthesized by T lymphocytes; $S1P_1$ and $S1P_2$ by mastocytes and macrophages; $S1P_5$ by dendritic and natural killer cells.

Receptor expression depends on cell differentiation stage and cell activation. Immature and mature dendritic cells mainly express S1PR4 and S1PR3, respectively. Changes in local S1P concentration could induce a switch from pro-inflammatory M1 to anti-inflammatory M2 macrophage subtypes. Signaling from S1PR1 regulates proliferation and maturation of T lymphocytes, as well as cytokine synthesis.

3.17.3.2 Sphingosine 1-Phosphate and Cell Exit from Lymphoid Tissues

The development of the immune system relies, in particular, on egress of developing T lymphocytes, or thymocytes, from the thymus to organs, where they can defend against microorganisms.

Sphingosine 1-phosphate operates in immunocyte function and migration to ensure the body's immunity [431]. Receptor S1PR1 intervenes in B- and T-lymphocyte exit from lymph nodes as well as mature thymocyte egress from thymus, IgG-secreting plasmocyte from spleen, and IgA-producing B lymphocyte from Peyers patches.

The thymus is used for T-cell development and tolerance induction. Double-negative (CD4− and CD8−) precursors develop into double-positive thymocytes in the thymic cortex. Cells selected for weak recognition of major histocompatibility complexes give rise to semi-mature, single positive (CD4+ or CD8+) thymocytes that localize to the medulla. On the other hand, strongly self-reactive, semimature, single-positive thymocytes are rejected. Cells that pass through the tolerance checkpoint undergo further maturation in the medulla.

In single-positive thymocytes, the transcription factor Krüppel-like factor KLF2 is upregulated. Therefore, KLF2 target genes are activated, such as those that encode $S1P_1$ and L-selectin (or CD62L). Perivascular channels that contain thymocytes in the thymus may serve as a thymocyte reservoir, in which pericyte- and vessel-derived S1P is protected from rapid degradation. Mature single-positive lymphocytes then exit the thymus.

Thymocyte exit from the thymus at corticomedullary junctions via blood vessels, rather than lymphatics (at least in mice), requires S1P detection by thymocytes [432]. Postcapillary thin-walled venules at the corticomedullary junction have relatively large caliber (10–50 μm). They possess a single layer of ensheathing α-smooth muscle actin-positive pericytes. Unlike most vascular beds, thymic blood vessels are ensheathed by neural crest-derived pericytes. These neural crest-derived pericytes release S1P for exiting thymocytes. These specialized pericytes hence promote reverse transmigration of cells across the vascular endothelium. Endothelial cells that also produce S1P can work in synergy with pericytes to elicit thymocyte egress. Lymphatics can also contribute to lymphocyte egress from the thymus.

3.17.3.3 Sphingosine 1-Phosphate and Inflammation

Chemotaxis mediated by S1P depends on S1P concentration and receptor type.[104] The S1P lipid increases inflammation, as it stimulates the production of inflammatory mediators, such as interleukin-1β and tissue factor.

3.17.3.4 Sphingosine 1-Phosphate Effects in Various Immunocyte Types

Sphingosine 1-phosphate activates S1PR3 on dendritic cells and S1PR2 in mastocytes, hence priming degranulation. Elevated circulating S1P concentration often raises helper-2 T-cell responses (without changing or possibly dampening T_{H1} cell response).[105]

Differentiating dendritic cell precursors as well as activated hematopoietic stem cells and mastocytes upregulate S1P receptors. The S1PR1 receptor regulates immunocyte migration (Table 3.22). Receptors S1PR3 and S1PR5 also regulate circulation and localization of dendritic cells and natural killer cells, respectively.

Ligands of S1PR1 sequester lymphocytes in secondary lymphoid organs. They hinder T-cell motility between medullary cords and lymphatic sinuses, mainly the migration through the endothelium of lymph nodes [433]. Expression of S1PR1 by lymphocytes is regulated via Krüppel-like factor-2, downregulated by high S1P levels and C-type lectin CLec2c (or CD69; a receptor in lymphocytes, natural killer cells, and platelets) upon lymphocyte activation.

3.18 Adipokines

Adipose tissue operates as an endocrine organ, as adipocytes secrete adipokines. Adipokines indeed function as hormones and cytokines. They are secreted not only by adipocytes, but also other cell types.

Adipocytes secrete numerous proteins that influence the activity of insulin receptor by auto-, para-, and endocrine mechanisms. Several adipocyte-derived hormones and cytokines, such as leptin, resistin, retinol-binding protein-4, tumor-necrosis factor-α and its soluble receptors sTNFR1 and sTNFR2, interleukin-1 and -6, plasminogen activator inhibitor-1, and lipocain-1 hamper insulin signaling.

On the other hand, apelin, adiponectin, chemerin, omentin, vaspin, and visfatin enhance signal transmission from insulin receptor. Elevated plasma level of leptin and resistin as well as lowered plasma concentration of adiponectin are correlated with risks of cardiovascular diseases (Table 3.23).

[104] In vitro, low and high S1P concentration promotes and reduces chemotaxis, respectively. In mastocytes, S1PR2 can inhibit migration via Rac1 GTPase-activating protein.

[105] In the presence of interleukins IL1β and IL6 and TGFβ, S1P causes differentiation into T_{H17} cells and reduces that of both T_{H1} and T_{H2} cells.

Table 3.22. Sphingosine 1-phosphate receptors (S1PRi or S1P$_i$) in immunocytes (Source: [431]). Receptor S1PR1 increases the chemotaxis of immunocytes, as well as S1PR5 in natural killer (NK) cells and S1PR2 in macrophages, although the latter reduces mastocyte migration.

Cell type	Receptor
Innate immunocytes	
Dendritic cell	S1PR1–S1PR5
Eosinophil	S1PR2–S1PR3
Macrophage	S1PR1/2
Mastocyte	S1PR1/2
NK cell	S1PR5
Adaptive immunocytes	
B lymphocyte	S1PR1/3
NKT cell	S1PR2/4
T lymphocyte	S1PR1/4

Table 3.23. Adipokines and cardiovascular risk (Source: [434]).

Hypoadiponectinemia	Visceral fat accumulation Hyperglycemia Hyperlipidemia Hypertension Endothelial dysfunction Inflammation
Hyperleptinemia	Activated sympathetic system Prothrombotic effect Endothelial dysfunction Vascular smooth muscle cell hypertrophy Cardiovascular remodeling Hypertension
Hyperresistinemia	Impaired glucose metabolism Inflammation Endothelial dysfunction

3.18.1 Adiponectin

Adiponectin[106] is synthesized mainly by adipocytes.[107] It forms homomultimers. Adiponectins exist as low-molecular-weight, full-length trimers, and

[106] Necto: to bind.

[107] The adiponectin gene ADIPOQ localizes to chromosomal band 3q27, a susceptibility locus for diabetes and cardiovascular disease [444]. It is also named C1q and collagen domain-containing protein, adipose most abundant gene transcript-1 (APM1), adipocyte complement-related 30-kDa protein (ACRP30), adiponec-

globular cleavage fragments. The trimer can dimerize to form a mid-molecular-weight hexamer. The latter can oligomerize to give rise to high-molecular-weight polymers, such as 12- and 18-mers. Full-length adiponectin requires post-translational modifications (e.g., hydroxylation and glycosylation) to be active. Low-, mid-, and high-molecular-weight adiponectin complexes bind to heparin-binding EGF-like growth factor (HBEGF); hexa- and polymers to platelet-derived growth factor PDGFb; and polymers to FGF2 fibroblast growth factor.

It is similar to collagen-8 and -10 as well as C1q complement factor (Sect. 3.21). Adiponectin specifically binds to collagen-1, -3, and -5 in intima of catheter-injured vessel wall [435]. Adiponectin is also produced in skeletal myocytes, hepatocytes, cardiomyocytes, and endothelial cells. Adiponectin circulates in blood at high concentrations (5–10 mg/ml). Men have lower plasma adiponectin levels than women.

Adiponectin is involved in the control of fat metabolism and insulin sensitivity, with anti-inflammatory, antidiabetic, and anti-atherogenic activities. In liver and other tissues, adiponectin raises fatty acid oxidation and reduces glucose synthesis. In myocytes, adiponectin-induced activation of cell energy sensor adenosine monophosphate-activated protein kinase stimulates phosphorylation of acetyl coenzyme-A carboxylase, glucose uptake, and fatty acid oxidation. In hepatocytes, full-length and globular adiponectin stimulates AMP-activated protein kinase phosphorylation (activation). Globular adiponectin yields this effect in skeletal myocytes, cardiomyocytes, and hepatocytes. Low-molecular-weight adiponectin, but not high-molecular-weight adiponectin, reduces lipopolysaccharide-mediated interleukin-6 release in human monocytes and represses nuclear factor-κB activation [436]. Adiponectin synthesis by adipocytes is reduced in obesity, insulin resistance, and type-2 diabetes (metabolic syndrome; Vol. 6 – Chap. 7. Vascular Diseases).

Adiponectin possesses an adaptor, phosphotyrosine interaction, PH domain, and leucine zipper containing-protein APPL1,[108] as well as 2 G-protein-coupled receptors, AdipoR1 and AdipoR2 (Vol. 3 – Chap. 7. G-Protein-Coupled Receptors), that are mainly synthesized in skeletal muscle and liver, respectively. In addition, T-cadherin may contribute to adiponectin signaling. Activation of AdipoR1 and AdipoR2 by adiponectin leads to phosphorylation of AMP-activated protein kinase and its downstream target acetylCoA carboxylase.

Adiponectin promotes endothelial expression of vascular cell adhesion molecule VCAM1, intercellular adhesion molecule ICAM1, and pentraxin-3

tin serum level quantitative trait locus 1 (AdipQTL1), gelatin-binding protein-28 (GBP28), and ACDC.

[108] A.k.a. Deleted in colorectal cancer (DCC)-interacting protein-13α. It is involved in crosstalk between adiponectin and insulin signaling. It binds to many proteins, such as small GTPase Rab5a, transmembrane dependence (i.e., active upon ligand binding as well as unbound) receptor DCC, and PKB2 kinase and phosphoinositide 3-kinase catalytic subunit PI3KC1Cα (encoded by the PIK3CA gene).

Table 3.24. Effects of adipokines on glucose level. Adiponectin inhibits hepatic glucose production and enhances glucose uptake in muscle. High blood interleukin-6 level decreases insulin sensitivity.

Decrease	Increase
Adiponectin	Resistin
Leptin	RBP4
Omentin	TNFα, IL6
Visfatin	

[444].[109] Adiponectin acts as an anti-oxidant, as it decreases reactive oxygen species. It also augments endothelial production of nitric oxide.

Adiponectin stimulates a ceramidase activity associated with its receptors AdipoR1 and AdipoR2 in cells that express both receptors to promote insulin sensitivity and cell survival and decrease inflammation [437]. It enhances ceramide catabolism and formation of anti-apoptotic sphingosine 1-phosphate. This effector explains the entire set of beneficial effects exerted by adiponectin.

3.18.1.1 Adiponectin and Cell Metabolism

Adiponectin affects gluconeogenesis and lipid catabolism. Adiponectin hinders atherosclerosis. Adiponectin favors insulin activity in the muscles and liver via activated AMPK (Table 3.24). Nuclear receptor NR1c3, or transcription factor PPARγ, upregulates adiponectin expression and reduces the plasma concentration of TNFα. Tumor-necrosis factor-α produced in adipose tissues, prevents adiponectin synthesis. It phosphorylates insulin receptors, and hence desensitizes insulin signaling.

3.18.1.2 Adiponectin and Myocardium

Epicardial adipose tissue as well as cardiomyocytes synthesize adiponectin. In cardiomyocytes, adiponectin causes glucose and fatty acid uptake and AMPK phosphorylation. Adiponectin expression is correlated with coronary function. Plasma adiponectin level in the great cardiac vein is significantly higher than that in the left coronary artery [438]. Therefore, adiponectin is locally produced in the coronary circulation and participates in the modulation of coronary blood flow in different regions of the myocardium.

[109] Long pentraxin Ptx3 is related to the pentraxin family (C-reactive protein and serum amyloid-P component in humans). It is prominently synthesized by endothelial cells, macrophages, and dendritic cells, in response to inflammatory signals (e.g., interleukin-1, tumor-necrosis factor, and lipopolysaccharides). It participates in the regulation of innate immunity against pathogens. Long pentraxin Ptx3 can be produced in endothelial cells, hepatocytes, and fibroblasts by IL1β and tumor-necrosis factor-α. It is also detected in cardiomyocytes.

Full-length adiponectin promotes cardiomyocyte survival via the AMPK pathway. It also protects from ischemia–reperfusion injury (Vol. 6 – Chap. 6. Heart Pathologies) via suppression of tumor-necrosis factor signaling owing to cyclooxygenase-2 [436]. Adiponectin stimulates production of nitric oxide via AMPK-mediated phosphorylation of NOS3 nitric oxide synthase. However, adiponectin inhibits expression of inducible nitric oxide synthase NOS2 and NADPH oxidase. It then blocks peroxynitrite formation and represses oxidative and nitrative stresses.

3.18.1.3 Adiponectin and Vascular Endothelium

Adiponectin suppresses endothelial cell apoptosis [439]. Adiponectin is involved in angiogenesis. It stimulates formation of capillary-like structures in vitro by umbilical vein endothelial cells via the AMPK–NOS3 axis [440]. Adiponectin promotes phosphorylation of AMP-activated protein kinase, stimulates the phosphatidylinositol 3-kinase–protein kinase-B pathway, and activates endothelial nitric oxide synthase NOS3 [436].

On the other hand, adiponectin impedes endothelial cell proliferation and superoxide production induced by oxidized low-density lipoproteins via NADPH oxidase [441]. Globular adiponectin also enhances NOS3 activity that is repressed by oxidized low-density lipoproteins. Globular adiponectin attenuates or suppresses proliferation and migration of coronary artery endothelial cells induced by vascular endothelial growth factor as well as related signaling effects, such as generation of reactive oxygen species, activation of protein kinase-B, kinases ERK1 and ERK2, and GTPase RhoA in vitro [442].

Adiponectin protects the cardiovascular system owing to its insulin-sensitizing and anti-inflammatory properties via AMP-activated protein kinases and the cAMP–PKA pathway. In endothelial cells, both globular and full-length adiponectins enhance the production of nitric oxide and limit synthesis by mitochondria and NADPH oxidases of reactive oxygen species induced by elevated glucose levels.

In addition, the cAMP–PKA pathway mediates adiponectin effects in endothelial cells against activity of tumor-necrosis factor and high glucose levels. Adiponectin then impedes expression of cell adhesion molecules (e.g., E-selectin and vascular cell adhesion molecule-1 [443]) induced by TNFα, hence leukocyte extravasation.

In human aortic endothelial cells, adiponectin suppresses TNFα-induced IκBα phosphorylation and subsequent NFκB activation. As it leads to cAMP accumulation, it allows crosstalk between cAMP–PKA and NFκB pathways. Globular and full-length adiponectin also protect against an increase in endothelial permeability induced by angiotensin-2 or tumor-necrosis factor [436].

3.18.1.4 Adiponectin and Inflammation

Adiponectin has dominant anti-inflammatory features, thus anti-atherogenic and antidiabetic effects (Table 3.25). Adiponectin regulates the expression

Table 3.25. Adipocytokines and their activity in inflammation and immunity (Sources: [444, 445]). Adiponectin antagonizes tumor-necrosis factor-α (TNFα), as it impedes its production in various cells, such as hepatocytes and macrophages, and counteracts its effects. It also inhibits endothelial nuclear factor-κB (NFκB) signaling via a cAMP-dependent pathway. Adiponectin may play a role in cell growth, angiogenesis, and tissue remodeling by binding and sequestering various growth factors with distinct binding affinities according to its complex type (tri-, hexa-, or polymer).

Adipokin	Inflammatory and immune effects
Adiponectin	Anti-inflammatory ↓ endothelial adhesion molecules, ↓ phagocytosis, T-cell responses, ↓ B-cell lymphopoiesis, ↓ NFκB, TNFα, IL6, Ifnγ, ↑ IL1RA, IL10 Pro-inflammatory ↑ CXCL8 in presence of lipopolysaccharide
Leptin	Pro-inflammatory ↑ TNFα, ROS, IL6, IL12, chemotaxis, neutrophil activation, thymocyte survival, lymphopoiesis, T-cell proliferation, NK-cell function, ↑ T_{H1} response, ↓ T_{H2} activity
Resistin	Pro-inflammatory ↑ endothelial adhesion molecules, ↑ NFκB, TNFα, IL1β, IL6, IL12
Visfatin	Pro-inflammatory ↑ IL8, IL6, ↓ Neutrophil apoptosis

of both pro- and anti-inflammatory cytokines. It suppresses the synthesis of tumor-necrosis factor-α, interleukins IL6 and IL8, and interferon-γ and favors the production of anti-inflammatory cytokines, such as interleukins IL10 and IL1-receptor antagonist (IL1RA) by monocytes, macrophages, and dendritic cells. Adiponectin increases the synthesis of tissue inhibitor of metallopeptidases in macrophages via IL10 [446].

Adiponectin inhibits the expression of adhesion molecules via inhibition of TNF and NFκB. Globular adiponectin suppresses TNF-induced activation of NFκB, prevents activation of IκB kinase, and leads to cAMP accumulation in endothelial cells [445]. This effect is blocked by inhibitors of adenylyl cyclase or protein kinase-A. Adiponectin also hampers foam cell formation. Adiponectin reduces vascular smooth muscle cell proliferation, migration, and apoptosis by attenuating synthesis of platelet-derived and fibroblast growth factors.

Table 3.26. Apelin effects on blood circulation (Source: [447]; EC: endothelial cell; SMC: smooth muscle cell). Its effectors comprise: (1) inhibitory G protein (Gi) that inhibits adenylyl cyclase; (2) Ca^{++}; and (3) ERK and S6K kinases and the PI3K–PKB axis. Neurons of the supraoptic and paraventricular nuclei of the hypothalamus synthesize apelin, apelin receptor, and vasopressin (or antidiuretic hormone). Water deprivation enhances production of apelin receptor and vasopressin in the supraoptic nucleus. Apelin and its receptor are also expressed in peripheral tissues with highest levels in heart, kidney, and lung.

Effect	Mediator
Vasoconstriction	ApJ receptor (SMC)
Vasodilation	Nitric oxide (ApJ receptor on EC)
Positive inotropy	Na^+–H^+ exchanger
Blood volume	Vasopressin

3.18.2 Apelin

Apelin is synthesized from a single gene in adipocytes, vascular and cardiac cells, some cells of the endocrine pancreas, gastric enterochromaffin-like cells, and colonic epithelial cells. Its production increases with elevated insulin level. On the other hand, the concentration of cardiac apelin is lowered by angiotensin-2.

Apelin circulates in blood and can then act as a hormone. Apelin regulates the cardiovascular function and fluid homeostasis (Table 3.26).

Apelin activates the G-protein-coupled receptor ApJ[110] that signals via inhibitory G protein (Gi; Vol. 4 – Chap. 8. Guanosine Triphosphatases). Consequently, it inhibits adenylyl cyclase.[111] On the other hand, activated ApJ stimulates extracellular-regulated kinases (Vol. 4 – Chap. 5. Mitogen-Activated Protein Kinase Modules) via protein kinase-C. In addition, apelin can activate S6K kinase (Vol. 4 – Chap. 4. Cytosolic Protein Ser/Thr Kinases).

3.18.2.1 Apelin Types

Apelin derives from a 77-amino acid *preproapelin* with a conserved 23-amino acid C-terminus (amino acids 55–77). Paired basic amino acid residues (Arg–Arg and Arg–Lys) are cleaving sites for endopeptidases. Many possible isoforms result from the maturation of a prepropeptide (preproapelin) into apelin peptides: $apelin_{36}$ (or $apelin_{(42--77)}$), the long form of apelin and shorter isoforms that correspond to a shorter C-terminus (Table 3.27). In addition, $apelin_{13}$ gives rise to a pyroglutaminated form ($^{pGlu}apelin_{13}$). These

[110] Previously termed angiotensin-2 receptor-like protein-1; Vol. 3 – Chaps. 6. Receptors and 7. G-Protein-Coupled Receptors).

[111] In humans, APJ mRNA is detected in spleen, intestine, thymus, prostate, testis, and ovary, but not adipocytes, neurons, pulmonary and renal epithelial cells, nor secretory cells of adrenal gland [447].

Table 3.27. Apelin isoforms. For example, $apelin_{36}$, the long form of apelin, and $apelin_{15}$ contain the C-terminal 36 and 15 residues of preproapelin, respectively. The 12 C-terminal residues (66–77) are the minimal structural core required for apelin activity.

Isoform	Other type of notation
$Apelin_{11}$	$Apelin_{(67--77)})$
$Apelin_{12}$	$Apelin_{(66--77)})$
$Apelin_{13}$	$Apelin_{(65--77)})$
$Apelin_{15}$	$Apelin_{(63--77)})$
$Apelin_{16}$	$Apelin_{(62--77)})$
$Apelin_{17}$	$Apelin_{(61--77)})$
$Apelin_{19}$	$Apelin_{(59--77)})$

isoforms differ by their tissue distribution and activity potency. Yet, these numerous apelin isoforms interact with a single receptor.

Apelin is a substrate of angiotensin-converting enzyme-related carboxypeptidase ACE2, another enzyme of the renin–angiotensin–aldosterone system, in addition to angiotensin-converting enzyme ACE1 [448].[112] The carboxypeptidase ACE2 cleaves (inactivates) apelin-36 and hydrolyzes apelin-13 [447].

Apelins were initially discovered as inducers of extracellular acidification. $Apelin_{13}$ and $^{pGlu}apelin_{13}$ have similar activities. $Apelin_{17}$ and $apelin_{36}$ are 8- and 60-fold less efficient, respectively [447]. However, elevation in extracellular acidification generated by $^{pGlu}apelin_{13}$ is transient, whereas that by $apelin_{36}$ is sustained. Repression of adenylyl cyclase is stronger by $apelin_{13}$ and $^{pGlu}apelin_{13}$ that have similar affinity for their substrate than by $apelin_{17}$ (slightly lower affinity) and $apelin_{36}$ (3 times lower affinity). In addition, $apelin_{36}$, $apelin_{17}$, and $apelin_{13}$ cause a sharp elevation in intracellular calcium concentration with little dose–response difference.

Apelin receptor is expressed on activated T lymphocytes. $Apelin_{36}$ is more effective in the inhibition of human immunodeficiency virus infection, as it blocks apelin receptor that serves as a HIV coreceptor [448].[113]

$Apelin_{13}$ and, to a much lesser extent, $apelin_{36}$, stimulate proliferation of gastric epithelial cells in vitro [447]. Apelin also acts as a mitogenic peptide for endothelial cells. $Apelin_{13}$, and to a lesser extent, $apelin_{36}$ operate as chemoattractants, especially during retinal angiogenesis. Apelin can also modulate cytokine production.

[112] Enzyme ACE2 cleaves C-terminus of angiotensin-1 and -2 into inactive $angiotensin_{(1--9)}$ or $angiotensin_{(1--7)}$, a functional antagonist of angiotensin-2 that acts predominantly as a vasodilator [448].

[113] Observation of inhibition of apelins on HIV entry shows that apelin-36 is 16- and 90-fold more efficient than apelin-17 and apelin-13, respectively [447].

3.18.2.2 Apelin Effects in the Cardiovascular Apparatus

Apelin has potent positive inotropic and vasorelaxant activity as well as anti-atherogenic and anti-aneurysmal properties. It can reduce expression and activation of inflammatory cytokines and chemokines, although it neither markedly changes intimal adhesion molecule expression nor medial and adventitial cell cytokine production [449].

Apelin acts as a potent endothelium-independent vasoconstrictor and endothelium-dependent vasodilator by targeting its receptors on vascular smooth muscle and endothelial cells, respectively. $Apelin_{12}$ is more potent for reducing blood pressure than $apelin_{13}$, a strong regulator of the cardiovascular function, whereas $apelin_{11}$ remains inactive.

Apelin receptor is produced in vascular endothelial cells that line small intramyocardial, endocardial, renal, pulmonary, and adrenal vessels as well as large vessels, such as coronary arteries and saphenous veins [448]. Although both the receptor and its ligand can be expressed by endothelial cells, subsets of endothelial cell population can produce either apelin or its receptor for paracrine signaling (but not autocrine regulation) [447].

Apelin receptor expression increases during formation of retinal vessels and diminishes after vessel stabilization. Lower ApJ levels are also observed in vascular smooth muscle cells as well as cardiomyocytes.

Apelin and its receptor are highly expressed in the heart of various species, but more abundantly in the right atrium than in the left ventricle. Apelin is a positive inotropic agent, as it promotes activity of Na^+–H^+ exchangers. Cardiac apelin level is downregulated by angiotensin-2.

In humans, apelin plasma level is significantly elevated by insulin as well as in obesity. During hypoxia such as in ischemic cardiomyopathy, apelin level is heightened, whereas it decays in atrial fibrillation and chronic heart failure [444]. In addition, apelin expression is significantly attenuated in ventriculomyocytes of experimental models of chronic pressure overload [447]. In humans with idiopathic dilated cardiomyopathy, apelin level decays, but atrial and ventricular apelin receptor level rises. Apelin plasma level in the early stage of heart failure can also rise and then fall.

3.18.2.3 Apelin Effects in the Central Narvous System

Messenger RNAs that encode apelin and its receptor abound in the central nervous system. In humans, apelin is expressed in the caudate nucleus, hippocampus, thalamus, hypothalamus, basal forebrain, frontal cortex, corpus callosum, amygdala, substantia nigra, and pituitary, as well as spinal cord [448].

Apelin and its receptor are coexpressed in neurons of the supraoptic and paraventricular nuclei of the hypothalamus that synthesize vasopressin. It hence participates in the regulation of fluid homeostasis.[114]

At least in the hypothalamus, apelin released by a neuron can activate presynaptic apelin receptors on the same neuron as well as those on nearby neurons, i.e., for both auto- and paracrine regulations [447]. In addition, acute stress increases apelin receptor production in the paraventricular nucleus. Repeated stresses cause a sustained upregulation of apelin receptors.

3.18.3 Leptin

Leptin (λεπτος: thin) is encoded by the human LEP gene (murine obese gene [OB]). The expression of the leptin gene is regulated by food intake and energy status, as well as several hormones and inflammatory mediators (e.g., tumor-necrosis factor, interleukin-1, and leukemia inhibitory factor). This non-glycosylated peptide is mainly produced by adipocytes in response to high lipid levels. Yet, leptin can be produced by various tissues, especially those of the cardiovascular system, such as blood vessel wall cells and cardiomyocytes that also synthesize leptin receptors (Vol. 3 – Chap. 6. Receptors).

Leptin modulates via the hypothalamus satiety and fat storage. In the hypothalamus, leptin also contributes to the control of the concentration of growth hormone, thyroxin, and sex steroids [444]. Leptin participates in the regulation of pancreatic islet cell activity, hematopoiesis, angiogenesis, wound healing, osteogenesis, and gastrointestinal function.

Leptin interacts with 6 types of cognate receptors (LepRa–LepRf) that belong to the class-1 cytokine receptors. Signal transduction involves the JaK–STAT pathways (Vol. 4 – Chap. 3. Cytosolic Protein Tyrosine Kinases) and AMP-activated protein kinase (Vol. 1 – Chap. 4. Cell Structure and Function).

Leptin circulates in the blood (at a concentration of a few ng/ml) and in the cerebrospinal fluid, crossing the blood–brain barrier in order to regulate food intake by the hypothalamus. Leptin concentration is proportional to insulin concentration and inversely proportional to glucocorticoid level [444].

Testicular and ovarian steroids decrease and increase leptin concentration, respectively [444]. Angiotensin-2 stimulates leptin production in adipose tissue by activating angiotensin-2 type 1 receptor and extracellular signal-regulated kinases ERK1 and ERK2 [73] (Vol. 4 – Chap. 5. Mitogen-Activated Protein Kinase Modules).

[114] Axonal transport translocates vasopressin to the posterior pituitary, where it is released in response to osmotic stimuli sensed by hypothalamic neurons to regulate water and sodium uptake in kidneys as well as vascular tone.

3.18.3.1 Leptin Resistance

Obese humans develop leptin and insulin resistance characterized by endoplasmic reticulum stress during which un- and misfolded proteins accumulate in the endoplasmic reticulum.[115] Endoplasmic reticulum stress blocks leptin ability to activate its signaling pathways [450]. Reduced endoplasmic reticulum stress sensitizes cells to leptin.

Elevated leptin level is able to induce leptin resistance via an increase in concentration of suppressor of cytokine signaling SOCS3 that blocks leptin receptors. Increased serum leptin concentration is often associated with insulin resistance, hypertension, endothelial dysfunction, and inflammation, as well as myocardial infarction and stroke. Conversely, administration of leptin stimulates the sympathetic nervous system, natriuresis, and nitric oxide-dependent vasodilation.

3.18.3.2 Leptin Activity in the Central Nervous System

In the brain, leptin influences the cortex, hippocampus, and hypothalamus activity. Leptin acts on hypothalamic neurons to repress food intake and promote energy consumption.[116] Leptin favors energy expenditure especially for the body's growth, reproduction, and immunity. In the hypothalamus, leptin controls not only appetite, but also level of sex steroids, thyroxin, and growth hormone. Leptin can be coexpressed with growth hormone in somatotrope neurons of the anterior pituitary.[117]

Leptin and Satiety in the Hypothalamus

Leptin receptor LepRb is found in the hypothalamus, especially in the satiety center.[118] Leptin hampers the activity of neurons in the arcuate nucleus expressing neuropeptide-Y and agouti-related peptide (AgRP). It favors the activity of neurons expressing α-melanocyte-stimulating hormone

---

[115] RNA-dependent protein kinase-like endoplasmic reticulum-bound eIF2α kinase (PERK) that is a resident endoplasmic reticulum-stress inducible kinase and inositol-requiring kinase IRE1 are major participants of the unfolded protein response. Their activity rises during development of obesity. Conversely, transcription factors X-box binding protein XBP1 and Activating transcription factor ATF6 favor endoplasmic reticulum adaptive capacity.

[116] Adipose tissues aim at storing energy. Adipocytes saturated with lipids can lead to lipid accumulation in other tissues, reducing their functioning. Increased leptin activity in the arcuate nucleus in the hypothalamus inhibits the production of neuropeptide-Y in the paraventricular nucleus, thereby reducing feeding.

[117] Somatotrope neurons have leptin receptors. Leptin can thus be an autocrine or paracrine regulator.

[118] Leptin receptor is expressed at low levels in manifold tissues, but at high levels in the mediobasal hypothalamus, particularly in the arcuate, ventromedial, dorsomedial, and ventral premamillary hypothalamic nuclei [451]. Expression of LepRb is moderate in the periventricular region and posterior hypothala-

derived from pro-opiomelanocortin and cocaine- and amphetamine-regulated transcript in the lateral arcuate nucleus. Melanin-concentrating hormone and orexins expressed in lateral hypothalamic area are inhibited indirectly by leptin. In addition, leptin increases catecholamine (adrenaline and noradrenaline) secretion by acting on the ventromedial hypothalamus, but not on hypothalamic arcuate, paraventricular, and dorsomedial nuclei, to prime a sympathetic activation [453].

Leptin controls cell metabolism via its actions in the hypothalamus and peripheral organs. Leptin regulates energy balance between intake and expenditure and glucose metabolism by activation of long-form leptin receptor LepRb in hypothalamic leptin-responsive neurons via mechanisms that depend or not on LepRb Tyr phosphorylation [454]. Leptin-bound phosphorylated LepRb (Tyr1138) activates Janus kinase JaK2 and recruits STAT3 signal transducer and activator of transcription. Phosphorylated LepRb (Tyr985) mediates cellular events that involve cytosolic phosphatase PTPn11, extracellular signal-regulated kinase, and SOCS3 suppressor of cytokine signaling. Additional mechanisms independent on LepRb Tyr phosphorylation control food intake, physical activity, adaptive thermogenesis, and glucose metabolism.

Leptin and Synaptic Adaptivity in the Hippocampus

Leptin influences synaptic adaptivity (plasticity). It stimulates the presence of ionotropic glutamate receptor-1 (GluR1 [AMPAR]; Vol. 3 – Chap. 2. Membrane Ion Carriers) at hippocampal synapses and increases synaptic strength, as it inhibits the phosphatase PTen [455]. Phosphatidylinositol (3,4,5)-trisphosphate, the level of which then rises, actually enables GluR1 exocytosis. Leptin enhances the amplitude of excitatory postsynaptic currents in the hippocampus.

3.18.3.3 Leptin Activity in the Cardiovascular System

In the heart, leptin regulates the contractility, metabolism, production of extracellular matrix components by cardiomyocytes, and contributes to the cardiomyocyte size [434].

Leptin receptors are widely distributed on endothelial cells and vascular smooth muscle cells. Leptin stimulates mitogen-activated protein kinases and

mic nucleus and low in the paraventricular nucleus and lateral hypothalamic area. Receptor LepRb also localizes to the nucleus tractus solitarius, lateral parabrachial nucleus, as well as motor and sensory nuclei and brainstem areas that are not associated with energy balance. Activation of leptin receptors in the hypothalamus: (1) represses orexigenic pathways that involve neuropeptide-Y and agouti-related peptide and (2) stimulates anorexigenic pathways that are linked to pro-opiomelanocortin and cocaine- and amphetamine-regulated transcript [452].

phosphatidylinositol 3-kinase. Leptin induces smooth muscle cell proliferation and migration [456]. Leptin also favors platelet aggregation. It promotes angiogenesis [457] (Vol. 5 – Chap. 10. Vasculature Growth). In addition, leptin intervenes in hematopoiesis.

Leptin upregulates the expression of vascular endothelial growth factor via activation of NFκB and PI3K [452]. Leptin also supports the production of nitric oxide synthase NOS2, thereby that of reactive oxygen species.

Plasma leptin action on arterial blood pressure regulation depends on soluble leptin receptor, nitric oxide metabolism, and blood flow behavior, as well as leptin-mediated hypothalamic activation centers of sympathetic neuronal regulation [458]. Effects of sympathetic neuronal activation differ according to organs. Kidneys belong to main targets of the sympathetic nervous system. Leptin's excitation modifies glomerular perfusion and sodium excretion.

Impaired leptin function in vascular cells can facilitate onset of hypertension. In vitro, leptin stimulates the proliferation and hypertrophy of vascular smooth muscle cells and their production of matrix metallopeptidase MMP2. It can then alter arterial wall remodeling. Leptin also stimulates the production of cytokines, such as tumor-necrosis factor-α and interleukin-6, that can alter blood vessel wall functioning. It also promotes the secretion of pro-atherogenic lipoprotein lipase by cultured macrophages [434]. In aortic endothelial cells, leptin causes superoxide and CCL2 production, hence favoring monocyte recruitment.

3.18.3.4 Leptin Activity in Kidney

By its short- and long-term effect in the kidney, leptin influences blood pressure by 2 opposing processes: (1) via renal sympathetic activation and (2) via nitric oxide synthesis. Kidneys express high amounts of leptin receptors. In addition to its direct effect on the renal functioning via leptin receptors, leptin has indirect activity via nitric oxide in renal tubules.

Leptin stimulates overall sympathetic nerve activity, especially in kidneys, adrenal glands, and brown adipose tissue [459]. However, leptin does not increase arterial pressure and heart rate. In the renal medulla, leptin inhibits Na^+–K^+ ATPase that is responsible for tubular sodium reabsorption, hence augmenting natriuresis [460]. On the other hand, leptin does not influence cortical Na^+–K^+ ATPase. Leptin-induced nitric oxide impedes renal sodium reabsorption, as it attenuates the activity of both apical transporters and basolateral Na^+–K^+ ATPases [461]. In particular, quantity of Na^+–Cl^- cotransporters is related to nitric oxide production.

Leptin stimulates nitric oxide production and release in vascular endothelial cells by activating a PI3K-independent PKB–NOS3 pathway, thus inducing vasorelaxation [462]. Nitric oxide synthesis in the renal medulla enhances medullary blood flow, hence preventing sodium retention and blood pressure elevation [463]. Leptin-induced nitric oxide activity hence opposes pressor and antinatriuretic effect of leptin-primed sympathetic activation.

3.18.3.5 Leptin Activity in Lung

Lung genesis depends on endodermal sonic Hedgehog signaling to mesodermal Wnt–β-catenin pathway (Vol. 3 – Chap. 10. Morphogen Receptors), followed by *parathyroid hormone-related protein* (PTHrP) action from endoderm to mesoderm [464]. Intra-uterine lung development is driven by fluid-caused distension. Distension (maturation) of fetal rat lung explants upregulates PTHrP signaling, as well as fibroblast-specific adipocyte differentiation-related protein (ADRP) and peroxisome proliferator-activated receptor-γ, but downregulates sonic Hedgehog and Wnt–β-catenin signaling. Parathyroid hormone-related protein and cAMP repress sonic Hedgehog and Wnt–β-catenin axes. Lung genesis indeed culminates with pulmonary surfactant production by epithelial type-2 cells. Differentiation of alveolar epithelial and mesenchymal cells is promoted by parathyroid hormone-related protein that is a stretch-sensitive molecule produced by alveolar type-2 pneumocytes.

Developing lung lipofibroblasts express leptin [465]. Leptin and its receptor are mutually expressed exclusively by fetal lung fibroblasts and type-2 pneumocytes that can therefore interact in a paracrine manner. Leptin thus participates in the parathyroid hormone-related protein paracrine stimulation of fetal lung maturation. In fetal lung acini, leptin is produced in alveolar interstitial fibroblasts that are subjected to parathyroid hormone-related protein secreted by formative alveolar epithelium under moderate stretch [466]. Leptin then acts back on leptin receptors of alveolar type-2 pneumocytes to prime surfactant synthesis.

Leptin also stimulates synthesis of surfactant phospholipids and proteins in adult human airway epithelial cells. When alveolar lipofibroblasts and epithelial type-2 cells are simultaneously stretched in vitro, surfactant synthesis increases (5-fold) [466]. Lipofibroblast stretching upregulates adipose differentiation-related protein expression and enhances parathyroid hormone-related protein binding (2.5-fold) and triglyceride uptake (15–25%). Alveolar type-2 pneumocyte stretching augments leptin stimulation of surfactant synthesis (3-fold). Paracrine leptin control hence is also stretch sensitive.

3.18.3.6 Leptin and Bone Remodeling

Leptin bound receptors in neurons attenuate bone formation and accrue osteoclast differentiation, whereas in osteoblasts they do not influence bone remodeling [467]. However, leptin via mediator cocaine- and amphetamine-regulated transcript (CART) inhibits bone resorption. Leptin regulates bone mass via sympathetic activity independently of its effect on energy metabolism. Leptin activation of the sympathetic tone could occur after binding to its receptor on hypothalamic neurons.

3.18.3.7 Leptin Activity in Cell Metabolism

Leptin stimulates AMP-activated protein kinase that decreases ATP-consuming anabolism and increases ATP-manufacturing catabolism. Leptin decreases insulin levels by inhibition of proinsulin synthesis and reduction of secretion.

In myocytes, leptin improves insulin sensitivity and reduces intracellular lipid levels by direct activation of AMP-activated protein kinase combined with indirect inputs to the central nervous system. In hepatocytes, leptin also enhances insulin sensitivity.

3.18.3.8 Leptin Activity in Inflammation

Leptin has dominant pro-inflammatory effects (Table 3.25). It heightens immune responsiveness by directly acting on immune cells. Leptin binds to its receptor LepRb and activates: (1) the mitogen-activated protein kinase pathway (P38MAPK and ERK) and (2) signal transducer and activator of transcription STAT3, thus producing pro-inflammatory cytokines TNFα and interleukin-6 and -12 in monocytes and macrophages.

Leptin favors activities of monocytes, macrophages, and natural killer cells [445]. Leptin stimulates the chemotaxis of neutrophils and the production by neutrophils of reactive oxygen species. It also stimulates the production of IgG2a by B lymphocytes. Moreover, it increases secretion by T lymphocytes.

3.18.4 Resistin

Resistin[119] (Retn) causes resitance of tissues to insulin (hence its name "resist + in") is synthesized by adipocytes and other cells of adipose tissues, as well as by myocytes, pancreatic cells, and macrophages. Primary resistin sources can differ between mammals: adipocytes in rodents and stromal vascular cells in human adipose tissue. Resistin prepeptide size also varies between mammals (slightly shorter in humans than in rodents). Resistin circulates mostly as high-molecular-weight hexamers and also more active low-molecular weight complexes [445].

Resistin synthesis is influenced by insulin, endothelin-1, and adrenaline, as well as pituitary, thyroid, and steroid hormones. Resistin can functionally interact with other adipokines, such as adiponectin and leptin.

Resistin activates phosphatidylinositol 3-kinase and members of the mitogen-activated protein kinase modules: extracellular signal-regulated protein kinase, P38MAPK, and Jun N-terminal kinase.

Resistin can reduce glucose uptake by muscles, adipose tissues, and liver, hence affecting insulin sensitivity. Therefore, in cardiomyocytes, resistin can

[119] Resisto, ere: to resist. A.k.a. adipose tissue-specific secretory factor (ADSF), C/EBPϵ-regulated myeloid-specific secreted Cys-rich protein, Cys-rich secreted protein A12α-like protein-2.

impair glucose metabolism. Resistin has dominant pro-inflammatory features. Resistin increases the production of tumor-necrosis factor and interleukins IL1β, IL6, and IL12 by various cell types via NFκB [444]. These cytokines upregulate the resistin expression. In human vascular endothelial cells, resistin increases the synthesis of pro-inflammatory agents and provokes the release of endothelin-1 and production of adhesion molecules and chemokine ligands. Resistin actually heightens the production of intercellular (ICAM1) and vascular cell (VCAM1) adhesion molecule-1, as well as chemokine ligand CCL2 by endothelial cells [444].

High plasma level of resistin correlates with pro-atherogenic inflammatory markers, metabolic syndrome, endothelial dysfunction, and elevated cardiovascular risk. In human coronary artery endothelial cells, clinically relevant concentrations of resistin (40 or 80 ng/ml) administered during 24 h reduce the production of endothelial nitric oxide synthase (NOS3) and elevate the cellular levels of reactive oxygen species via P38MAPK and JNK [468].

3.18.5 Visfatin

Visfatin, or nicotinamide phosphoribosyltransferase (NAmPT or NAmPRTase)[120] is produced by adipocytes and lymphocytes. Therefore, this protein acts as both an enzyme and adipokine that can circulate in the blood stream.

Plasma level of visfatin is negatively correlated to endothelial function. Visfatin is able to upregulate the activity of interleukin-6 and tumor-necrosis factor-α [434]. It also favors production of transforming growth factor-β1, plasminogen activator inhibitor-1, and collagen-1. In addition, visfatin fosters the maturation of vascular smooth muscle cells and inhibits neutrophil apoptosis [444].

Visfatin has an insulin-like activity because it binds to and activates insulin receptor at a binding site different from that of insulin [469]. It thus favors glucose uptake, as it enhances synthesis and membrane incorporation of glucose transporter GluT1.

Vasodilator visfatin initiates [470]: (1) phosphorylation (Ser1177) and dephosphorylation (Thr495) of nitric oxide synthase NOS3; (2) phosphorylation (Ser473) of protein kinase-B; and (3) phosphorylation of vasodilator-stimulated phosphoprotein (Ser239). Furthermore, visfatin has angiogenic activity via extracellular signal-regulated protein kinase-1 and -2 and fibroblast growth factor-2 [471]. Moreover, it acts on endothelial cells by an auto- and/or paracrine mechanism. It indeed upregulates the expression of chemokine CCL2

[120] A.k.a. pre-B-cell colony-enhancing factor (PBEF1). Its systematic name is nicotinamide-nucleotide:diphosphate phosphoα^{D}ribosyltransferase, also named nicotinamide mononucleotide (NMN) pyrophosphorylase, NMN pyrophosphorylase, and NMN synthase. It catalyzes the condensation of nicotinamide with 5-phosphoribosyl 1-pyrophosphate to yield nicotinamide mononucleotide that is further processed into nicotinamide adenine dinucleotide.

and its receptor CCR2 via phosphatidylinositol 3-kinase and nuclear factor-κB [472].

Visfatin released by perivascular adipose tissue not only inhibits vascular smooth muscle cell contraction, but also stimulates proliferation of these cells [473]. This effect relies on the nicotinamide phosphoribosyltransferase activity that produces nicotinamide mononucleotide needed for mitogen-activated protein kinases P38MAPK and extracellular signal-regulated kinases ERK1 and ERK2.

Leptin increases visfatin production in adipose tissue via phosphatidylinositol 3-kinase and mitogen-activated protein kinase [474]. High-glucose level also elevates visfatin synthesis in mesangial cells that reside around renal blood vessels [475].

On the other hand, angiotensin-2 binds to the nuclear receptor NR1c3 (or peroxisome proliferator-activated receptor-γ) in adipocytes, skeletal myocytes, and endothelial cells. It then prevents the release of visfatin [476].

3.18.6 Other Adipokines and Adipocyte Products

3.18.6.1 Adipsin

Adipsin[121] is mainly synthesized in monocytes and resident macrophages in adipose tissue. It intervenes in the rate-limiting step in the complement activation alternative pathway. It could also generate an acylation-stimulating protein that increases adipocyte triglyceride production [444].

3.18.6.2 Angiopoietin-like Peptide-4

Angiopoietin-like peptide-4 (AngptL4),[122] induced by peroxisome proliferator-activated receptors (nuclear receptors NR1c) especially in the liver and adipose tissue, increases the triglyceride level [444]. Its concentration correlates with those of lipoproteins. It inhibits lipoprotein lipase as well as blood triglyceride clearance.

This glycosylated, secreted protein is, in particular, produced under hypoxia by endothelial cells [479]. During hypoxia, full-length AngptL4 accumulates in the subendothelial extracellular matrix via heparan sulfate proteoglycan. Therefore, in the extracellular matrix, AngptL4 exists either as a matrix-bound, immobilized, full-length protein or soluble form. This survival factor for vascular endothelial cells is involved not only in lipid metabolism, but also the regulation of glucose homeostasis and angiogenesis.[123]

[121] A.k.a. complement factor-D and C3-convertase activator.

[122] A.k.a. angiopoietin-like protein-4 and fasting-induced adipose factor. The class of angiopoietin-like proteins of the angiopoietin family includes 7 known proteins (AngptL1–AngptL7).

[123] The matrix-associated form of AngptL4 limits the formation of stress fibers and focal contacts in endothelial cells, thereby decreasing the motility of endothelial cells and angiogenesis via an autocrine pathway [479].

3.18.6.3 Chemerin

Chemerin, or retinoic acid receptor responder RARRes2,[124] is a chemoattractant. Chemerin is an adipokine that serves as a regulator in adipogenesis or adipocyte function.[125] It operates in auto- and paracrine signaling for adipocyte differentiation.

Chemerin is secreted as an inactive prochemerin that undergoes extracellular serine peptidase cleavage of its C-terminus to generate an active chemerin that can circulate in blood (estimated concentration 3.0–4.4 nM in humans). Chemerin concentration increases when body mass, blood pressure, and triglyceride level rise [444].

Chemerin targets the G-protein-coupled receptor CmkLR1 chemokine-like receptor. Chemerin receptor defects impair cell differentiation into adipocytes and decrease adiponectin and leptin expression [444].

3.18.6.4 Hepcidin

Hepcidin is mainly synthesized in the liver as a hepcidin preprohormone and then prohormone, before being secreted. It was first described as a urinary antimicrobial peptide.

This hormone regulates iron metabolism, as it prevents iron efflux from enterocytes and macrophages. This peptide actually inhibits ferroportin involved in iron export as well as iron absorption and secretion by enterocytes and transport across the placenta [444]. Iron release from macrophages is also prevented by ferroportin inhibition.

Hepcidine concentration correlates with levels of C-reactive protein and interleukin-6 [444]. Interleukin-6 that is synthesized by white adipose tissue is a potent stimulator of hepcidin production and secretion by the liver. It is also produced in adipose tissue, particularly in obese subjects. Its expression rises during iron overload and inflammation. During adipocyte hypoxia, hepcidin concentration can either decay or heighten [477].

Acute inflammation, especially during cardiac surgery, can affect the regulation of hepcidin expression in subcutaneous adipose tissue, but not in epicardial adipose tissue, thereby contributing to inflammation-induced systemic changes of iron metabolism [478].[126]

---

[124] A.k.a. tazarotene-induced gene product TIG2.

[125] Retinoids inhibit cell differentiation. They are then used in the treatment of hyperproliferative dermatological diseases.

[126] When iron metabolism is dysregulated, iron is sequestered in the reticuloendothelial system. A subsequent hypoferremia occurs, i.e., a limited availability for erythropoiesis, that causes anemia observed in chronic inflammatory diseases.

3.18.6.5 Omentin

Omentin[127] is a galactofuranose binding-lectin. It indeed acts as a receptor for lactoferrin as well as bacterial arabinogalactans. In addition, it is an insulin sensitizer made by vascular stromal cells within fat pads that enhances glucose uptake [452]. It is encoded by 2 genes that are selectively expressed in visceral adipose tissue.

Omentin concentration decays in obesity and insulin resistance [444]. On the other hand, omentin concentration rises when high-density lipoprotein and adiponectin concentration heightens.

3.18.6.6 Retinol-Binding Protein-4

Retinol-binding protein-4 (RBP4) of the lipocalin family is devoted to the delivery of retinol, a form of vitamin A. It is actually the specific blood carrier of retinol. It delivers retinol from the liver store to tissues. It is also secreted by adipocytes. In plasma, the RBP-retinol complex interacts with transthyretin that prevents its filtration through the renal glomerulus.

This adipokine impairs insulin action on the liver and muscles [452]. Retinol-binding protein-4 contributes to insulin resistance.

3.18.6.7 Vaspin

Visceral adipose tissue-derived serine peptidase inhibitor (serpin) yields the portmanteau vaspin (the noun serpin is also a portmanteau). Vaspin or Serpin-A12 suppresses the production of tumor-necrosis factor, leptin, and resistin [445]. However, this extracellular adipokine improves insulin sensitivity.

The concentration of insulin-sensitizing vaspin decays in physically fit subjects and rises in obese humans, especially those with impaired glucose tolerance [444].

3.18.6.8 Insulin Resistance

Transient increases in non-esterified fatty acid levels, such as acute changes after a meal, enhance insulin secretion, whereas chronic elevations associated with insulin resistance reduce insulin secretion by the pancreas. Insulin resistance is actually linked to lipolysis and release of non-esterified fatty acids into the circulation [452]. Circulating non-esterified fatty acids reduce glucose uptake by adipocytes and myocytes, and promote glucose release by hepatocytes.

---

[127] A.k.a. endothelial lectin, intelectin-1 (Itln1), and intestinal lactoferrin receptor (IntL).

3.19 Colony-Stimulating Factors

Colony-stimulating factors (CSF) are devoted to hematopoietic (blood) cells. They are able to generate mature myeloid cells from bone marrow precursors. They can also act on mature myeloid cells, especially during immune responses. Colony-stimulating factors include granulocyte colony-stimulating factor (gCSF), or colony-stimulating factor CSF3, granulocyte–macrophage colony-stimulating factor (gmCSF), or CSF2, and macrophage colony-stimulating factor (mCSF), or CSF1.

Colony-stimulating factors intervene in wound healing, as well as placental and fetal development. Smooth muscle, endothelial and epithelial cells, neurons, and keratinocytes express receptors for gmCSF and mCSF [480].

Colony-stimulating factors act as pro-inflammatory cytokines that are associated with tumor-necrosis factor and some interleukins, such as IL1, IL17, and IL23 [480]. In vitro, gmCSF favors production of pro-inflammatory cytokines, such as tumor-necrosis factor and interleukins IL6, IL12, and IL23. Cultured cells stimulated by mCSF tend to manufacture interleukin-10 and CCL2 chemokine ligand.

3.19.1 Granulocyte–Macrophage Colony-Stimulating Factor (CSF2)

Granulocyte–macrophage colony-stimulating factor (gmCSF or CSF2) is synthesized upon stimulation, especially by interleukin-1, tumor-necrosis factor, and lipopolysaccharides. Secreted gmCSF is a single glycosylated polypeptide chain. In vitro, gmCSF promotes survival and activation of macrophages, neutrophils, and eosinophils, as well as maturation of dendritic cells. Moreover, gmCSF is required for maturation of alveolar macrophages and invariant natural killer T cells [480].

The gmCSF receptor (CSF2R) is a heterodimer composed of a specific ligand-binding CSF2Rα and a signal-transduction CSF2Rβ subunit. The latter is also a component of interleukin IL3R and IL5R receptors. Activated gmCSF receptor stimulates 3 known pathways: (1) the Janus kinase–signal transducer and activator of transcription (JaK–STAT); (2) mitogen-activated protein kinase (MAPK); and (3) phosphoinositide 3-kinase (PI3K) axis.

In vitro, gmCSF and mCSF preferentially target cells with antigen-presenting and phagocytic features, respectively. Stimulation of immature cells of the macrophage lineage with gmCSF alone promotes dendritic cell-like phenotype, whereas stimulation with both mCSF and gmCSF leads to a stronger macrophage phenotype. A costimulus such as that caused by lipopolysaccharides and gmCSF or mCSF is usually required to activate monocytes and macrophages.

3.19.2 Macrophage Colony-Stimulating Factor (CSF1)

Macrophage colony-stimulating factor (mCSF or CSF1) is constitutively produced by several cell types, such as endothelial and smooth muscle cells, fibroblasts, stromal cells, macrophages, and osteoblasts. Several mCSF types exist: cell-surface protein, secreted glycoprotein, and proteoglycan. The glycoprotein and proteoglycan isoforms circulate throughout the body.

The mCSF receptor (CSF1R) is a homodimeric type-3 receptor Tyr kinase. It elicits survival, proliferation, differentiation, and possibly activation of cells of the monocyte–macrophage lineage. Furthermore, mCSF is crucial for maintenance of several macrophage-lineage populations associated with tissue integrity.

3.19.3 Granulocyte Colony-Stimulating Factor (CSF3)

The blood concentration of granulocyte colony-stimulating factor (gCSF or CSF3) rises under stress such as infection. The gCSF receptor belongs to the type-1 cytokine receptors. The colony-stimulating factor-3 receptor (CSF3R) reaches its highest expression on neutrophils. Activation of CSF3R triggers numerous signaling cascades using cytosolic kinases.

3.20 Tumor-Necrosis Factor Superfamily

Tumor-necrosis factors (TNF) belong to the class of cytokines that includes chemokines, interleukins, interferons, and others molecules (monokines, lymphokines, stem cell factor, colony-stimulating factor, glucose phosphate isomerase [GPI],[128] osteopontin, etc.).

Tumor-necrosis factors act via TNF receptors, TNFR1 and TNFR2, that are also called tumor-necrosis factor receptor superfamily member-1A (TNFRSF1a) and -1B (TNFRSF1b).

Tumor-necrosis factor irreversibly triggers the caspase cascade, hence apoptosis. The TNFR receptors actually associate with procaspases via adaptors to cleave other inactive procaspases (Chap. 4). Tumor-necrosis factors also launch a pro-inflammatory program in immune cells.

Tumor-necrosis factors are able to interact with receptors on endothelial cells to increase vascular permeability, thereby favoring leukocyte extravasation to reach the site of infection.

The tumor-necrosis factor superfamily (TNFSF) comprises numerous members, the so-called *TNF ligands*, among which tumor-necrosis factor, more precisely TNFα, or tumor-necrosis factor superfamily member-2 (TNFSF2)[129] and tumor-necrosis factor-β, or TNFSF1.[130]

[128] A.k.a. autocrine motility factor (AMF).

[129] A.k.a. cachexin or cachectin.

[130] A.k.a. lymphotoxin-α. Lymphotoxin-β is a type-2 membrane protein that heterotrimerizes with lymphotoxin-α (predominantly 1 lymphotoxin-α and

Tumor-necrosis factor-α acts synergistically with EGF and PDGF growth factors. Agent TNFSF2 induces expression of several interleukins. Tumor-necrosis factor-β is able to kill different cell types and induce terminal differentiation in others. Protein TNFSF1 inhibits lipoprotein lipase in endothelial cells. It is mainly synthesized by T lymphocytes, in particular CD8+ cytotoxic T lymphocytes.

Ligands of the TNFSF superfamily are detected on B lymphocytes, dendritic cells, macrophages, CD4+ and CD8+ T lymphocytes, NK and NKT cells, mastocytes, and smooth muscle and endothelial cells during inflammation. They include homotrimers TNFSF4 and TNFSF15 as well as TNFSF7 and TNFSF9 [481] (Tables 3.28, 3.29, and 3.30).

The superfamily of TNF ligands also contains numerous other members, such as TNFSF6 that is synthesized by monocytes, dendritic cells, and bone marrow stromal cells, TNFSF10, TNFSF11 that is expressed by helper T cells, and TNFSF13b. Member TNFSF12–TNFSF13 is a hybrid protein composed of the cytoplasmic and transmembrane domains of TNFSF12 fused to the C-terminus of TNFSF13.

The tumor-necrosis factor receptor superfamily (TNFRSF) is constituted by: TNFRSF1a/b, TNFRSF3–6, TNFRSF6b, TNFRSF7–9, TNFRSF10a–10d, TNFRSF11a/b, TNFRSF12a, TNFRSF13b/c, TNFRSF14, TNFRSF16–19, TNFRSF21, TNFRSF25, and TNFRSF27 (Table 3.31). In particular, TNFRSF4, TNFRSF7, TNFRSF9, and TNFRSF25 associate with TNF ligands TNFSF4, TNFSF7, TNFSF9, and TNFSF15, respectively.

Members of the TNFRSF superfamily also link to adaptors *TNFR-associated factors* (TRAF) that can bind to and prevent activity of inhibitors of nuclear factor-κB (Vol. 4 – Chap. 9. Other Major Signaling Mediators). In CD4+ and CD8+ T lymphocytes, TNFRSF4, -7, and -9 increase the expression of anti-apoptotic molecules [481]. Receptors TNFRSF4, -7, -9, and -25 are rapidly expressed on human natural and inducible CD4+, CD25+, FoxP3+ regulatory T cells. Interactions between TNFSF and TNFRSF can then modulate development and function of these regulatory T cells.

Upon immunocyte activation, certain interactions between ligands and receptors[131] mediate crosstalk between T lymphocytes and other cell types to modulate immune response. Expression of TNF ligands and receptors is

2 lymphotoxin-β or 2 lymphotoxin-α and 1 lymphotoxin-β) to anchor it to the plasma membrane.

[131] Interactions between TNFSF4 and TNFRSF4, TNFSF7 and TNFRSF7, TNFSF9 and TNFRSF9, and TNFSF15 and TNFRSF25 can: (1) stimulate conventional T lymphocytes and antigen-presenting cells; (2) mediate communication between CD4+ and CD8+ T lymphocytes, NK and T cells, NKT and antigen-presenting cells, and T lymphocytes and other types of immune or tissue cells; (3) promote immune responses. Other costimulatory interactions between TNF ligands and receptors are also important, such as those based on TNFRSF1b (or TNFR2), TNFRSF3, TNFRSF5, TNFRSF8, TNFRSF13b, TNFRSF14, and TNFRSF18 [481].

Table 3.28. Members of the tumor-necrosis factor (ligand) superfamily (TNFSF; AITRL: activation-inducible TNF-related ligand; ApoL: apoptosis antigen ligand; APrIL: A proliferation-inducing ligand; BAFF: B-cell activating factor; BLyS: B-lymphocyte stimulator; CD: cluster of differentiation; DTL: dendritic cell-derived TNF-like molecule; FasL: Fas antigen ligand; GITRL: glucocorticoid-induced TNF-related ligand; HVEM: herpes virus entry mediator; Light: ligand highly induced in activated T lymphocytes and macrophages homologous to lymphotoxins; LT: lymphotoxin; ODF: osteoclast differentiation factor; OPgL: osteoprotegerin ligand; RANKL: receptor activator for nuclear factor-κB ligand; TL: TNF-like factor [e.g., TL1a: endothelial cell-derived TNF-like factor]; TALL: TNF-and ApoL-related leukocyte-expressed ligand; TRAIL: TNF-related apoptosis-inducing ligand; TRANCE: TNF-related activation-induced cytokine; TRDL: TNF-related death ligand; TWeak: TNF-related weak inducer of apoptosis; VEGI: vascular endothelial growth inhibitor).

Type	Common name	Other aliases
TNFSF1	TNFα	
TNFSF2	TNFβ	LTα
TNFSF3	LTβ	
TNFSF4	OX40L	CD134L, CD252
TNFSF5	CD40L	CD154
TNFSF6	FasL	CD95L
TNFSF7	CD27L	CD70
TNFSF8	CD30L	CD153
TNFSF9	4-1BBL	CD137L
TNFSF10	TRAIL	CD253, Apo2L, TL2
TNFSF11	RANKL	CD254, ODF, OPgL, TRANCE
TNFSF12	TWeak	Apo3L
TNFSF13	APrIL	TALL2, TRDL1
TNFSF13b	TNFSF20, TALL1	CD257, BAFF, BLyS, DTL, THANK
TNFSF14	Light	CD258, HVEML
TNFSF15	VEGI	TL1a
TNFSF18	GITRL	TL6, AITRL

enhanced by T lymphocytes, and antigen-presenting, natural killer (NK), NKT, and activated endothelial cells, as well as other cell types.

Receptors TNFRSF4, -7, -9, and -25 can cooperate with T-cell receptors to promote T-cell division [481]. Members of the TNFSF and TNFRSF superfamilies enable interactions between antigen-presenting cells (dendritic cells, B lymphocytes, and macrophages) and CD4+, CD8+ T lymphocytes, respectively. These interactions are elicited by activating signals from TNFRSF5–TNFSF5 pairs and Toll-like and T-cell receptors (Vol. 3 – Chap. 11. Receptors of the Immune System), or cytokine receptors bound to their specific cytokines (TNF, interleukins, and thymic stromal lymphopoietin) [481]. These interactions initiate intracellular signaling in TNFR-loaded cells that

Table 3.29. Members of the TNF ligand superfamily and their cellular sources (Source: [482]; EdA: ectodysplasin-A). TNF ligands are synthesized as type-2 membrane proteins. They can be cleaved as soluble homotrimers, except membrane-bound TNFSF2 and -3 and TNFSF13 and -13b, that can also form heterotrimers. The soluble forms can act as agonists or antagonists of membrane-bound forms. TNFSF ligands have a limited or no activity in soluble form. However, the interactions of at least 2 trimers that create hexamers or higher-order clusters, render the soluble species active.

Type	Cell source
TNFSF1	Macrophage, NK, T, and B cell, adipocyte
TNFSF2/3	Macrophage, dendritic, NK, T, and B cell
TNFSF4	Macrophage, dendritic, T, and B cell
TNFSF5	T and B cell
TNFSF6	Activated splenocyte, thymocyte
TNFSF7	NK, T, and B cell
TNFSF8	Monocyte, T cell
TNFSF9	Macrophage and dendritic and B cell
TNFSF10	Dendritic, NK, and T cell
TNFSF11	Osteoblast, activated T cell
TNFSF12	Monocyte
TNFSF13	Macrophage, lymphoid cells
TNFSF13b	Monocyte, macrophage, and dendritic and T cell
TNFSF14	Granulocytes, monocytes, and dendritic and T cell
TNFSF15	Endothelial cell
TNFSF18	Macrophage and dendritic and B cell
EDA1/2	Skin

stimulates nuclear factor-κB1 and -κB2,[132] as well as other mediators, such as kinases (phosphoinositide 3-kinase, protein kinase-B, and extracellular-signal-regulated and Jun N-terminal kinase), and transcription factors (e.g., nuclear factor of activated T cells). These interactions also prime intracellular signaling in TNF ligand-loaded cells that generates secretion of pro-inflammatory cytokines by antigen-presenting cells (TNF and interleukins IL1, IL6, and IL12) and favors cell proliferation.

[132] Activated nuclear factor-κB1 causes cell division and enhances cell survival. Nuclear factor-κB1 indeed promotes cell proliferation via survivin, aurora-B kinase, cyclin–cyclin-dependent kinase complexes, as well as survival via anti-apoptotic proteins (B-cell lymphoma BCL2, BCL2-related protein-A1, and BCLxL) and/or inhibition of pro-apoptotic proteins (e.g., BCL2-interacting mediator of cell death). It can also contribute to the production of cytokines (interleukins IL2, IL4, and IL5, as well as interferon-γ.

Table 3.30. Superfamilies of tumor-necrosis factor ligands (TNFSF) and receptors (TNFRSF) and their effects (EdA: ectodysplasin-A; ILi: interleukin-i).

Type	Receptors	Function
TNFSF1	TNFRSF1a/b	Apoptosis, survival, cytotoxicity, inflammation, production of IL1 and IL6
TNFSF2	TNFRSF1a/b TNFRSF3	Cell proliferation, genesis of secondary lymphoid tissue, inflammation
TNFSF3	TNFRSF3	Lymphoid cell genesis
TNFSF4	TNFRSF4	Survival of T cells and population expansion, T-cell activation
TNFSF5	TNFRSF5	B-cell costimulation, proliferation, differentiation, germinal center development
TNFSF6	TNFRSF6b	Apoptosis, cytotoxicity, angiogenesis
TNFSF7	TNFRSF7	Survival of T cells and population expansion, T-cell activation
TNFSF8	TNFRSF8	T_{H2} cell-mediated processes
TNFSF9	TNFRSF9	T-cell costimulation, survival of T cells and population expansion
TNFSF10	TNFRSF10a–c TNFRSF11b	Apoptosis, activation of dendritic, NK, and cytotoxic T cells
TNFSF11	TNFRSF11a	Dendritic cell division (survival factor for activated dendritic cells), maintenance of immune tolerance, osteoclast differentiation and activation, lactation
TNFSF12	TNFRSF12a	Stimulation of cell growth, angiogenesis, apoptosis, induction of inflammatory cytokines
TNFSF13	TNFRSF17	B/T-cell proliferation, T-cell survival, T-cell-independent type-2 antigen response
TNFSF13b	TNFRSF13c, TNFRSF17	B-cell development (survival and maturation factor)
TNFSF14	TNFRSF3 TNFRSF6b	Initiation of T-cell costimulation
TNFSF15	TNFRSF25	Inhibition of endothelial cell proliferation, angiogenesis, apoptosis
TNFSF18	TNFRSF18	T-cell regulation,
EDA1/2		hair follicle and sweat gland development

3.21 C1q and Tumor-Necrosis Factor Superfamily

Complement C1q molecule is composed of 18 polypeptidic chains (6 A, 6 B, and 6 C chains). Agent C1q is a component of the classical pathway of

Table 3.31. Aliases of members of the tumor-necrosis factor receptor superfamily (DR: death receptor).

Type	Alias
TNFRSF1a	TNFR1, CD120a
TNFRSF1b	TNFR2, LTαR, CD120b
TNFRSF3	CD18, TNFR3, LTβR
TNFRSF4	CD134, OX40, TxGp1R, Act35
TNFRSF5	CD40
TNFRSF6a	CD95, FasR, Apo1, Apt1
TNFRSF6b	DcR3, TR6
TNFRSF7	CD27
TNFRSF8	CD30
TNFRSF9	CD137, 4-1BB
TNFRSF10a	CD261, Apo2, DR4, TRAILR1
TNFRSF10b	CD262, DR5, KILLER, TRICK2, TRAILR2
TNFRSF10c	CD263, DcR1, TRID, TRAILR3
TNFRSF10d	CD264, DcR2, TrunDD, TRAILR4
TNFRSF11a	CD265, RANK, ODFR, TRANCER
TNFRSF11b	Opg, OCIF, FDCR1, TR1
TNFRSF12a	CD266, TWEAKR
TNFRSF13b	CD267, TNFRSF14b
TNFRSF13c	CD268, BAFFR, BR3
TNFRSF14	CD258, TR2, ATAR, HVEM, LIGHTR
TNFRSF16	NGFR
TNFRSF17	CD269, BCMA
TNFRSF18	AITR, GITR
TNFRSF19	TAJ, TRADE, TRoy
TNFRSF21	DR6
TNFRSF25	TNFRSF12, DR3, DDR3, TR3, Apo3, LARD, TRAMP, WSL1, WSL-LR
TNFRSF27	EdA2R, XEDAR, TR14

complement activation for antimicrobial defense. It connects classical pathway-driven innate immunity to IgG- or IgM-mediated acquired immunity.[133]

Molecule C1q possesses many partners (Table 3.32), hence operating in several processes (Table 3.33). C1q receptor corresponds to the complex formed by calreticulin (Crt or Crtc)[134] and low density lipoprotein receptor-related protein LRP1[135] The C1q–CaBP3–LRP1 complex activates

[133] Binding of C1q to IgG- or IgM-containing immune complexes leads to the autoactivation of C1r serine peptidase proenzyme that, in turn, activates C1s, another Ser propeptidase. The C1q–C1r–C1s complex is the C1 agent. Activated C1 complex subsequently activates the components (C2–C9) of the classical pathway.

[134] A.k.a. cC1qR, calregulin, CRP55, Ca^{++}-binding protein CaBP3, and calsequestrin-like protein.

[135] A.k.a. α2-macroglobulin receptor and CD91.

Table 3.32. Ligands of C1q (Source: [482]). Molecule C1q interacts via its globular domain. This globular domain pertains also to various proteins, such as adiponectin, collagen-8 and -10, multimerin, precerebellin, and elastin microfibril interface-located protein (emilin).

β-Amyloid
C-reactive protein (CRP)
Decorin
IgG, IgM
Lipopolysaccharides
Pentraxins (Ptx)
Phospholipids (e.g., cardiolipin)
Serum amyloid protein (SAP)
Viral proteins

P38MAPK and NFκB and stimulates the production of pro-inflammatory cytokines and chemokines in macrophages.

On the other hand, members of the TNF superfamily are involved in inflammation, adaptive immunity, apoptosis, energy homeostasis, and tissue regeneration. Both C1q and TNFα are most often inducers of pro-inflammatory activators.[136] Members of the C1q and TNF superfamily are active as self-assembling trimers.

3.22 Superfamily of Interleukins

Among cytokines, interleukins (IL) can be grouped into several families (Table 3.34; Vol. 3 – Chap. 11. Receptors of the Immune System). Interleukins are involved in immunity and hematopoiesis (Table 3.35; Vol. 5 – Chap. 2. Hematopoiesis). Interleukins bind to their cognate receptors (Table 3.36). Interleukin-8 is the CXCL8 chemokine; the interleukin-8 G-protein-coupled receptors IL8Rα and IL8Rβ that are encoded by the IL8RA and IL8RB genes correspond to the CXCR1 and CXCR2 chemokine receptors.

3.22.1 Interleukin-1 Family

The interleukin-1 family (IL1F) comprises 11 members: IL1α, IL1β, IL1 receptor antagonist (IL1RA), IL18 (or IL1F4),[137] IL33 (or IL1F11), and IL1F5

[136] Agent C1q can suppress lipopolysaccharide-induced production of interleukin-12β subunit and TNFα in bone marrow-derived dendritic cells. It precludes the LPS-induced MyD88-dependent pathway, thereby reducing NFκB activity and delaying MAPK phosphorylation [482]. Adiponectin can also suppress mature macrophage function.

[137] A.k.a. interferon-γ-inducing factor (IGIF).

Table 3.33. Proteins that contain globular, trimeric C1q (gC1q), their cellular sources and functions (Source: [482] CORS26: 26-kDa collagenous repeat-containing sequence protein; CTRP5: C1q- and tumor-necrosis factor-related protein-5; emilin: elastin microfibril interface-located protein). Except in precerebellin and multimerin, the gC1q domain is always located at the C-terminus of a collagen-like sequence. The gC1q structure is homo- (adiponectin, collagen-8 and -10, and multimerin) or heterotrimeric (C1q).

Protein	Source	Function
Adiponectin	Adipocyte	Antidiabetic and anti-atherogenic adipokine, insulin sensitivity, energy homeostasis, lipid and glucid catabolism
C1q	Hepatocyte	Clearance of apoptotic cells, phagocytosis of bacteria, neutralization of retroviruses, modulation of activity of fibroblasts, and dendritic and B cells
Collagen-8	Endothelial cell	Vascular development and remodeling
Collagen-10	Chondrocyte	Lattice formation
CORS26	Prechondrocyte, fibroblast, adipocyte	Skeletal development
CTRP5	Epithelial cell, hepatocyte	Lattice formation
Emilin-1	Ubiquitous	Smooth muscle adhesion to elastic fibers, elastogenesis
Emilin-2	Lung	Development of heart chambers
Emilin-3 (multimerin-2)	Endothelial cell	Angiogenesis
Emilin-4 (multimerin-2)	Platelet, endothelial cell	Cell adhesion
Precerebellin	Purkinje cell	Development and stability of synapse

to IL1F10. Members IL1F5, IL1F6, IL1F8, IL1F9, and IL33 are active as full-length molecules, but their activity is less efficient than that of isoforms without their complete N-terminus [483]. Translated precursors proIL1α, proIL1β, and proIL1RA are processed by calpain, caspase-1, and signal peptidase into active cytokines, respectively [483].

Interleukin-1 family members are synthesized by and act on innate immune cells, such as basophils, eosinophils, neutrophils, mastocytes, and natural killer cells, to contribute to their survival and activity. Natural killer cells promote immune response from type-1 helper T cells by secreting interferon-γ. In addition, IL1 family members target lymphocytes to reinforce certain adaptive immune responses; IL18 and IL33 mainly activate T_{H1} and T_{H2} cells,

Table 3.34. Interleukin families (ILiF [i: integer]; CNTF: ciliary neurotrophic factor; CT1: cardiotrophin-1; LIF: leukemia inhibitory factor; OSM: oncostatin-M).

Family	Members
IL1F	IL1α, IL1β, IL1RA, IL18 (IL1F4), IL33 (IL1F11), IL1F5–IL1F10
IL2F	IL2, IL3, IL4, IL7, IL9, IL13, IL15, IL21
IL6F	IL6, IL11, IL27, IL30, IL31, LIF, OsM, CNTF, CT1
IL10F	IL10, IL19, IL20, IL22, IL24, IL26
IL12F	IL12, IL23, IL27, IL35
IL17F	IL17a–IL17d, IL17e (IL25), IL17f1/2

respectively, and IL1, as well as IL18 and IL33, act in T_{H17} cell differentiation and response maintenance [483].[138]

3.22.1.1 Interleukin-1

Interleukin-1[139] increases B-cell proliferation caused by mitogens. Therefore, IL1 family members promote innate and adaptive immune responses, as they enhance activities of B lymphocytes and helper T cells.

Interleukin-1 is secreted by macrophages, neutrophils, endothelial and smooth muscle cells, B and T lymphocytes, and fibroblasts in particular. Interleukin-1α and -β rapidly increase expression of hundreds of genes in multiple cell types. Positive and negative feedbacks amplify and terminate IL1 signaling.

Interleukin-1β is secreted and circulates, whereas IL1α is generally connected to the plasma membrane of the producing cell and acts locally [483]. Interleukin-1β is mainly produced by monocytes and macrophages, whereas IL1α is more widely expressed. The latter is, in particular, synthesized by endothelial cells. Interleukins IL1α and IL1β are agonists, whereas IL1RA is a specific antagonist.

Interleukin-1 controls the expression of numerous genes. Interleukin-1 increases the synthesis of [484]: (1) several cytokines, including IL1RA and itself; (2) enzymes, such as cyclooxygenase and nitric oxide synthase; (3) growth factors (fibroblast, keratinocyte, hepatocyte, nerve, and insulin-like growth factor); (4) clotting factors; (5) neuropeptides, (6) extracellular matrix molecules,

[138] Receptor IL18R is uniquely produced by T_{H1} cells in response to IL12 stimulation [483]. Interleukin-18 amplifies proliferation and Ifn-γ production in T_{H1} cells. Interleukin-33 enhances the cytokine production by T_{H2} cells that express IL1RL1 receptor.

[139] A.k.a. leukocyte endogenous mediator, hematopoietin-1, endogenous pyrogen, catabolin, and osteoclast-activating factor.

Table 3.35. Interleukins associated with vascular and blood cells. Neutrophils (Nφ), monocytes (Mo), and macrophages (Mφ) are important sources of IL1 family cytokines. They also respond to IL1, IL18, and IL33. Dendritic cells (DC) synthesize IL1 and IL18. All IL1 family members are secreted by epithelial cells [483]. They represent the main IL33 source. Eosinophils (Eφ) are particularly influenced by IL33. Among IL1 family members, only IL1 is produced by B (B Lφ) and T (T Lφ) lymphocytes. IL1 family members also strongly influence mastocytes (MC) that mediate innate allergy.

Interleukin	Sources	Targets and functions
IL1	Mo, Mφ, Nφ, B and T Lφ, dendritic and mast cells, epi- and endothelial cells,	Granulocytes, Mo, DC, MC, FB, EC Hematopoiesis, inflammation Proliferation, chemotaxis Collagen synthesis
IL2	T_{H1} and NK Lφ	B, T, and NK Lφ
IL3	T Lφ	Hematopoietic progenitor cells
IL4	T_{H2} and mast cells	B Lφ, Eφ, mastocytes
IL5	T_{H2} and mast cells	Eφ
IL6	T_{H2} and B Lφ, Mφ	B and T Lφ, thrombopoiesis
IL7	Thymic and marrow cells	B and T lymphopoiesis
IL8	Mφ, Nφ	Nφ, T Lφ
IL9	T Lφ	Hematopoiesis, thymopoiesis
IL10	T_{H2}, CD8+ T, and B Lφ, Mφ	B and mast cells Inhibition of cytokine production, anti-inflammatory
IL11	Stromal cells	Hematopoiesis, thrombopoiesis
IL12	B Lφ, Mφ	NK Lφ
IL13	T_{H2}	B Lφ, Eφ, mastocytes
IL18	Mo, Mφ, Nφ, DC, epithelial cells	Bφ, Nφ, Mo, Mφ, MC NKC, NKTC
IL33	Epithelial cells	Bφ, Eφ, Nφ, Mo, MC, NKC, NKTC

etc.; as well as density of plasmalemmal receptors for IL2, IL3, IL5, and gm-CSF in particular. Interleukin-1 increases the production of colony-stimulating factors and stem cell factors. It enhances the activation of T lymphocytes in response to antigens. This activation leads to increased T-cell production of IL2 that raises T-cell activation (*autocrine loop*). It also induces expression of interferon-γ by T lymphocytes. This T cell activation by IL1 is mimicked by tumor-necrosis factor-α secreted by activated macrophages.

Liganded IL1 receptor primes a sequence of phosphorylation and ubiquitination that activates NFκB and JNK and P38MAPK pathways. These pathways cooperate to stimulate the expression of canonical IL1 target genes, such as those that encode cytokines IL6 and IL8, chemokine CCL2, enzymes cyclooxygenase COx2, MAPK phosphatase DUSP1, inhibitor of NFκB-α, and IL1α and -β [485].

Table 3.36. Cytokine receptor families (Sources: KEGG, [511]). Cytokines can also be classified into type-1 cytokines (IL2, IL12, Ifnγ, TGFβ, TNFβ [or TNFSF2]) that promote cellular immune response and type-2 cytokines (IL4, IL5, IL6, IL9, IL10, and IL13) that favor humoral immune (antibody) response. Type-1 and -2 cytokines then refer to cytokines produced by CD4+ type-1 (T_{H1}) and -2 (T_{H2}) helper T cells, respectively. Some of these type-1 and -2 cytokines are cross-regulators: interferon-γ and interleukin-12 decrease the concentrations of type-2 cytokines, whereas IL4 and IL10 reduce the levels of type-1 cytokines. Naive CD8+ T cells can polarize into CD8+ and CD4+ type-1 or -2 effectors that produce type-1 or -2 cytokines, such as CD8+ cytotoxic T_{C1} and T_{C2} cells.

	Class-1 cytokines
IL1 receptor family	Interleukin-1α, -1β, -18, -33
	Interleukin-1 receptors IL1R1 and IL1R2
	IL1 receptor-related proteins IL1RL1 and IL1RL2
	IL1 receptor accessory protein (IL1RAcP)
IL2 receptor family	Interleukin-2, -4, -7, -9, -13, -15, -21
	Thymic stromal lymphopoietin
IL3 receptor family	Interleukin-3, -5
	Granulocyte-macrophage colony-stimulating factor
IL6 receptor family	Interleukin-6, -11, -12A/B, -23
	Oncostatin-M, leukemia inhibitory factor,
	ciliary neurotrophic factor, cardiotrophin-1,
	B-cell-stimulating factor-3, leptin,
	granulocyte colony-stimulating factor (CSF3)
Single-chain family	Erythropoietin, growth hormone, prolactin
	Class-2 cytokines
IL10 receptor family	Interleukin-10, -19, -20, -22, -24, -26
	(IL10R1–IL10R2, IL20R1–IL10R2, IL20R1–IL20R2
	(IL22R1–IL10R2, IL22R1–IL20R2)
Type-I interferons (IfnαR1–IfnαR2)	Interferon-α, -β, -κ, -ω
Type-II interferons (IfnγR1–IfnγR2)	Interferon-γ
Type-III interferons (IL28R1–IL10R2)	Interferon-λ (interleukin-28 and -29)

3.22.1.2 Interleukin-18

Interleukin-18 is expressed by macrophages and dendritic cells as well as epithelial cells.[140] Molecule ProIL18 is constitutively produced, but needs to be processed by caspase-1 to be activated [483]. On the other hand,

[140] Many subpopulations of dendritic cells, but not all, produce IL1 and IL18. They also respond to IL1 and IL18, often in combination with other cytokines [483].

T lymphocytes, NK cells, and macrophages are targeted by IL18, which promotes the production of interferon-γ and tumor-necrosis factor [486].

Both IL1 and IL18 enhance the secretion by mastocytes of IL3, IL5, IL6, IL13, and tumor-necrosis factor, but only in the presence of IgE or IL3. Interleukin-18 promotes the production of IL4 to IL6 and IL13 by basophils when combined with another stimulus, particularly IL3, hence T_{H2} immune response [483]. Interleukin-18 as well as IL1 and IL33 cooperate with IgE to improve histamine release by basophils. Interleukin-18 excites natural killer cells for Ifnγ production and enhances their cytolytic action, as it raises the production of perforin and TNFSF6 [483]. Both IL18 and IL33 influence activity of natural killer T cells. They enhance production by NKT cells of IL4, IL5, IL13, gmCSF, and TNF upon T-cell receptor stimulation.

3.22.1.3 Interleukin-33

Interleukin-33 abounds in many tissues. It particularly acts on eosinophils to enhance their survival and adhesion, as well as their production of superoxide (O_2^-) and CXC-chemokine ligand CXCL8. Moreover, it amplifies eosinophil-mediated immune responses, as it operates synergistically with 3 important eosinophil-targeting cytokines IL3, IL5, and granulocyte–macrophage colony-stimulating factor.

In addition, IL33 stimulates cytokine production by mastocytes in the absence of additional signals (unlike IL18). It also increases mastocyte maturation and survival. Moreover, it enhances the effects of other stimuli of mastocyte activation, such as thymic stromal lymphopoietin or IgE [483]. Once stimulated by IL12, IL33 promotes Ifnγ production by natural killer cells.

3.22.1.4 Group of Interleukins-36 (IL1F6 and IL1F8–IL1F9)

Interleukins IL36α, IL36β, and IL36γ, also called IL1 family members IL1F6, IL1F8, and IL1F9, respectively, are encoded by distinct genes. They target the same receptor complex composed of interleukin-1 receptor-like protein IL1RL2[141] and IL1R accessory protein (IL1RAcP). These pro-inflammatory cytokines prime similar signaling axes.

IL36 Receptor antagonist (IL36RA) corresponds to IL1 Family member IL1F5. It binds to IL1RL2 and antagonizes cytokine ligands similarly to IL1RA that counteracts IL1α and IL1β.

3.22.1.5 Interleukin-37 (IL1F7)

Interleukin-37, or IL1F7, produces anti-inflammatory effects in macrophages and epithelial cells, as it prevents the production of inflammatory cytokines induced by Toll-like receptor agonists as well as that of IL1 and tumor-necrosis factor.

---

[141] A.k.a. interleukin-1 receptor-related protein IL1RRP2.

3.22.1.6 Interleukin-1 Receptor Family

Interleukin-1 receptor family includes type-1 and -2 interleukin-1 receptors (IL1R1 and IL1R2) and interleukin-1 receptor-like-1 and -2 (IL1RL1 and IL1RL2). Plasmalemmal receptor IL1RL1 is a marker of type-2 helper T lymphocytes. Receptor IL1RL1 (IL1RL1b or ST2) can be secreted (soluble IL1RL1 form [IL1RL1a]). Activity of IL1 is controlled by both the receptor antagonist IL1RA and IL1R2 decoy receptor.

Interleukins IL18 and IL33 are inhibited by the binding partner IL18BP and soluble IL1RL1a, respectively. Activities of IL1F6, IL1F8, and IL1F9 are regulated by the receptor antagonist IL1F5 [483]. Therefore, some IL1 family members are counteracted by a corresponding receptor antagonist and others by a soluble receptor or binding protein.

Expression of IL1 receptor member IL1RL1b can be primed by mechanical stimulus in cardiomyocytes. Concentration of soluble (plasmatic) form IL1RL1a increases 1 day after myocardial infarction and can serve as a biomarker for myocardial injury [487].

3.22.2 Interleukin-2 Family

The IL2 family of cytokines include IL2, IL4, IL7, IL9, IL13, IL15, IL21, and thymic stromal lymphopoietin (TSLP).

3.22.2.1 Interleukin-2

Interleukin-2 is responsible for T-cell proliferation. It also acts on macrophages as well as B-, NK-, and lymphokine-activated killer cells [484]. Natural killer cells secrete TNFα, Ifnγ, and gmCSF in response to IL2 that activates macrophages. Interleukin-2 is secreted by activated CD4+ helper T cells.

3.22.2.2 Interleukin-3

Synthesized by activated T cells and NK cells, IL3 acts on hematopoietic progenitors, in combination with other cytokines. Activated CD4+, CD8+ T cells, basophils, and mastocytes produce IL4 and IL13. IL4 is a T-cell growth factor and promotes the differentiation of T_{H0} precursors toward the T_{H2} lineage [484]. Interleukin-3 elicits many actions of IL4. Interleukin-5, manufactured by activated CD4+, CD8+ T cells and mastocytes, influences eosinophil and basophil production and functioning. It enhances IL2-dependent differentiation and proliferation of T cells.

3.22.2.3 Interleukin-7

Interleukin-7 is the major thymopoietic cytokine that is produced by bone marrow and thymic stromal cells. IL7 stimulates stem cells to form lymphoid

Table 3.37. Cellular sources of interleukin-9 and transcription factors involved in IL9-producing cell development (Source: [492]; NKT: natural killer T cell).

Sources	Production Transcription factors
T_{H2}	IL4/5/9/10/13; GATA3, IRF4, MAF, STAT6
T_{H9}	IL9; PU1, IRF4
T_{H17}	IL9/17; BATF, IRF4, RORα, RORγ2, STAT3
T_{Reg}	IL9/10; FoxP3, SMAD2–SMAD4, STAT5
Mastocyte	IL9; IL1β and histamine promote further IL9 production
NKT	IL4/5/9/13

progenitors, which lead to B and naive T cells. In particular, IL7R plays a role on the common lymphoid progenitor, the source of all lymphoid lineages [488]. IL7 is compulsory for T-cell but not B-cell development. IL7 also stimulates the proliferation of mature T cells.

3.22.2.4 Interleukin-9

Interleukin-9 acts on erythroid progenitors, in the presence of Epo, and T and B cells [489]. It is also involved in immunity against helminths. It recruits mastocytes to the infection site. Transforming growth factor-β is required for IL9-producing T-cell differentiation from type-2 helper T cells [490]. Among committed T lymphocytes, interleukin-9 is synthesized in T_{H2}, T_{H9}, T_{H17}, and regulatory T cells (Table 3.37). The transcription factor PU1 of the ETS family[142] is required for the development of IL9-secreting subsets of CD4+ helper T cells [491].

Interleukin-9 receptor (IL9R) that is composed of the cytokine-specific IL9R α and γ chain promotes the cross-phosphorylation of JaK1 and JaK3 Janus kinases. This cross-phosphorylation enables activation of signal transducer and activator of transcription complexes, specifically STAT1 homodimers, STAT5 homodimers, and STAT1–STAT3 heterodimers [492]. Target cells of IL9 include mast, regulatory T, T helper-17, and antigen-presenting cells (Table 3.38).

[142] A.k.a. spleen focus-forming virus (SFFV) proviral integration proto-oncogene product SPI1.

Table 3.38. Cellular targets of interleukin-9 (Source: [492]; APC: antigen-presenting cell [monocytes and macrophages]).

Target cell	Function
Mastocyte	Mastocyte proliferation, synthesis of IL1β/5/6/9/10/13 (IL5 and IL13 elicit eosinophilia and airway mucus production)
T_{H17}	Autocrine growth factor (cell proliferation)
T_{Reg}	Enhancement of cell regulation
APCs	Synthesis of TGFβ

Interleukin-15

Interleukin-15 is expressed in heart, skeletal muscle, lungs, liver, and kidneys, as well as in activated monocytes, macrophages, endothelial cells, fibroblasts, etc. It triggers the proliferation of activated B lymphocytes and their immunoglobulin production, as well as the proliferation of NK cells and activated CD4+CD8+ T cells. IL16 synthesized by CD4+CD8+ T cells, eosinophils, and mastocytes is a chemokine for CD4+ cells, monocytes, and eosinophils.

3.22.2.5 Thymic Stromal Lymphopoietin

Thymic stromal lymphopoietin (TSLP) is a type-1 cytokine that promotes the activation of B and dendritic cells (Table 3.39). It binds to specific receptors on CD11c+ (α_X integrin) dendritic cells. It activates dendritic cells to produce IL4, IL5, IL13, and TNFα. In addition, TSLP promotes T_{H2} cell responses [493].

Signal TSLP on dendritic cells improves cell survival, upregulates major histocompatibility complex class-2 molecules, costimulators CTLA4 counter-receptor CD86 and TNFRSF5, various chemokines, notably CCL17 and CCL22 that target CCR4 [493]. The TSLP receptor contains TSLPR and IL7Rα.[143] Receptor engagement can activate STAT5 transcription factor.

3.22.3 Interleukin-6 Family

3.22.3.1 Interleukin-6

Fibroblasts, endothelial cells, activated B cells, monocytes, and helper T cells produce IL6. The latter promotes development and functioning of both B and T lymphocytes and megakaryocyte maturation [494]. Interleukin-6 acts

[143] Receptor IL7R contains IL7Rα and common cytokine receptor γ-chain [γc]).

Table 3.39. Cellular targets of thymic stromal lymphopoietin (Source: [493]; ↑: increase; ↓: decrease; CRTH2: chemoattractant receptor homologous molecule expressed on T_{H2} cells; NKT: natural killer T cell).

Target cell	Effect
Dendritic cell	Survival ↑, production of IL12/23 ↓, production of TNFSF4 ↑, activation of CD4+ T cells, T_{H1}/T_{H2} differentiation, maintenance and further polarization of CRTH2+ T_{H2} effector-memory cells
Mastocyte	Production of IL13 ↑
Basophil	Enhanced response, promotion of T_{H2} response
Eosinophil	Eosinophil recruitment, upregulation of α_M integrin and ICAM1
B Lymphocyte	Progenitor proliferation
CD4+ T cell	IL4 expression ↑, T_{H2} differentiation
CD8+ T cell	Cytotoxicity ↑
NKT cell	Production of IL13 ↑

in synergy with IL1 and tumor-necrosis factor. It enhances the production of immunoglobulin and glucocorticoid synthesis.

Skeletal myocytes also produce IL6 during exercise. Interleukin-6 acts as an autocrine factor that upregulates its mRNA levels via Ca^{++}–calmodulin-dependent kinase kinase (Vol. 1 – Chap. 4. Cell Structure and Function), thereby supporting AMPK activation during exercise. It is rapidly released into the circulation following exercise. It increases fatty acid oxidation, basal and insulin-stimulated glucose uptake, and translocation of GluT4 to the plasma membrane. Interleukin-6 and other myokines may defend against type-2 diabetes[144] with insulin-sensitizing effect. Interleukin-6 increases lipolysis

[144] Diabetes mellitus, or simply diabetes, is characterized by altered metabolism with hyperglycemia caused by low production of insulin by β cells of the pancreas with or without resistance to insulin. The classical triad of diabetes symptoms is polyuria (frequent urination), polydipsia (increased thirst and fluid intake), and polyphagia (increased appetite). Type-1, -2, and gestational diabetes have different causes. Type-1 diabetes is usually due to autoimmune destruction of the pancreatic β cells. Type 2 diabetes results from insulin resistance in target tissues. Gestational diabetes occurs when pregnancy hormones are responsible for insulin resistance in genetically predisposed women.

and fatty acid oxidation in patients with type-2 diabetes independently of growth hormone and/or cortisol [495].

Skeletal muscles can act as endocrine organs. Interleukin-6 stimulates the production of anti-inflammatory cytokines, such as IL1 receptor antagonist and IL10. Interleukin-10 impedes the production of IL1a, IL1b, IL8, and TNFα. Long-term effect of exercise can hinder chronic diseases associated with inflammation, such as type-2 diabetes. However, IL6 activates suppressor of cytokine signaling (SOCS) in the liver ($\sim$25 fold), leading to hepatic insulin resistance [496]. Interleukin-6 increases SOCS3 expression 2-fold in muscle and concomitantly glucose uptake. Negative effects of IL6 on SOCS3 can be overridden by the positive effects on AMPK [497].

3.22.3.2 Interleukin-11

Interleukin-11 is expressed by fibroblasts, and endothelial cells among others. In synergy with IL3, IL4, IL7, IL12, IL13, SCF, and gmCSF, IL11 stimulates the proliferation of primitive stem cells as well as commitment and differentiation of multi-lineage progenitors [498]. IL11 acts synergistically to stimulate: (1) megakaryocytopoiesis and thrombopoiesis with IL3, Tpo, or SCF; (2) erythropoiesis with IL3, SCF, or Epo; and (3) myeloid colony formation with stem cell factor.

3.22.3.3 CT1, CNTF, LIF, and OsM

Cytokines of the IL6 family include cardiotrophin-1 (CT1), ciliary neurotrophic factor (CNTF), leukemia inhibitory factor (LIF), and oncostatin-M (OsM). All IL6s that homo- or heterodimerize protein GP130 (i.e., complexes GP130–GP130, GP130–LIFR, and GP130–OsMR) activate kinases JaK1, JaK2, and, to a lesser extent Tyk2 (Vol. 4 – Chap. 3. Cytosolic Protein Tyrosine Kinases). However, JaK1 is required for GP130-mediated signaling.[145]

Cardiotrophin-1 is produced by cardiomyocytes and cardiac fibroblasts subjected to mechanical stress overload and/or exposed to an excessive amount of angiotensin-2 [499]. Secreted CT1 interacts with heterodimeric receptor formed by GP130 and leukemia inhibitory factor receptor (LIFR) to initiate exaggerated cardiomyocyte growth, reduce calsequestrin expression, and impede formation of longitudinal bundles of cardiomyocytes.

Pleiotropic cytokine oncostatin-M signals via plasmalemmal receptors that contain protein GP130 on various cell types. Type-1 and -2 oncostatin-M receptors depend on the type of β-receptor subunit of the GP130 complex, i.e., GP130–LIFRβ and GP130–OsMRβ complexes, respectively, as some cell types stimulated by OsM are unresponsive to LIFR, but respond to OsMR.

[145] Upon homodimerization, GP130 phosphorylates (activates) both STAT1 and STAT3. Protein GP130 preferentially activates STAT3.

Specific OsM activity is mediated by type-2 receptor, whereas common functions of LIF and OsM depend on type-1 receptor. Receptors LIFR or OsMR can homodimerize in embryonic stem and tumor cells.

3.22.4 Interleukin-10 Family

3.22.4.1 Interleukin-10

Interleukin-10, an anti-inflammatory cytokine, limits the immune response to pathogens, thereby preventing damage to the body's cells. In humans, IL10 is a homodimer. Interleukin-10 is expressed by many immune cells, CD4+ T cells (T_{H0}, T_{H1}, T_{H2}, and T_{H17} clones), regulatory T lymphocytes, CD8+ T cells, monocytes, and macrophages [484], as well as B cells, eosinophils, and mastocytes [500], and dendritic and natural killer cells, eosinophils, and neutrophils [501] (Table 3.40).

The synthesis of interleukin-10 is regulated by chromatin structure (epigenetic control), in addition to the transcription control and post-transcriptional regulation [501].[146] Production of IL10 is often triggered by pro-inflammatory cytokines. Extracellular signal-regulated kinases ERK1 and ERK2 contribute in many cell types to IL10 synthesis. In macrophages, positive and negative feedback loops regulate IL10 production [501].[147]

Interleukin-10 hinders the synthesis of T_{H1}-derived (IL2, Ifnγ, and gmCSF) and monocyte-derived (IL1α and -β, IL6, IL8, TNFα, gmCSF, and gCSF) cytokines, but induces IL1Ra production by macrophages. Interleukin-10 can prevent monocyte differentiation into type-1 dendritic cells, the most important antigen-presenting cells. In the presence of monocytes and/or macrophages, IL10 precludes the proliferation of resting T cells, the reduced proliferation being only partially due to decreased IL2 production. Interleukin-10

[146] Transcription factors Specific protein SP1, Activating transcription factor ATF1, CCAAT–enhancer binding protein C/EBPβ, cAMP-responsive-element-binding protein (CREB), Ifn-regulatory factor IRF1, nuclear factor-κB, and signal transducer and activator of transcription STAT3 transactivate the Il10 gene in macrophages and T-cell lines [501]. In addition, 2 cofactors of the homeobox (HOX) family, pre-B-cell leukemia transcription factor PBx1 and PBX-regulating protein PReP1, cause IL10 expression in mouse macrophages. In T_{H1} cells, mothers against decapentaplegic homolog SMAD4 and musculoaponeurotic fibrosarcoma proto-oncogene homolog (MAF) promote IL10 synthesis. In T_{H2} cells, MAF, Jun and GATA-binding protein GATA3 elicits IL10 expression.

[147] Signaling pathways based on kinases P38MAPK and ERKs that provoke IL10 production are controlled by interferon-γ and IL10 itself. The latter stimulates dual-specificity protein phosphatase DUSP1 that impedes P38MAPK action. On the other hand, IL10 upregulates mitogen-activated protein kinase kinase kinase MAP3K8 to amplify its own production. In addition, Ifnγ can also block the PI3K–PKB pathway, thereby relieving the inhibtion on glycogen synthase kinase GSK3. The latter suppresses IL10 expression via cAMP response element-binding protein (CREB) and Activator protein-1.

Table 3.40. Immune cell sources of interleukin-10 (Source: [501]; ND: not described; ERK: extracellular signal-regulated protein kinase). Interleukin-10 is not expressed by plasmacytoid dendritic cells. Ii can be synthesized in macrophages and myeloid dendritic cells after activation of Toll-like and other receptors (e.g., C-type lectin domain family members CLec4l and CLec7a).

Source	Mediators of IL10 production
Macrophage	ERK1/2
Myeloid dendritic cell	ERK1/2
B lymphocyte	ND
Mastocyte	ND
Eosinophil	ND
Helper T cells	ERK1/2
T_{H1}	IL12, STAT4
T_{H2}	IL4, STAT6
T_{H17}	TGFβ, IL6/21/27, STAT3
Regulatory T cells	ERK1/2, TGFβ

can also enhance the cytotoxic activity of CD8+ T cells. It stimulates B cells and mastocytes [502] (Table 3.41).

3.22.4.2 Interleukin-10 Family Members

In humans, 5 molecules are structurally related to IL10: IL19, IL20, IL22, IL24, and IL26 (Table 3.42). Interleukin-22 is a member of the IL10 family that is involved in inflammation and wound healing. It is mainly produced by T_{H17} and T_{H22} cells.

3.22.5 Interleukin-12 Family

Interleukin-12 is produced by macrophages, activated monocytes, and neutrophils. A positive feedback exists between IL12 and Ifnγ, this loop being controlled by IL10, TGFβ, IL4, and IL13, which downregulate IL12 production and the ability of T and NK cells to respond to IL12 [484]. Interleukin-12 activates NK cells and enhances via Ifnγ the phagocytic activity. It favors T_{H1} cell differentiation and functioning and inhibits T_{H2} cell differentiation. It synergizes with other hematopoietic factors to promote proliferation of early multipotent hematopoietic progenitors and lineage-committed precursors [503].

3.22.6 Interleukin-17 Family

Most of interleukin-17 released during inflammation is produced by innate immune cells (Table 3.43) rather than T_{H17} cells and CD4+ memory T cells. Innate immune, IL23-dependent, IL17-producing cells include macrophages

Table 3.41. Effect of IL10 on immune cells (Source: [500]). Interleukin-10 has potent effects on numerous cell populations, in particular circulating and resident immune cells as well as epithelial cells. It is immunoregulatory rather than simply immunosuppressive and anti-inflammatory. It indeed stimulates some functions of innate immunity, such as NK cell activity, non-inflammatory removal of particles, cells, and microbes by stimulating phagocytosis, and T_{H2}-related immunity, but suppresses directly and indirectly T_H response and pro-inflammatory cytokine secretion by macrophages. It acts on dendritic cells and macrophages (autocrine inhibition) to prevent the development of T_{H1} responses. It can suppress T_{H2} and allergic responses. On the other hand, IL10 enhances the differentiation of IL10-secreting T_{Reg} cells. In some circumstances, IL10 activates mastocytes and improves the activity of CD8+ T, NK, and B cells.

Cell type	Effects
B cells	Growth, IgE synthesis (+)
Dermal dendritic cells	CD86 expression, antigen presentation (−)
Endothelial cells	Expression of E-Selectin (+)
Eosinophils	Release of IL8, gmCSF (−)
Keratinocytes	Secretion of TNFα, IL6 (−)
Langerhans cells	Antigen presentation (−)
Mastocytes	Production of TNFα (−), growth (+), antigen-induced histamine liberation (+)
Monocytes, macrophages	Production of TNFα, IL1/6/8/12 (−), expression of MHC class-2, CD86, CD54, CD40 (−), antigen presentation (−), production of IL1Ra (+), soluble TNF receptors (+)
Neutrophils	Production of TNFα, IL1/8 (−), production of IL1Ra (+)
NK cells	Cytotoxicity (+)
T cells	Production of IL2 and Ifnγ (−), mitogen-induced proliferation (−)

Table 3.42. Properties of IL10 homologs (Source: [500]).

Type	Source	Effects
IL19	Activated monocytes	
IL20	Monocytes	Enhancement of IL10-induced expression of inflammation-related genes
IL22	Activated T cells, Stimulated mastocytes, Mesangial cells (renal glomerulus)	Induction of acute phase response, inhibition of IL4 production in T_{H22}, activation of STAT3 in hepatocytes

Table 3.43. Innate IL17-producing cells (Source: [504]; βG: β-glucan; AHR, aryl hydrocarbon receptor; BP: bacterial product; CLec7a: C-type lectin domain family-7 member-A, or dectin-1; GL: glycolipid; ID2: inhibitor of DNA-binding-2; IL: interleukin; IRF4:interferon regulatory factor; Ly6g: lymphocyte antigen 6 complex, locus G; MICA: MHC class-1 polypeptide-related sequence-A; NCR: natural cytotoxicity triggering receptor [NCR1 a.k.a. natural killer cell P46-related protein (NKp46)]; NKG2d: natural killer group-2 member-D; NOD2: nucleotide-binding oligomerization domain protein-2; RAE1: retinoic acid early transcript-1; ROR: retinoic acid receptor-related orphan receptor; Runx: Runt-related transcription factor; SCA: stem cell antigen; STAT: signal transducer and activator of transcription; TLR: Toll-like receptor; TNF: tumor-necrosis factor). Interleukin-17A is synthesized at its highest levels in thymus-dependent lymphocytes, such as adaptive αβ T- and innate γδ T cells, as well as invariant natural killer T (iNKT) and lymphoid-tissue inducer-like (LTi) cells. Cluster of differentiation CD90, or Thy1, is a glycosyl-phosphatidylinositol (GPI)-anchored cell surface antigen originally discovered as a thymocyte antigen.

Cell type	Trigger	Location	Transcription factors
CD3+, CD27− γδ	IL23–IL23R, IL1–IL1R, RAE1/MICA–NKG2D, βG–CLec7a, BP–TLR	Gut, skin	RORγ2, Runx1, AHR, IRF4
CD1d+, CD3+, NK1.1− iNKT	IL23–IL23R, GL–CD1d	Lung, skin, liver	RORγ2,
CD3−, NCR1+	IL23–IL23R, IL15–IL15R, RAE1/MICA–NKG2D,	Gut, skin	RORγ2, AHR, IRF4, ID2
CD4+, SCFR+, CD90+, CD3− LTi-like	IL23–IL23R, IL7–IL7R, BP–TLR	Spleen, lamina propria	RORγ2, AHR, ID2, STAT3
CD90+, SCA1+, CD3−, CD4−, SCFR−	IL23–IL23R, IL7–IL7R, BP–TLR	Lamina propria	RORγ2, TBx21
Paneth cell	TNF–TNFR, BP–NOD2	Intestine	
Ly6g+, CD11b+	BP–TLR	Lung, kidney	

and dendritic cells that predominantly reside in the skin and mucosae, where they serve as sentinels of the immune system [504]. They possess IL23R to amplify the inflammatory response.

Table 3.44. Interleukin-17 family (Source: [504]; iNKT: invariant natural killer T cell; LTi: lymphoid-tissue inducer).

Type	Cellular sources
IL17a	TCRαβ and TCRγδ T cells, iNKT and LTi-like cells
IL17b	Intestinal and pancreatic cells
IL17c	Thymic and splenic cells
IL17d	T, smooth muscle, and epithelial cells
IL17e	Mast and epithelial cells
IL17f	T_{H17} cells

Interleukin-17 (IL17 or IL17a) is secreted by CD4+, CD45RA+, CD45RO+ activated memory T cells [505]. Interleukin-17 stimulates T and endothelial cells as well as fibroblasts and macrophages to express various cell-specific cytokines [506]. It also exhibits indirect hematopoietic activity by enhancing the capacity of fibroblasts to sustain the proliferation of CD34+ hematopoietic progenitors and their differentiation into neutrophils [507].

Interleukin-17 causes the production of many cytokines (gCSF, gmCSF, IL1β, IL6, TGF-β, TNF-α), chemokines (IL8, CXCL1,[148] and CCL2), and prostaglandins (e.g., PGE2) in fibroblasts, endothelial and epithelial cells, and macrophages.

Interleukin-17 is the founding member of the IL17 family (IL17F) that includes IL17a to IL17d, IL17e (or IL25), and IL17f1 and -f2 isoforms (Table 3.44). Interleukin-17a[149] is both an innate and adaptive cytokine. It is produced from 4 to 8 h after microbial infection to enhance neutrophil chemotaxis, as it promotes IL6, gCSF, and CXCL8 (or IL8) production [504].

3.22.7 Interleukin-34

Interleukin-34 is expressed in various tissues (heart, brain, lung, liver, kidney, thymus, testis, ovary, small intestine, prostate, and colon), especially in the spleen, in particular by sinusoidal endothelium in its red pulp. It regulates myeloid cell growth and differentiation. Interleukin-34 binds to macrophage colony-stimulating factor receptor (or colony-stimulating factor receptor CSF1R) [508].

3.22.8 Interleukin-8 (CXCL8)

Stimulated monocytes, neutrophils, T and NK cells, fibroblasts, and endothelial cells secrete interleukin-8. Interleukin-8 is chemoattractant for fibroblasts,

---

[148] A.k.a. IL8-related chemotactic cytokine growth-regulated oncogene GROα.
[149] Initially termed cytotoxic T-lymphocyte antigen CTLA8.

Table 3.45. Interferons (Nφ: neutrophil). Interferon-α and -β are secreted by many cell types, such as lymphocytes (Lφ; B, NK, and T cells), plasmacytoid dendritic cells, macrophages (Mφ), fibroblasts, endothelial cells, osteoblasts, among others.

Interferon	Main sources	Principal targets
Ifnα, -β	Mφ, Nφ	NK Lφ, Mφ
Ifnγ	T_{H1}-, NK Lφ	Mφ, Nφ, NK Lφ

neutrophils, basophils, T lymphocytes, and endothelial cells [509].[150] It also activates integrins (Vol. 1 – Chap. 7. Plasma Membrane) in monocytes and eosinophils for adhesion to endothelial cells.

3.23 Interferons

Type-1 (interferon-I), -2 (interferon-II), and -3 (interferon-III) interferons (Table 3.45) have antiviral, antiproliferative, and immunomodulatory effects. They activate the Janus kinase–signal transducer and activator of transcription signaling, mitogen-activated protein kinase P38MAPK, and phosphatidylinositol 3-kinase cascades [510]. Interferons induce the expression of multiple genes. Certain genes are regulated by Ifn I and -II, whereas others are selectively regulated by distinct Ifns.

Type-1 interferons (Ifnα, Ifnβ, Ifnδ, Ifnϵ, Ifnζ, Ifnκ, Ifnω, and Ifnτ),[151] produced by many cell types, such as T and B cells, bind to a specific receptor (interferon-α and -β receptors [IfnAR]) to yield antiviral activities. They stimulate both macrophages and natural killer cells. Interferon-ω is released by leukocytes at the site of viral infections or tumors.

Type-2 interferon, or interferon-γ, is secreted by CD8+ T cells. Interferon-γ is involved in the regulation of the immune and inflammatory responses. It potentiates the effects of Ifnα and Ifnβ. It promotes the presentation of antigen to CD4+ helper T cells. It also stimulates macrophages.

Type-3 interferons, i.e., members of the interferon-λ class, include interleukin-28a (Ifnλ2), -28b (Ifnλ3), and -29 (Ifnλ1). They pertain to the class-2

[150] During extravasation, leukocytes migrate stimulated by IL8 concentration gradient and accumulate at locations of high concentration.

[151] Type-1, class-α interferons are produced by leukocytes against viral infection. They include 13 subtypes (Ifnα1–Ifnα2, Ifnα4–Ifnα8, Ifnα10, IfnαFNA13–Ifnα14, Ifnα16–Ifnα17, and Ifnα21, in addition to Ifnα pseudogene IfnαP22). Type-1, class-β interferons that are synthesized by fibroblasts have also an antiviral activity. Two Ifnβ subtypes exist (Ifnβ1 and Ifnβ3; Ifnβ2 corresponds to interleukin-6). In humans, another isoform exists: Ifnκ (but not Ifnδ, Ifnϵ, Ifnζ, and Ifnτ). In humans, interferon-ω released by leukocytes comprises a single functional form (Ifnω1) and several pseudogenes (IfnωP2, -P4, -P5, -P9, -P15, -P18, and -P19).

cytokines that comprise interleukin-10 family (IL10, IL19, IL20, IL22, IL24, and IL26) as well as type-1 and -2 interferons.

Interferons binds to interferon receptors (Table 3.36). Interferon-I receptors are heterodimers composed of IfnαR1 and IfnαR2 subunits encoded by the IFNAR1 and IFNAR2 genes. Kinases JaK1 and TyK2 (Vol. 4 – Chap. 3. Cytosolic Protein Tyrosine Kinases) associate with IfnAR1 and IfnAR2c subunits, respectively. Interferon-II (interferon-γ) receptor is also a heterodimer with IfnγR1 and IfnγR2 subunits encoded by the IFNGR1 and IFNGR2 genes. Members of the interferon-λ class target the receptor dimer made of IL28Rα (IfnλR1) and IL10Rβ (IL10R2) encoded by the IL28RA and IL10RB genes.

4

Cell Survival and Death

Some hormones and growth factors, such as insulin-like growth factor IGF1 and growth hormone, determine cell longevity. When a cell experiences a stress, survival, which can involve cellular self-digestion (autophagy), and death compete until a decision is made according to the context, i.e., when signaling related to a given fate becomes dominant.

... Has some misfortune fallen to your lot?
This too, will pass away – absorb the thought, ...
Are you upon earth's heights? No cloud in view?
Go read your motto once again: This too,
Shall pass away; ...

(This Too, Shall Pass Away, Ella Wheeler Wilcox [1850–1919])

4.1 Cell Fate and Homeostasis

Homeostasis refers to the regulation of synthesis of molecules within the cell as well as the organization and function of the cell and tissue (collection of specialized cells embedded within an extracellular matrix) to maintain a condition that fluctuates within a determined range owing to a dynamical adjustment.

Homeostasis without tissue growth implies that cell duplication is compensated by elimination of another cell. Pathologies are often associated with excessive or deficient cell death. Cell loss occurs acutely in infarction and progressively in degenerative diseases. There are several cell death modes.

Protein homeostasis is controlled by proteasomal turnover of unstable proteins via the ubiquitin system and lysosomal clearance of the majority of stable proteins through macroautophagy, most often referred to as autophagy.

4.1.1 Dependence Receptors

Most transmembrane receptors are inactive in the absence of their cognate ligand and initiate downstream signal transduction only upon ligand binding (*positive signaling*). The classical trophic theory states that, under normal conditions, cells die unless a trophic (growth) factor is present. The survival of cells in the presence of a trophic factor results from the binding of the trophic factor with its receptor Tyr kinase. Liganded RTKs foster cell survival via the PI3K–PKB and Ras–Raf–MAP2K–ERK pathways. The absence of growth factor provokes RTK inactivation, hence loss of survival signal transduction and, by default, cell death.

However, some receptors are also active in the absence of their ligand (*negative signaling*). In particular, *dependence receptors* prime cell death in the absence of ligands. The family of dependence receptors consists of 15 known members, such as netrin-1 receptor Deleted in colorectal carcinoma (DCC) and Hedgehog receptor Patched [512].

Netrin-1 receptor Uncoordinated Unc5h2 that acts in patterning of epithelial tissues and developmental angiogenesis pertains to the family of dependence receptors. Overexpression of Unc5h2 causes apoptosis via activation of death-associated protein kinase (DAPK), the activity of which is inhibited by autophosphorylation, as well as elevated activity of protein phosphatase-2 [513]. Agent Unc5h2 recruits different PP2 regulators to produce either a survival or death signal. Scaffold 65-kDa PP2 regulatory subunit A isoform β PPP2R1β[1] couples the catalytic subunit of protein phosphatase-2 ($PP2_c$) with regulatory subunits to dephosphorylate DAPK or cancerous inhibitor of PP2 (CIP2). In the absence of netrin-1, PPP2R1β fosters cell survival and decreases PP2 activity. Netrin-1 prevents and disrupts the interaction of Unc5h2 with PP2 and PPP2R1β, but not DAPK. It favors the association of PP2 with CIP2 inhibitor.

Insulin and insulin-like growth factor IGF1 that bind to their cognate receptors promote cell growth and survival via the PI3K–PKB and MAPK–ERK pathways. On the other hand, insulin and IGF1 receptors can foster apoptosis [514]. Both IR and IGF1R have thus a dual role in cell fate and pertain to the family of dependence receptors. Liganded IR and IGF1R preclude apoptosis, whereas unbound IR and IGF1R support apoptosis using multiple pathways.

In addition to IR and IGF1R, neurotrophin Tyr kinase receptors NTRK1 and NTRK3 operate as trophic and dependence receptors [515]. Ligand availability primes cell survival, differentiation, proliferation, and migration. In the absence of cognate trophic ligand, they trigger apoptosis in a kinase-independent manner.

Receptors Rearranged during transfection (ReT) with its coreceptors GDNF family receptor-α (GFRα1–GFRα4) tethered by members of the

[1] A.k.a. PR65β.

glial cell line-derived neurotrophic factor (GDNF) family, ephrin-B3 receptor EPHa4, and pleiotrophin[2] receptor Anaplastic lymphoma kinase can also cause apoptosis [512] (Vol. 3 – Chap. 8. Receptor Kinases).

4.1.2 Stress, Errors, Quality Control, and Autophagy

Cells are endowed with *quality controllers* that enable them to sense possible errors that can occur during experienced environmental stresses, then to choose between repair and degradation according to the mistake type and required energetic cost. Molecular and cellular turnover, i.e., molecule synthesis and removal as well as organelle and/or cell birth, growth, and death ensure homeostasis. Kinetics of interaction of proteins with chaperones and nucleic acids with repair machineries can serve as timers in quality control to detect abnormal molecules.

Cell organelle generation, growth, and clearance are controlled by quality control mechanisms. During autophagy, part of the cytoplasm containing potentially toxic proteins or damaged organelles that are targeted for clearance is surrounded by a cisternal membrane, the *phagophore* or *isolation membrane*. This membrane closes to form a double-membrane vacuole, the *autophagosome*. The autophagy system orchestrates the engulfment of some part of the cytoplasmic content into the autophagosome and coordinates fusion of the latter with lysosome, into which input materials are degraded.

Among organelles, mitochondrial dynamics (fusion, fission, and *mitophagy*, i.e., mitochondrial autophagy) and endoplasmic reticulum-associated degradation and unfolded protein response control the cell energy source and the secretory pathway that produces proteins for partial exocytosis and export in the extracellular medium, respectively.

Under nutrient-rich conditions, large amounts of ribosomal subunits are assembled. Ribosomal proteins have increased turnover kinetics compared with other cytoplasmic proteins. In response to starvation, ribosomes in excess accumulate in autophagy vesicles to be degraded (*ribophagy*). Ribosome degradation involves a selective autophagy pathway, in addition to non-selective autophagy [516]. A specific ubiquitin–proteasome module is involved.

4.1.3 Autophagy and Apoptosis

Autophagy (self-eating) is a cell adaptation mechanism to stress that protects a cell against death or leads to death according to the stress condition.

[2] Pleiotrophin pertains to the neurite growth-promoting factor [NEGF] family with midkine. It is indeed called neurite outgrowth-promoting factor NeGF1 as well as heparin affinity regulatory peptide (HARP), heparin-binding brain mitogen (HBBM), heparin-binding growth associated molecule (HBGAM), heparin-binding growth factor HBGF8, and osteoblast-specific factor OSF1. During embryo- and fetogenesis and early postnatal development, pleiotrophin is expressed in the nervous system as well as other organs, notably lungs and kidney, among others.

On the other hand, *apoptosis* (self-killing or type-1 cell death) is the first identified genetically programmed death process.[3]

When a cell experiences a stress, it initiates a response that causes cell survival or death. When cells are unable to withstand stress, death proteins, such as BAX and BAK of the B-cell lymphoma protein-2 (BCL2) family (Sect. 4.2.1) permeabilize mitochondrial membrane and provoke cell apoptosis.

On the other hand, autophagy favors stress adaptation and avoids cell death. However, in some conditions, autophagy can lead to *autophagic cell death*, or type-2 (non-apoptotic) cell death. Therefore, according to circumstances, autophagy serves as a cell survival pathway and suppresses apoptosis or serves as a helper or collaborator of apoptosis. Regulators of both pathways are indeed connected. Moreover, numerous death stimuli can activate either pathway. Last, but not least, both pathways share several genes required for their respective execution.

4.2 Main Death Mediators

4.2.1 B-Cell Lymphoma Protein-2 Family

The BCL2 gene was discovered in B-cell follicular lymphomas, in which transcription by immunoglobulin heavy-chain gene promoter and enhancer on chromosome 14 becomes excessive [519]. BCL2 family proteins identified in B-cell lymphoma (leukemia) tumor cells are classified into 3 classes: (1) anti-apoptotic proteins that encompass BCL2, BCL2a1,[4] BCLb,[5] BCLw, BCL extra-large protein (BCLxL), and BCL2-related myeloid cell leukemia sequence protein MCL1; (2) pro-apoptotic proteins that comprise BCL2-antagonist–killer (BAK), BCL2-associated X protein (BAX), and BCL2-related ovarian killer (BOK);[6] and (3) BH3-only proteins that include BCL2 antagonist of cell death (BAD), BH3-interacting domain death agonist (BID), BCL2-interacting killer (BIK), BLK, or NBK, BCL2-like molecule (BIM or BOD), BCL2-modifying factor (BMF), Harakiri (Hrk or death protein DP5), Noxa (Latin: damage), and P53-upregulated modulator of apoptosis (PUMA or BBC3).

Cytoprotection is ensured by anti-apoptotic BCL2 proteins because they antagonize pro-apoptotic proteins BAK and BAX and repress mitochondrial outer membrane permeability, thus preventing apoptosis. Anti-apoptotic BCL2 proteins also preclude autophagy. On the other hand, other BCL2 family members or modulators activate autophagy via class-3 phosphatidylinositol 3-kinase (PI3KC3) with catalytic vacuolar protein sorting VPS34 and

[3] The notion of "programmed cell death" was introduced by Lockshin in 1964 [517]. In 1972, a peculiar type of cell death was dubbed "apoptosis" [518].

[4] A.k.a. A1 and BFL1.

[5] A.k.a. BCL2L10.

[6] A.k.a. MTD.

Table 4.1. Subcellular location of proteins of the B-cell lymphoma (leukemia; BCL2) family (Source: [519]; BAD: BCL2 antagonist of cell death; BAK, BCL2-antagonist/killer-1; BAX: BCL2-associated X protein; BCLxL: BCL extra-large protein; BID: BH3-interacting domain death agonist; MCL1: BCL2-related myeloid cell leukemia sequence protein-1).

Type	Healthy cell	Apoptotic cell
	Anti-apoptotic members	
BCL2	Endoplasmic reticulum, nuclear and outer mitochondrial membrane	Endoplasmic reticulum, mitochondria
BCLxL	Cytosol, mitochondria, endoplasmic reticulum	Endoplasmic reticulum, mitochondria
BCLw	Cytosol, mitochondria	Mitochondria
MCL1	Cytosol, mitochondria, endoplasmic reticulum	Mitochondria
	Pro-apoptotic members	
BAK	Outer mitochondrial membrane, endoplasmic reticulum	Mitochondria, endoplasmic reticulum
BAX	Cytosol	Endoplasmic reticulum, mitochondria
BOK	Cytosol, mitochondria	Mitochondria
BID	Cytosol	Cytosol, mitochondria

regulatory VPS15 subunits (encoded by the Pik3c3c and Pik3r4 genes, respectively).[7]

BH3-only inactivators that respond to cell-death stimuli bind to anti-apoptotic BCL2 family proteins, hence releasing BH3-only activators of pro-apoptotic BCL2 family proteins [521]. Proteins of the BCL2 family have their subcellular locations (Table 4.1). They interact to promote or prevent cell apoptosis.

Protein kinase-B and ribosomal S6 kinase RSK2 phosphorylate BAD that then dissociates from BCL2, hence freeing BCL2 that can inhibit pro-apoptotic BAX and BAK proteins [520].

On the other hand, the cytosolic protein Tyr phosphatase PTPn5 contributes to BAK dephosphorylation (activation; Tyr108) in response to apoptotic stimuli [522]. It is controlled by extracellular signal-regulated kinases ERK1 and ERK2 that inhibit PTPn5 and promote cell survival.

[7] Kinase PI3KC3 is required for autophagosome formation (Sect. 4.4). It constitutes the *PI3K complex* with beclin-1, ultraviolet wave resistance-associated gene product (UVRAG), Activating molecule in beclin-1-regulated autophagy protein (AMBRA1), and endophilin B1 family member BAX-interacting factor BIF1 (a.k.a. endophilin-B2 and SH3 domain GRB2-like endophilin-B1 [SH3GLB1]) [520].

4.2.2 Reactive Oxygen Species

Dioxygen is required for cell life. In addition, it produces reactive oxygen species, such as superoxide, peroxides, hydroxyl radicals, and other related species that are involved in cell signaling as well as aging and diseases.

Reactive oxygen species are produced by mitochondria (2–5%), by various enzymes (xanthine oxidase, cyclooxygenases, lipoxygenases, myeloperoxidases, heme oxygenase, monoamine oxidases, aldehyde oxidase, and cytochrome P450-associated enzymes), and mainly by NADPH oxidases.

At physiological concentrations, reactive oxygen species can act as second messengers, particularly in cell growth and proliferation. During growth factor stimulation as well as in cell adhesion, ROS regulate the cytoskeleton organization and cell motility, mainly via oxidation of and/or indirect influence on adhesion or cytoskeletal proteins, such as Src kinase, β-actin, focal adhesion kinases, and several protein Tyr phosphatases.

However, these oxidants can be toxic, as they interact with other chemical species and produce derived products, which can cause oxidative damage. Therefore, the cell has anti-oxidant defenses that encompass a set of enzymes as well as other proteins such as peroxisome proliferator-activated receptor-γ coactivator-1. Under normal conditions, ROS are cleared from the cell by anti-oxidant enzymes, such as *superoxide dismutase* (SOD), *glutathione peroxidase* that degrades hydrogen peroxide, peroxiredoxins that decompose peroxides and hydrogen peroxide, and thioredoxin reductase, in addition to non-enzymatic anti-oxidants.

The increased production of reactive oxygen species raises oxidative protein damage and lipid peroxidation. In particular, under pathological conditions, elevated ROS levels damage polyunsaturated fatty acids in membrane lipids, and DNA. Such damages lead to cell dysfunction and death. The balance between beneficial and deleterious effects also depends on cellular localization and possibly ROS sources.

The mitochondrial respiratory chain can generate reactive oxygen species. The ROS level increases when electron flow through the respiratory chain is impeded by respiratory inhibitors or altered by uncoupling electron transport from oxidative phosphorylation. When cells experience hypoxia, oxidative stress rises. Reactive oxygen species stimulate hypoxic nuclear genes, but another activation source exists. Mitochondria actually produce nitric oxide in a nitrite (NO_2^-)-dependent, nitric oxide synthase-independent manner, using cytochrome-C oxidase that acts as a nitrite reductase [523].

ROS and Insulin Resistance

Increased production of reactive oxygen species and resulting chronic oxidative stress causes insulin resistance via Jun N-terminal kinase of the mitogen-activated protein kinase modules. Jun N-terminal kinase phosphorylates insulin receptor substrate IRS1 that is then degraded, hence repressing insulin signaling. Signaling that activates JNK involves stress-induced

MAP3K4 and MAP3K7 as well as scaffold JNK-interacting protein (JIP). Activated JNK dimerizes and moves into the nucleus, where it can phosphorylate Jun transcription factor to stimulate stress responses and apoptosis. On the other hand, JNK-specific phosphatase MKP7 dephosphorylates (inactivates) JNK enzyme. Inhibition of activated JNK in the cytosol, especially at the cell cortex, prevents IRS1 degradation and promotes insulin signaling and insulin-dependent glucose uptake.

Skeletal muscles are responsible for about 80% of the peripheral glucose uptake. In skeletal myocytes, oxidative phosphorylation in mitochondria that results from exercise heightens ROS production and activates JNK, but increases insulin sensitivity. Acute oxidative stress raises insulin-dependent phosphorylation of protein kinase-B.

Whereas acute oxidative stress leads to JNK activation in the nucleus and improves insulin signaling, chronic oxidative stress in skeletal myocytes activates JNK in the cytoplasm and provokes insulin resistance [524]. Acute oxidative stress is able to restore insulin signaling in insulin-resistant cells. Acute oxidative stress redistributes MKP7 from the nucleus to cytoplasm, even in insulin-resistant cells.

4.2.3 Caspases

Caspases[8] have several functions, including inflammation, cell proliferation, differentiation, and survival/apoptosis [525] (Table 4.2). All caspases are produced as inactive zymogens. Initiation of caspase activation needs adaptor complexes, such as apoptosomes and inflammasomes.

Activated caspase substrates include cytokines, kinases, transcription factors, and polymerases. Many caspase inhibitors control their activities. In addition to their catalytic activities, caspases have non-proteolytic functions for the recruitment of adaptors and effectors that modulate NFκB activation.

Caspase-1 mainly activates inflammatory cytokines (Chap. 3).[9] Inflammatory caspases (group-1 caspases) also include caspase-4 and -5 in humans [527]. They are activated by members of the NOD-like receptor (NLR) family, such

[8] Caspases are cysteinyl aspartate peptidases, i.e., cysteine peptidases, which cleave substrates targeting aspartate residues.

[9] Caspase-1, also called interleukin-1β-converting enzyme, targets interleukin-1β precursor (proIL1β, the production of which is induced by lipopolysaccharides) to generate active interleukin-1β. Caspase-1 is activated by $P2X_7$ ion channel–receptor stimulated by extracellular ATP. Interleukin-1β initiates and amplifies various events during microbial invasion and tissue injury. Caspase-1 also activates interleukin-18. Interleukin-18 stimulates the production of interferon-γ by splenocytes and pro-inflammatory cytokines, upregulates adhesion molecules, and stimulates natural killer lymphocytes. Both interleukin-1β and -18, activated by inflammasomes, activate their specific receptors (IL1R and IL18R) for signaling. Caspase-1 can also activate NFκB.

Table 4.2. Caspase (cysteine aspartic acid-specific peptidases) functions (Sources: [525, 527, 528]). Caspases are produced as zymogens that need to be activated. They cleave their substrates after specific tetrapeptide motifs. The caspase family can be subdivided into initiators that auto-activate and initiate proteolysis (activation) of other caspase effectors. Caspase-1, -4, -5, and -12 are activated during innate immune responses for regulation of inflammatory cytokine processing (e.g., interleukin-1β and -18). Caspase-2, -3, and -6 to -10 are stimulated during cell apoptosis.

Caspase	Functions
Caspase-1	Cytokine maturation, inflammation, NFκB activation
Caspase-2	NFκB activation, stress-induced apoptosis initiation, erythrocyte formation
Caspase-3	Apoptosis, erythrocyte and platelet formation
Caspase-4	Inflammation
Caspase-5	Inflammation
Caspase-6	Apoptosis
Caspase-7	Apoptosis
Caspase-8	Apoptosis initiation, proliferation and activation of T, B, and NK cells, maturation of monocytes into macrophages, NFκB activation
Caspase-9	Apoptosis initiation
Caspase-10	Apoptosis initiation
Caspase-11	Inflammatory cell migration, caspase-1 activation
Caspase-12	Inflammation, autophagy

as NAIP (neuronal apoptosis inhibitory protein), NALP (NACHT [NAIP domain], LRR [leucine-rich repeat], and PYD [pyrin domain] containing proteins), and IPAF (IL1β-converting enzyme peptidase-activating factor), that form inflammasomes. Inflammatory caspases are controlled at the inflammasome level by inhibitors to avoid excessive cytokine production.

Executioner caspases (caspase-3, -6, and -7) conduct apoptosis. Apoptotic caspases are activated by *initiator caspases* (caspase-2, -8, -9, and -10) that are recruited to adaptor complexes [525]. Caspase-9 is the main component of the mitochondrial intrinsic apoptotic pathway regulated by BCL2 protein family, because of its higher catalytic activity.

Caspase-2 can also be involved in apoptosis. It also elicits MAPK and NFκB signaling. Caspase-2 activation requires the assembly of a proteic complex that comprises adaptor RAIDD and P53-induced protein with a death domain (PIDD), the *piddosome*.

Cytochrome-C release from the mitochondria favors the formation of septameric apoptosomes, composed of apoptotic peptidase-activating factor-1 (APAF1) that activates caspase-9. Apoptosome-bound caspase-9 cleaves (activates) caspase-3.

Caspase-8 recruited to the *death-inducing signaling complex* (DISC) is implicated in the extrinsic apoptotic pathway triggered by tumor-necrosis factor and death receptor TNFRSF6a. Caspase-8 and -10 mediate NFκB-dependent inflammation in response to virus invasion. Caspase inhibitors for apoptosis regulation include flice-inhibitory proteins (FlIP), upregulated by the TNFα–NFκB pathway, and inhibitors of apoptosis proteins (IAP).[10]

Catalytic and non-proteolytic functions of caspases are implicated in cell differentiation, proliferation, and NFκB activation [528]. Although caspase-8 can initiate apoptosis, it also acts in the proliferation of immunocytes (T, B, and NK cells), according to the caspase-8 adaptor recruitment level, hence to caspase-8 activation magnitude. T-cell receptor stimulates the recruitment of caspase-8 and its adaptor Fas-associated death domain-containing protein (FADD) to a suitable proteic complex required for TNF-induced NFκB activation.

Caspase-3 targets substrates associated with cell cycle control, such as cyclin-dependent kinase inhibitor and transcription factor nuclear factor for activated T cells. Caspase-2, -3, and -9 are transiently activated during erythroblast differentiation. Caspase-3 and -9 activation is involved in platelet formation from megakaryocytes. Both differentiation types lead to enucleated cells (erythrocytes and platelets). In addition, caspase-8 is implicated in the maturation of monocytes into macrophages.

Protein nitrosylation regulates cell activity. Caspase-3 is inhibited by nitrosylation. It is denitrosylated by thioredoxin Trx1 in resting lymphocytes and by Trx2 after activation of cell death-promoting receptor TNFRSF6a [529]. Caspase-3 denitrosylation requires thioredoxin reductase.

4.2.4 Cytochrome-C

Cytochrome-C is a water-soluble, heme-containing protein. Two known isoforms — testicular and somatic — of cytochrome-C exist. Cytochrome-C participates in ATP synthesis. It is also required in apoptosome assembling that involves a core of 7 apoptotic peptidase-activating factor-1s.

Cytochrome-C is a mitochondrial protein that is allocated to cristae of the inner mitochondrial membrane, where it can associate with membrane phospholipid cardiolipin. It is a component of the mitochondrial electron-transport chain. The electron-transport chain comprises 4 membrane-bound

[10] Member IAP1 of the inhibitor of apoptosis family with ubiquitin ligase activity is also termed Birc2. It is required for endothelial cell survival [526]. Protein IAP1 favors the formation of tumor-necrosis factor receptor complex-1 in endothelial cells that enhances the transcriptional activity of NFκB activation.

proteic complexes (complex-1–complex-4). The electron-transport chain uses the electron carriers ubiquinones, cytochromes, and iron-sulfur proteins to transfer electrons. Electron transfer is used for proton outflux from the mitochondrial matrix. This flux generates the inner mitochondrial membrane potential. A passive H^+ flux into the mitochondrial matrix occurs through a proton pore that is associated with ATP synthase. Cytochrome-C uses its heme as a redox intermediate to carry electrons between complex-3 and -4 of the electron-transport chain. Cytochrome-C accepts electrons from complex-3 for cytochrome oxidase (complex-4). Electrons are donated to O_2 to form H_2O.

When a cell receives an apoptotic signal, cytochrome-C is released into the cytosol to trigger the programmed cell death pathway. In the cytosol, cytochrome-C commits apoptotic peptidase-activating factor APAF1 and forms the apoptosome that activates caspase-9. Cytochrome-C release is controlled by many agents, especially members of the B-cell lymphoma protein-2 family.

4.2.5 Other Mediators of Cell Survival or Death

4.2.5.1 NFκB Transcription Factor and MAPK Enzyme

Cell exposure to inflammatory cytokines, pathogens, and DNA damages activates nuclear factor-κB (NFκB; Vol. 4 – Chap. 9. Other Major Signaling Mediators), mitogen-activated protein kinases (Vol. 4 – Chap. 5. Mitogen-Activated Protein Kinase Modules), and transcription factors interferon regulatory factors (IRF) that upregulate anti-apoptotic proteins and cytokines, and cause autophagy or apoptosis.

4.2.5.2 RIPK Signaling via MAPK and Members of TNFSF and TNFRSF

Receptor-interacting protein RIPK1 is an adaptor kinase implicated in cell fate [530]. Kinase RIPK1 forms integrator complexes of signaling pathways triggered by T-cell receptor (TCR), Toll-like receptors TLR3 and TLR4 (Vol. 3 – Chap. 11. Receptors of the Immune System),[11] as well as death receptors, such as members of the tumor-necrosis factor receptor superfamily (Sect. 3.20) TNFRSF1a (or TNFR1), TNFRSF10a, and TNFRSF10b.

Protein RIPK1 also binds to death domain-containing adaptor proteins, such as tumor-necrosis factor receptor-associated death domain- (TRADD) and Fas receptor (TNFRSF6a)-associated death domain (FADD)-containing proteins, as well as with other types of adaptors in order to recruit kinases, such as MAP3K1, MAP3K3, and RIPK3.

[11] Toll-like receptors TLR3 and TLR4 can activate the NFκB and MAPK pathways. Stimulated TLR3 and TLR4 also activate interferon regulatory factor IRF3 to produce type-1 interferons.

When tumor-necrosis factor-α (or TNFSF1) binds to its receptor TNFR1 (or TNFRSF1a), TNFR1 recruits RIPK1 and TNFR-associated factor TRAF2 in plasmalemmal *complex 1*. Several pathways are then activated, such as NFκB, P38MAPK, and JNK (Fig. 4.1). Activation of NFκB prevents apoptosis, its blockade causes apoptosis due to TNFα. Receptor endocytosis allows the recruitment by TRADD of FADD and caspase-8, which form *complex 2*, leading to apoptosis. Moreover, caspase-8 favors RIPK1 cleavage. RIPK1 ubiquitination in membrane rafts favors the recruitment of TGFβ-activated kinase TAK1 (MAP3K7)-binding protein TAB2 (MAP3K7IP2) and formation of the TAK1–TAB1–TAB2 (MAP3K7–MAP3K7IP1– MAP3K7IP2) complex. The latter activates IKK and NFκB.[12] In addition, MAP3K7 also binds and phosphorylates MAP3K3 that activates NFκB. Ligand TNFSF6 can also activate NFκB via caspase-8 (Fig. 4.1), but it is mainly aimed at eliciting apoptosis.

4.2.5.3 GSK3β Kinase

Glycogen synthase kinase-3β phosphorylates β-catenin and promotes its degradation. Activated P38MAPK phosphorylates (inhibits) GSK3β and supports cell survival at least in some cells, such as thymocytes and neurons [531].[13]

4.2.5.4 Protein Kinase-B

Survival kinase protein kinase-B also phosphorylates GSK3β. Moreover, PKB phosphorylates (inactivates) FoxO transcription factors as well as Ub ligase double minute DM2 that targets P53 to reduce the expression of several pro-apoptotic genes [520]. Protein kinase-B also impedes the activity of several kinases that act upstream from the JNK module to prevent activation of Jun N-terminal kinase and its pro-apoptotic function.

In addition, protein kinase-B and ribosomal S6 kinase RSK1 phosphorylate cAMP-responsive element-binding protein to promote the expression of anti-apoptotic genes, such as Bcl2 and Bclxl [520].

4.2.5.5 Protein Ser/Thr Kinase DAPK

Calcium calmodulin-regulated death-associated kinase (DAPK; Vol. 4 – Chap. 4. Cytosolic Protein Ser/Thr Kinases) associates with the cytoskeleton. It can cause membrane blebbing and contribute to stress fiber formation,

[12] The NFκB inhibitory protein TNFαIP3 (or A20) recruited to the TNFR1 complex impedes RIPK1 ubiquitination.

[13] Enzyme P38MAPK is phosphorylated (activated) by MAP2K3 and MAP2K6. The P38MAPK pathway controls cell differentiation, blocks cell proliferation, and induces apoptosis. P38MAPK participates in G2–M cell cycle checkpoint. Enzymes of the P38MAPK subfamily can also promote cell survival.

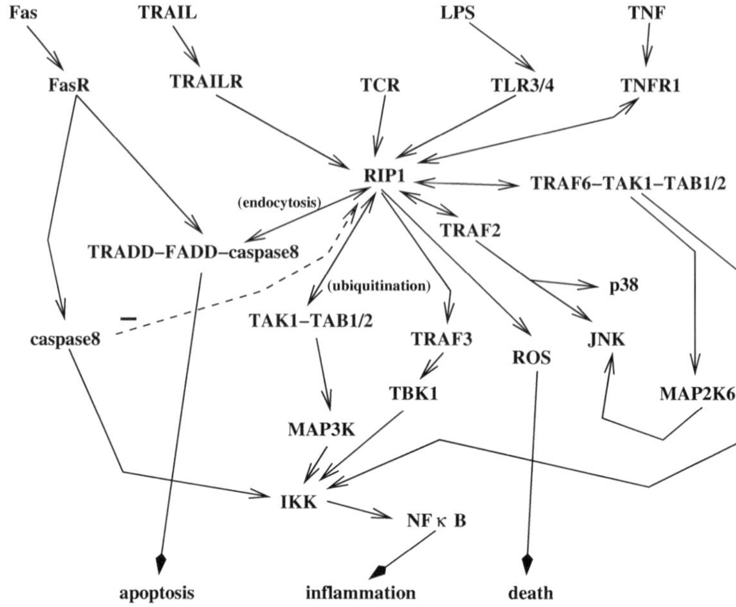

Figure 4.1. Kinase RIPK1 and cell fate (Source: [530]). Enzyme RIPK1 integrates numerous signals to initiate different cellular responses, according to the type of RIPK1-containing complexes, complex subcellular location, and targeted substrates. Agent TNFSF1 (TNFα) binds to its receptor TNFRSF1a (TNFR1) and activates either pro- or anti-apoptotic pathways. Activated TNFR1 recruits RIPK1 and TRAF2 that forms plasmalemmal complex-1. Ubiquitination of RIPK1 in membrane rafts favors the formation of the TAK1–TAB1–TAB2 complex that activates IKK and NFκB. The complex composed of TNFSF10 (TRAIL) and its receptors TNFRSF10a (TRAILR1) and TNFRSF10b (TRAILR2) undergoes endocytosis and triggers most often apoptosis or NFκB activation. Agent TNFSF10 can stimulate enzymes of the mitogen-activated protein kinase module, such as JNK and P38MAPK, with cooperation of RIPK1 and TRAF2 on the one hand and IKK using RIPK1 on the other, but at a lower magnitude and speed than TNFSF1. Receptor TNFRSF6a recruits FADD and caspase-8 to the plasma membrane and forms the death-inducing signaling complex (DISC). Once recruited to DISC, caspase-8 is activated and leads to apoptosis. Activated Toll-like receptor TLR3 recruits RIPK1, TRAF6, TAK1, TAB1, and TAB2. This complex activates IKKβ and MAP2K6, and subsequently JNK enzyme. Under TNF stimulation, RIPK1 translocates to the mitochondrion and reduces the interaction between adenine nucleotide translocase and cyclophilin-D, thus reducing ATP synthesis and enhancing production by mitochondria of reactive oxygen species (ROS) that are involved in apoptosis.

as it phosphorylates myosin-2 regulatory light chain. This tumor suppressor and antimetastatic agent operates in early P53-dependent transformation checkpoint. It is able to modulate both apoptosis and autophagic cell death. Kinase DAPK phosphorylates beclin-1, thereby reducing beclin-1–BCLxL binding [520]. In addition, DAPK enhances P53 activity via the CDKN2A–ARF locus product p14ARF.

Kinase DAPK is phosphorylated (activated) by extracellular signal-regulated protein kinase. A feedback loop allows DAPK to target ERK for its cytosolic sequestration [520]. Therefore, nuclear effects of ERK, such as prosurvival signaling that counters apoptosis, are eliminated.

4.2.5.6 P53 Transcription Factor

The transcription factor P53 upregulates pro-apoptotic genes, such as Bax, Puma, and Noxa, and downregulates anti-apoptotic genes such as Bcl2. Cytoplasmic P53 also enhances mitochondrial outer membrane permeability induced by BAX [520]. Moreover, P53 may inhibit the target of rapamycin (TOR) pathway. On the other hand, P53 activates damage-regulated modulator of autophagy (DRAM), a lysosomal membrane protein. Therefore, P53 is an activator of apoptosis and possesses a dual role in autophagy regulation. Nuclear P53 favors autophagic cell death; cytoplasmic P53 represses prosurvival autophagy.

4.2.5.7 Cyclin-Dependent Kinase Inhibitors

The transcription factor P53 elicits cyclin-dependent kinase inhibitor CKI1a (encoded by the gene CDKN1A) that inhibits the activity of Ccn–CDK2 and Ccn–CDK4 complexes. In humans, the alternate reading frame of the CDKN2A–ARF locus encodes CKI2a (P14ARF or P19ARF in mice).[14] Factor P14ARF is a stabilizer of P53, as it antagonizes P53's inhibitor DM2. Ubiquitin ligase DM2 indeed binds to P53 and promotes its degradation (Chap. 2). Consequently, P14ARF functions in the P53 pathway to stimulate apoptotic genes and cause autophagy [520]. Protein P14ARF can also localize to the outer membrane of mitochondria, where it binds to BCL2 and BCLxL.

4.2.5.8 Short Mitochondrial Alternate Reading Frame

The P14ARF mRNA can also produce a second, small isoform that lacks the N-terminal domain necessary for nuclear localization and DM2 binding

[14] The cyclin dependent kinase inhibitor gene CDKN2A encodes 3 alternatively spliced CKI2a variants. Two isoforms are inhibitors of CDK4 kinase. The third gene transcript contains an alternate open reading frame (ARF) that produces a protein that sequesters DM2 in the nucleolus. Ubiquitin ligase DM2 targets P53 and thus prevents P53 transcriptional activation. Short mitochondrial form of alternate reading frame (smARF isoform) that is produced after removal of nucleolar functional domains, translocates to mitochondria to reduce the mitochondrial transmembrane potential.

and localizes to mitochondria. This short mitochondrial form of alternate reading frame (smARF) reduces the mitochondrial transmembrane potential and causes autophagic cell death [520].[15] Mitochondrial smARF, unlike nucleolar, full-length P14ARF is able to provoke P53-independent autophagy.

4.2.5.9 Transcription Factors of the E2F Family

Members of the E2F family of transcription factors are involved in DNA repair as well as cell survival, proliferation, differentiation, and death. The CDKN2A–ARF locus is an E2F1 target gene. Factor E2F1 thus triggers apoptosis via P14ARF and P53 [520]. In addition, E2F1 upregulates the expression of the autophagy genes Atg8 (or mammalian microtubule-associated protein-1 light chain-3 [Map1lc3]), Atg1, Atg5, and damage-regulated modulator of autophagy (Dram). Therefore, E2F1 links apoptosis and autophagy.

4.3 Cell Degradation and Death

Cells permanently experience various exogenous and endogenous stress types. Damage responses range from complete recovery to cell death. Cell death contributes to the formation of tissues and organs and regulates tissue homeostasis by eliminating unwanted cells. Proliferating cells can adopt a state of permanent cell cycle arrest (cellular *senescence*). Multiple mechanisms of cell death exist.

4.3.1 Autophagy and Autophagic Cell Death

Apoptosis enables elimination of unwanted cells, especially during embryo/fetogenesis and immune processes. Apoptosis actually corresponds to a line of defense against infection.

Autophagy is primarily aimed at protecting cells when cells experience nutrient deprivation or any other stress (Sect. 4.4). This survival mechanism indeed starts up the defense against stress. Autophagy also controls the protein turnover during tissue development and regulates cell proliferation. It involves the engulfment of parts of the cytoplasm inside autophagosomes. Beclin-1 (mammalian ortholog of yeast AtG6) and autophagy-related gene product AtG5 are involved in autophagy.

In some conditions, autophagy and apoptosis are mutually exclusive. However, autophagy is also linked to cell death. Consequently, according to the scenario (nature of stress and cell environment as well as cell type), both processes either act in synergy or counteract each other. In particular, AtG5 also

[15] Within a peptidase K-resistant compartment of the mitochondria, smARF interacts with P32 mitochondrial protein that stabilizes smARF and increases its ability to cause mitochondrial membrane dissipation and autophagy [520].

regulates apoptosis.[16] Coordination and crosstalk thus exist between different cell death modalities. *Autophagic cell death* that accompanies autophagy corresponds to a genetically programmed self-elimination.

4.3.2 Criteria of Cell Death Types

Different types of cell death are defined by morphological criteria, but cell death can be classified according to enzymological criteria (with and without induction by nucleases or peptidases, such as caspases, calpains, cathepsins, and transglutaminases), functional aspects (programmed, accidental, immunogenic, or non-immunogenic) [532].

In the absence of a defined biochemical event that corresponds to a point-of-no-return, a cell can be considered dead when [532]: (1) the plasma membrane has lost its integrity; (2) the cell and its nucleus have undergone fragmentation; and (3) the cell or its fragments have been engulfed by an adjacent cell.

Biochemical cascades are triggered by cell stresses whatever the cell death mode [533]. Mechanisms that prime transient growth arrest, senescence, or apoptosis in response to cellular stress rely on a set of mediators, such as P53 and CKI2a.

Morphological, biochemical, or immunological data alone cannot accurately define the cell death type (Tables 4.3, 4.4, and 4.5). (1) Apoptosis is characterized by caspase activation and degradation by surrounding cells. (2) Massive autophagic vacuolization can induce cell death. (3) Cell necrosis, currently defined by ruptures of the plasma membrane that trigger inflammation, particularly involves calpains and cathepsins. (4) Mitotic catastrophe occurring during or shortly after a failed cell division results from a combination of deficient cell cycle checkpoints and cellular damages, which activates self-destruction. Permeabilization of both mitochondrial membranes usually corresponds to the point of no-return in the cascade of events leading to cell death.

Cell death has been subdivided into regulated and unregulated mechanisms. Regulated *apoptosis* (Sect. 4.6) is a type of programmed cell death defined by nuclear and chromatin condensation (*pyknosis*) followed by nuclear fragmentation (*karyorrhexis*).

Apoptosis is carried out by a coordinated reaction cascade with initiator and executioner caspases. Caspases are activated via 2 main pathways — extrinsic and intrinsic — by distinct apoptotic stimuli. The B-cell lymphoma-2 family encompasses 2 main pro-apoptotic proteins: BAX and BAK. Members of the death-associated protein kinase family are common to both apoptosis and autophagy.

[16] Upon apoptotic stimuli, AtG5 is cleaved by calpain. Its N-terminal product then translocates to mitochondria, where it interacts with BCLxL and promotes cytochrome-C release and caspase activation.

Table 4.3. Cell fate under stress and corresponding morphological features (**Part 1**; Source: [532]; BCL: B-cell leukemia/lymphoma protein; BCLxL: BCL extra-large protein). Apoptosis that can occur without oligonucleosomal DNA fragmentation often means programmed cell death. Cell death can occur in the context of autophagy.

Cell fate	Features
Apoptosis	Rounding-up of the cell, retraction of pseudopods, reduction of cellular and nuclear volume (pyknosis), nuclear fragmentation (karyorrhexis), chromatin condensation, plasma membrane rupture, uptake by phagocytes, absence of modifications of cytoplasmic organelles, mitochondrial membrane permeabilization, mitochondrial transmembrane potential dissipation, activation of pro-apoptotic BCL2 family proteins, activation of caspases
Autophagy	Absence of chromatin condensation, absence or little association with phagocytes, vacuolization of the cytoplasm, accumulation of autophagic vacuoles beclin-1 dissociation from BCL2, BCLxL, dependency on autophagy-related genes

Non-apoptotic cell death pathways, such as autophagic cell death, necroptosis, and poly^{ADP}ribose polymerase PARP1-mediated cell death, drive cellular responses to stress, genotoxic damage, and infection [534]. These multiple pathways of non-apoptotic cell death are regulated.

Unregulated *necrosis* (Sect. 4.9) is caused by overpowering stresses that are incompatible with cell survival. Necrosis is associated with inflammation. *Necroptosis* occurs when death receptors are activated by their ligands, whereas caspases are inhibited, leading to a shift from apoptosis to non-apoptotic cell death with features of autophagy and necrosis. *Parthanatos* is a caspase-independent cell death modality characterized by nuclear condensation.

In *entosis*, a cell invades another cell, even when it is still alive. It corresponds to heterophagy. In vitro, epithelial cells that lose attachments to the extracellular matrix, even in the absence of death signal, can be ingested by other normal cells. Cell internalization is reversible, but most ingested cells die. Entosis could eliminate mislocalized epithelial cells.

4.4 Cell Autophagy

Autophagy is initiated by the formation of double membrane-enclosed vesicles, the so-called *autophagosomes*, that engulf cytoplasmic proteins and

Table 4.4. Cell fate under stress and corresponding morphological features (**Part 2**; Source: [532]; HMGB: high-mobility group HMG-box domain-containing protein; PARP: poly^{ADP}ribose polymerase; RIPK: receptor-interacting protein kinase). Anoikis is induced by the loss of attachment to matrix and/or other cells. Death domain and Toll-like receptors elicit necrosis. Necroptosis refers to regulated necrosis in contrast with accidental necrosis. Paraptosis that is triggered by insulin-like growth factor receptor-1 leads to extensive cytoplasmic vacuolization and mitochondrial swelling, without any other morphological hallmark of apoptosis. Pyroptosis is a type of cell death induced by bacteria that involves caspase-1 activation by inflammasome. Adaptor PYD and CARD domain-containing protein (PYCARD; a.k.a. Apoptosis-associated speck-like protein containing a CARD domain [ASC] and caspase recruitment domain-containing protein CARD5) and caspase-1 form a pyroptosome. Pyronecrosis is necrotic cell death of macrophages. Entosis is a cell death modality primed by cell engulfment.

Cell fate	Features
Anoikis	Absence of cell adhesion
Necrosis	Gain in cell volume (oncosis), rupture of plasma membrane, swelling of cytoplasmic organelles, lysosomal membrane permeabilization, mitochondrial alterations (nitroxidative stress, mitochondrial membrane permeabilization), increase in the cytosolic calcium concentration lipid degradation, activation of non-caspase peptidases, moderate chromatin condensation activation of calpains and cathepsins, reactive oxygen species overgeneration, HMGB1 release, PARP1 hyperactivation, RIPK1 phosphorylation and ubiquitination

organelles, then fuse with lysosomes to become *autolysosomes*. Therefore, autophagosomes carry proteins and organelles destined for autophagy to lysosomes. In particular, autophagy sequesters cytosolic constituents in autophagosomes in nutrient-deprived cells.

During periods of starvation, autophagy yields amino acids and fatty acids to maintain metabolism and energy compatible with cell survival. Cell autophagy also allows the cell to survive against metabolic stresses. Cell autophagy also occurs during cell differentiation and growth, as well as in the innate and adaptive immune responses, thus avoiding accumulation of microbes. Under normal conditions, autophagy eliminates damaged organelles and potentially toxic proteic aggregation.

Two stages of autophagosome genesis can then be defined [535]: (1) formation and (2) maturation and fusion with lysosomes. Microtubules and microtubule-associated proteins participate in the formation and maturation

Table 4.5. Cell fate under stress and corresponding morphological features (**Part 3**; Source: [532]). Cornification is a specific form of programmed cell death in epidermis to create a barrier function. Cornified skin ensures mechanical resistance, water repulsion, and structural stability. Excitotoxicity in neurons subjected to excitatory amino acids (e.g., glutamate) leads to the opening of ^{N}methyl ^{D}aspartate Ca^{++} channels, subsequent cytosolic Ca^{++} overload, and activation of lethal pathways that include features of apoptosis and necrosis. Wallerian degeneration refers to neuron or axon partial degeneration.

Cell fate	Features
Cell death occurring during metaphase	Micronucleation (uneven distribution of chromosomes)
Cell death preceded by multinucleation	Deficient nucleus separation
Cornification	Formation of corneocytes (dead keratinocytes) Elimination of cytosolic organelles Modification of plasma membrane Accumulation of lipids in F- and L-granules Extrusion of lipids in the extracellular space

of autophagosomes. Intermediate and actin filaments may be involved in the early stages of autophagosome production. Most of the autophagosomes receive material from the endocytic compartments (early and late endosomes as well as multivesicular bodies) before they fuse with lysosomes. The *amphisome* that results from these fusions is more acidic than the autophagosome and acquires hydrolytic enzymes [535]. The endoplasmic reticulum may be the main source of lipids such as phosphatidylinositol 3-phosphate in the autophagosomal membranes, as a direct connection exists between the endoplasmic reticulum and the *isolation membrane*, or *phagophore* [535].

Autophagy is associated with sequestration, degradation, and recycling of cellular macromolecules and organelles (e.g., endoplasmic reticulum and mitochondria) by lysosomes. The autolysosomal content is degraded by lysosomal hydrolases. The ubiquitin–proteasome module is required to remove protein aggregates and damaged or excess organelles when fast adaptation is needed. Moreover, catabolism of cell organelles saves energy that allows cell survival.

Although autophagy constitutes a stress adaptation that counteracts apoptosis, under certain circumstances, it is combined with apoptosis (Sect. 4.7). The composition of autophagosome may determine its prosurvival function rather than triggering a prodeath mechanism. Yet, autophagy in dying cells represents a failed attempt of cytoprotection.

4.4.1 Mediators of Autophagy

Autophagic pathways include macro-, micro-, and chaperone-mediated autophagy. Autophagy is controlled by: (1) the status of cellular energy and its controller such as AMP-dependent protein kinase (AMPK); (2) the uptake of nutrients, particularly amino acids, and its regulators such as target of rapamycin (TOR); and (3) the stimulation by growth factors such as insulin. Target of rapamycin senses nutrient status, regulates cell growth, and precludes autophagy, acting on autophagy-execution proteins. The kinase TOR is inactivated by cell starvation. Autophagic TOR-dependent and -independent pathways degrade aberrant molecules and organelles and allow clearance of toxic proteins. Massive, coordinated autophagy can result from nutrient deprivation. Ammonia, a product of glutaminolysis, is a diffusible factor that stimulates autophagy [536].

4.4.1.1 AMPK Pathway and Cell Starvation

Prolonged autophagy induces programmed cell death that differs from apoptosis. An increased cellular AMP/ATP ratio occurs during deprivation of nutrients and energy, such as hypoxia, ischemia, and glucose impoverishment. Metabolic stresses thus lead to AMP accumulation that activates the energy sensor adenosine monophosphate-activated protein kinase (Vol. 4 – Chap. 4. Cytosolic Protein Ser/Thr Kinases).

Sensing of nutrient and energy availability, especially of cellular levels of adenosine monophosphate, is linked to the LKB1–AMPK pathway. The cyclin-dependent kinase inhibitor CKI1b is an effector of this energy-sensing pathway.[17] Once CKI1b is phosphorylated (stabilized) by AMP-activated protein kinase, it promotes cell autophagy in response to metabolic stress rather than apoptosis [537].

The energy sensor AMPK inhibits the nutrient sensor target of rapamycin complex TORC1.[18] During amino acid starvation, AMPK phosphorylates (inactivates) Raptor, a TORC1 subunit, and (activates) TSC2, a RHEB GTPase-activating protein (Sect. 2.3). Under nutrient-rich conditions, TORC1 suppresses the activity of uncoordinated-51-like kinase subunit ULK1.[19] Under

[17] Cyclin-dependent kinase inhibitor CKI1b impedes the activity of cyclin-dependent kinases, thus causing cell cycle arrest. Both CDK2 and CDK4 can act in case of CKI1b depletion. On the other hand, depletion of CDK2 and CDK4 promote autophagy. Small interfering RNAs that downregulate CKI1b cause apoptosis.

[18] In normal conditions, amino acid availability activates TORC1 via Rag guanosine triphosphatase. Glucose availability stimulates TORC1 via the phosphorylation (inactivation) of TSC2, a RHEB GTPase-activating protein.

[19] In mammals, the 2 orthologs of autophagy gene product AtG1 (identified in *Saccharomyces cerevisiae* [yeast]), ULK1 and ULK2 (related to Unc51 in *Caenorhabditis elegans* [nematode]), act as catalytic subunits of the AtG1 kinase complex (Sect. 4.4.1.2) that initiates autophagy.

nutrient-poor and low-energy conditions, AMPK phosphorylates ULK1 to trigger autophagy [538].

4.4.1.2 Autophagy-Related Genes and Autophagosome Formation

Autophagy is governed by *autophagy-related genes* (Atg) and their products (AtG). Several autophagy-related gene products are required for the formation of autophagosomes. At least 18 different AtG proteins are recruited to the isolation membrane (Table 4.6).

Autophagy-related gene products are regulated by between-protein interactions, acetylation, oxidation, phosphorylation, and proteolytic cleavage [535]. Starvation-induced autophagy is associated with oxidation of AtG4a and AtG4b, deacetylation of AtG5, AtG7, AtG12, and MAP1LC3. Proteins AtG5, AtG7, and beclin-1 are calpain substrates; AtG4d and beclin-1 are caspase substrates. Cleavage of AtG5 by calpain-1 and beclin-1 by caspase-3 prevents autophagy.

Four conserved AtG signaling modules control the major steps of autophagy [539]: (1) the uncoordinated-51-like kinase (ULK1–ULK2) complex or AtG1 kinase complex (AtG1–AtG13–AtG17) that controls the initiation of autophagosome formation; (2) class-3 phosphatidylinositol 3-kinase PI3KC3 complex (PI3KC3–beclin-1 or vacuolar protein sorting dimer VPS34–VPS15) that controls the production of phosphoinositide signals that facilitate assembly of the autophagosome; (3) ubiquitin-like protein (UbL) conjugation axes composed of activase AtG7, 2 conjugases AtG3 and AtG10, and 2 ubiquitin-like proteins (AtG8 and AtG12) for autophagosome maturation and cargo recruitment; (4) recycling complex that contains AtG9, AtG2, AtG18, and AtG21 participates in the transfer and recycling of components from the isolation membrane to the growing autophagosome.

Under nutrient-rich conditions, activated TOR kinase inhibits AtG1. Under starvation, the activity of TOR kinase is suppressed. Protein AtG1 inhibits S6 kinase by blocking its phosphorylation [540]. Otherwise, activated S6K promotes cell growth. Reactive oxygen species, especially H_2O_2, intervene in autophagy induced by amino acid deprivation [541]. B-cell lymphoma protein-2 inhibits PI3KC3 that is activated by UV wave resistance-associated gene protein (UVRAG) [534]. Elevation in AtG8a expression in aging neurons increases cell lifespan by more than 50% and resistance to oxidative stress [542].

Autophagosome Initiation

During starvation, autophagy is primed by the ULK1–ULK2 complex (mammalian orthologs of yeast AtG1 kinase; ULK: uncoordinated-51-like kinase) under the control of the TORC1 complex [535]. Target of rapamycin complex TORC1 that contains TOR kinase phosphorylates (inactivates) both ULK1 and ULK2. On the other hand, both ULK1 and ULK2 associate with

Table 4.6. Autophagy-related gene products (AtG) and their orthologs and functions (Barkor: beclin-1-associated autophagy-related key regulator [a.k.a. Atg14-like protein (Atg14L)]; FIP200: 200-kDa focal adhesion kinase family-interacting protein; MAP1LC3: microtubule-associated protein-1 light chain-3 [a.k.a. LC3]; PtdEtn: phosphatidylethanolamine; RB1CC: RB1-inducible coiled-coil domain-containing protein; ULK: uncoordinated-51-like kinase; UVRAG: ultraviolet wave resistance-associated gene product; WIPI: WD repeat domain-containing phosphoinositide-interacting protein). The AtG4 subfamily is composed of 4 members (AtG4a–AtG4d). The human MAP1LC3 family is composed of 6 AtG8 orthologs: MAP1LC3a to MAP1LC3c and 3 MAP1LC3 paralogs, the $GABA_A$ receptor-associated proteins GABARAP1, GABARAPL1, and GABARAPL2 (or GATE16, GABARAP1, and AtG8L) with partially redundant roles in autophagy. The AtG9 member and its mammalian ortholog (mAtg9 or Atg9L1) are the only transmembrane AtG proteins. The AtG12–AtG5 dimer behaves as a ligase-like protein. Phospholipid-binding WIPI1 and WIPI2 of the human WIPI family act as a PI3P scaffold in the early stages of autophagy.

Type	Ortholog	Function
AtG1	ULK1/2	S6K inhibitor, kinase
AtG2	AtG2L	Recycling complex, vesicle assembly
AtG3	AtG3L	Conjugase-like protein
AtG4	AtG4L	Cysteine peptidase
AtG5	AtG5L	AtG8 lipidation (PtdEtn ligase)
AtG6	Beclin-1	Component of the PI3KC3 (VPS34–VPS15) complex
AtG7	AtG7L	Activase-like protein
AtG8	MAP1LC3	Conjugation to phosphatidylethanolamine
	AtG8L	Organizer of regulation of autophagosome formation Receptor or regulator of material engulfment (?)
AtG9	AtG9L1	Recycling complex
AtG10	AtG10L	Conjugase-like protein
AtG11	AtG11L	
AtG12	AtG12L	Ubiquitin-like protein
AtG13	AtG13L	AtG1 activator
AtG14	Barkor	Assistance of beclin-1 in early autophagosome formation and late autophagosome–lysosome fusion
AtG16	AtG16L	Phosphatidylethanolamine ligase component
AtG17	FIP200	Autophagosome initiation
AtG18	WIPI	Phosphoinositide interactor; scaffold
AtG21		Recycling complex
AtG101	C12orf44	TOR-regulated RB1CC1–AtG13–AtG101 regulatory complex

heat shock protein HSP90, a chaperone for kinase stabilization. In addition, ULK1 and ULK2 link to AMPK [539].

Autophagosome initiation and formation is controlled by the ULK–AtG13–FIP200[20] In mammals (but not yeast), the ULK–AtG13–FIP200 complex is a stable complex that is not regulated by nutrition conditions. Mammalian AtG13-binding protein AtG101 located at the phagophore stabilizes AtG13 [535].[21] Inactivation of TOR by starvation activates ULKs and causes phosphorylation of AtG13 and FIP200 [535].

Autophagosome Maturation

The beclin-1–VPS34–AtG14L complex (or beclin-1–PI3KC3–Barkor), i.e., the so-called PI3KC3 complex, AtG9L1, and WIPI proteins (human orthologs of yeast AtG18) contribute to the assembly of the initial isolation membrane and maturation of autophagosomes [535]. Phosphatidylinositol 3-phosphate generated by PI3KC3 is an important component for the formation of autophagosomes.

The activity of the beclin-1–PI3KC3–Barkor complex is controlled by activators, such as Vacuole membrane protein-1 (VMP1), Activating molecule in beclin-1-regulated autophagy (AMBRA1), myeloid differentiation factor-88 (MyD88), and inhibitors, such as members of the BCL2 family. In addition, phosphoinositide 3-phosphatases of the myotubularin family degrade PI3P [535].

In normal conditions, AtG9L1 travels between the trans-Golgi network and late endosomes. During starvation, AtG9L1 resides peripherally and colocalizes with microtubule-associated protein-1 light chain-3 (MAP1LC3; mammalian ortholog of yeast AtG8) and small GTPase Rab7. The transfer of AtG9L1 depends on mitogen-activated protein kinase P38MAPK and its binding protein P38MAPKIP. Agent AtG9L1 might carry lipids and/or serve as a recruitment platform.

Elongation of the isolation membrane of autophagic vesicles depends on 2 ubiquitin-like conjugation systems based on AtG12 and MAP1LC3 that act sequentially [543]. The human MAP1LC3 family is composed of members MAP1LC3a to MAP1LC3c and the MAP1LC3 paralogs GATE16, GABARAP1, and AtG8L. The MAP1LC3 conjugation system also intervenes to close the autophagosome [535].

The first ubiquitin-like conjugation for elongation of the autophagosome membrane involves AtG5 protein. The conjugation of AtG5 to AtG12 occurs soon after their synthesis owing to activase AtG7 activase and AtG10 conjugase. The AtG12–AtG5 dimer then binds AtG16L to be recruited to the membrane of the forming autophagosome.

---

[20] The AtG1–AtG13–AtG17 complex in yeast (FIP200: 200-kDa focal adhesion kinase family-interacting protein).

[21] The interaction between AtG101 and the ULK–AtG13–FIP200 complex is stable, whatever the nutritional status.

The second conjugation associates phosphatidylethanolamine, a component of the phospholipid bilayer, with AtG8 (or MAP1LC3) by the sequential action of AtG4, AtG7, and AtG3. The peptidase AtG4b of the ATG4 subfamily first cleaves MAP1LC3 that is then conjugated to PtdEtn ($MAP1LC3^{PtdEtn}$) [535]. Agents AtG7 and AtG3 act as activase and conjugase for the conjugation of MAP1LC3 to phosphatidylethanolamine. The AtG12–AtG5–AtG16L trimer operates as a ligase for MAP1LC3 lipidation, thereby converting MAP1LC3 from its soluble cytoplasmic form (sMAP1LC3)[22] to membrane-bound, autophagosome-associated form (mMAP1LC3[23] for outer and inner membrane expansion of the autophagosome. Members of the ATG8 subfamily MAP1LC3, GABARAP, GABARAPL1, and GABARAPL2 connect to PtdEtn-conjugation enzymes AtG3, AtG5, and ATG7 [539].

The WD repeat-containing phosphoinositide-interacting proteins WIPI1 and WIPI2 act as a PI3P scaffold in the early stages of autophagy. Protein WIPI1 also associates with MAP1LC3+ structures. Protein WIPI2 tethers to AtG2a via heat shock protein HSP40 (and AtG2a links to AtG2b). Protein WIPI4 binds to AtG2a and AtG2b.[24]

Tectonin β-propeller repeat-containing protein TecPR1 interacts with the AtG12–AtG5–AtG16L complex and the related protein TecPR2 connects to ATG8 subfamily members [539]. Protein TecPR1 also binds to human ortholog of lipid transfer regulator Foie gras-1 involved in autophagosome formation, coatamer-ϵ component TTC15, and components of the TRAPP3 vesicle-tethering complex. The latter includes TRS85, TRAPPc2L (or TRS20), TRAPPc3, TRAPPc4 (or TRS23), and TRAPPc5 (or TRS31) [539]. The TRAPP3 vesicle-tethering complex is required for assembly of AtG8 onto pre-autophagosomes and serves as a guanine nucleotide-exchange factor for Rab1 GTPase. Depletion of TecPR1 and its associated protein TTC15 increases autophagosome number [539]. Moreover, depletion of TTC15, but not TecPR1, alters autophagosome flux.

When the autophagosome is formed, the AtG12–AtG5–AtG16L complex leaves the autophagosome. The pool of MAP1LC3 associated with the autophagosomal outer surface is cleaved from phosphatidyl ethanolamine by AtG4 peptidase and recycled [535].

Amphisome and Autolysosomes

After their formation, autophagosomes can merge with endosomes (early, late, and multivesicular endosomes) to form *amphisome*, before they fuse with lysosomes. Beclin-1 interactors Rubicon and UVRAG participate in the regulation of the maturation of autophagosomes. The beclin-1–PI3KC3–UVRAG–

[22] A.k.a. MAP1LC3-1.

[23] A.k.a. MAP1LC3-2.

[24] Depletion of WIPI1 or WIPI4 increases the autophagosome number, whereas that of WIPI2 decreases the autophagosome density [539].

Rubicon complex may attenuate endocytosis, whereas the beclin-1–PI3KC3–UVRAG complex stimulates the maturation of autophagosomes and endocytic transfer [535].

Beclin-1 (AtG6) Interactome

During nutrient-poor conditions, beclin-1 interacts with class-3 phosphatidylinositol 3-kinase (VPS34) that induces autophagy. The constitutive BCL2– and BCLxL–beclin-1 association is disrupted by autophagic signals. Jun N-terminal kinases phosphorylate BCL2, thereby triggering its release from beclin-1. Beclin-1 is also phosphorylated by death-associated protein kinase (DAPK), hence its affinity for BCLxL is reduced. Autophagy is incompatible with cell division. Cyclin-dependent kinases CDK1 and CDK5 phosphorylate PI3KC3, thereby impeding its interaction with beclin-1 [544].

In addition, BH3-only member BNIP3 of the BCL2–adenovirus E1B 19-kDa protein-interacting protein family that is upregulated during hypoxia competes for association of beclin-1 with anti-apoptotic BCL2 and BCLxL, hence leading to dissociation of BCL2– and BCLxL–beclin-1 complexes [520].

Among many beclin-1-interacting proteins, anti-apoptotic factors BCL2 and BCLxL inhibit autophagy only when they lodge in the endoplasmic reticulum, but not mitochondrial-resident BCL2 and BCLxL [545]. Beclin-1 dissociates from BCL2 for autophagy. Nutrient-deprivation autophagy factor-1 (NAF1)[25] localizes to both the endoplasmic reticulum and mitochondria. It binds to the Ca^{++} channel and inositol trisphosphate receptor IP_3R1 as well as BCL2 at the endoplasmic reticulum. It promotes beclin-1–BCL2 connection. Therefore, this regulator of beclin-1–BCL2 interaction prevents beclin-1-mediated autophagy.

Other beclin-1 interactors, such as UV radiation resistance-associated gene product (UVRAG) or AtG14L, promote PI3KC3 activity and autophagosome formation and maturation. Endophilin-B1 family member BIF1 (endophilin-B2 or SH3GLB1) interacts with beclin-1 via UVRAG to regulate PI3KC3 (VPS34–VPS15). It regulates apoptosis, as it connects to and activates BAK and BAX. Beclin-1-associated autophagy-related key regulator (Barkor) competes with UVRAG to link to beclin-1. Both Barkor and UVRAG sequentially assist beclin-1 in early autophagosome formation as well as late autophagosome–lysosome fusion [545]. Activating molecule in beclin-1-regulated autophagy (AMBRA1 or DCAF3) serves as a substrate adaptor for the cullin-related Cul4–DDB1–DDA1 ubiquitin ligase [539]. The Cul4–DDB1–DDA1–AMBRA1 complex that may cause ubiquitin-dependent proteolysis links to the PI3KC3–beclin-1 complex. In the central nervous system, AMBRA1 binds to beclin-1 and favors autophagosome formation [545]. On the

[25] A.k.a. CDGSH iron–sulfur-binding domain-containing protein CISD2, ZCD2, Noxp70, and Miner1.

other hand, RUN domain beclin-1 interactor and cysteine-rich containing protein Rubicon precludes autophagy. When Rubicon associates with the beclin-1–PI3KC3–UVRAG complex, it impedes autophagosome maturation [545].

Cardiomyocyte Autophagy

Autophagy and ubiquitin–proteasome pathways particularly protect cardiomyocytes. Altered autophagy has been observed in various heart diseases. In the absence of AtG5, heart hypertrophy and failure occur. Conversely, autophagy plays a beneficial role in hearts in response to pressure overload or β-adrenergic stresses.

4.4.1.3 Other Mediators of Autophagy

Role of Small GTPases

Small GTPase Rab5 (Vol. 4 – Chap. 8. Guanosine Triphosphatases and Their Regulators) regulate the conjugation of AtG12 at autophagosomal membrane via PI3KC3 [535]. Small GTPase Rab33b may contribute to the localization of the AtG12–AtG5–AtG16L complex to the autophagosomal membrane. Small GTPase Rab7 is required for autophagosome maturation and fusion of autophagosomes with late endosomes and lysosomes. Small GTPase Rab11 contributes to the fusion of autophagosomes and multivesicular bodies. Small GTPases Rab22 and Rab24 might also intervene in autophagy.

Proteins of the ATG8 subfamily interact with multiple Rab regulatory monomeric GTPases. They may thus contribute to the control of localized activation of these Rab GTPases and vesicle dynamics [539].

In response to amino acids, Ras-related small GTPases (Rag) promote TORC1 redistribution to subcellular loci that contain its activator small GTPase Ras homolog enriched in brain (RHEB).

Actors of Intracellular Transfer of Molecules

Endosomal sorting complexes required for transport (ESCRT) operate in the sorting of proteins during endocytosis and genesis of multivesicular endosomes (Vol. 1 – Chap. 9. Intracellular Transport). The ESCRT3 complex is needed for fusion between autophagosomes and multivesicular bodies and lysosomes that generate amphisomes and autolysosomes [535].

Hepatocyte growth factor-regulated Tyr kinase substrate (HRS) acts in endosomal sorting upstream from ESCRT complexes. It binds to phosphatidylinositol 3-phosphate and facilitates the maturation of autophagosomes [535].

Among vesicular soluble ^{N}ethylmaleimide-sensitive factor attachment protein receptors, vesicle associated membrane proteins VAMP3 and VAMP7 contribute to the control of the fusion between autophagosomes and multivesicular endosomes and amphisomes with lysosomes, respectively [535].

Lysosomal-associated membrane proteins LAMP1 and LAMP2 as well as ATPases are membrane constituents of the endo- and lysosomes. The role of a LAMP family member differs according to the cell type (e.g., fibroblast vs. hepatocyte) [535]. Vacuolar ATPases are necessary for the lysosomal degradation of autophagic cargos. The ATPases associated with various cellular activities (AAA) are also committed in intracellular membrane fusion. ^{N}Ethylmaleimide sensitive factor (NSF) is an AAA ATPase that binds to and disassembles SNARE complexes for their recycling.

Damage-regulated autophagy modulator (DRAM) is a transmembrane protein target of P53. It also controls the formation and maturation of autophagosomes [535].

Cytoskeleton

Microtubules (Vol. 1 – Chap. 6. Cell Cytoskeleton) intervene in the maturation of autophagosomes, especially the fusion between autophagosomes and lysosomes [535]. Autophagosomes move bidirectionally along microtubules. Their centripetal movement is carried out by the nanomotors dynein and kinesin. Phosphatidylinositol 3-phosphate-binding FYVE[26] and coiled-coil domain-containing protein FYCO1, a Rab7 effector and AtG8-binding protein, resides at the autophagosome exterior and may tether autophagosomes to plus-end-directed microtubule nanomotors of the kinesin class KIF5b and KIF23 [539]. Protein KIF5b links FYCO1+ autophagosomes to microtubules to maintain cortical localization.

In addition, FYCO1 may serve as a platform for assembly of vesicle fusion and trafficking factors, such as vacuole fusion factors Mon1a and Mon1b and small GTPase Rab7a. Another PI3P-binding ankyrin repeat and FYVE domain-containing protein AnkFY1 connects to both PI3P and Rab5 on endosomes and stimulates endosomal fusion.

Kinases

Many kinases target cytoplasmic and nuclear regulators of autophagy (Table 4.7). In addition, Jun N-terminal kinases control the expression of beclin-1 via transcription factor Jun. The Hippo kinase complex that is implicated in cell growth control is composed of protein Ser/Thr kinases STK3 (or MST2) and STK4 (or MST1), in addition to inhibitory Ras interaction/interference protein RIN1, afadin, and Ras association domain-containing family proteins RASSF1, RASSF2, and RASSF5 and stimulatory scaffold protein WW45 [539]. Kinases STK3 and STK4 are linked to cargo recruitment and vesicle trafficking.

Cell uptake of ^{L}glutamine, the rate-limiting controller of TORC1, is carried out by solute carrier superclass member SLC1a5. Loss of SLC1a5 activates

[26] Domain found in Fab1, Yotb, Vac1, and EEA1.

Table 4.7. Kinase regulators of autophagy (4eBP1: inhibitory eIF4e-binding protein; AMPK: AMP-activated protein kinase; BCL: B-cell lymphoma protein; CKI: cyclin-dependent kinase inhibitor; DAPK: death-associated protein kinase; eIF: eukaryotic translation initiation factor; FIP200: 200-kDa focal adhesion kinase family-interacting protein; FoxO: forkhead transcription factor; IRS: insulin receptor substrate; JNK: Jun N-terminal kinase; PERK: protein kinase-like endoplasmic reticulum kinase; PKB: protein kinase B; S6K: P70 ribosomal S6 kinase; TORC: target of rapamycin complex; ULK: uncoordinated-51-like kinase).

Kinase	Substrate	Effect
AMPK	CKI1b	↑ Autophagy
DAPK	Beclin-1	↓ Affinity for BCL2 and BCLxL
JNK1	BCL2	Beclin-1 release
PERK	eIF2α	Conversion of sMAP1LC3 to mMAP1LC3
PKB	FoxO1/3/4	FoxO inactivation
S6K	IRS1	IRS1 inhibition
STK3/4		Cargo recruitment, vesicle trafficking, Hippo kinase complex, cell growth control
TORC1	AtG13, ULK1/2	ULK inactivation (↓ autophagy)
	S6K, 4eBP1	Cell growth
TORC2	PKB	↓ FoxO3 activity (↓ autophagy)
ULK1	ULK1, AtG13, FIP200	Autophagy initiation

autophagy and inhibits cell growth [535]. Transporter SLC3a2 that carries ^{L}leucine and essential amino acids into cells contributes also to ^{L}glutamine sensitivity. The TORC1 complex is incorporated into the ULK1–AtG13–FIP200 complex via ULK1 in normal conditions. Under starvation, TORC1 dissociates from this complex, and ULK1 is activated. Activated ULK1 autophosphorylates and phosphorylates both AtG13 and FIP200 to initiate autophagy [535].

Transcription Factors

Transcription factors FoxO1 and FoxO3 stimulate autophagy, as they upregulate the expression of several Atg genes, such as Atg4b, Atg12, Gabarapl1, Map1lc3, Ulk1, as well as PIK3C3 and BNIP3. Transcription factor E2F1 causes autophagy, as it raises the expression of autophagy genes Map1lc3, Ulk1, Atg5, and Dram. Transcription factor NFκB controls the expression of beclin-1. Yet, NFκB prevents autophagy in macrophages as well as TNFα- and starvation-dependent autophagy [535]. Transcription factor P53 regulates the expression of autophagy facilitator Dram and attenuator TP53-induced glycolysis and apoptosis regulator (TIGAR).[27] Hypoxia-inducible factor HIF1 is

[27] Agent TIGAR can indirectly contribute to the fall in the intracellular ROS level.

a transcription factor that targets BNIP3 and its related protein BNIP3L and causes hypoxia induced mitochondrial autophagy. Hypoxia-inducible BH3-only proteins BNIP3 and BNIP3L dissociate the beclin-1–BCL2 complex. Transcription factors ATF4 and DNA-damage-inducible transcript DDIT3 (or CCAAT/enhancer-binding protein (C/EBP)-homologous protein CHOP) are regulated by protein kinase-like endoplasmic reticulum kinase (PERK) to increase the expression of Map1lc3b and Atg5, respectively.

Ubiquitination

Three cullin (Cul)-related Ub ligases — the Cul2–EloB–EloC–KlhDC10,[28] Cul3–KBTBD6–KBTBD7,[29] and Cul4–DDB1–DDA1[30] complexes — intervene in autophagy [539]. Moreover NEDD4 Ub ligase connects to multiple ATG8 subfamily proteins for lysosomal degradation. Yet, NEDD4 can intervene in AtG8 lipidation.

On the other hand, ubiquitin-specific peptidase USP10 and its adaptor G3BP1 associate with the ULK complex and limit the autophagosome number [539].

Non-Canonical Autophagosome Formation

Non-canonical formation of functional autophagosomes employs only a part of the repertoire of AtG proteins. An autophagy form independent of AtG5 and AtG7 involves monomeric GTPase Rab9 located on the trans-Golgi network and endosomes [535]. Another form of autophagy depends on AtG5 and AtG7, but not beclin-1 and PI3KC3. In the 2 non-canonical autophagosome formations, ULKs are required as initiators [535].

4.4.2 Clearance of Protein Aggregates

Autophagy represents a clearance mechanism for intracellular protein aggregates such as those observed in certain degenerative diseases (Huntington's and Parkinson's diseases). Autophagy primed by the target of rapamycin pathway causes clearance of these toxic aggregates.

Histone deacetylase HDAC6 promotes fusion between autophagosomes and lysosomes, as it recruits ^{F}actin, but not during starvation (selective autophagy) [546]. Both HDAC6 and ^{F}actin are required for proper clearance of misfolded proteins as well as damaged or potentially toxic substrates from the cell, but not organelles.

---

[28] Elongin-B (EloB) and -C (EloC) are also called transcription elongation factor-B polypeptides TCEb2 and TCEb1, respectively (KlhDC: kelch domain-containing protein).

[29] KBTBD: kelch repeat and BTB [POZ] domain-containing protein.

[30] DDB: damage-specific DNA-binding protein; DDA: de-etiolated homolog (DET1) and DDB1-associated protein.

Table 4.8. Autophagy during ischemia–reperfusion events involves many processes and factors (Source: [548]; AtG: autophagy-related gene product). Decrease in ATP concentration activates AMP-activated protein kinase (AMPK) that inhibits target of rapamycin, as it phosphorylates tuberous sclerosis complex TSC2, as well as protein synthesis, as it phosphorylates eukaryotic elongation factor eEF2. Hypoxia-inducible factor-1α (HIF1α) stimulates mitochondrial autophagy, as it upregulates BNIP3. Beclin-1 (mammalian ortholog of yeast AtG6) is part of the class-3 phosphatidylinositol 3-kinase complex that induces autophagy. Expression of Beclin-1 rises during reperfusion following organelle damage caused by hypoxia. Reactive oxygen species production in mitochondria is strongly augmented by an ROS-induced ROS release mechanism when electron transport resumes during reperfusion. Endoplasmic reticulum stress activates RNA-dependent protein kinase-like endoplasmic reticulum kinase (PERK; a.k.a. eukaryotic initiation factor-2α kinase eIF2αK3) to stimulate autophagy.

Agent	Mediator
ATP depletion	AMPK
Hypoxia	HIF1α, BNIP3
Organelle damage	Beclin-1 (AtG6)
Oxidative stress	AtG4
Endoplasmic reticulum stress	PERK (eIF2αK3)

On the other hand, cyclic adenosine monophosphate increases intracellular calcium concentration, then favors the activity of peptidase calpain that cleaves G protein, thereby activating G-protein subunit $G\alpha_s$ that stimulates adenylyl cyclase. Activators RapGEF3 and -4 leads via Rap2b and phospholipase-C to inositol triphosphate formation. This process prevents autophagy and protein aggregate clearance. Reduction in intracellular calcium levels and cAMP formation using calcium channel Ca_V1 blockers, $K_{IR}6.2$ channel ($i_{K_{ATP}}$ current) opener, and $G\alpha_i$ activator induces autophagy, hence avoiding degenerative diseases associated with protein aggregate formation [547].

4.4.3 Autophagy and Ischemia–Reperfusion Events

In the cardiovascular system, autophagy protects cardiomyocytes from pressure overload [548]. Cardiac remodeling involves increased rate of cardiomyocyte death and precedes heart failure. Autophagy primes cell survival during ischemia–reperfusion events (Table 4.8), but cell death can occur despite autophagy.

4.4.4 Tumor Cell Death and Autophagy

Autophagy can eliminate damaged organelles and limit the effects of cellular stresses that promote cellular transformation. In precancerous stage, the cell that has been exposed to mutagens, pathogens, and/or other cellular stresses that cause DNA damage as well as mitochondrial alterations with ROS release is the site of gene mutations that disrupt the cell cycle control, DNA repair, and cell death. When the DNA damage cannot be resolved by autophagy, the cell can still avoid transformation owing to a death response. Transformed proliferating cells can still trigger cell death. Once cancer cells form a not yet vascularized, solid tumor, hypoxia can trigger apoptotic cell death [520]. Autophagy can limit the extent of further genomic damage. Autophagy that consumes neuropilin-1 can also be anti-angiogenic.

Oncogenic signals (ERK, class-1 PI3K, PKB, and RSK1) suppress autophagy. Conversely, tumor suppressors, such as inhibitors of the TOR pathway, such as tuberous sclerosis complexes TSC1 and TSC2, phosphatase and tensin homolog deleted on chromosome 10 (PTen), DAPK, beclin-1, p14ARF and smARF, UVRAG, and P53 via AMPK, activate autophagy.

Tumor-killing cellular and viral proteins can selectively kill tumor cells. These include apoptin with apoptin-interacting partners, human α-lactalbumin made lethal to tumor cells (Hamlet), TNFSF10,[31] melanoma differentiation-associated gene product MDA7, E4orf4, parvovirus minute virus of mice encoded protein NS1, and brevinin-2R [549].

Suppression of autophagy can promote tumorigenesis, but autophagy helps tumoral cell survival. Autophagy thus operates in both tumor suppression and survival. Oncogenes (e.g., phosphatidylinositol 3-kinase, protein kinase-B, and anti-apoptotic proteins of the BCL2 family) inhibit autophagy, whereas proteic tumor suppressors (e.g., BH3-only proteins, death-associated protein kinase-1, phosphatase and tensin homolog, tuberous sclerosic complexes-1 and -2), except tumor suppressor P53, stimulate autophagy [550].

On the other hand, P53 pertains to the DNA-damage response, a cell defense mechanism, that leads to cell cycle arrest or apoptosis. Following DNA damage, arginine methylation of P53 by protein arginine methyltransferase-5 (Vols. 1 – Chap. 5. Protein Synthesis and 3 – Chap. 1. Signal Transduction) promotes the transcription of cell cycle arrest mediators and represses apoptosis, as it favors cyclin-dependent kinase inhibitor CKI1a and 14-3-3σ that provoke cell cycle arrest rather than P53-dependent cell death regulators Noxa, APAF1, and PUMA that elicit apoptosis (Sect. 4.6).

4.5 Cell Senescence

Aging can be defined as a progressive decline in cell function, ultimately resulting in cellular senescence (senex: old, senior; senesco: aging; senescence most often is a synonym of aging). Senescence, the permanent arrest of cell division, describes the functioning deterioration that follows development and maturation. Cellular senescence is a feature of dividing cells that can proliferate and spend long periods in reversible quiescence, resuming division in

[31] A.k.a. tumor-necrosis factor-related apoptosis-inducing ligand TRAIL.

response to proliferation signals. Cell senescence arises as soon as the cell loses its ability to divide and maintain its genetic material. Progressive loss in gene regulation modulates aging pattern and rate. Senescent cells have altered expression of hundreds of genes.

Senescent cells are found in many renewable tissues, such as vasculature and hematopoietic niches. They are also markers of age-related pathology such as atherosclerosis, as well as benign dysplastic or preneoplastic lesions. High-mobility group AT-hook domain-containing regulator HMGA2 in neural stem cells promotes cell renewal in young mice, as it maintains low concentrations in tumor suppressors, such as CKI2a and CKI2d. Production of HMGA2 decreases as neural stem cells age and levels of tumor suppressors increase. With aging, the cell's strategy is to avoid cancer, not to maintain youth.

The permanent suspension of division can be induced by: (1) strong expression of H-Ras GTPases that involves the P14ARF–P53 pathway; (2) chromatin disturbances; or (3) shortened telomeres that also implicate P53 transcription factor [551].

4.5.1 Oncogenic Events

Oncogenes are mutant genes that can transform cells in conjunction with additional mutations. During potent oncogenic events, cells can irreversibly enter into growth arrest and often become resistant to apoptotic signals [552]. Their gene expression widely changes, especially those that encode cell cycle inhibitors or activators. Many senescent cells overexpress genes that encode secreted proteins acting on their environment. Senescence induced by oncogenes can be associated with that generated by chromatin disturbances or shortened telomeres [552]. Oncogene-induced senescence does not occur in all cell types. It suppresses tumorigenesis in vivo. Tumorigenesis indeed requires additional mutations, such as in p53 or Cki2A genes.

4.5.2 Telomere Shortening

DNA damages that cannot be repaired, but that do not lead to apoptosis, such as shortened telomeres, are responsible for cell senescence.[32] Cumulative oxidative stresses may also induce telomere shortening. Short telomeres are assumed to trigger the onset of senescence. Aging is also characterized by a decaying ability to respond to stress and to retain cell homeostasis.

Telomerase can synthesize new telomeric repeats and restore telomere length. Telomerase activity is regulated by protein kinase-B, protein kinase-C, and extracellular signal-regulated kinases ERK1 and ERK2. Nitric oxide can prevent senescence of endothelial cells by activating telomerase [553].

[32] Telomeres are repetitive DNA sequences at the ends of chromosomes that shorten as the number of cell divisions increases.

4.5.3 Chromatin Disturbances

Cell senescence can also be caused by chromatin perturbation. Inhibition of histone deacetylase that promotes euchromatin formation induces senescence. Senescence can be triggered both by heterochromatin disruption and downregulation of histone acetyltransferase that elicits heterochromatin formation [552].

4.5.4 Defective Repair and Reactive Species

Cell senescence can be accelerated by unusual stress levels with defective nuclear or mitochondrial DNA repair enzymes. Oxygen metabolized by mitochondria is converted to superoxide ions, and subsequently to hydrogen peroxide, hydroxyl radicals, and other reactive species. These molecules can generate free radicals that damage DNA and structural proteins. Metal ions can participate. Glucids can react with amino acids and DNA bases and produce carbohydrate adducts (glycation). These compounds can form reactive species that target proteins or DNA, impairing their functions.

The DNA-damage response is characterized by activation of sensor kinases (ATMK, ATRK, DNAPK), formation of DNA-damage foci that contain activated histones H2a.x, and cell cycle arrest by activated checkpoint proteins, particularly P53 and CDK inhibitors. This signaling pathway contributes to G0-arrest stability. A positive feedback loop between the DNA-damage response and elevated mitochondrial production of reactive oxygen species in senescing cells sets a necessary and sufficient condition for long-term maintenance of growth arrest, until a point of no return is reached during establishment of senescence [554]. High CKI1a concentrations stimulate growth arrest and DNA damage-induced protein GADD45, P38MAPKα (MAPK14), GRB2, TβR2, and TGFβ, cause mitochondrial dysfunction, and increase ROS production. Augmented ROS production, in turn, provokes additional DNA damage that maintains the DNA-damage response in a permanent active state. In summary, the sustained activation of CKI1a by the DNA-damage response induces mitochondrial dysfunction and reactive oxygen species production via the CKI1a–GADD45a–MAPK14–TGFβ pathway in senescing human cells [554].

4.5.5 P53 Transcription Factor and Cyclin-Dependent Kinase Inhibitors

Growth arrest is achieved during cell senescence by the transcription factor P53 and cyclin-dependent kinase inhibitors such as CKI2a. These cell type-specific pathways respond to different stimuli, but interact.

The P53 pathway is regulated by P14ARF that inhibits ubiquitin ligase DM2 that targets P53 for degradation. The CKI2a–RB (retinoblastoma protein) pathway promoted by small GTPase Ras that generates senescence-associated heterochromatin foci (SAHF) is committed either primarily or secondarily to engagement of the P53 pathway [552]. However, senescence can occur independently of these pathways.

Tissue aging can be at least partially caused by stem cell depletion, especially in mitotically active tissues. Stem cell abundance or proliferative potential decreases with age. Expression of CKI2a increases with age in stem and progenitor cells of brain, hematopoietic niches of bone marrow, and pancreas [552]. Hyperactive CKI2a and P53 avoid tumor occurrence, but lead to accelerated aging.

4.5.6 Nuclear Factor-κB

The gene expression database AGEMAP ("atlas of gene expression in mouse aging project") has been aimed at cataloging aging-related changes in mice [555]. Nervous, vascular, and steroid-responsive tissues undergo distinct aging processes, suggesting that these tissues contribute strongly to the body's decline. However, aging is coordinated in time among tissues with rapid or slow aging patterns for multiple tissues. In addition, age is regulated differently in separate mammal species, but genes involved in the electron-transport chain show common age regulation in all explored species. Age-related alterations in gene expression of 9 tissue types are related to a set of 14 transcription-factor binding motifs [556]. The motif most strongly associated with aging was that of transcription factor nuclear factor-κB. Mammal species share regulatory mechanisms of aging. NFκB is required for expression of many genes that are upregulated in epidermal aging.

4.5.7 Jun N-Terminal Kinases

Jun N-terminal kinases[33] regulate components of AP1 transcription factor and prevent P53-mediated senescence, as they phosphorylate (activate) components Jun and JunD [557].[34]

4.5.8 Interferon-β

Sustained signaling by certain anti-proliferative cytokines such as interferon-β yields senescence [552]. Acute interferon-β stimulation reversibly stops cell growth, but chronic activation augments intracellular oxygen radicals and causes P53-dependent senescence. Chronic transforming growth factor-β activity also provokes senescence.

[33] A.k.a. mitogen-activated protein kinases MAPK8–MAPK10.

[34] JNK phosphorylates various AP1 and AP1-related transcription factors, such as Jun, JunB, JunD, and ATF2. Moreover, JNK stimulates AP1 activity by upregulating Jun mRNA and controlling the half-life of Jun.

4.5.9 Klotho

α-Klotho is associated with aging and short lifespan. Overexpression of Klotho in mice extends their lifetimes. Klotho operates as a circulating hormone that binds to a plasmalemmal receptor and represses intracellular signals of insulin and insulin-like growth factor IGF1 [558]. Agent Klotho also interacts with TRP channels and fibroblast growth factor receptor. α-Klotho influences calcium homeostasis. It binds to Na^+–K^+ ATPase at the plasma membrane. The lower the α-Klotho level, the smaller the calcium concentration and the greater the amount of Na^+–K^+ ATPases in the choroid plexi and the kidneys [559]. Klotho interacts with various Wnt family members, particularly morphogen Wnt3, to suppress Wnt activity (Vol. 3 – Chap. 10. Morphogen Receptors).

4.5.10 Secreted Molecules

Secreted molecules, such as secreted phospholipase-A2, insulin-like growth factor-binding protein IGFBP7, and chemokines that regulate cell senescence act via type-1 (multifunctional M-type) transmembrane glycoprotein phospholipase-A2 receptor of the C-type lectin family (CLec13c), chemokine receptors CXCR2, and interleukin-6 receptor. Receptor PLA2R elicits cell senescence via ROS production, as it controls the production of cytokines rather than releasing arachidonic acid and exciting the MAPK pathway. Moreover, PLA2R activates the P53 pathway [560].

Some substances secreted by a senescent cell have detrimental effects on the surrounding tissue, and might contribute to potential risk of cancer and chronic inflammation with age. Factor P53 promotes cell senescence and keeps in check products secreted by senescent cells. Upregulation of genes that encode extracellular matrix-degrading enzymes, inflammatory cytokines, and growth factors, can affect the behavior of neighboring cells or even distant cells.

4.5.11 Diet

Restricted dietary intake without malnutrition favors longevity. Certain genes and corresponding pathways increase survival under conditions of dietary restriction. Factor FoxO can be implicated. It influences survival of the whole body, particularly controlling hormone production.

Moderate magnesium deficiency accelerates cellular senescence. Fibroblasts cultured in magnesium-deficient conditions exhibit a decreased replicative lifespan and symptoms of accelerated senescence without loss in viability [561].[35]

[35] Acute and severe magnesium deficiencies in cultured cells reduce oxidative stress defense, cell cycle progression, and cellular viability, but elevate activity of proto-oncogenes Fos and Jun, as well as NFκB.

4.6 Cell Apoptosis

Apoptosis represents a modality of programmed cell death among other types of programmed cell death that can be distinguished by morphological observation. Apoptosis is characterized by morphological changes, and cell shrinkage leading to small round cells with nuclear pyknosis (chromatin condensation) and karyorhexis (nuclear fragmentation). The Golgi stack, endoplasmic reticulum, and mitochondria also undergo fragmentation during apoptosis. In particular, numerous proteins are released from the mitochondrial intermembrane space. Intermembrane space cytochrome-C triggers the apoptosome assembly.

Apoptosis pathways conduct fast cell death. Apoptosis also contributes to phagocyte cleaning and prevents unwanted immune responses. Membrane alterations are detected by macrophages that engulf apoptotic cells.

Many apoptosis regulators exist with functional specialization and compensatory control (redundancy in apoptosis execution) to increase the efficiency of the apoptosis signaling pathways that respond to distinct extra- and intracellular stimuli.

Apoptosis results from 2 mechanisms: the extrinsic and intrinsic pathway (Fig. 4.2). (1) The *receptor-mediated extrinsic pathway*, implicated in the elimination of unwanted cells during development, is initiated by ligand-induced activation of the death receptors at the plasma membrane. (2) The *mitochondrion-mediated intrinsic pathway* is triggered by stresses (e.g., DNA damage). Proteins that participate in apoptosis are associated with mitochondrial outer-membrane permeabilization and caspase activation. These mediators are not exclusively dedicated to self-destructive events, as they exert some function in normal (death-unrelated) processes.

Both proteolytic pathways during cell degradation involve the activation of cysteinyl aspartate-specific peptidases, the so-called caspases (Tables 4.9 and 4.10). Both initiator and executioner caspases are required. Initiator caspases are inactive monomers activated by dimerization [563]. Caspases target several hundred proteins for controlled and restricted proteolysis that does not disturb neighboring cells and avoids the release of immunostimulatory molecules.

4.6.1 Intrinsic Pathway

The mitochondrial intrinsic pathway is activated by stimuli that permeabilize the outer mitochondrial membrane and release proteins from the mitochondrial intermembrane space. The cytosol contains inactive procaspases and auto-inhibited monomeric apoptotic peptidase-activating factor APAF1. In normal conditions, cytochrome-C and second mitochondria-derived activator of caspase (SMAC) remain sequestered in the mitochondrial intermembrane space. Inhibitors of apoptosis proteins (IAP) inhibit caspases, but are inhibited by SMACs. Mitochondrial permeability transition pores (PTP) are

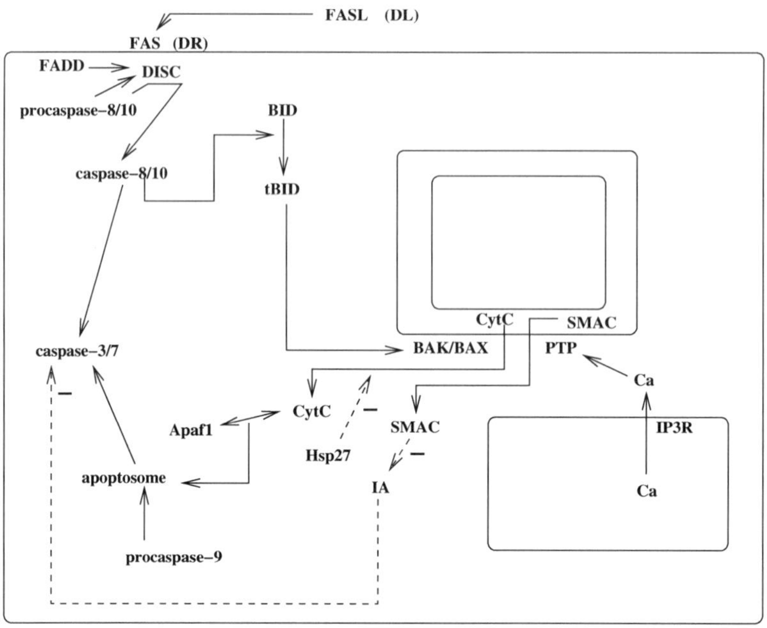

Figure 4.2. Intrinsic and extrinsic apoptotic pathways (Source: [562]; BCL2: B-cell lymphoma protein-2). An extracellular death ligand (DL) stimulates a death receptor (DR) to initiate the extrinsic apoptotic pathway. Engagement of the DR TNFRSF6a (Fas) by the DL TNFSF6 (FasL) leads to the assembly of the death-inducing signaling complex (DISC) formed by FAS-associated death-domain protein (FADD) and initiator procaspase-8 or -10. The DISC complex favors the auto-activation of caspase-8 or -10. Initiator caspase then triggers the apoptotic cascade. In normal conditions, pro-apoptotic members of the BCL2 family are inhibited by anti-apoptotic members of this family. Cytochrome-C (CytC) and second mitochondria-derived activator of caspase (SMAC) remain sequestered in the mitochondrial intermembrane space (MIS). Inhibitors of apoptosis proteins (IA) inhibit caspases, but are inhibited by SMACs. The mitochondrial intrinsic pathway is activated by stimuli that permeabilize the outer mitochondrial membrane (OMM) and release proteins from MIS. In the cytosol, CytC binds to APAF1 to form apoptosome. Agent APAF1 binds to and triggers auto-activation of initiator procaspase-9. Activated caspase-9 then activates executioner caspase-3 and -7. The piddosome that contains P53-inducible, death domain-containing protein (PIDD) activates procaspase-2. In addition, Ca^{++} released from the endoplasmic reticulum (ER) via IP_3Rs enables mitochondrial permeability transition pore (PTP) opening. Opening of PTPs promotes molecular flux into mitochondrion and osmotic imbalance that can cause OMM rupture. Protein BID that is cleaved (activated) into truncated $_t$BID by caspase-8 is a converging point of both pathways. Activated BID primes BAK and BAX oligomerization to form OMM pores.

closed. Pro-apoptotic members of the BCL2 family are inhibited by anti-

Table 4.9. Apoptosis mediators. (**Part 1**) B-cell lymphoma-2 (BCL2) proteins (Sources: [519, 534, 564]). The BCL2 family comprises 3 sets of proteins that contain 1 to 4 BCL2 homology (BH) domains (BAD: BCL2 antagonist of cell death; BAK: BCL2-antagonist/killer-1; BAX: BCL2-associated X protein; BID: BH3-interacting domain death agonist; BIK: BCL2-interacting killer; BIM, BCL2-like molecule-11; BOK: BCL2-related ovarian killer; BMF: BCL2-modifying factor; Hrk: harakiri [or death protein-5]; ERK, extracellular signal-regulated kinase; JaK–STAT: Janus kinase–signal transducer and activator of transcription; MCL1: BCL2-related myeloid cell leukemia sequence protein-1 [or BCL2L3]; PUMA: P53-upregulated modulator of apoptosis, or BCL2-binding component-3). Anti-apoptotic BCL2 family members can be cleaved by caspases, then losing BH4 domain so that they exhibit pro-apoptotic rather than anti-apoptotic activity. They can prevent mitochondrial damage by pro-apoptotic BCL2 family members. BH3-only death sentinels that antagonize anti-apoptotic BCL2 family members of the outer mitochondrial membrane (OMM) bind on BAX and BAK to relieve their inhibition. At apoptosis onset, BAK and BAX undergo conformational changes and increase OMM permeability that leads to mitochondrial fragmentation. A set of BH3-only factors might directly induce formation of BAX–BAK channels that lead to release of cytochrome-C and other mitochondrial intermembrane space proteins. Various BH3-only proteins interact with specific subsets of anti-apoptotic BCL2 family members, yielding various pathways toward apoptosis. In some cases, BAX and BAK interact with voltage-dependent anionic channel of the mitochondrial permeability pore.

Type	Features
BCL2	**Anti-apoptotic** (BCL2, BCL2a1, BCLb, BCLw, BCLxL, and MCL1)
	BH4 domain-containing proteins
	BCLxL induction by growth factors via JaK–STAT
	BCL2 prevention of oligomerization of BAX and BAK
	Pro-apoptotic (BAK, BAX, and BOK)
	Multidomain (BH1, BH2, and BH3)-containing proteins
	Oligomerization of BAX and BAK leads to cytochrome-C efflux
	BH3-only (BAD, BID, BIK, BIM, BMF, Hrk, Noxa, and PUMA)
	BH3 domain-containing proteins
	Upregulation of Noxa, PUMA, and BID by P53
	BID activation by caspase-8
	BAD inactivation by growth factor receptor
	BIK activation in response to inhibited protein synthesis
	BIM tethered to microtubules
	BIM stimulation by forkhead box transcription factor FoxO3a
	(growth-factor deprivation) and CCAAT enhancer binding protein
	CEBPα and CEBP homologous protein
	(endoplasmic reticulum stress)
	BIM phosphorylation and degradation by ERK kinase
	BMF tethered to actin filaments
	BMF activation by release from actin–myosin complexes

Table 4.10. Apoptosis mediators (**Part 2**; Sources: [519, 534, 564]; Mitochondrial damage might also be caused by intramitochondrial K^+ influx or action of caspase-2 on mitochondria. Adaptors include nucleotide-binding and oligomerization domain-like receptor (NLR), apoptotic peptidase-activating factor-1 (APAF1), and P53-induced protein with a death domain (PIDD).

Type	Features
Caspases	Peptidases
NLR	Caspase-1 activatation during infection
APAF1	Caspase-9 activatation following cytochrome-C release
PIDD	Caspase-2 during genotoxic stress

apoptotic members of this family. In addition, heat shock protein HSP27 prevents cytochrome-C release.

In the intrinsic pathway, the apoptotic stimulus triggers the assembly of the *apoptosome*. Cytochrome-C has 2 main functions: electron transport and caspase activation. Once released in the cytosol from mitochondria, cytochrome-C is sensed by apoptosis-peptidase-activating factor APAF1, binds to APAF1 to form heptamers, and is recruited to the apoptosome. Cytochrome-C then activates APAF1 in the apoptosome to recruit and stimulate procaspase-9 into caspase-9[36] and caspase-3, leading to cell apoptosis.[37]

Mitochondrial dysfunction disrupts energy production and can trigger apoptosis. Protein BID that is cleaved (activated) into truncated $_t$BID by caspase-8 is a converging point of both pathways. Activated BID primes BAK and BAX oligomerization to form outer mitochondrial membrane pores, in addition to mitochondrial permeability transition pores and other mitochondrial carriers such as voltage-dependent anion channels (porins).

Mitochondrial outer membrane permeabilization releases mitochondrial intermembrane proteins, such as cytochrome-C. The sequential activation of caspases is controlled both positively and negatively by B-cell lymphoma protein-2 family members. Cytochrome-C liberation from mitochondria requires the activation of cell death mediator BAK of the BCL2 family that is located in the mitochondrial outer membrane, where it remains inactive unless activated. Another member of the BCL2 family, pro-apoptotic BCL2-associated X protein (BAX) is also needed to activate caspases. BH3-only members initiate apoptosis by engaging multiple prosurvival relatives in guarding BCL2-antagonist killer (BAK) and BAX [565]. BCL2 proteins are also implicated in embryogenesis and adult tissue homeostasis (Table 4.11).

[36] Caspase-9 activation requires the ATPase activity of APAF1. Caspase-9 and its downstream effectors are inhibited by X-linked inhibitor of apoptosis (XIAP).

[37] Mitochondrial release of cytochrome-C can lead to caspase activation for cell differentiation, such as monocyte-to-macrophage differentiation.

Table 4.11. Targets of BCL2 proteins (Source: [519]). Members of the BCL2 family are involved from early embryogenesis to adult tissue homeostasis. BCL2-like proteins not only are involved in events linked to apoptosis, but also operate in normal, death-unrelated phenomena such as synaptic remodeling.

Type	Target cells
	Anti-apoptotic members
BCL2	Lymphocytes, renal epithelium, melanocyte progenitors
BCLxL	Erythroid progenitors, neurons
BCLw	Sperm cells
BCLa1	Granulocytes, mastocytes
MCL1	Lymphoid cells, hemopoietic stem cells
	Pro-apoptotic BAX–BAK family members
BAK	Thrombocyte turnover
BAX	Neuronal development, sperm cell differentiation
	Pro-apoptotic BH3-only proteins
BAD	Cell sensitive to deprivation of epidermal or insulin GF
BID	Cell types sensitive to Fas-induced apoptosis
BIK	
BIM	Lymphoid and myeloid cells
HRK	Neuron types sensitive to deprivation of nerve growth factor
Noxa	Fibroblasts, keratinocytes
PUMA	Many cell types

Pro-apoptotic BH3-only proteins BID and BIM bind to and activate cytosolic pro-apoptotic protein BAX [566]. The interaction site for direct activation differs from the binding groove for anti-apoptotic proteins that confers survival functionality. Once activated, BAX self-oligomerizes to disrupt mitochondrial integrity by assembling within the outer mitochondrial membrane, and initiates apoptosis.

Activated caspases cleave many constituents of the cell cytoskeleton: (1) components of actin and myosin filaments (actin, myosin, spectrins, α-actinin, gelsolin, and filamin); (2) microtubular proteins (tubulins, microtubule-associated protein τ, dynein, etc.); (3) intermediate filament proteins (vimentin and keratins); and (4) lamins. Proteolysis of cytoskeletal constituents contributes to cell retraction in the early stage of apoptosis, as well as membrane blebbing [564].

Nuclear fragmentation relies on the disintegration of lamin-A, -B, and -C by caspases. The actin cytoskeleton and RoCK1 participate in nuclear fragmentation, as the actin cytoskeleton is attached to the nuclear membrane. The microtubule network permits dispersal of nuclear fragments.

Transcription factors, such as nuclear factor of activated T cells NFAT1 and NFAT2, nuclear factor-κB (NFκB1 and RelA), Activator protein AP2α, and basic transcription factor BTF3, and translation factors, such as eukaryotic translation initiation factors eIF2a, eIF3, eIF4b, eIF4e, eIF4g, and eIF4h, are degraded by caspases during apoptosis, as are ribosomal RNA and proteins [564]. Within the dying cell, caspase-activated endonucleases cleave chromatin at internucleosomal sites. Kinase STK4 (or mammalian sterile-20 kinase MST1) activated by caspase-3 phosphorylates histone-2B, hence leading to chromatin condensation. DNA degradation avoids immune system activation. The activated program leads to DNA fragmentation, cytoplasm shrinkage, membrane changes, and cell death without damage to neighboring cells.

Whereas the plasma membrane of viable cells displays markers that inhibit phagocytosis, apoptotic cells lose these plasmalemmal molecules and exhibit other markers such as phosphatidylserine for phagocytosis. Chemoattractant lysophosphatidylcholine is released by caspase-3. Several integrins, such as $\alpha_M\beta_2$ and $\alpha_X\beta_2$ integrins (or complement receptors CR3 and CR4), are implicated in the recognition of dying cells by bridging thrombospondin and complement component C3b [564].

Attracted phagocytes release anti-inflammatory molecules, such as interleukin-10 and TGFβ. Moreover, phagocytes behave differently with apoptotic cells than with pathogens, as they minimize the production of pro-inflammatory cytokines such as TNFα. Certain kinds of apoptotic cells display plasmalemmal calreticulin that can bind CD91 to trigger cell engulfment. They are targeted by dendritic cells for T-cell-mediated immune response.

4.6.2 Extrinsic Pathway

The extrinsic or death-receptor apoptosis pathway is primed by an extracellular ligand, i.e., a pro-apoptotic and -inflammatory cytokine, such as TNFSF1 (or tumor-necrosis factor-α), TNFSF6, and TNFSF10, that binds to a transmembrane receptor of the death-receptor family. Death receptors are members of the tumor-necrosis factor receptor superfamily (Chap. 3 and Vol. 3 – Chap. 11. Receptors of the Immune System) that contain an intracellular death domain and can recruit caspase-8 via adaptor Fas (TNFRSF6)-associated death domain-containing protein (FADD).[38] The ligand-dependent, plasmalemmal *death-inducing signaling complex* is then formed and activates caspase-8 and -10.

Caspase-8 activation is prevented by flice-inhibitory protein. The catalytic domain of caspases includes large and small units. Cleavage of these dimer units leads to caspase activation. Caspase-9 and -8 cleave procaspase-3 and -7. Cleaved, active caspases achieve apoptosis. Activated caspase-8 stimulates downstream caspase effectors, such as caspase-3, -6, or -7, without BCL2 involvement.

[38] Protein with a FADD domain is also named MORT1.

In some cells such as hepatocytes, the extrinsic pathway can intersect the intrinsic pathway via cleavage and activation by caspase-8 of pro-apoptotic BH3-only protein BID [519]. Truncated BID (tBID) then translocates to mitochondria and activates caspase-9 and its caspase effectors caspase-3, -6, and -7 via the intrinsic pathway.

4.6.3 Apoptosis Features and Role

In the early stages of apoptosis, the cell retracts and detaches from neighboring cells and the extracellular matrix. Caspases dismantle cell–cell adhesion complexes and cell–matrix focal adhesion sites to facilitate the subsequent apoptotic cell removal by phagocytes. Caspases target focal adhesion kinase, BCAR1 (or CAS), and tensin, as well as β- and γ-catenins of adherens junctions, cadherins, and desmosome-associated proteins.

Apoptosis is required during normal body development, especially in the central nervous system where genetic deficiency of apoptotic genes, such as caspase-9, apoptotic adaptor peptidase-activating factor APAF1, BAX, or BAK, results in abnormalities. Apoptosis also maintains tissue homeostasis, especially in the immune system. Inappropriate activation of apoptosis leads to acute and chronic diseases. Dysregulation of apoptosis leads to cancer and autoimmunity.

4.6.4 Other Apoptosis Mediators

Nucling is a short-lived regulator of apoptosis. It contributes to the formation of the cytochrome-C–APAF1–caspase-9 apoptosome. It interacts with galectin-3 that participates in apoptosis [568]. It prevents galectin-3 synthesis, as it inhibits NFκB nuclear translocation and activity. In addition, its expression is regulated by nuclear factor-κB and tumor-necrosis factor-α that stimulates the canonical NFκB activation pathway [569]. Nucling undergoes both proteasomal degradation and proteolytic cleavage by caspases.

Nitric oxide is able to prime apoptosis. Diverse apoptotic stimuli indeed activate neuronal (NOS1) or inducible (NOS2) nitric oxide synthase. Nitric oxide nitrosylates glyceraldehyde 3-phosphate dehydrogenase (GAPDH), abolishing its catalytic activity. Nitrosylated GAPDH is then able to bind to ubiquitin ligase SIAH1 that translocates to the nucleus to trigger cell death [567]. Nuclear GAPDH is acetylated by nuclear acetyltransferases P300 and cAMP responsive element-binding protein (CREB)-binding protein (CBP) to stimulate its acetylation and catalytic activity on targets such as P53 transcription factor.[39] Tumor suppressor P53 is able to activate survival pathways. However,

[39] Glyceraldehyde 3-phosphate dehydrogenase is able to bind to both P300 and CREB-binding protein, the latter in the presence of nitric oxide. Nitric oxide influences interaction between GAPDH and P300 and CBP as it facilitates GAPDH nuclear translocation.

P53 can also trigger apoptosis, particularly upon oxidative stress, via PUMA in the presence of nitric oxide.

4.6.5 Anoikis

Anoikis is a particular type of apoptosis due to loss in integrin-mediated contact between the cell and the extracellular matrix. Reactive oxygen species produced via small GTPase Rac1 activated by integrin engagement ensure cell escape from anoikis [570]. Reactive oxygen species activate Src kinase that phosphorylates epidermal growth factor receptor and subsequently stimulates the Raf–ERK and PI3K–PKB signaling pathways. The latter phosphorylates and degrades via ubiquitin pro-apoptotic protein BIM of the BH3-only family, therefore suppressing anoikis.

In adherent cells, BIM is sequestered in a cytoskeletal motor complex. Upon cell detachment, BIM is released and translocated into mitochondria, where it interacts with BCLxL, hence neutralizing its prosurvival function. Both BIM sequestration by cell adhesion and EGFR signaling due to integrin commitment in cell adhesion protect from anoikis.[40]

4.7 Interplays between Apoptosis and Autophagy

Crosstalk between autophagy and apoptosis relies on interaction of beclin-1 with B-cell leukemia/lymphoma protein BCL2 at the endoplasmic reticulum. Three different types of interplay exist between autophagy and apoptosis [520]: (1) autophagy antagonizes apoptotic cell death and promotes cell survival; (2) autophagy acts as an enabler of apoptosis, as it participates in some events during apoptotic cell death without causing death; and (3) apoptosis and autophagy cooperate to provoke cell death.

According to circumstances, autophagy either avoids or leads to cell death, i.e., apoptosis and autophagy triggered by similar stimuli combine or are mutually exclusive [572]. Certain pathways drive cytoprotective autophagy that prevents apoptosis. Cyclin-dependent kinase inhibitor CKI1b causes autophagy and impedes apoptosis in response to nutrient depletion. Activated AMP-dependent kinase phosphorylates CKI1b, thus allowing cells to survive. Autophagy can also kill cells either by destroying large proportions of cytosol

[40] Integrin engagement in cell adhesion causes ligand-independent activation of many growth factor receptors, such as epidermal, platelet-derived, hepatocyte, and vascular endothelial growth factor receptor as well as insulin receptor. Integrin–EGFR crosstalk is mediated by Tyr kinase Src and involves the recruitment of adaptor BCAR1 [571]. Integrins induce EGFR phosphorylation (Tyr845, Tyr1068, Tyr1086, and Tyr1173, but not on Tyr1148, a major site of phosphorylation in response to EGF). Integrins also increase the amount of EGFRs expressed in the plasma membrane.

and organelles or stimulating pro-apoptotic signals. Autophagy is also able to cause cell necrosis [572].

Several pathways triggered by stresses generate both autophagy and apoptosis [572]. Reactive oxygen species stimulate apoptosis via mitochondrial outer membrane permeabilization and autophagy via cysteine peptidase autophagy-related gene product AtG4.

Ceramide, a sphingolipid, induces apoptosis via the intrinsic pathway. Ceramide can also prime autophagy. Ceramide can raise the expression of beclin-1 owing to Jun N-terminal kinases that phosphorylate Jun transcription factor that targets the gene encoding beclin-1. On the other hand, sphingosine 1-phosphate causes autophagy, but inhibits ceramide-induced apoptosis.

Calcium overload in mitochondria leads to apoptosis. Calcium can generate autophagy possibly via calmodulin-dependent kinase kinase-β. Calcium also stimulates calpains that could contribute to autophagy or apoptosis.

Transcription factor P53 yields apoptosis in cell exposed to genotoxic stress. It also activates damage-regulated autophagy modulator.

Beclin-1 is a BH3-only protein that triggers apoptosis, as it activates BAX or BAK or it inhibits anti-apoptotic members of the BCL2 family. Beclin-1 is also a tumor suppressor that operates via short mitochondrial form (smARF) of alternate reading frame of the CDKN2A gene (P14ARF). It stimulates autophagy and caspase-independent cell death and nucleolar isoform P14ARF that activates P53 by hindering P53 inhibitor DM2 ubiquitin ligase. The link of beclin-1 to UV irradiation resistance-associated tumor suppressor gene product (UVRAG) and class-3 phosphatidylinositol 3-kinase PI3KC3 causes autophagy.

Calcium–calmodulin-regulated death-associated protein kinase DAPK1 favors either type-1 apoptosis or type-2 autophagic cell death in response to various stimuli (interferon-γ, tumor-necrosis factor-α, transforming growth factor-β, oncogenes, and detachment from the extracellular matrix) [572]. Tumor suppressor DAPK1 activates nucleolar P14ARF. DAPK-related protein kinase DAPK2 and ZIP-kinase DAPK3 also control both apoptosis and autophagic cell death.

In contrast, cross-inhibitions can occur between apoptosis and autophagy. Beclin-1 is inhibited by BCL2, BCLxL, and MCL1. Autophagy AtG12–AtG5–AtG16 complex can be cleaved by a calpain to produce truncated AtG5 that translocates to mitochondria and induces mitochondrial outer membrane permeabilization. Short AtG5 loses autophagic activity and acquires a pro-apoptotic function.

4.7.1 Apoptosis as an Antagonist of Autophagy

Under some circumstances, autophagy impedes apoptosis and favors cell survival. Endoplasmic reticulum stress can be resolved by autophagy of protein aggregates and misfolded proteins, hence avoiding apoptosis subsequent to the endoplasmic reticulum stress response. Autophagy promotes cell survival during periods of starvation. It can also protect cells from anoikis [520]. Furthermore, autophagy maintains genomic integrity in the presence of metabolic stress, drug treatment, or radiation damage via mitophagy, i.e., scavenging of depolarized mitochondria that are sources for genotoxic reactive oxygen species.

4.7.2 Autophagy as an Enabler of Apoptosis

In certain conditions, autophagy assists the apoptotic program, but does not directly cause cell death. Autophagy enables the execution of particular components of apoptotic cell death. During nutrient deprivation, autophagy saves ATP and then favors ATP-dependent phosphatidylserine exposure that yields a phagocytic signal to neighboring phagocytes.

4.7.3 Apoptosis and Autophagy as Partners

Apoptosis and autophagy cooperate to ensure efficient cell death, especially when one cell death program is inhibited [520]. Apoptosis, autophagy, or both are used to eliminate the cell. Coordination between the cell death pathways can result from simultaneous activation of both pathways by common inducers, such as ceramide and TNFRSF10b. Moreover, autophagy can improve or even be necessary for caspase-dependent cell death. Inhibition of autophagy is able to prevent apoptosis initiated by death signals such as TNFα combined with suppression of NFκB [520]. Under other circumstances, autophagy is restrained by the apoptotic pathway. Autophagic cell death mediated by receptor-interacting protein kinase and Jun N-terminal kinase is activated on caspase-8 inhibition.

4.8 Non-Apoptotic Cell Death

Non-apoptotic cell death, or type-2 cell death, is used in rare circumstances when the apoptotic machinery is genetically unavailable, especially in developmental events. Cell necrosis is an accidental cell death. The regulated, or programmed, necrosis is called *necroptosis*. Its execution involves the active disintegration of mitochondrial, lysosomal, and plasma membranes. In fact, non-apoptotic cell death includes necroptosis and *PARP1-mediated necrotic death*. These processes serve specific functions in response to pathogen infection, nutrient and energy deprivation, and DNA damage. Death-domain receptor engagement by cognate ligands can lead to non-apoptotic cell death (using morphological criteria), because of the activity of caspase inhibitors.

Non-apoptotic cell death is characterized by accumulation of double membrane-enclosed vesicles that are characteristic of autophagy. Autophagosomes fuse with lysosomes and the autophagosome content is degraded by lysosomal enzymes.

4.8.1 Necroptosis

Necroptosis is controlled by numerous initiators, effectors, and modulators, such as death receptors of the tumor-necrosis factor receptor superfamily (TNFRSF1a and -1b, TNFRSF6a, and TNFRSF10a and -10b), receptor-interacting protein kinases RIPK1 and RIPK3, caspase inhibitors, ubiquitin ligases, deubiquitinases, and reactive oxygen species) [573]. In addition, necroptosis can be ignited by pathogen recognition receptors, such as Toll-like, NOD-like, and retinoic acid-inducible gene I-like receptors, as well as DNA damage. It is activated by the same stimuli that initiate apoptosis, but causes pathological disturbances (fast mitochondrial dysfunction, organelle swelling, and plasmalemmal permeabilization, without nuclear fragmentation) [534].

Necroptosis is primed by TNFα that recruits and activates the intracellular signaling complex made of adaptor kinase RIPK1 (but not TRADD–RIPK1 complex for apoptosis).[41] In addition, necrostatins inhibit necroptosis, but not TNFα-induced apoptosis.[42] Kinase RIPK1 translocates into mitochondria and disrupts the association of ADP–ATP translocase with peptidylprolyl isomerase-D (or cyclophilin-D), a component of the mitochondrial permeability transition pore.

Activated RIPK1 increases the production of reactive oxygen species from the mitochondrial respiratory chain and the RIPK1–Rac1–NADPH oxidase complex and activates Jun N-terminal kinase. Autophagy can be activated during necroptosis in some cell types. Phospholipase-A2, lipoxygenases, and acid sphingomyelinase are also activated. In opposition to overwhelming stress-induced activation that leads to necrosis, necroptosis deals with activation mechanisms regulated by internal signaling.

Enzymes RIPKs are involved in 2 distinct complexes, the TNFR complex-1 and -2 [573]. Ligand-bound TNFR1 triggers the assembly of a cytoplasmic complex that includes TNFR-associated death domain adaptor (TRADD), TNFR-associated factor TRAF2 and TRAF5, cellular inhibitor of apoptosis IAP1 and IAP2, and receptor-interacting protein kinase RIPK (Table 4.12). In this *TNFR complex-1*, RIPK1 can be ubiquitinated by IAPs and then recruit transforming growth factor-β-activated kinase TAK1 (MAP3K7), TAK1-binding proteins TAB2 and TAB3 (MAP3K7IP2 and -3) to ignite the canonical nuclear factor-κB pathway. Ubiquitinated RIPK1 promotes the activation of nuclear factor-κB. Riboflavin kinase bridges TNFR1 to NADPH oxidase such as NOx1 that contributes to TNFα induced necroptosis due to the overproduction of reactive oxygen species. RIPK1 can also be deubiquitinated by cylindromatosis (Cyld) and tumor-necrosis factor-α-

[41] Kinase function of RIPK1 is not involved in the activation of nuclear factor-κB that leads to apoptosis.

[42] Necrostatin-1 and -3 do not influence RIPK1-mediated activation of NFκB as well as the activity of P38MAPK and JUN N-terminal kinase JNK1. Several other necrostatins suppress necrosis, but they inhibit RIPK1 indirectly by interfering with upstream mediators.

induced protein TNFαIP3, as well as OTU domain-containing protein-7B and ubiquitin-specific peptidase USP21. Deubiquitinated RIPK1 acts as a cell death-inducing kinase. When TNFR1 is endocytosed, the *TNFR complex-2* is built with TRADD, Fas (TNFSF6)-associated protein with a death domain (FADD), caspase-8, and RIPK1 and RIPK3 kinases. Normally, caspase-8 is activated in the TNFR complex-2 to initiate apoptosis. In the presence of a caspase-8 inhibitor, RIPK1 interacts with RIPK3 to generate a necroptosis-inducing complex, the *necrosome*.[43] The RIPK1–RIPK3 necrosome: (1) stimulates glycogenolysis and glutaminolysis, as it enhances the activity of glycogen phosphorylase, glutamate–ammonia ligase, and glutamate dehydrogenase-1; (2) activates JNK-mediated degradation of ferritin, thus increasing the labile iron pool, and fosters sphingomyelinase-induced generation of ceramide that is converted into sphingosine by ceramidase to foster a cytosolic Ca^{++} wave that activates calpains and cytosolic phospholipase-A2. The latter triggers lipid peroxidation via arachidonic acid. Sphingosine, calpains, and lipid hydroperoxides cause lysosome membrane permeabilization. In addition, reactive oxygen species generated by the mitochondrial respiratory chain and NADPH oxidase-1, ceramide metabolism, and labile iron pool elevation initiate a mitochondrial uncoupling and lipid peroxidation and promote the opening of the permeability-transition pore complex. The permeabilization of mitochondrial membranes provokes the translocation of cytotoxic proteins such as apoptosis-inducing factor from the mitochondrial intermembrane space to the cytosol.

Necroptosis inhibitor necrostatin delays the opening of the mitochondrial permeability transition pore in isolated cardiomyocytes at reperfusion after 30-mn ischemia. During ischemia–reperfusion events, cardioprotector necrostatin-1 impedes the activity of peptidylprolyl isomerase-D, hence reducing mitochondrial permeability transition pore opening during reperfusion [574].

4.8.2 PARP1-Mediated Cell Death

The cell death mediated by PARP1 includes 2 pathways: (1) energy collapse and (2) apoptosis-inducing factor translocation. However, PARP1 overactivation can lead to caspase-independent cell death. Poly^{ADP}ribose polymerase PARP1 is involved in initiating DNA repair. Enzyme PARP1 is indeed rapidly activated by DNA strand breaks and recruits DNA-repair factors by attaching ^{ADP}ribose units to chromatin-associated proteins.

Activated PARP1 also leads to the release of inflammatory mediator High-mobility group HMG-box domain-containing protein HMGB1 that can alert immunocytes to the presence of dangerous cells with damaged DNA [575].[44]

---

[43] When caspases are inhibited, RIPK1 and RIPK3 are phosphorylated.

[44] Necrotic cells release inflammatory mediators such as HMGB1 that activate cytokine production from innate immunocytes. Chromatin-associated protein HMGB1 is sequestered at condensed chromatin during apoptosis. Protein

Table 4.12. The TNFR1 interactome in necroptosis (Source: [573]; (+): stimulation; (−): inhibition; AA: arachidonic acid; Cyld: cylindromatosis; ECM: extracellular medium; FADD: Fas (TNFSF6)-associated protein with a death domain; IAP: inhibitor of apoptosis; JNK: Jun N-terminal kinase; LMP: lysosomal membrane permeabilization; LOx: lipoxygenase; NFκB: nuclear factor κ light chain-enhancer of activated B cells; NOx: NADPH oxidase; OTUD7b: OTU domain-containing [zinc finger] protein-7B [a.k.a. Cezanne]; cPLA2: cytosolic phospholipase-A2; PM: plasma membrane; RFK: riboflavin kinase; RIP: receptor-interacting protein; ROS: reactive oxygen species; SMase: sphingomyelinase; TNF: tumor-necrosis factor; TNFR: TNF receptor; TNFαIP: tumor-necrosis factor-α-induced protein; TRADD: TNFR-associated death domain-containing protein; TRAF: TNFR-associated factor; USP: ubiquitin-specific peptidase).

Factor	Location	Effect
Caspase-8	Cytoplasm	TNFR complex-2 member, cleavage (−) of RIPK1 and RIPK3
Ceramidase	Mitochondrion, PM	Conversion of ceramide into sphingosine, lysosomal membrane permeabilization
Cyld	Cytoplasm, PM	RIPK1 deubiquitination, NFκB (−), necroptosis (+)
FADD	Cytoplasm	Adaptor in TNFR complex-2
IAP	Cytoplasm, PM	RIPK1 ubiquitination, NFκB (+)
JNK1	Cytoplasm, mitochondrion	Degradation of ferritin, ROS production
LOX	Cytoplasm	Conversion of AA into lipid hydroperoxides, lysosomal membrane permeabilization
NOX1	PM	ROS production
OTUD7b	Cytoplasm, PM	RIPK1 deubiquitination, NFκB (−), necroptosis (+)
cPLA2	Cytoplasm	Production of arachidonic acid, LMP
RFK	Cytoplasm, PM	TNFR complex-1 member, coupling of TNFR1 to NOx1
RIPK1	Cytoplasm, PM	Component of the necrosome
RIPK3	Cytoplasm, PM, mitochondrion	Component of the necrosome
SMase	Lysosome, PM	Transformation of sphingomyelin into ceramide, ROS production, lipid peroxidation, LMP
TNF	ECM, PM	Necroptosis (in the absence of caspase action)
TNFαIP3	Cytoplasm, PM	RIPK1 deubiquitination, NFκB (−), necroptosis (+)
TNFR2	PM	Facilitation of RIPK1 activation
TRADD	Cytoplasm, PM	Adaptor
TRAF2/5	Cytoplasm, PM	NFκB (+), necroptosis (−)
USP21	Cytoplasm, PM	RIPK1 deubiquitination, NFκB (−), necroptosis (+)

PARP1 regulates the translocation of HMGB1 from the nucleus to the cytosol in cells that lose plasma membrane integrity during necrosis. Inflammatory mediator HMGB1 activates macrophages and TNFα production.

Enzyme PARP1 also mediates cell death induced by secondary DNA damage following cell injury that leads to the translocation of poly^{ADP}ribose polymers into the cytosol. This process triggers translocation of mitochondrial oxidoreductase apoptosis-inducing factor (AIF) from mitochondria to the nucleus, where it mediates cell death [576]. The oxidoreductase activity of apoptosis-inducing factor is required for mitochondrial complex-1 function. Glycohydrolase that degrades PAR polymer prevents PARP1-dependent AIF release. Cells with reduced AIF level are resistant to PARP1-dependent cell death and PAR polymer cytotoxicity.

In addition, PARP1 mediates cell death following DNA damage in another process that involves TNF receptor-associated factor TRAF2–RIPK1-dependent activation of Jun N-terminal kinase-1, which contributes to mitochondrial dysfunction and necrotic death.

4.9 Cell Necrosis

Mild insults to the cell cause apoptosis, whereas intense stress generates uncontrollable necrosis. Apoptosis is a regulated cell death program, that eliminates the cell without generating any inflammatory response. In contrast, necrosis is related to catastrophic cell fate with organelle swelling, mitochondrial dysfunction, massive oxidative stress, and rapid plasma membrane permeabilization. During necrosis, the rapid loss of membrane integrity releases the cellular content into the extracellular space. This release damages neighboring cells and activates the immune system.

Alarmins, or danger-associated molecular patterns (DAMP), released by necrotic cells that target pattern-recognition receptors include high-mobility group protein HMGB1, uric acid, some heat shock proteins, single-stranded RNA, and DNA [564]. Release of alarmins stimulates macrophages, neutrophils, dendritic cells, and natural killer cells. After necrosis, cell fragments are taken up by phagocytes that produce pro-inflammatory cytokines. The subsequent inflammatory response can damage neighboring cells.

When tissues are not sufficiently oxygenated, concentrations in transcription factor HIFα and HIF prolyl hydrolase PHD2 rise. Both molecules avoid large tissue necrosis by favoring angiogenesis and contribute to repair damaged tissues after a slight, initial hypoxia to prevent later, more important lesions.

Death receptors, such as TNFRSF1a (TNFR1), TNFRSF6a, TNFRSF25 (a.k.a. TNFRSF12 and death receptor DR3), as well as TNFRSF10a (a.k.a. DR4) and TNFRSF10b (a.k.a. DR5), can induce necrotic cell death following the generation of reactive oxygen species by activated NADPH oxidases (Vol. 4 – Chap. 9. Other Major Signaling Mediators). The latter form complexes with adaptors, such as FADD, TRADD, and RIPK1. Adaptors RIPK1 and TRADD form a complex with NOx2 and recruit NOx1 and Rac1 to form a superoxide-producing complex. The effect of reactive oxygen species on necrotic cell death is likely mediated via sustained JNK activation.

When some great sorrow, like a mighty river,
Flows through your life with peace-destroying power
And dearest things are swept from sight forever,
Say to your heart each trying hour:
"This, too, shall pass away."...

When fortune smiles, and, full of mirth and pleasure,
The days are flitting by without a care,
Lest you should rest with only earthly treasure,
Let these few words their fullest import bear:
"This, too, shall pass away."...

(This, Too, Shall Pass Away, Lanta Wilson Smith [1851–1921])

5 Circadian Clock

Day full-blown and splendid-day of the immense sun, action, ambition, laughter,
The Night follows close with millions of suns, and sleep and restoring darkness.

(Youth, Day, Old Age and Night, Walt Whitman [1819–1892])

Many cellular events happen at specific times of the day. Day–night (light–darkness) alternation has trained organisms[1] to anticipate periodic environmental changes and organize metabolism and behavior accordingly. Biological daily clocks coordinate the activity of the body's organs and tissues. Many mammalian cells contain a molecular clock, i.e., a set of genes of which the cyclical expression controls self-sustained oscillators that are reset every day. The clock can keep ticking after the removal of all external stimuli.

Circadian rhythms that, in particular, organize metabolic and energetic cycles (e.g., glycolytic and oxidative cycles) result from transcription and translation feedback loops. Clock proteins are transcriptional activators that support the synthesis of repressors that inhibit these activators, which repress their own production.

Nevertheless, in human anucleated erythrocytes, periodic cycles can occur independently of transcription [577].[2] Therefore, non-transcription-based

[1] In 1729, Jean-Jacques d'Ortous de Mairan (1678–1771) devised an experiment that proved the existence of a circadian rhythm in plants. He tested whether the rhythmic movement of mimosa leaves required exposure to sunlight or resulted from an internal clock. The circadian cycle persisted in the absence of daylight, as a plant in a dark box for several days maintains leaf motion.

[2] Most other organelles, especially mitochondria, suppliers of chemical energy, also lack in erythrocytes. Anti-oxidant peroxiredoxins are among the most abundant proteins in erythrocytes. The non transcriptional control of circadian rhythms regulates the activity of peroxiredoxins. These enzymes are characterized by an entrainable (i.e., tunable by environmental stimuli), temperature-compensated (i.e., clock does not work slower at lower temperatures or quicker at higher

rhythms couple with transcription-based oscillators to control circadian signals.

5.1 Intracellular Transcription-Based Clocks

Circadian rhythm[3] drives about 20% of gene expression. Resulting cellular rhythmicity relies on several interlinked transcriptional and translational feedback loops that generate oscillations entrained by the circadian clock.[4]

The circadian clock is composed of a cerebral pacemaker and a network of interconnected oscillators in most peripheral tissues. The hypothalamic suprachiasmatic nucleus (SCN) acts as the master clock. This neuronal conductor orchestrates all molecular, cell-autonomous, slave circadian clocks in peripheral cells. Day–night cycle is the predominant cue for the cerebral pacemaker.[5] In mammals, only pacemaker neurons of the suprachiasmatic nucleus respond to light. The peripheral oscillators are synchronized owing to nervous and endocrine, as well as paracrine communications.

Daily rhythm in electrical behavior and gene expression is produced by the circadian clock in both diurnal and nocturnal organisms with phase reversal between diurnality or nocturnality. Besides, feeding behavior and the body's metabolism also strongly influence circadian rhythms. Cerebral clock drives rest–activity and fasting–feeding cycles, whereas subsidiary peripheral clocks that are mostly set by daily feeding prepare and adapt the control of the body's metabolism.[6]

Hundreds of tissue-specific genes that regulate diverse biological processes (intracellular transport, signal transduction, and cell metabolism, cycle, and survival, etc.) are controlled by the body's clocks. The central clock entrails peripheral clocks to synchronize via neurohormonal signals. The master clock

---

values) redox cycle (duration ~24 h). In erythrocytes, dimeric hemoglobin generates peroxide as well as tetrameric hemoglobin, but to a lesser extent. Hemoglobin evolve between tetramers and dimers according to a circadian rhythm. Another redox element — the NADH–NADPH couple — experiences circadian oscillations. Agent NADPH is a reduced form of NADP that serves as a cofactor, which couples energy flux with changes in the oxidation of peroxiredoxin and hemoglobin. Peroxiredoxins protect red blood cells from peroxides, as they oxidize when reactive oxygen species accumulate and subsequently dimerize. Conversely, their reduced form is monomeric. When ROS concentration rises, peroxiredoxins oligomerize.

[3] Circa: about and diem: day (i.e., period ~24 h).

[4] When clocks are stably synchronized to a pacemaker, they are said to be entrained. Circadian oscillations can be tuned by external signals and reset when they do not coincide. Cellular oscillators can be entrained by various stimuli, such as light, temperature, and feeding schedules.

[5] Free running of circadian rhythm refers to circadian gene expression in the absence of external cues, such as the day–night cycle.

[6] In the absence of circadian clocks, activity rhythms as well as energy balance and glucose homeostasis are impaired.

continuously maintains phase coherence among the body's cells, whereas peripheral oscillators without crosstalk are not synchronized in the absence of a central timekeeper.

Circadian clocks modulate numerous behaviors. Groups of cells that display cyclic changes are defined as biological oscillators. Pacemakers exhibit sustained rhythms and regulate the functioning of other cells. Self-sustained oscillators determine the period of a given activity, whereas damped oscillators require inputs to sustain their rhythmicity. Circadian pacemakers are self-sustained oscillators.

Circadian oscillators consist of autoregulatory negative feedback loops in which activators stimulate transcription of inhibitors that suppress the activity of activators (Sect. 5.4). In the latter part of the day, transcription mediated by transcription activators CLOCK (circadian locomotor output cycles kaput) and BMAL1 (brain and muscle ARNT-like protein) is inhibited. Certain components of involved regulatory networks are connected and operate coordinately. Post-translational modification of clock proteins regulates clock function (Sect. 5.5).

A simple negative feedback loop of interacting genes or proteins can generate a biological oscillator. Activators initiate the production of repressors, which in turn, after a delay, inhibit activators and slow their own production and activity. As repressors decline in level and/or activity, activators start the cycle again. The cycle duration of an oscillator can be modulated by intracellular signals and protein modifications as well as extracellular signals (e.g., neurotransmitters and ionic fluxes that either depolarize or hyperpolarize the cell) to adjust to environmental timing cues. Factors that modulate amplitude and phase of circadian rhythms must synchronize circadian cells.

The frequency of a negative-feedback oscillator most often cannot be adjusted without altering its amplitude [578]. Therefore, oscillators such as cell cycle clock that include both negative and positive feedback loops, can achieve tunable frequency over a wide frequency range without substantial changes in oscillator amplitude. Positive-plus-negative feedback oscillators also appear to be more robust, as oscillations occur over a much broader range of enzyme concentrations and kinetic constant values.

Attributes that describe an oscillation are periodicity, robustness, and entrainment. Therefore, clock components must operate on clock period, amplitude, and phase. Many mammalian clock genes have been identified (Tables 5.1 to 5.3).

A nuclear complex is constituted by 3 Per (Per1–Per3) and 2 Cry (Cry1–Cry2) proteins, with RNA-binding non-POU domain-containing octamer-binding protein (NonO), histone methyltransferase subunit WD repeat domain-containing protein WDR5, splicing factor proline and glutamine-rich (SFPQ)[7] and the scaffold Sin3a homolog that acts as a corepressor that

---

[7] A.k.a. polypyrimidine tract-binding protein [PTB]-associated splicing factor (PSF).

Table 5.1. Clock and clock-associated genes in mammals (**Part 1**; Source: [579]; BMAL: brain and muscle ARNT-like protein; CLOCK: circadian locomotor output cycles kaput; EGFR: epidermal growth factor receptor; GR: glucocorticoid receptor; ID2: inhibitor of DNA binding-2 [bHLH-containing transcriptional repressor]; MAOa: monoamine oxidase-A; MPG: ^{N}methylpurine (^{N}methyladenine)-DNA glycosylase; NAmPT: nicotinamide phosphoribosyltransferase; Opn4: opsin-4, or melanopsin; Per: period; PKG: cGMP-dependent protein kinase; RasD1: Ras, dexamethasone-induced-1 [a.k.a. Activator of G-protein signaling AGS1 and dexamethasone-induced Ras-related protein DexRas1]; SCN: suprachiasmatic nucleus; TGFA: transforming growth factor-α gene; VIP: vasoactive intestinal peptide [neuropeptide]; VIPR2: vasoactive intestinal peptide receptor-2 [G-protein-coupled receptor]). Circadian rhythms are oscillators generated by feedback loops in which 3 Per (Per1–Per3) and 2 Cry (Cry1–Cry2) proteins operate in a nuclear complex with RNA-binding non-POU domain-containing octamer-binding protein (NonO), histone methyltransferase subunit WD repeat domain-containing protein WDR5, splicing factor proline and glutamine-rich (SFPQ; or polypyrimidine tract-binding protein [PTB]-associated splicing factor [PSF]), and scaffold Sin3a homolog (a transcription corepressor as a subunit of the histone deacetylase complex) to inhibit the transcriptional activity of the CLOCK–BMAL1 dimer that represses the production of its own constituents. The turnover of Per and Cry proteins relieves the inhibition of the CLOCK–BMAL1 activity to trigger a new cycle.

Gene	Function
	Input genes
Gr	Resets circadian phase in peripheral tissues
Id2	Regulates photic entrainment
OPN4	Photoreceptor; light input signals to SCN
Pkg2	Light entrainment via Per1 and Per2 expression
RASD1	Potentiation of photic and suppression of non-photic entrainment
Vip	Neuronal clock synchronizer in the SCN
Vipr2	Neuronal clock synchronizer in the SCN
	Output genes
Egfr	Regulation of cell proliferation
MaoA	Mediation of CLOCK influence on mood
Mpg	Mediation of genotoxic stress
Nampt	NAD^+ biosynthesis (rate-limiting enzyme)
TGFA	Regulation of cell proliferation

complexes with histone deacetylases HDAC1 and HDAC2 [580]. This large PER complex prevents the transcriptional activity of the CLOCK–BMAL1 dimer that, itself, represses the production of its constituents. The PER complex, indeed, rhythmically associates with DNA-bound CLOCK-BMAL1 at the Per1 promoter and recruits the Sin3–HDAC complex that deacetylates

Table 5.2. Clock and clock-associated genes in mammals (**Part 2**: pacemaker; Source: [579]; AMPK: AMP-dependent protein kinase; bHLHb2(3): bHLH domain-containing transcription factor-B2(3) [a.k.a. Hairy and enhancer-of-split-related protein HRT2(1) and differentially expressed in chondrocytes DEC1(2); transcriptional repressors]; BMAL: brain and muscle ARNT-like protein; CIPC: CLOCK-interacting protein, circadian; CLOCK: circadian locomotor output cycles kaput; CBP: CREB-binding protein [transcription coactivator]; Cry: cryptochrome; CSNK1D, CSNK2A: casein kinases CK1δ and CK2α genes; DBP: albumin D-element-binding protein [member of proline and acidic amino acid-rich region (PAR) family (family-1A4 according to the structural classification) of the basic region leucine zipper (bZIP) transcription factors (PAR/bZIP); forms the heterodimer DBP–HLF or DBP–TEF]; FBx: F-box protein of Ub ligase SKP1–cullin-1–F-box (SCF) complex FBxW: F-box and WD repeat domain-containing subunit of Ub ligase SCF complex; Gsk3B: glycogen synthase kinase-3β gene; HLF: hepatic leukemia factor [PAR/bZIP family member; forms the heterodimer DBP–HLF and TEF–HLF]; TEF: thyrotroph embryonic factor [a.k.a. vitellogenin gene-binding protein (VBP); PAR/bZIP family member]; Tim: timeless homolog; bHLH: basic helix–loop–helix).

Gene	Function
Ampk	Phosphorylation of Cry1 for degradation
bHLHb2/3	Repress CLOCK–BMAL1-mediated transcription
Cbp	Activation of CLOCK–BMAL1-mediated transcription
Cipc	Repression of CLOCK–BMAL1-mediated transcription
CSNK1D	Phosphorylation (destabilization) of Per
CSNK2A	Phosphorylation (destabilization) of Per
Dbp–Tef–Hlf	Activation of D-box-containing clock genes
FBX13	Mediation of Cry degradation, regulation of Cry and Per expression
FBXW11	Mediation of CLOCK turnover
Gsk3B	Phosphorylation of Tim, regulation of nuclear localization of Tim and Per

histone-3 and -4, thereby precluding transcription, whereas CLOCK preferentially acetylates $H3K_9$.[8]

5.2 Intracellular Energy Metabolism-Based Clocks

Circadian fluctuations not only result from oscillations in gene transcription managed by self-sustained molecular clocks that exist in all nucleated cells and are synchronized by the central pacemaker (*genetically based oscillators*), but also on non-transcriptional mechanisms such as the energy availability

[8] The Sin3–HDAC complex includes HDAC1 and HDAC2, the histone-binding proteins (nucleosome-remodeling factor subunit) retinoblastoma-binding proteins RBBP4 (or RBAP48) and RBBP7 (or RBAP46), and 18-kDa and 30-kDa Sin3a-associated proteins SAP18 and SAP30 [581].

Table 5.3. Clock and clock-associated genes in mammals (**Part 3**: pacemaker; Source: [579]; BMAL: brain and muscle ARNT-like protein; CLOCK: circadian locomotor output cycles kaput; NFIL3: nuclear factor, interleukin-3-regulated [E4 promoter-binding protein E4BP4; transcription repressor]; NonO: non-POU domain-containing octamer-binding protein [nuclear RNA- and DNA-binding protein]; NPAS: neuronal bHLH and PAS [Per (Period circadian protein), ARNT (aryl hydrocarbon receptor nuclear translocator protein), and Sim (single-minded protein)] domain-containing transcription factor; Per: period; PGC: PPARγ transcriptional coactivator; PRKCA: protein kinase-Cα gene; RACK1: receptor of activated protein kinase-C1; RorA/B/G: retinoid-related orphan receptor-α/β/γ genes; SIRT: silent information regulator-two [sirtuin; class-3 histone deacetylase]; RRE: Rev response element (retinoic-acid responsive element [BMAL1 is partly regulated by a competition between ROR activators and reverse erythroblastoma-α(Rev-ErbA) repressor at its promoter region]); WDR5: WD repeat-containing protein-5 [subunit of histone methyltransferase]).

Gene	Function
Nfil3	Repression of D-box-containing genes
NONO	Modification of CLOCK–BMAL1 activity
Npas2	Dimerization with BMAL1 to activate E box-containing genes
Nr1D1/2	Represses RRE box-containing clock genes
PGC1A	Modification of ROR function
PRKCA	Phosphorylation of CLOCK–BMAL1
Rack1	Phosphorylation of CLOCK–BMAL1
RorA/B/G	Activates RRE box-containing clock genes (e.g., Bmal1)
Sirt1	Association with Per2, repression of CLOCK–BMAL1-mediated transcription
Wdr5	Association with Per1, repression of CLOCK–BMAL1-mediated transcription

that is nonetheless entrained by alternating cycles of light and darkness and temperature oscillations (*metabolically based oscillators*). Genetic and metabolic oscillators are coupled. The metabolic cofactor nicotine adenine dinucleotide (phosphate) NAD(P) influences the activity of some clock proteins [582]. Energetic cycles, such as glycolytic and oxidative cycles, have a transcription-dependent circadian periodicity.

Yet, rhythmic cycles in the activity of some substances can occur independently of transcription and depend on the turnover in cellular energy [582]. Among these substances, peroxiredoxin enzymes are involved in cycles of oxidation and reduction and can inactivate reactive oxygen species [583].[9]

[9] Class-2 peroxiredoxins are oxidized and dimerize, or even oligomerize, when reactive oxygen species accumulate. In the liver, the expression of peroxiredoxins undergoes a circadian rhythm.

In humans, erythrocytes that lack a nucleus and mitochondria exhibit a rhythmic functioning. In these enucleated capsules, i.e., in the absence of gene transcription, 2 main circadian features can be observed [583]: (1) a self-sustained, temperature-independent, circadian oligomerization and (2) synchronization of peroxiredoxin oxidation cycles with temperature rhythm (entrainment). Moreover, in nucleated blood cells, inhibitors of transcription and translation do not interfere with the peroxiredoxin oxidation rhythm.

Last, but not least, oligomeric transitions of hemoglobin (dimer–tetramer transition) obey a rhythm. Hence, the oxygen-carrying capacity of the erythrocyte displays a circadian variation. Glycolysis, the single source of energy in erythrocytes, as well as the concentrations of ATP and NAD(P)H oscillate.

5.3 Nervous Control of Circadian Rhythm

The hypothalamic suprachiasmatic nucleus[10] is the master circadian clock that is synchronized to the daily light–darkness cycle detected from retinal cues via the retinohypothalamic tract [585]. It synchronizes the phase of circadian clocks in peripheral tissues.

Daily rhythms in electrical activity, metabolism, and gene expression of the suprachiasmatic nucleus are similarly set up in both nocturnal and diurnal species. Light input from the retina entrains central circadian oscillators of the suprachiasmatic nucleus. The retina contains 3 photoreceptor types (rods, cones, and melanopsin-containing retinal ganglion cells) that send light information to the cerebral pacemaker. The intrinsic response is driven by melanopsin and the extrinsic response corresponds to rod and cone light inputs to melanopsin-containing retinal ganglion cells that project to the master clock via the retinohypothalamic tract. Extrinsic light signals from rods and cones are able to compensate for several circadian light functions, whereas cues from melanopsin-containing retinal ganglion cells are mandatory for circadian light information. Intensity-dependent retinal light-based signaling rather than neural mechanisms downstream from the suprachiasmatic nucleus determine diurnality or nocturnality, as light intensity change can produce temporal-niche switching [586].

All cell types are capable of exhibiting intrinsic circadian oscillations, as they have clock genes. Connections between circadian cells and the body's tissues include nervous and hormonal signals. Neurons of the suprachiasmatic

[10] The suprachiasmatic nucleus comprises about 20,000 neurons. Spontaneous action potentials in the suprachiasmatic nucleus drive circadian timing of behavior. Ionic fluxes control the spontaneous firing rate. Large conductance Ca^{++}-activated K^+ channel (BK) in the suprachiasmatic nucleus is regulated by the intrinsic circadian clock [584]. Cultured neurons of the suprachiasmatic nucleus exhibit daily rhythms in protein synthesis, metabolism, firing rate, and neuropeptide secretion. The suprachiasmatic nucleus is composed of bilateral nuclei.

Table 5.4. Downstream connections of the suprachiasmatic nucleus (Source: [585]). The SCN central region receives light-derived cues via the retinohypothalamic tract. The circadian-regulated hormone melatonin (^{N}acetyl 5-methoxytryptamine) is expressed in many metabolic tissues. Melatonin is secreted into the blood stream from the epiphysis to contribute to the regulation of the circadian rhythm. It is also produced by bone marrow cells, lymphocytes, and epithelial cells. It targets G-protein-coupled melatonin receptors that lodge in the brain and some peripheral organs. The MT_1 subtype, a.k.a. Mel_{1A} and MtnR1a, resides mainly in the pituitary gland and suprachiasmatic nucleus. The MT_2 subtype, a.k.a. Mel_{1B} and MtnR1b, localizes principally to the retina.

Neural region	Effect
Medial preoptic region via subparaventricular zone	Thermoregulation
Dorsomedial nucleus via paraventricular nucleus	Hormone secretion (melatonin, corticosteroids)
Lateral hypothalamus and ventrolateral preoptic nucleus	Sleep–wake cycle Feeding–fasting cycle

nucleus project to many regions for timing of hormone release, feeding behavior, and body temperature regulation (Table 5.4). The effect of the suprachiasmatic nucleus is predominantly mediated by the sympathetic nervous system and its neurotransmitter noradrenaline. In addition, the suprachiasmatic nucleus secretes circulating regulators to control rhythms in the brain and peripheral organs, such as transforming growth factor-α, cardiotrophin-like cytokine,[11] and prokineticin-2 [585].

Several brain regions (dorsal medial hypothalamus, retina, olfactory bulb, etc.) act as circadian pacemakers. Certain neurons operate as circadian pacemakers for various activities, such as locomotion, olfaction, and learning [587]. The hypothalamic dorsomedial nucleus synchronizes daily behavior with expected mealtime and regulates many food anticipatory rhythms, although it is not the food entrainable oscillator. It integrates circadian and energy information and processes sleep–wake cycle, body temperature, and certain locomotor activities. The dorsomedial nucleus is connected to the: (1) parabrachial nucleus that is required for food entrainment; (2) preoptic ventrolateral nucleus and lateral hypothalamus that regulate wakefulness; (3) subparaventricular zone, medial preoptic area, and raphe pallidus nucleus that control body temperature; (4) hypothalamic paraventricular nucleus in charge of corticosteroid secretion; and (5) suprachiasmatic nucleus.

[11] Cardiotrophin-like cytokine factor CLCF1, a.k.a. novel neurotrophin NNT1 and B-cell-stimulating factor BSF3, is a cytokine of the interleukin-6 family. It is closely related to cardiotrophin-1 and ciliary neurotrophic factor.

In the central pacemaker, genes involved in protein synthesis (from translation initiation to folding and post-translational modification) and transport are controlled by the circadian clock.

The core body temperature undergoes daily rhythms (~2-K fluctuation).[12] The suprachiasmatic nucleus of mammals that is synchronized with the daily light–dark cycle remains insensitive to temperature change, although, in many organisms, the daily cycle driven by the environmental temperature serves as a signal that participates in the synchronization of their circadian clock. However, in mammalian peripheral organs, such as lungs, pituitary, liver, kidneys, and olfactory bulb, minor fluctuations in temperature cause phase shifts in Per2 expression [588].[13]

In mammals, SCN pacemaker neurons exhibit circadian-regulated oscillations in membrane potential mediated by voltage-gated calcium channels Ca_V1 [589]. Calcium is an important signal transducer in pacemaker cells that couples membrane potential with the circadian clock. Mammalian SCN pacemaker neurons express both large- (BK or $K_{Ca}1$) and small-conductance (SK or $K_{Ca}2$) calcium-activated potassium channels under the control of the circadian rhythm. Voltage-gated sodium and calcium channels of involved neurons may impede sensitization of the suprachiasmatic nucleus to temperature variations [590]. The thermoresistance of the suprachiasmatic nucleus requires the existence of both dorsal and ventral regions of the nucleus. Intercellular communications thus permit a collective resistance.

On the other hand, temperature-dependent entrainment of peripheral clocks depends on the heat shock response pathway. A sudden temperature jump can actually induce the oligomerization of heat shock transcription factor HSF1 that primes the synthesis of heat shock proteins to stabilize proteins in stress conditions. Compounds that specifically inhibit HSF1 also prevent temperature-induced phase shifts of peripheral clocks. Besides, several clock proteins possess a heat shock element (HSE), i.e., a binding site for HSF1 [591].[14]

---

[12] Ectotherms control their body temperature via environmental heat sources. On the other hand, endotherms maintain a certain body temperature despite large variations in external temperature.

[13] Effects of the activity, sleep, feeding, and metabolic state must be separated from those of temperature. Restricted feeding regimens that alter circadian gene expression in the liver provoke a deep depression of body temperature when food is unavailable.

[14] Glucocorticoids that reset peripheral circadian oscillators cause an inhibition of the HSF1-mediated transcription [588]. Both HSP90 and HSP70 act as molecular chaperones for inactive glucocorticoid receptors. Redox state that fluctuates with the circadian rhythm also regulates HSF1 activation.

5.4 Transcriptional Regulators of the Circadian Clock

The circadian clock contains regulatory loops that rely on expression of combinations of transcriptional activators and repressors that target many clock-controlled genes. The transcriptional network of circadian clocks includes at least 3 clock-controlled DNA elements: morning (E-box/E′-box), day (D-box), and night (retinoic acid responsive element [RRE]) that are influenced by at least 11, 4, and 5 transcription factors, respectively (Table 5.5).[15] In addition to 3 basic circadian phases (morning, day, and night), circadian oscillations comprise supplementary phases (dawn, noon, dusk, and late night). Morning activation and night repression of genes drive day transcription mediated by D-box, whereas day activation and morning repression yield night transcription regulated by RRE [592].

Many disparate physiological activities (sleep–wake cycle, blood circulation, feeding, body temperature, hormonal secretions,[16] memory formation, and glucose, lipid, and mitochondrial oxidative metabolisms) follow cycles of repeated oscillations, being regulated by the body's clock, exhibiting mainly circadian rhythms that match the earth's rotation. Endogenous rhythms are coordinated with the environment via circadian clock entrainers.[17]

The body clock is controlled by an interplay of numerous molecular pathways, with feedback loops leading to gene transcription and protein translation. The clock regulator consists of both transcriptional activators and repressors (Table 5.6). Several genes are involved in the circadian rhythm, such as Clock, Bmal1, PER, CRY, and nuclear receptor-1D1 (NR1d1).[18]

Transcription factors that belong to the bHLH-PAS[19] class, the circadian locomotor output cycles kaput/neuronal PAS domain containing protein-2

---

[15] D-box is associated with transcriptional activators Dbp, Tef, and Hlf and repressor E4bp4. Dbp, Tef, and Hlf are expressed during the morning under the control of E-box, whereas E4bp4 is produced during the night under the supervision of RRE segment. Retinoic acid responsive element, responsible for night expression, is regulated by activators RORα, RORβ, and RORγ and repressors nuclear receptors NR1d1α and NR1d1β (a.k.a. Rev-erbAα and Rev-erbAβ). RORα and RORβ are expressed during the day under the control of D-box, whereas NR1d1α and NR1d1β are regulated by E-boxes.

[16] The expression of estrogen receptor ERβ is subjected to a circadian regulation that determines the cell response to estrogens [593].

[17] These external cues are also called zeitgebers (zeitgeber: time-giver, synchronizer; index of the circadian cycle time).

[18] A.k.a. Rev-Erbα and thyroid hormone receptor-α-like factor (THRAL).

[19] bHLH-PAS stands for basic helix–loop–helix–Per-ARNT-Sim (Per: Period; ARNT: aryl hydrocarbon receptor nuclear translocator; Sim: single-minded). The bHLH-PAS class (class-7 of basic helix–loop–helix proteins) of transcription factors includes regulatory proteins like hypoxia-inducible factor and endothelial PAS domain-containing protein EPAS1 that regulate the cellular response to low oxygen concentration.

Table 5.5. Bulk regulation of the circadian rhythm ($\oplus \longrightarrow$: stimulation; $\ominus \longrightarrow$: inhibition; $\downarrow$: repression; E-box: enhancer box sequence of DNA). Transcriptional activators CLOCK and BMAL1 dimerize and bind to E-box-containing gene sequences (e.g., promoters of PER [PER1–PER3] and CRY [CRY1–CRY2], among other clock-controlled genes). The synthesized proteins enter the nucleus. Once accumulated in the nucleus, transcriptional repressors Per and Cry dimerize and, after a specific delay, inhibit the expression of Clock and Bmal1 genes, thereby repressing their own production during the late afternoon. Afterward, Per and Cry are gradually phosphorylated and degraded; their inhibition is then relieved to initiate the following cycle. Two other feedback loops involve nuclear retinoid-related orphan receptors (ROR), transcriptional activators, and dimeric Reverse (Rev)-Erbα repressors that connect to RRE DNA elements (short feedback loop controlling Bmal1 transcription) on the one hand and D-box-binding protein (DBP) and E4 promoter-binding protein E4BP4 that tether to the D-Box gene sequence on the other. Peroxisome proliferator-activated receptor PPARα and coactivator PPARγ coactivator PGC1α also modulate Bmal1 transcription.

Time of the day	Involved gene sequence	Events
Morning (dawn)	E-box	BMAL1–CLOCK–E-box linkage $\oplus \longrightarrow$ Per, Cry (negative feedback loop) $\oplus \longrightarrow$ ROR, Rev-Erb $\oplus \longrightarrow$ DBP
Day (late morning, noon, and afternoon)	D-box	DBP $\oplus \longrightarrow$ D-box Per, Cry $\ominus \longrightarrow$ E-box Rev-Erbα $\ominus \longrightarrow$ RRE (BMAL1) Per, Cry synthesis $\downarrow$
Night (dusk, midnight, and late night)	RRE	$\oplus \longrightarrow$ E4BP4 E4BP4 $\ominus \longrightarrow$ D-box Progressive degradation of Per and Cry ROR $\oplus \longrightarrow$ RRE (BMAL1) (positive feedback loop) BMAL1–CLOCK $\oplus \longrightarrow$ E-box

(CLOCK/NPAS2) and brain and muscle ARNT-like-1 (BMAL1) heterodimerize to regulate the circadian clock. The transcription factor NPAS2 is a substitute for CLOCK in the suprachiasmatic circadian clock [594]. Factors CLOCK and NPAS2 can coordinately function and independently heterodimerize with BMAL1 in the suprachiasmatic nucleus to maintain rhythmicity. The CLOCK–BMAL1 heterodimer has DNA-binding ability and targets E-box motifs in gene promoters (Fig. 5.1).

Table 5.6. Main circadian transcriptional activators and repressors. Three basic circadian phases exist: morning, day, and night. Genes with promoters that have E-box (morning), D-box (day), and retinoic acid responsive element (RRE; night) sequences are regulated by the circadian clock. Transcription activators CLOCK and BMAL1 heterodimerize and induce the expression of genes for Period and Cryptochrome that contain E-box elements in their promoters.

Agent	Activity
Brain and muscle Arnt-like protein-1 (BMAL1)	bHLH-PAS transcription activator that dimerizes with CLOCK to stimulate E-box-containing promoters
Circadian locomotor output cycles kaput (CLOCK)	bHLH-PAS transcription activator that dimerizes with BMAL1 to activate E-box-containing promoters
Neuronal PAS domain protein-2 (NPAS2)	Transcription factor similar to CLOCK that is highly expressed in forebrain and dimerizes with BMAL1 to activate gene transcription
Cryptochrome proteins (Cry)	Transcriptional repressors that dimerize with Per to inhibit CLOCK–BMAL1-mediated gene transcription
Period proteins (Per)	PAS domain-containing proteins that dimerize with Cry to inhibit CLOCK–BMAL1-induced gene transcription

The rhythm molecules include the major transcriptional repressors Cryptochromes Cry1[20] and Cry2 and Periods Per1 and Per2. Heterodimers of transcription factors BMAL1[21] and either CLOCK or NPAS2 activate the production of Per and Cry proteins by PER and CRY genes. In turn, Per and Cry inhibit their own expression by repressing CLOCK–BMAL1 (negative feedback loop). Therefore, at low Per and Cry concentrations, PER and CRY are activated by heterodimers BMAL1–CLOCK and/or BMAL1–NPAS2. Once their levels reach a threshold, Cry and Per bind to BMAL1–CLOCK/NPAS2, thus forming multimers that translocate to the nucleus to subsequently inhibit transcription of their own and other clock-controlled genes, especially canceling BMAL1–CLOCK/NPAS2 activity. During the night, the Per-Cry repressor complex is degraded. The CLOCK–BMAL1 complex can then activate a new cycle of transcription.

Circadian oscillators are resilient to fluctuations in transcription rates and temperature. Rhythmicity is maintained even when the control of gene expression is impaired. The circadian rhythm also has a compensatory mechanism

[20] Transcriptional repressor Cry1 is expressed during evening.

[21] A.k.a. aryl hydrocarbon receptor nuclear translocator-like factor.

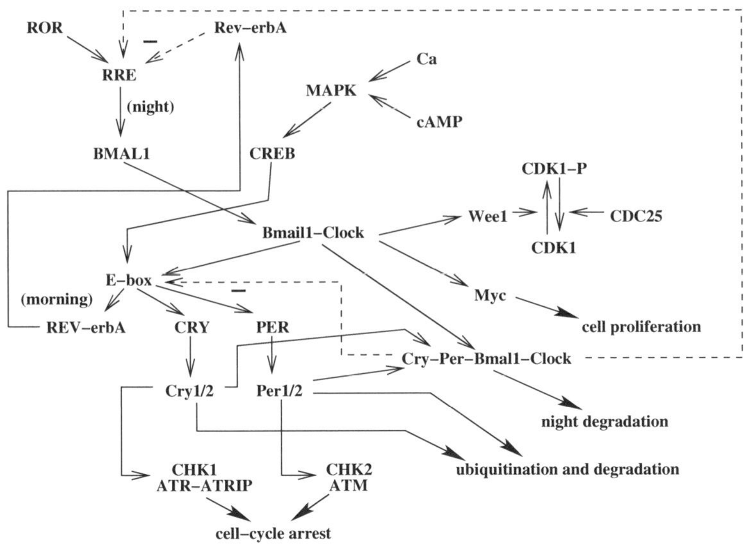

Figure 5.1. Circadian clock and its feedback loops (Source: [585]). The circadian clock involves a primary negative feedback loop that includes circadian locomotor output cycles kaput (CLOCK) and its paralog neuronal PAS domain protein-2 (NPAS2), brain and muscle ARNT-like-1 (BMAL1), Period homologs Per1 and Per2, Cryptochrome Cry1 and Cry2. Per and Cry form a heterodimer that translocates to the nucleus and interacts with the CLOCK–BMAL1 complex to inhibit transcription of their own and other clock-controlled genes. Afterward, Per–Cry repressor complex is degraded and CLOCK–BMAL1 complex can then activate a new cycle of transcription. Another feedback loop is composed of NR1d1 that is expressed upon stimulation by CLOCK–BMAL1 complex. Nuclear receptor NR1d1 represses Bmal1 gene transcription, thereby competing with retinoic acid-related orphan receptor (ROR) that binds to ROR response elements (RRE) in Bmal1 promoter. The CLOCK–BMAL1 complex also regulates the transcription of myelocytomatosis oncogene (Myc) and Wee1. Protein Wee1 phosphorylates (inactivates) CDK1 that causes G2 arrest. Cyclin-dependent kinase-1 is dephosphorylated by cell division cycle-25 for cell cycle progression from G2 phase to mitosis. Protein Per1 interacts with ataxia telangiectasia mutated protein (ATMK) and checkpoint kinase-2 and Cry with checkpoint kinase-1 and the complex formed by ataxia telangiectasia and Rad3-related kinase (ATRK) and its interacting protein (ATRIP). Post-translational modification and degradation of circadian clock proteins also regulate the circadian clock. Proteins Per and Cry are phosphorylated for polyubiquitination and degradation. Cyclic adenosine monophosphate (cAMP) and calcium (Ca) determine amplitude, phase, and period of the transcriptional feedback loops of the circadian pacemaker via mitogen-activated protein kinase (MAPK) and cAMP response element binding (CREB).

when temperature change occurs in the physiological range. Factor Per1 intervenes in transcription and temperature compensation [595]. Cultured mouse fibroblasts, but not cells lacking Per1, exhibit a 26-h period at 310 K (37 °C) and a 24-h period at 304 K (31 °C).

Furthermore, NR1d nuclear orphan receptors, retinoid orphan receptors, and Differentially expressed in chondrocytes factors (DEC)[22] operate in the rhythmic activity of the pacemaker [593]. Nuclear hormone receptor NR1d1 that is targeted by the BMAL1–CLOCK complex and represses Bmal1 transcription constitutes a second (subsidiary) negative feedback loop to add robustness to the circadian clock.

The circadian clock is governed not only by transcriptional and post-translational feedback loops that control periodical activity of circadian genes, but also by small molecules, such as cyclic adenosine monophosphate and calcium that determine amplitude, phase, and period of the transcriptional feedback loops of the circadian pacemaker [596]. Concentration of cAMP not only is regulated by the circadian pacemaker, but also influences the activity of the pacemaker. Agent cAMP hence is a component of the central oscillator. Rhythmic cAMP signaling associated with the transcriptional oscillator sustains progression of the transcriptional timing of the suprachiasmatic nucleus as well as peripheral cells. In addition, microRNAs and RNA-binding proteins that influence RNA stability also participate in core clock timing.

5.5 Modifications and Degradation of Circadian Clock Proteins

Post-translational modifications (acetylation, methylation, phosphorylation, sumoylation, and ubiquitination) regulate subcellular location, stability, activity, and turnover of circadian clock proteins. Robustness of feedback loops of the circadian clock increases due to post-translational modifications of Per and Cry repressors, especially reversible protein phosphorylation governed by kinases and phophatases. Regulator BMAL1 is acetylated, phosphorylated, and sumoylated. These distinct post-translational modifications can finely tune the regulation of the circadian clock.

Mammalian circadian clock proteins undergo a daily cycle of accumulation followed by phosphorylation and then degradation. Phosphorylation of circadian clock proteins thus contributes to circadian timing. The accuracy of the duration of the circadian phases depends, at least partly, on the regulation of nuclear import and export of clock components. Changes in subcellular localization of clock components can result from phosphorylation. Protein phosphorylation (Table 5.7) in circadian negative feedback is not only required for protein degradation, but also nucleocytoplasmic transport [597].

[22] Proteins DECs belong to the class-B of basic helix–loop–helix domain-containing transcription factors. They are potent transcriptional repressors that bind to DNA at E-box.

Table 5.7. Phosphorylation of components of the circadian clock (Source: [598]; CamK: calcium–calmodulin-dependent kinase; CK: casein kinase; GSK: glycogen synthase kinase; MAPK: mitogen-activating protein kinase; PKA(C): protein kinase-A(C)). Protein fate depends on the phosphorylation site. In Drosophila, CamK2 phosphorylates CLK, the CLOCK ortholog.

Circadian factor	Kinase
	Positive components
BMAL1	CK2α
CLOCK	(nuclear translocation of CLOCK, BMAL1)
	PKA (CREB-induced Per1 activation)
	PKCα
	MAPK
	Negative elements
Per1–Per3	CK1δ, CK1ϵ
Cry1–Cry2	(nuclear localization of Per1, Per2)
	(Per2 degradation)
	GSK3β
	(nuclear localization of Per2)

In particular, Per2 is destabilized when it is phosphorylated by casein kinases. Conversely, Per2 may be stabilized by binding protein phosphatase-1. Post-translational modifications and degradation of circadian clock proteins also determine circadian periodicity. Per and Cry concentrations actually reach repeatedly low values due to feedback loops and also degradation; a new cycle of Cry and Per production then occurs. In addition, the Per–Cry complex that translocates to the nucleus is resistant to ubiquitin-dependent degradation.

5.5.1 Mitogen-Activated Protein Kinases

Mitogen-activated protein kinase activated by the Ras-Raf1-MAP2K pathway phosphorylates (inhibits) BMAL1, a positive regulator of the autoregulatory feedback loop of the circadian oscillator, to repress the CLOCK–BMAL activity [599]. In addition, the Ca^{++}-sensitive cAMP–MAPK–CREB cascade may be involved. A cAMP-responsive element (CRE) site indeed exists in the promoter of PER genes that binds CRE-binding protein (CREB).

5.5.2 Casein Kinases

Levels of Per and Cry depend on phosphorylation by casein kinases CK1δ and CK1ϵ,[23] and then on sumoylation.

[23] Casein kinases CK1ϵ and CK1δ are encoded by genes CSNK1D and CSNK1E, respectively. Mutations in genes PER2 and CSNK1D cause the familial advanced

Phosphorylation of circadian clock proteins leads to polyubiquitination and degradation by 26S proteosome. Cry forms a complex with F-box and leucine-rich repeat-containing protein FBxL3, a subunit of the ubiquitin ligase complex SKP1–cullin–F-box (SCF), for degradation by proteasomes [600]. A mutation in FBxL3 induces longer circadian rhythms of about 27 h in homozygotes, delaying Cry degradation rate and prolonging the duration of negative feedback [601].

The Per2 degradation rate that is regulated by CK1δ and CK1ϵ is also subject to chemical perturbations, but not temperature. Some pharmacological compounds can affect CK1δ- and CK1ϵ-dependent phosphorylation and lengthen the period of the clock from circadian to circabidian (48 h) [602].

Casein kinase CK2 is constituted by catalytic CK2α and regulatory CK2β subunits. Casein kinase CK2 can be considered as a holoenzyme, as CK2α monomer exists as an active form in vivo. Casein kinase CK2α alone phosphorylates circadian regulator BMAL1 (Ser90), whereas CK2β precludes BMAL1 phosphorylation in a dose-dependent manner [603].

5.5.3 Protein Kinase-C

Receptor for activated C kinase RACK1, a scaffold that recruits protein kinase-Cα and its substrates, causes association of some proteins with BMAL1 at the time of day when CLOCK–BMAL1 transcription activator is inhibited [604]. Both RACK1 and PKCα are recruited into a nuclear BMAL1 complex during the negative feedback phase of the cycle. Scaffold RACK1[24] stimulates phosphorylation of BMAL1 by PKCα. Therefore, the Ca^{++}-sensitive PKC pathway not only relays external signals, but also participates in the circadian feedback loop.

5.5.4 Ubiquitination

F-box and WD repeat-containing proteins FBxW1[25] and FBxW11,[26] also called β-transducin repeat-containing proteins βTrCP1 and βTrCP2, are subunits of the SCF complex (SCF–βTrCP). This ubiquitin ligase complex binds to Per1 and Per2 for degradation (robustness) [605]. βTrCP1 and βTrCP2 are located in the nucleus and cytosol, at least in certain cell types, respectively. Their activity differs according to their residence site. Moreover, although βTrCP1 and βTrCP2 have similar targets (Per1 and Per2, as well as IκBα, NFκB1, and CDC25a), they also link to distinct proteins.

---

sleep-phase syndrome. Factors CLOCK and BMAL activate the transcription of PER and CRY genes. Factors Per and Cry heterodimerize in the cytoplasm following phosphorylation and enter the nucleus, where they inhibit CLOCK–BMAL transcriptional activation of Period (Per) and Cryptochrome (Cry) genes.

[24] A.k.a. guanine nucleotide-binding (G) protein-β polypeptide 2-like 1 (GNB2L1).

[25] A.k.a. FBxW1a.

[26] A.k.a. FBxW1b and Homologous to Slimb protein (HOS).

5.5.5 Histone Post-Translational Modifications

Circadian transcription of Clock and Bmal1 genes is associated with circadian changes in histone-H3 acetylation and methylation, as well as chromatin remodeling. Protein CLOCK is both a transcription activator and an enzyme. CLOCK indeed is a histone acetyltransferase that acetylates not only histone-H3, but also non-histone substances such as its dimerization partner BMAL1. Histone deacetylase NAD^+-activated sirtuin Sirt1 transduces signals originating from cellular metabolites to the circadian clock circuitry [606]. Sirt1 contributes to circadian control via histone-H3 acetylation. Sirt1 is recruited to the Clock–Bmal1 chromatin complex at circadian promoters. Moreover, Sirt1 is required for strong circadian transcription of several core clock genes (Bmal1, RORγ, PER2, and CRY1) [607]. Sirt1 linked to the CLOCK–BMAL1 complex under circadian control promotes deacetylation and degradation of Per2.

Chromatin remodeling with different types of post-translational modifications (acetylation, methylation, phosphorylation, and ubiquitination)[27] is needed for gene transcription. Circadian gene transcription is assisted by histone acetylation. Conversely, histone methyltransferase mixed lineage (myeloid–lymphoid) leukemia MLL1[28] and its product histone $H3K_4me^3$ undergo circadian cycle at clock-responsive promoters [608].

5.6 Other Circadian Clock Regulators

Circadian clock genes are regulated by glucocorticoids. Conversely, their secretion from the adrenal gland bears a circadian rhythm. Multiple clock genes possess a glucocorticoid response element (GRE). Among them, several genes such as PER2 are directly regulated by the glucocorticoid receptor [609].

The master clock in the suprachiasmatic nucleus can synchronize peripheral clocks (Sect. 5.7). However, in many organs, the feeding–fasting rhythm yields the dominant timing signal (zeitgeber). Nutrient-responsive adenosine monophosphate-activated protein kinase triggers phosphorylation (Ser71 and Ser280) of the clock component Cryptochrome-1 and its subsequent degradation. Enzyme AMPK is phosphorylated (activated) by protein kinases, such as liver kinase-B1 (LKB1) and calcium–calmodulin-dependent protein kinase kinase-β. In mouse livers, AMPK activity and nuclear localization are rhythmic and inversely correlated with Cry1 abundance in the nucleus [610]. Thus, cryptochrome serves as partner to chemical energy sensors in mammals that can transduce nutrient signals to the body's clocks.

[27] In general, acetylation fosters transcriptional activation. On the other hand, methylation of histones leads to activation or repression, depending on the sites of modification. Lys4 (K4)-based trimethylation of histone-H3 triggers transcriptional activation.

[28] Mammalian homolog of Drosophila trithorax.

Melanoma antigen family member-D1 (MAGeD1)[29] binds to the RAR-related orphan receptor RORα (NR1f1) to regulate core clock genes, such as Bmal1, REV-ERBα (NR1D1), and interleukin-3-regulated nuclear factor (NFIL3),[30] via Rev-Erbα–ROR responsive elements (RRE) [611]. The circadian regulator MAGeD1 is a non-rhythmic protein that enhances rhythmic input and removes irrelevant, perturbing noise from the circadian clock.

5.7 Peripheral Circadian Clocks

Circadian clock genes are ubiquitously expressed. Circadian oscillators consist of positive and negative transcriptional feedback loops. The majority of the body's cells exhibit cell-autonomous, self-sustained, circadian oscillations in protein synthesis. Peripheral oscillators are entrained by the SCN pacemaker for a coordinated activity, but can act independently of the suprachiasmatic nucleus. Although they are synchronized by the master clock, they are weakly coupled to the suprachiasmatic nucleus. The suprachiasmatic nucleus resets rapidly to shifts in light cycle, but resynchronization of peripheral oscillators takes a much longer time.

Peripheral circadian oscillators are controlled by signals arising from central and proximal signals. The suprachiasmatic nucleus serves as a coordinator by centrally regulating rhythms of body activity, feeding, and temperature to synchronize the phase of peripheral oscillators. Circadian behavioral rhythms are regulated by the central pacemaker, but tissue-specific rescue can attenuate pathophysiological consequences of BMAL1 deficiency [585]. Disruption of circadian clocks in hepatocytes eliminates cycling of the majority of genes for which hepatocyte control is dominant, but other genes (e.g., PER2, Nocturnin, and members of the heat shock protein family) continue to respond to rhythmic signals generated by the master clock.

Circadian clock genes influence multiple processes by: (1) rhythmically regulating pathway components and (2) directly intervening in cell activities under circadian regulation, such as cell metabolism, cell divison cycle, and DNA-damage control. Conversely, cell metabolism as well as environmental modulators of metabolic state, such as diet type, influence circadian clocks. Circadian clocks and metabolism are linked to adapt to the periodic environment.

At least 50 different transcriptional regulators are under circadian regulation. The liver, among other peripheral organs, can be considered as a damped circadian oscillator. Liver and heart clocks are delayed by 4 to 6 h relative to the SCN clock. Few clock-controlled genes, such as NR1d1 and NR1d2 that encode orphan nuclear receptors (transcription repressors), are common in

---

[29] A.k.a. neurotrophin receptor-interacting MAGe homolog.

[30] A.k.a. E4 promoter-binding protein E4BP4.

different tissues [612]. Moreover, few clock-controlled genes are strictly regulated by the circadian clock, as most are probably controlled by numerous pathways.

5.8 Circadian Clock and Metabolism

The circadian molecular clock genes coordinate the regulation of behavior, energy balance, and metabolism. The RNA deadenylase *nocturnin*, an effector of the clock-controlled genes, participates in the coregulation of nutrient uptake, metabolism, and storage [613]. Nocturnin expression strongly increases in the early evening hours in the liver. The rhythmic transcription of nocturnin in several peripheral tissues modulates lipid accumulation in the liver. In addition, levels of metabolic hormones (glucagon, insulin, ghrelin, leptin, and corticosterone) as well as multiple metabolism-associated receptors, carriers, enzymes, and cofactors (e.g., glucagon receptor, glucokinase, glucose transporter, glucose 6-phosphate transport protein, pyruvate kinase, pyruvate dehydrogenase, etc.) show circadian oscillations.

The circadian control of energy metabolism involves peroxisome proliferator-activated receptor-γ coactivator PGC1α that is required for glucose, lipid and energy homeostasis. Transcription coactivator PGC1α is stimulated by many environmental cues (temperature, nutritional status, and physical activity). Factor PGC1α is rhythmically produced by the liver and skeletal muscles in mice [614]. Factor PGC1α connects the activity of circadian clock genes to the regulation of energy metabolism via nuclear receptors RORα of the RAR-related orphan nuclear receptor (ROR) family. It is rhythmically expressed and deacetylated by Sirt1 in an NAD^+-dependent fashion. Activity of SIRT1 also oscillates.

Coupling between circadian rhythms and periodic changes in metabolic activity involves heme, an iron-containing porphyrin that serves for enzymes involved in oxidative metabolism and transcription factors of the circadian rhythm regulation. Orphan nuclear receptor NR1d1 (or Rev-ErbAα) that controls transcription of the gene encoding BMAL1 acts as a heme sensor to coordinate the cellular clock and glucose and energy metabolism in hepatocytes [615]. Heme reversibly binds circadian core clock repressor NR1d1 to regulate its interaction with a nuclear receptor corepressor complex. Heme also suppresses hepatic gluconeogenic gene expression via NR1d1 repressor.

The liver adjusts the body's metabolism to daily feeding–fasting cycles by appropriate expression of many liver genes involved in the metabolism of lipids, proteins, and carbohydrates. Moreover, rate-limiting steps in hepatocyte metabolism are regulated by circadian clocks. Glucose plasma concentration must be kept nearly constant whatever the time of the day to provide a constant source of fuel at least for neurons and erythrocytes.[31] Inactivation

[31] During the absorptive phase, blood glycemia is adjusted by insulin. During the postabsorptive phase, glucose is produced in the liver by glycogenolysis or

of BMAL1 specifically in hepatocytes is associated with normal rest–activity and feeding–fasting rhythms, as the master clock remains fully functional, but maintains GluT2 expression at constitutively low levels, thereby causing hypoglycemia despite adequate gluconeogenesis and glycogenolysis rates [616].[32] On the other hand, BMAL1 deficiency in all cells does not significantly disturb glucose blood level, as arrhythmic organisms ingest small food amounts throughout the day that balance the absence of adaptive GLUT2 production. Loss in liver rhythm rather than BMAL1 disruption generates impaired glucose homeostasis. Therefore, in addition to acute hepatic responses to circulating glucose, the hepatic circadian clock at least maintains a daily rhythm of hepatic glucose export.

In mice, both food availability and the temporal pattern of feeding determine the repertoire, phase, and amplitude of the circadian transcriptome in the liver. During prolonged fasting, only a small subset of transcripts continues to display a circadian pattern [617]. In addition, temporally restricted feeding restores rhythmic transcription of numerous genes in oscillator-deficient liver.

Glucagon is the main regulatory hormone during fasting that maintains normal blood glucose concentration by stimulating hepatic gluconeogenesis, the process of glucose synthesis under fasting. In hepatocytes, glucagon binds to glucagon receptors and triggers the cAMP pathway, hence phosphorylation of cAMP-responsive element-binding protein and dephosphorylation of the CREB-regulated transcription coactivator CRTC2 to prime the transcription of gluconeogenic genes that encode phosphoenolpyruvate carboxykinase (PEPCK) and glucose 6-phosphatase. In addition, hepatic gluconeogenesis is also regulated by the circadian clock [618]. Glucagon exerts its strongest action in the morning, when Cry concentration is high. Activity of CREB during fasting is modulated by Cry1 and Cry2 Cryptochromes. Phosphorylation of CREB in response to glucagon decays due to the inhibition by Cryptochrome repressor at the level of plasmalemmal $G\alpha_s$-coupled receptors, thereby precluding GPCR-stimulated cAMP production [618]. Therefore, Cryptochrome in the liver controls glucose levels during fasting and facilitates insulin sensitivity. Insulin is also rhythmically produced by pancreatic β cells.

In mitochondria, NAD^+ is used in oxidation–reduction reactions that maintain energy balance in cells. Agent NAD^+ operates as a hydrogen carrier and a coenzyme for dehydrogenases involved in ATP-generating oxidative phosphorylation. Molecule NAD^+ also serves as a substrate for ADP ribosylation. The circadian clock influences NAD^+ synthesis. As circadian

---

gluconeogenesis and released into the blood stream via GluT2 transporter, GluT2-mediated export being the rate-limiting step.

[32] Hepatocyte oscillators control the cyclic expression of glycogen synthase, glycogen phosphorylase, phosphoenolpyruvate carboxykinase, glucokinase, glucose-6-phosphatase, and GluT2 glucose transporter. Transporter GluT2 devoted to glucose export has a peak expression during the fasting phase, whereas carriers and enzymes involved in glucose import and storage (e.g., glucokinase) are highly produced during the feeding phase.

Table 5.8. The circadian clock and metabolism according to time of day (Source: [620]; ↑: increase; ↓: decrease).

	Wake and feeding	Sleep and fasting
Nervous system	Sympathetic tone ↑	Melatonin secretion ↑ Growth hormone secretion ↑
Adipose tissue	Lipogenesis ↑ Adiponectin secretion ↑	Lipid catabolism ↑ Leptin secretion ↑
Liver	Glycogen synthesis ↑ Cholesterol synthesis ↑ Bile acid synthesis ↑	Gluconeogenesis ↑ Glycogenolysis ↑ Mitochondrial genesis ↑
Muscle	Fatty acid uptake ↑ Glycolysis ↑	Oxidative metabolism ↑
Pancreas	Insulin secretion ↑	Glucagon secretion ↑
Adrenal gland	Glucocortocoid ↑	

clock mediators regulate transcription and translation of NAD^+ production enzymes, they participate in cellular metabolism [619]. Whereas NAD^+ precursor nicotinamide inhibits sirtuin-1, NAD^+ activates sirtuin-1. Sirtuin-1 regulates production of NAD^+ synthesizing enzyme nicotinamide phosphoribosyl transferase (NAMPT) in association with the CLOCK–BMAL1 dimer. Therefore, NAD^+ conjoins the circadian clock and cell metabolism. Furthermore, circadian rhythms can be controlled by cell metabolism. For example, cyclic ^{ADP}ribose that is produced from NAD^+ by ADP-ribosyl cyclase modulates circadian oscillations in calcium release via ryanodine receptors.

The circadian clock crosstalk with metabolic transcription networks via nuclear hormone receptors or directly via clock activators or repressors [620] (Table 5.8). In particular, Per2 directly links to promoters of certain metabolic genes. In addition, Cry repressor modulates gluconeogenesis via glucagon and inhibition of cAMP signaling. The circadian clock also regulates triglyceride packaging into chylomicrons that transport dietary lipids. In the liver, the circadian clock via its mediator nocturnin prevents lipid accumulation. On the other hand, nuclear hormone receptors modulate the circadian clock according to changes in environmental conditions, especially the nutritional status. Variations in the concentration of glucocorticoid hormone, retinoic acid, heme, and fatty acids affect glucocorticoid, retinoic acid, Rev-Erbα, and peroxisome proliferator-activated receptors, as well as neuronal PAS domain-containing protein NPAS2 [620].

5.9 Circadian Control of the Renal Function

Renal excretion of water and electrolytes exhibits a circadian rhythm. Secretion and reabsorption capacities of the distal nephron segments, such as distal convoluted tubule, connecting tubule, and cortical collecting duct undergo circadian changes [621]. These segments possess an intrinsic circadian control system. The circadian clock of nephron cells relies on circadian core clock genes (Clock, Bmal1, Npas2, PER, CRY, Nr1d1) and clock-controlled transcriptional factors, such as *albumin D-box-binding protein* (DBP), *hepatic leukemia factor* (HLF), and *thyrotrophic embryonic factor* (TEF). In particular, the renal circadian clock influences the production of regulators of water or sodium balance, such as vasopressin receptor V_2, aquaporin-2 and -4, and epithelial sodium channel (ENaC).[33]

Furthermore, secretion of hormones that control the transport of water and electrolytes in the distal nephron and collecting duct also depends on the circadian clock. The adrenal clock controls circadian rhythm of glucocorticoid synthesis and secretion [622]. Aldosterone that regulates sodium reabsorption in the nephron shares a common synthesis pathway with glucocorticoids. Aldosterone secretion may thus also experience a circadian rhythm.

5.10 Circadian Control of the Cell Cycle

The cell cycle (Chap. 2) undergoes circadian modulation. Replication of DNA is restricted to night [585]. The circadian clock controls the transcription of transcription factor Myc (myelocytomatosis oncogene homolog) and Wee1 kinase. Transcription of Wee1 is influenced by the CLOCK–BMAL1 complex. In regenerating liver, Wee1 expression is coregulated with Per1. The cell cycle entry into M phase is suppressed during the day when the transcription of Per and Wee1 is high. Protein Wee1 phosphorylates (inactivates) CDK1 that causes G2 arrest.[34] The activity of CcnB1–CDK1 complex is also regulated by the circadian clock.

The circadian clock regulates the cell division cycle not only by controlling gene expression, but also by post-translational modifications. Protein Per1 interacts with checkpoint proteins ataxia telangiectasia mutated (ATMK) and checkpoint kinase-2 and Cry and protein Timeless with checkpoint kinase-1 and the complex formed by ataxia telangiectasia and Rad3-related kinase (ATRK) and its interacting protein (ATRIP).

[33] In the cortical collecting duct, the maximal expression of mRNA for vasopressin receptor V_2 happens at zeitgeber time ZT20 and that of aquaporin-2 and -4 at ZT18, respectively. In distal convoluted and connecting tubules, the expression of vitamin-D receptor and sodium–calcium exchanger (SLC8a1) transcripts reaches its maximum at ZT20 and ZT23, respectively [621].

[34] Cyclin-dependent kinase-1 is dephosphorylated by cell division cycle-25 for cell cycle progression from G2 phase to mitosis.

The central circadian clock in the hypothalamic suprachiasmatic nucleus that controls behavioral rhythm is reset by light signals mediated by Per1 and Per2. Peripheral clocks are reset according to feeding and humoral signals. The circadian clock is also reset by external time cues that are independent of light signals for synchronization to environmental changes. Transforming growth factor-β activates activin receptor-like kinase to reset cellular clock via SMAD3 and the gene Dec1 [623]. Deleted in esophageal cancer DEC1 operates independently or in cooperation with PER for clock entrainment. As it acts on E-box gene, transcriptional activator and suppressor DEC1 upregulates the CRY1 gene expression and downregulates NR1d1 expression.

5.11 Circadian Rhythm Influence on Stem Cells

Circadian activity regulates the exit and entry of hematopoietic stem cells (Vol. 5 – Chap. 2. Hematopoiesis) out of and into the bone marrow, i.e., the transfer into and out of the flowing blood (circulating blood-cell progenitors peak ~5 h after light initiation and nadir, ~5 h after darkness in mice).

Many molecules control the motion of hematopoietic stem cells between the bone marrow and blood stream, such as granulocyte colony-stimulating factor (CSF3) that is able to mobilize hematopoietic stem cells (colony-forming units and SCA1+, SCFR+ cells) toward circulating blood via CXCL12 proteolysis and suppression of osteoblast activity.

The chemokine CXCL12 that sequesters blood-cell precursors into the bone marrow experience cyclic increases and decreases in concentration. Concentration of CXCL12 in the bone marrow niches indeed oscillates with periods of light and dark. Circulating hematopoietic stem cells and their derived progenitors fluctuate in antiphase with CXCL12 expression [624]. When CXCL12 diminishes and fails to anchor hematopoietic stem cells to the bone marrow, a small number of these blood-cell precursors are able to enter into the blood circulation.

Hematopoietic stem cell niches[35] are innervated by the sympathetic nervous system, thereby being controlled by adrenergic neurotransmitter noradrenaline.[36] Noradrenaline activates β2-adrenergic receptors on osteoblasts and β3-adrenergic receptors of stromal cells in the bone marrow. Osteoblasts

[35] In the bone marrow, hematopoietic stem cell resides in specific microenvironments. They are associated with sinusoids and nerve terminals, as well as with osteoblasts and CXCL12+ reticular cells. In humans, CD146+ (CD146: melanoma cell adhesion molecule) stromal cells that are precursors of both sinusoidal adventitial reticular cells and osteoblasts share similarities with mouse CXCL12+ reticular cells. Efferent nerve terminals are connected by gap junctions (Vol. 1 – Chap. 7. Plasma Membrane) to sinus adventitial reticular and other stromal cells of the bone marrow to form neuroreticular complexes.

[36] Lipopolysaccharides also are mobilizing agents of hematopoietic stem cells, as they induce noradrenaline release by bone marrow-resident macrophages.

that have β2-adrenoceptors (but not β3-adrenoceptors)[37] possess clock genes that inhibit bone formation. Activated β3-adrenoceptors of osteoblasts enhance hematopoietic stem cell egress. Activated β3-adrenoceptors of bone marrow stromal cells impede phosphorylation of transcription factor SP1 that binds to Cxcl12 gene promoter and thus reduces concentration of CXCL12 anchor.[38] Transient increase in circulating hematopoietic stem cells anticipates their occupancy in newly remodeled niches.

Hematopoietic stem cell circulation can then be synchronized with circadian remodeling of hematopoietic stem cell niches, as skeletal mass regulation by hormones, such as parathyroid hormone and leptin, and bone formation also bear a circadian rhythm.

5.12 Circadian Rhythm and Immune System

The immune system exhibits a rhythmic activity under the control of the circadian clock. In the immune system, lymphocyte proliferation, natural killer cell activity, humoral immune response, density of circulating leukocytes, cytokine concentration, and cortisol undergo a circadian rhythm. Susceptibility to infection and course of asthma also experience a daily variation. Both systemic pathways and cell-autonomous operators can relay timing information. On the one hand, circadian-controlled humoral factors such as cortisol and innervations by the autonomic nervous system may regulate the gene expression. On the other, local clocks in immune cells can directly control cell immune functions.

At least in mice, macrophages of the spleen, lymph nodes, and peritoneum display an intrinsic circadian rhythm that operates autonomously even ex vivo, as demonstrated by TNFα and IL6 secretion by stimulated macrophages [625]. More than 8% of the macrophage transcriptome oscillates in a circadian fashion, particularly regulators for pathogen recognition and cytokine secretion. Hence, the strength of pro-inflammatory cytokine production of macrophages in response to bacterial endotoxin is determined by the circadian phase of the macrophage clock rather than by bulk circadian modulators such as cortisol. However, systemic time signals such as glucocorticoids, melatonin, or catecholamines can intervene to synchronize or shape the immune response.

Lipopolysaccharides can also cause dephosphorylation and degradation of transcription factor Sp1 that activates CXCL12 expression.

[37] Fibroblastic reticular and preosteoblast cell lines express both ARβ2 and ARβ3 (Vol. 3 – Chap. 7. G-Protein-Coupled Receptors).

[38] Phosphorylation of transcription factor SP1 by protein kinase-A elicits its DNA binding. β3-Adrenoceptors can couple to both Gs and Gi/o proteins (Vol. 4 – Chap. 8. Guanosine Triphosphatases and their Regulators), thereby increasing and decreasing intracellular cAMP level and, subsequently, inducing or preventing PKA activity, respectively.

In macrophages, the circadian clock controls many functions, such as phagocytosis and antigen presentation. The genes that bear a circadian regulation encompass those that [625]: (1) encode chaperonins involved in the stress response (e.g., HSPa1, HSPa5, HSPc1, HSPd1, HSP110);[39] (2) factors of immune regulation (protectin [or CD59], C-type lectin CLec2c [or CD69], T-lymphocyte activation coregulator CD86, membrane glycoprotein CD200 receptors CD200R1 and CD200R4);[40] (3) components involved in phagocytosis, such as vesicle-associated membrane protein VAMP8 (or endobrevin); members of the solute carrier superclass (SLC2a9, i.e., glucose transporter GLuT9; SLC7a8, i.e., large neutral amino acid transporter subunit LAT2; SLC9a8 and SLC9a9, i.e., sodium–hydrogen exchangers NHE8 and NHE9; SLC9a3R2 regulator, or NHERF2; organic cation carrier SLC22a15; mitochondrial citrate carrier SLC25a1; SLC27a1, or fatty acid transport protein FATP1; SLC29a1, or equilibrative nitrobenzylmercaptopurine riboside-sensitive nucleoside transporter ENT1; SLC39a1, a.k.a. zinc–iron-regulated transporter-like protein ZIRTL and zinc transporter ZIP1; and MgtE-like magnesium transporter SLC41a3); immune receptor TLR1; lectins (C-type lectins CLec2i, Clec4d, and CLec5a; mannose-binding lectins LMan1 and LMan2; galactoside-binding, soluble lectin LGalS9; and sialic acid-binding Ig-like lectin SIgLec1); and integrins (α_5 and αFC GAP repeat-containing integrin [ItFG]);[41] and (4) compounds involved in wound healing and extracellular matrix homeostasis, such as MMP9 and prolyl 4-hydroxylase, α polypeptides P4Hα1 and -2.

[39] Chaperonins are proteic complexes that assist the folding of nascent polypeptides into functional molecules.

[40] Protectin is also called 20-kDa homologous restriction factor HRF20 and membrane attack complex inhibitory factor (MACIF). This GPI-anchored glycoprotein is expressed on all leukocytes, erythrocytes, endothelial cells, neurons, Schwann cells, glial cells, ependymal cells, and certain epithelial cells, such as nephron cells and bronchial epithelial cells [626]. Protein CD59 prevents the formation of a membrane attack complex formed by activated terminal complement proteins C5b to C9 to protect the cell from complement-mediated lysis. Agent CD69 (a.k.a. Activation inducer molecule [AIM], early antigen EA1, early T-cell activation antigen P60, and very early activation [VEA]) is expressed on activated T and B lymphocytes, macrophages, platelets, neutrophils, and natural killer cells. Agent CD86 (a.k.a. B-lymphocyte activation antigen B7-2 and CTLA4 counter-receptor B72), is expressed on B and T lymphocytes, dendritic cells, endothelial cells, fibroblasts, fibrocytes, eosinophils, monocytes, macrophages, NKT cells, pneumocytes, among others. Membrane glycoprotein CD200 targets its receptors CD200R1 and CD200R2 (or CD200R1L) to inhibit hematopoetic cells, especially myeloid cells. The CD200–CD200R complex also controls the activity of monocytes and macrophages. Receptors of CD200 are inhibitors of myeloid cells via docking protein DOK2 and activator RasGAP [627].

[41] Integrin ItFG1 is also called T-cell immunomodulatory protein (TIP).

5.13 Circadian Cycle Disorders

Circadian rhythms in different body tissues can lose synchrony. In addition, travelers can bear jet lag and have difficulty resetting the circadian rhythm. Special chronotypes include larks (morning-oriented, with activity from ~ 4 to ~ 19h) and owls (evening-oriented, i.e., individuals who wake up and go to sleep late). Larks and owls that follow normal schedules experience insomnia and sleepiness as well as possible health problems. Circadian cycle disturbances generated by genetic or environmental insults indeed cause diverse metabolic and behavioral disorders. A subset of insomnias (advanced and delayed sleep phase syndromes, non-24-hour sleep–wake syndrome, and irregular sleep–wake pattern) is linked to circadian alterations.

Familial advanced sleep-phase syndrome (FASPS) is caused by mutations in gene PER2 that increases nuclear PER2 turnover [585]. Delayed sleep-phase syndrome can result from abnormal PER3 gene. Clock mutations could be associated with obesity and metabolic syndrome and Bmal1 mutations to hypertension and type-2 diabetes. Circadian gene mutations can cause other behavioral dysfunctions [585]. Per2 mutation enhances alcohol consumption. Mutation of the Npas2 gene is implicated in seasonal affective disorder.

Because circadian clocks are cell autonomous and distributed throughout the body, both central and peripheral circadian oscillators can be targeted by therapy. In addition, toxicity of anti-cancer agents varies with administration time.

Mathematical models of the regulation of the circadian clock and cell division cycle by several key dynamical pathways address the control of cell proliferation, especially tumor growth. The circadian clock that regulates cell metabolism, growth, and DNA-damage control modulates responses to drugs. Mathematical models that incorporate drug metabolism aim at predicting optimal chronotherapy to tailor drug delivery according to body rhythms.

5.14 Modeling of the Circadian Rhythm

The circadian clock regulates the gene expression and protein activation in a periodic manner. Oscillating systems are characterized by their period, phase, and amplitude. Negative feedback loops control these properties of the oscillator.

The sensitivity analysis aims at studying the dependence of steady-state behavior on internal and external conditions. The stability analysis aims at characterizing long term behavior (bistability, oscillations, etc.). The bifurcation analysis aims at investigating the dependence of the dynamical behavior on internal and external conditions.

The 3-state, 6-parameter Goodwin oscillator is the simplest model. It consists of a negative feedback loop within a single gene expression pathway, i.e., describes the regulation of a translated protein that inhibits its own transcription. It is represented by a set of non-linear ordinary differential equations for concentrations of mRNA ([mRNA]), a circadian clock protein ([P]), and transcriptional repressor ([R]):

$$\begin{aligned}\frac{d}{dt}[mRNA] &= \kappa_1 \frac{1}{1+[R]^p} - \kappa_2[mRNA],\\ \frac{d}{dt}[P] &= \kappa_3[mRNA] - \kappa_4[P],\\ \frac{d}{dt}[R] &= \kappa_5[P] - \kappa_6[R]. \end{aligned} \tag{5.1}$$

Modeling of the circadian rhythm relies on kinetic equations for the major involved transcription factors (BMAL1, CLOCK, Cry, or Per; concentration $[TF_x]$) synthesized from the corresponding gene (Bmal1, Clock, CRY, or PER; concentration $[mRNA_x]$). It involves a set of parameters, such as the transcription rates (v_{tc}; production of the $mRNA_x$), translation rate (v_{tl}; production of the transcription factor TF_x), degradation rates of the mRNA ($v_{d_{mRNA_x}}$) and protein ($v_{d_{TF_x}}$), translocation rates from the nucleus to the cytosol ($v_{t_{TF_x}}$), rates of phosphorylation ($v_{TF_x^{P(i)}}$, $i = 1, 2$: first and second phosphorylations) and dephosphorylation, and association and dissociation rate of the transcriptional complexes Per–Cry and CLOCK–BMAL1. The negative feedback exerted by the nuclear transcriptional complex is described by an equation of the Hill type with a given degree of cooperativity and thresholds for repression.

A model of the mammalian circadian clock can be based on the intertwined positive and negative regulatory loops that involve the transcription of the PER, CRY, Bmal1, and Clock genes. It is described by a system of 16 kinetic equations that yields the temporal gradient of the concentrations of [628]: (1) PER, CRY, and Bmal1 mRNAs, (2) non-phosphorylated and phosphorylated transcription factors Per and Cry in the cytosol, (3) non-phosphorylated and phosphorylated Per–Cry complex in the cytosol and nucleus, (4) non-phosphorylated and phosphorylated transcription factor BMAL1 in the cytosol and nucleus: and (5) complexes Per–Cry and CLOCK–BMAL1 in the nucleus.

The cell division cycle in the cell populations controlled by a periodic input can be described as an eigenvalue problem. A population of cells structured by age a can be represented by the following set of equations [629]:

$$\begin{aligned}\partial_t n(t,a) + \partial_a n(t,a) + [\mathsf{d}(t,a) + \mathsf{P}(t,a)]n(t,a) &= 0,\\ n(t,0) &= 2\int_0^\infty \mathsf{P}(t,\alpha)n(t,\alpha)\,d\alpha,\\ n(0,a) &= n^0(a), \end{aligned} \tag{5.2}$$

where d and P are the T (1 d)-periodic death rate and cell division rate. Such systems of periodic differentiable equations are characterized by a dominant eigenvalue. When the circadian control exerts an effect only on the death rate (P does not depend on time), the dominant Floquet eigenvalue (associated with division $[\mathsf{P}(t,a)]$ and death $[\mathsf{d}(t,a)]$ rates) is higher than the dominant Perron eigenvalue (associated with period-averaged division $[(1/T)\int_0^T \mathsf{P}(\tau,a)\,d\tau]$ and death $[(1/T)\int_0^T \mathsf{d}(\tau,a)\,d\tau]$ rates) leads to a higher proliferation rate. Therefore, the periodicity has a strong effect on cell proliferation. A probabilistic approach can be based on the assumption that cells jump from one generation to the next, instead of dividing, with a time t_i of arrival in generation i and number of divisions $N_t(a)$ a cell of initial age a experiences at time t.

The functioning of cellular clocks was studied using signal processing [630]. Temporal oscillations results from feedback loops with a given period, duration, and amplitude. If $s(t)$ and $r(t)$ are 2 given signals, $s(t)$ can be decomposed into 2 parts, the first being a function of $r(t)$, and the second the derivative of a function of $r(t)$.

The night is come, but not too soon;
And sinking silently,
All silently, the little moon
Drops down behind the sky.

(The Light of Stars, Henry Wadsworth Longfellow [1807-1882])

6

Cell Motility

A noiseless patient spider,
I mark'd where, on a little promontory, it stood isolated,
Mark'd how to explore the vacant, vast surrounding,
It launch'd forth filament, filament, filament, out of itself.
Ever unreeling them, ever tirelessly speeding them.

(A Noiseless Patient Spider, Leaves of Grass, 1855,
Walt Whitman [1819–1892])

Cell migration is an integrated multistep process that occurs during embryo- and fetogenesis as well as, after birth, postnatal angiogenesis, immune response, and wound healing. Cell motility also contributes to tissue adaptation to environmental cues and remodeling. Furthermore, cell movements are responsible for progression of certain diseases, such as atherosclerosis and intimal hyperplasia, as well as cancer invasion and tumor metastasis. Concentration fields of regulators of cell activity generate a spatially distributed cell response (Sect. 6.2).

Mesenchymal cells, such as fibroblasts and smooth muscle cells, migrate more slowly than hematopoietic cells that shuttle between tissular compartments. Neutrophils and lymphocytes can reach velocities of about 20 µm/mn, whereas fibroblasts usually migrate at about 0.5 µm/mn.

Endothelial cells use *podosomes* that form rosettes upon TGFβ stimulation to create paths in the vessel wall. During their transmigration through the endothelium, leukocytes build podosomes to remodel their environment. Podosomes (in vitro lifetime 2–3 mn) have a dense actin core surrounded by a peripheral mesh of ^{F}actin cables.

6.1 Random and Guided Cell Movements

Cells move slowly on a matrix, or taxi, toward or away from a spatial intensity gradient of a given stimulus. The nature of the stimulus can be chemical (chemokinesis and chemotaxis), physical (thermo-, photo-, phono-, magneto-, and electrotaxis), or mechanical (rheotaxis in a moving medium or according to the matrix rheology).[1] *Chemokinesis* corresponds to a random cell migration primed by a motogenic factor applied symmetrically or not that does not determines the direction of migration. Motogens often are growth factors, such as platelet-derived (PDGF) and epidermal (EGF) growth factors. *Chemotaxis* results from the action of a soluble factor, the chemoattractant, applied asymmetrically that directs the direction of cell migration. The spatial organization and motion of cells subjected to chemical and physical signals from neighboring cells, surrounding fluid, and extracellular matrix determine the formation and maintenance of tissues.

Cell *taxis* refers to this guided displacement in response to a directional signal. In an organism, this innate behavioral response aims at shaping, remodeling, or repairing tissues. The main type of signals are chemoattractants or -repellents. Chemical sources are sensed by transmembrane receptors.

Necrotaxis is triggered by apoptotic and necrotic cells to attract macrophages. *Haptotaxis* corresponds to a motion directed by any chemoattractant bound to a matrix rather than soluble in the extracellular medium. (The cell moves toward tethered ligands, instead of soluble messengers that run toward cell-surface receptors.)

Chemotaxis, the motion of cells to reach a new functional site guided by a or many chemical gradients, follows cell reorganization with a cell front and back in the direction of the chemical gradient(s). Chemotactic pathways link the chemoattractant receptor to the pseudopod formation and cell motion. Cell movement can result from switches between attractive and repulsive migration in response to extracellular guidance cues. Directional migration is initiated by diverse types of extracellular cues, such as growth factors or chemokines.

Direction of cell migration depends on chemoattractant gradient as well as physical properties of the extracellular matrix (area availability and flexibility) and geometric shape of the cell colony. Cells extend from acute rather than obtuse angles formed by the cell population when geometric constraints are imposed by the extracellular matrix [632].

Cell motion is also controlled by the physical properties of the cell environment, especially its rigidity. In cell cultures, the support rigidity influences cell adhesion and cytoskeleton organization. The stability of focal adhesions depends on the support stiffness. In the presence of anisotropic stiffness, *haptotaxis* of epithelial cells occurs according to the direction of greatest support rigidity [633].[2]

[1] *Electrotaxis* is a directional displacement of cells in response to an electric field. Many signaling pathways are simultaneously involved during electrotaxis, such as guanylyl cyclases and phosphatidylinositol 3-kinases [631].

[2] Focal adhesion remodeling corresponds to a rate-limiting step in haptotaxis.

6.2 Cell Polarization and Remodeling during Displacement

The shape of migrating cells varies according to the number, size, and strength of adhesion foci from strongly spread cells that use many large adhesions (e.g., fibroblasts and endothelial cells) to highly protrusive cells that move using weak adhesions (e.g., lymphocytes) [634]. Cell–matrix focal adhesions correspond to spots that exert and detect mechanical forces. Migration modes depend also on the local composition, fiber orientation, and rheology of the extracellular matrix, electrochemical gradients, and intrinsic contractility of cells, among other factors.

6.2.1 Structural Polarity of Migrating Cells – Compartmentation

Cell migration relies on various repetitive processes that require a quick and precise regulation and a spatial and temporal integration of signals for the coordination of a set of displacements and deformations of cell regions. Various pathways partition the migrating cell into a functional front with a leading edge, a cell body, and a rear with a trailing edge (Fig. 6.1).

6.2.1.1 Cell Front – Lamellum, Lamellipodium, and Filopodium

The front compartment includes a *lamellum* (or lamella) and a *lamellipodium* or several lamellipodia that form the leading edge with possible filopodia. The cell front is characterized by regulated actin and focal adhesion dynamics in association with integrin signaling [635] (Fig. 6.2).[3] The cell rear deals with maintenance and nuclear functions (nucleic acid metabolism and cell cycle regulation).

Cell motion begins with cytoplasmic extensions, lamella and lamellipodia, and contraction of peripheral cytoplasmic sheets (cortex), attracted by chemotactic molecules (directional migration) or growth factors (random migration).

Lamellipodia (length 2–10 µm, thickness 0.1 µm) ruffle back and forth, forming transient pointwise contacts (attachment sites) with the substrate, slowly stretching and pulling the remaining part. When the elongated cell begins to move, the actin filaments organize into linear bundles[4] that span

[3] Integrin receptors sense changes in gradients of chemokines and physical agents in the extracellular matrix. The Ras–ERK pathway (Sect. 6.8.10.1 and Vol. 4 – Chap. 5. Mitogen-Activated Protein Kinase Modules) has particular substrates (e.g., myosin light chain kinase) and scaffolds (e.g., mitogen-activated protein kinase kinase-1-interacting protein-1) in the lamellipodium. Activity of ERK can also be hindered by EPHa2.

[4] Filament bundles observed in immotile cells yield tensile strength and structural support.

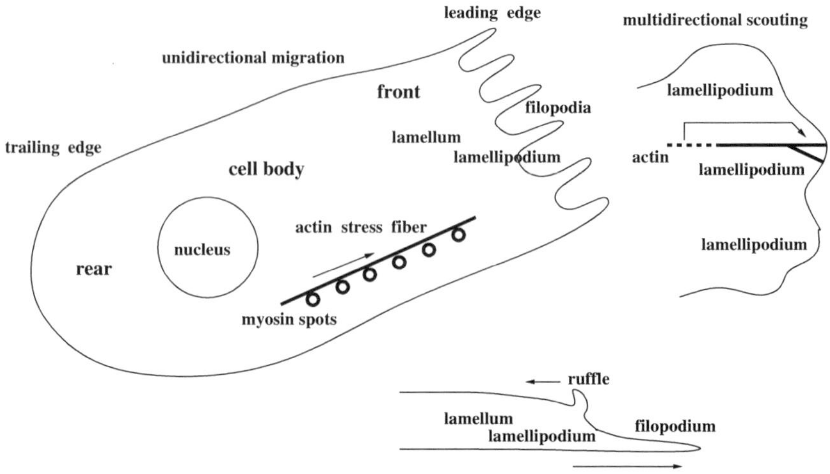

Figure 6.1. Migrating, polarized cell characterized by its leading and trailing edges. The front compartment comprises a lamellum (or lamella) and lamellipodium, or several lamellae and lamellipodia, as well as filopodia. A migrating cell has a dense mesh of cortical actin at its leading edge, with long, parallel, thrusting actin bundles extending into filopodia. Focal adhesion assembly–disassembly and cytoskeleton polymerization–depolymerization cycles allows cell motion on its matrix. Actin polymerization induces cell protrusions; contraction of actomyosin fibers causes protrusion and rear retraction. Actin undergoes a rearward flow in the lamellipodium due to membrane resistance at the leading edge, when it is already strongly stretched by actin polymerization, and myosin-2-induced contraction of actin filaments in the lamellum. Actin retrograde flow is achieved in particular by membrane ruffling. Tension developed by actomyosin fibers contributes also to disassembly of focal adhesions at both the front (prominently at the lamellum–lamellipodium border) and rear of the cell. Yet, Ca^{++}-activated peptidase calpain is a major agent of adhesion plaque disassembly in the retracting leading zone and expanding trailing region. The actomyosin cytoskeleton that relies mainly on myosin-2 causes the retraction of the rear of the migrating cell.

the whole cell in different directions with convergence to focal points and then rapidly disassemble to form a diffuse mesh leading to thin filopodia and lamellipodia.

Two actin networks in the lamellipodium and lamella occupy the front position of a migrating cell to control cell protrusion. The lamellipodium network is characterized by fast retrograde flow associated with actin polymerization localized in a band (width 1–3 µm) near the leading edge. The lamellar network is more stable with slow retrograde flow driven by contraction of actomyosin filaments that resides in a wider band (15 µm).

Growth and destruction of actin filaments in filopodia control formation and retraction of cell bulges during displacement. Filopodia are dynamic,

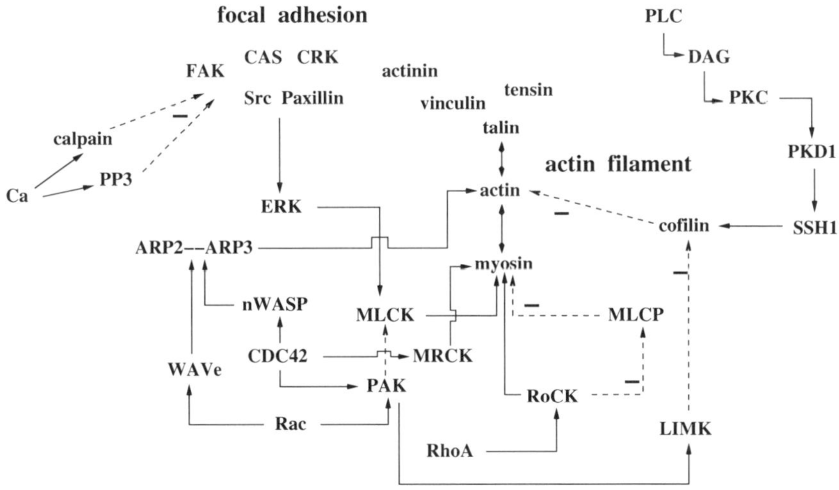

Figure 6.2. Linkage between focal adhesion and the cytoskeleton. During cell migration, protrusions of the front and translocation of the cell body is followed by the retraction of the rear of the cell. Cell body translocation that follows the formation of frontal protrusions results from a coordinated contraction of a set of actomyosin filaments and microtubular transfer, thereby depending on myosin-2 and microtubule nanomotors (dynein and kinesins). Myosin activity depends on 2 antagonist enzymes — myosin light-chain kinase (MLCK) and phosphatase (MLCP) —. Small GTPase RhoA elicits actin polymerization and remodeling in response to signals. RhoA activates Rho-associated, coiled-coil-containing protein kinase (RoCK) that phosphorylates (activates) myosin-2 as well as (inactivates) MLCP. Small GTPase Rac activates P21-activated kinase (PAK) that phosphorylates (inactivates) MLCK and myosin-2 as well as (activates) LIMK, which inhibits actin depolymerizer cofilin and microtubule remodeler stathmin. Several mechanisms support the disassembly of focal adhesions: (1) proteolytic cleavage of focal adhesion proteins, especially focal adhesion kinase (FAK) and talin, by calcium (Ca)-dependent calpain; (2) protein dephosphorylation (inactivation) by protein phosphatase-3 (PP3, PP2b, or calcineurin); (3) actomyosin contraction controlled by several pathways, among which the RhoA–RoCK axis that regulates the activity of myosin; (4) microtubule-induced adhesion disassembly; and (5) endocytosis of adhesion receptors (integrins) for recycling that depends on phosphoinositide-dependent kinase PDK1 and Rab GTPase. Focal adhesion kinase that remodels cell–matrix adhesions modulates the activity of Rac and Rho as well as the Src–ERK axis. Myotonic dystrophy kinase-related CDC42-binding kinase (MRCK) is a CDC42 effector that phosphorylates myosin to body cell translocation.

relatively long,[5] narrow (caliber 0.1–0.3 μm) actin-rich cell protrusions. Filopodia protrude from the lamellipodial actin network. Filopodia contain a core filled with parallel bundles of actin filaments with barbed ends located at the filopodial tip. Filopodial initiation and growth require actin polymerization

[5] Very short filopodia are often called microspikes.

at these barbed ends that involves multiple actin-associated proteins and regulators. Filopodial protrusion–retraction cycles are indeed caused by actin polymerization at the barbed ends and retrograde flow of actin filament bundle. In addition, interdigitated filopodia that protrude from opposing cells help to form intercellular adhesions.

Filopodia are involved not only in cell migration, but also in between-cell signaling. Filopodia actually have a sensory function. Filopodia contain receptors for diverse signaling and extracellular matrix molecules. Integrins and cadherins are located in the tips or along the shafts of filopodia. In macrophages, filopodia explore the environment.[6]

6.2.1.2 Cell Body

The microtubule-organizing center (MTOC) is relocalized between the nucleus and the leading edge in migrating cells. Yet, the nucleus can move away from the leading edge to reorient the microtubule-organizing center that remains stationary [636]. Rearward nuclear movement is coupled with actin retrograde flow. This motion is regulated by CDC42 GTPase, myotonic dystrophy kinase-related CDC42-binding kinase (MRCK), myosin, and actin. Myotonic dystrophy kinase-related CDC42-binding kinase is a CDC42 effector that phosphorylates myosin, thereby supporting the movement of the cell-body content, in particular the nucleus.

The monomeric GTPase CDC42 contributes to cell polarization as it localizes to the microtubule-organizing center (MTOC) and Golgi body in front of the nucleus toward the leading edge, thereby promoting the microtubule growth into the lamella and microtubule-mediated delivery of Golgi-derived vesicles to the leading edge.

On the other hand, dynein, partitioning-defective protein Par6, and PKCζ kinase contribute to the translocation of the microtubule-organizing center to maintain it at the cell barycenter, but do not participate in nuclear movement [636].

6.2.2 Functional Polarity of Migrating Cells

Several regulators contribute to the cell polarity that arises during cell migration (Table 6.1). Phosphatidylinositol 3-kinase and phosphatase and tensin homolog deleted on chromosome 10 (PTen) operate as amplifers of the stimulating chemoattractant gradient. Neutrophils that respond to small concentration gradients (1–2% change in chemoattractant concentration per cell length) amplify the asymmetrical stimulus internally via $PI(3,4,5)P_3$ that accumulates at the leading edge owing to activated RHO family GTPases — CDC42, Rac,

[6] After finding a pathogen, filopodia bind to it and then retract toward the macrophage body. The filopodia and lamellipodia then transform into an actin-based membrane structure, the phagocytic cup.

Table 6.1. Preferential localization of molecules and organelles in a polarized migrating cell. The phosphoinositide 3-kinase (PI3K) pathway activates Rac and CDC42 GTPases at the leading edge of migrating cells.

Cell front	Cell rear and side
Organelles	
MITOC, Golgi body	
Molecules	
PI3K	PTen
$PI(3,4,5)P_3$	PIP_2
Myosin-2	Actin
CDC42, Rac	Rho
Actin polymerization	Actin depolymerization
ARP2/3, WASP, WAVe,	Cofilin
profilin, ENA–VASP	
Crosslinkers, cappers	

and Rho — that stimulate PI3K and removal of phosphoinositide phosphatase PTen and SHIP at the trailing edge. Yet, PI3K-independent Rac activation exists in leukocytes via scaffold protein and Rac guanine nucleotide-exchange factor DOCK2 [637].

6.2.3 Adhesion–Deadhesion Cycle

Cell migration relies on the dynamic regulation of adhesion, deadhesion, and readhesion of cells on the extracellular matrix, especially regulated cycles of integrin-based connection–deconnection–reconnection of cells to the extracellular matrix.[7] Removal and recycling of focal adhesion happen alternately at the cell rear and front. They are associated with integrin endocytosis and translocation from the cell rear during its retraction to the front protrusions, and conversely. Integrins undergo several processes, such as heterodimerization, clustering, lateral diffusion in the plasma membrane, and interaction with the actin cytoskeleton and its partners that cause conformational changes to modify binding affinities for matrix components.

Cell adhesion turnover occurs in all compartments of the migrating cell (Table 6.2). Adhesion foci are created at the leading edge of protrusions and

[7] Nascent adhesions can turn over rapidly (~ 1 mn) or mature to focal complexes close to the leading edge at the lamellipodium–lamellum interface, where they persist for several minutes [634]. Focal complexes can mature into focal adhesions (size ~ 50 nm to ~ 1 µm) that are connected to stress fibers (large actomyosin bundles). Myosin-2 triggers stress fiber contraction (Sect. 6.8.5.4). Resulting tension can expose binding sites of adhesion constituents and assist in their recruitment. These constituents then assemble, thereby leading to adhesion maturation.

Table 6.2. Adhesion turnover (Source: [740]).

Adhesion site formation	Adhesion site removal and recycling
Integrin activation	FAK, Src, ERK, MLCK
PI3K, PKC, Rap1, talin	CAS, paxillin
Integrin clustering	
CDC42, Rac	

disassembled at the cell rear and base of protrusions. Kinases, such as focal adhesion (FAK), Src, extracellular signal-regulated (ERK), and myosin light-chain (MLCK) kinases (Vol. 4), as well as adaptors, such as BCAR1[8] and paxillin (Vol. 1 – Chap. 6. Cell Cytoskeleton), and talin-mediated linkage of actin filaments to focal adhesions regulate cell adhesion turnover in migrating cells [740].

6.2.4 Phases of Cell Migration

Cell motility is initiated by an actin-dependent protrusion of the leading edge. During cellular extension, new adhesions with the matrix form under the leading edge. Afterward, the nucleus is translocated forward through contraction forces developed by the actomyosin cytoskeleton (Table 6.3).

Afterward, retraction fibers pull the rear of the cell forward, adhesions at the cell rear disassemble, and, finally, the trailing edge retracts (Tables 6.4 and 6.5).

The rapid, reversible phosphorylation of mediators enables at least partly this control of the localization, timing, and duration of these processes as well as their coordination. Mediator phosphorylation state is controlled by kinases and phosphatases (Sect. 6.8.10).

The cytoskeleton reorganization (assembly and disassembly) of focal adhesions and junctions between adjacent cells and the cell and the extracellular matrix are required for cell motility and tissue formation and remodeling.

Cells can collectively migrate, especially during embryonic morphogenesis and cancer metastases. Chemoattractant cues are translated into directed cell movement. Cell cluster polarity is maintained because tip cells communicate with back cells to ensure coordinated, directed migration. Cell imaging measures cell shape changes and movements using labeled cytoskeletal components as well as other cellular elements without disturbing their function [643].

6.2.5 Cytoskeleton Remodeling

Cell motility is associated with changes in both cell configuration (filopodia, and lamellipodia) and mechanical state (sol–gel transition). When the

[8] A.k.a. CAS.

Table 6.3. Phases of cell displacement and phosphorylation–dephosphorylation control (**Part 1**; Source: [639]; ADF: actin depolymerizing factor; APC10: Anaphase-promoting complex subunit-10 [a.k.a. destruction of cyclin-B1 (DOC1)]; CAS: CRK-associated substrate [or BCAR1]; CRK: CT10 regulator of kinase; DOCK: dedicator of cytokinesis (GEF); ElMo: engulfment and cell motility adaptor; lmwPTP: low-molecular-weight protein Tyr phosphatase; PAK: P21-activated kinase; Pax: paxillin; PP: protein phosphatase; PPM: Mg^{++}–Mn^{++}-dependent protein phosphatase; PTen: phosphatase and tensin homolog deleted on chromosome 10; PTPn: protein Tyr phosphatase non-receptor; PTPR: protein tyrosine phosphatase receptor). Small GTPase Rac stimulates actin polymerization in lamellipodia. Focal adhesion kinase controls the focal adhesion turnover. It supports migration. Its phosphorylation by Src kinase and dephosphorylation by many phosphatases regulate its function. PTPn11 limits the phosphorylation degree of FAK and hence the number of focal adhesions to foster the migration. Yet, PTPn11 together with EPHa2 impedes cell migration.

Event	Stimulators	Inhibitors
Leading edge protrusion (phase 1)	Rac (lamellipodium) CDC42 (filopodium) RhoA (stress fiber) PPM1a, PP2 (via ADF–cofilin)	PTPn12, PTen (Rac inhibition) PTen (CDC42 inhibition) Rho–LIMK (via ADF–cofilin)
Formation of adhesion complexes (phase 2)	PTPRo, PTPn1b, PTPn11, PTPn11–RhoA Src–CAS–CRK lmwPTP (p190RhoGAP inactivation) Integrin–SHC–ERK Integrin–FAK–CAS CAS–CRK–DOCK1–ElMo–Rac Integrin–PTPn11–ERK PTPRo, PP2–Pax Pax–PAK FAK–PI3K–PIP_3 PTPn11, PPM1d–FAK	PTPRh, PTPn12, PP2 PTPRh, PTPn12–CAS Abl$\ominus \longrightarrow$ CAS–CRK interaction PTen$\ominus \longrightarrow$ CAS–CRK PTen–FAK PTPn11,PPM1d–FAK PTPn6–PI3K–CRKL–APC10

membrane is subjected to a local stimulus, the cortex gel undergoes pressure from the incompressible sol that generates a bulge. The protuberance content gels to form a stable lamellipodium. *Gelsolin*, activated by Ca^{++}, dissociates actin gel for actin–myosin interactions. Cell motility results from actin polymerization into filaments and depolymerization.

Cytoskeleton remodeling is regulated in space and time in response to external and internal signals to ensure cell adaptation to its environment as well as cell motion. The cell, indeed, displays a set of internal motions (cytoplasmic streaming, mitosis using mitotic spindle, vesicle-associated transport, etc.). Moreover, the cell is able to creep about and actively change its shape

Table 6.4. Phases of cell displacement and phosphorylation–dephosphorylation control (**Part 2**; Source: [639]; MLCK: myosin light-chain kinase; MLCP: myosin light-chain phosphatase; MRCK: myotonic dystrophy kinase-related CDC42-binding kinase; PP: protein phosphatase; PTPn12: protein Tyr phosphatase non-receptor type 12; RoCK: Rho-associated, coiled-coil-containing protein kinase). Cell migration can be decomposed into 4 main phases. (1) At the leading edge, a lamellipodium is created (broad, thin, actin-rich protrusion) and stabilized, from which filopodia are generated (smaller, finger-like, actin-rich extensions). Small GTPases of the RHO family regulate the actin remodeling in lamellipodia and filopodia. (2) Focal complexes are formed close to the leading edge and serve as precursors of larger focal adhesions. (3) The cell contracts owing to stress fibers and its content moves forward. (4) At the cell rear, adhesion sites are degraded and membrane receptors are recycled from the rear to the front, and then the cell rear retracts. The ezrin–radixin–moesin (ERM) family of crosslinkers between the actin cytoskeleton and plasmalemmal proteins intervenes as signal transducer in cell responses that involve cytoskeletal remodeling. Moesin is involved in Rho- and Rac-dependent assembly of actin filaments. It can be upregulated in migrating cells, although it not expressed in the resting state such as in vascular smooth muscle cells. Dephosphorylation of moesin (Thr558) precede tail retraction.

Event	Stimulators	Inhibitors
Cell body translocation (phase 3)	MLCK RoCK CDC42–MRCK–myosin	MLCP
Rear retraction (phase 4)	PP–moesin Rac, PTPn12–RhoA	MRCK–moesin

Table 6.5. Regulators of trailing edge retraction (phase 4; Sources: [640–642]; CAPNS1: calpain (calcium-activated peptidase, neutral) small subunit-1; PPP1R14b: protein phosphatase-1, regulatory [inhibitory] subunit-14B).

Pathway	Effect
Ca^{++}–calpain	Adhesion plaque disassembly (talin)
Ca^{++}–CAPNS1	Mechanical stress detection
Ca^{++}–PP3	Rear retraction
Rho–PP1, PP2	Moesin dephosphorylation, retraction
PPP1R14b–PP1	Retraction
RhoA–RoCK–MLCP	Contraction of stress fibers
Dynamin–FAK	Focal adhesion disassembly (microtubules)
Ca^{++}–MLCK	Contraction of stress fibers
ERK–MLCK	Contraction of stress fibers
Rac–PAK–MLCK	Relaxation of stress fibers

and migrate with frequent direction changes. Cell motility is necessary for blood coagulation, wound healing, and immune responses. Cell motion is triggered by signal transduction cascades from stimulated receptors that ensure signal transmission down to cell deformation.

During cell travel, the cell membrane continuously moves from the progression front to the tapered cell back as tracked vehicles owing to an endocytosis–exocytosis cycle [644]. Cell migration has long time scale directional persistence,[9] whereas lamellipodial dynamics have short time scales. The directional persistence of the cell path and lamellipodial dynamics has been studied using epidermal growth factor stimulation on fibronectin-coated substrata [645]. The short time scale kinetics of adhesion complex formation may modulate the directional persistence. High resolution[10] and molecular sensitivity of photoactivated localization microscopy allow the study of the dynamics of adhesion complexes in photon-tolerant cell lines [646]. In addition, gain and loss of paxillin among the set of involved proteins[11] at attachment junctions between cortical cytoskeleton and cell matrix can be measured during the initiation, maturation, and dissolution of adhesion complexes.

6.3 Cell Ensemble Migration

Collective migration of cohesive cells occurs during embryo- and fetogenesis, wound repair, and cancer invasion. Cell clusters can move as a 2-dimensional sheet of a cell monolayer, a three-dimensional strand with generation of a tube structure (e.g., duct and gland formation, such as angiogenesis and tracheobronchial tree generation), or a poorly organized mass [647].

Three features characterize collective cell migration: (1) cells remain connected with preservation of intercellular junctions during cell motion, in particular adherens junctions that maintain a coordinated organization; (2) the supracellular organization of actin cytoskeletons that generate force for migration and maintain intercellular junctions; (3) the cell ensemble that modifies the tissue structure along the migration path.

Collective migration differs from various types of multicellular displacements, such as those occurring in some morphogenetic events [647]:

---

[9] Migrating cells sustain a nearly constant direction of locomotion over time scales ranging from minutes to hours.

[10] High spatial resolution (~ 60 nm), but modest frame rates (~ 25 s per frame) characterize photoactivated localization microscopy. Time resolution suits slow migration.

[11] More than 90 proteins are implicated in thin adhesion plaque at the cell–substratum interface.

(1) *invagination*,[12] (2) *intercalation*,[13] (3) *expansive growth*,[14] (4) *emboli*,[15] and (5) *cell streaming*.[16]

Adherens junction proteins, such as cadherins and integrins that are connected to the actin and/or intermediate filament cytoskeleton provide cell cohesion. Many collectively migrating cells are epithelial cells. Homophilic intercellular adhesion (i.e., symmetrical adhesions with same components in connected cells) are mediated by epithelial, neural, or vascular endothelial cadherins (Vol. 1 – Chap. 7. Plasma Membrane). Integrins couple cells via intercellular matrix components (Vol. 1 – Chap. 7). In branching epi- and endothelia, desmosomal proteins (Vol. 1 – Chap. 7) participate in collective cell behavior. Gap junction connexins (Vol. 1 – Chap. 7) contribute to epi- and endothelium sprouting. Rear cells often have tight junctions (Vol. 1 – Chap. 7), but not necessarily leading cells [647].

Classical cadherins are major intercellular adhesion molecules at adherens junctions that contribute to morphogenesis (cell polarity, sorting, and proliferation, as well as neurite extension and fasciculation of axons). Some members of the cadherin class (desmosomal cadherins, kidney-specific [Ksp]- and liver–intestine [LI]-cadherin in the renal and intestinal epithelium, respectively, Fat1, etc.) are not only involved in mechanical adhesion, but also in signal transduction. Protocadherins are members of the cadherin class that constitute families, such as the PCDH2 and PCDH3 families. They either are widely expressed or highly specific to nervous tissues.[17]

Tissue morphogenesis requires that both individual cells and cell groups migrate in coordination. Joint migration of interacting cells requires transmembrane *protocadherin-10*,[18] homophilic cell adhesion molecules, that bind NCK-associated protein NAP1 to regulate WAVe-dependent rearrangement of the actin cytoskeleton [648]. Protocadherin-10 causes concomitant reorganization of ^{F}actin and N-cadherin at cell junctions to accelerate the motion

[12] Embryonic tissues can fold due to a shifting of cells together with surrounding tissues in response to change in shape of other cells. Cell displacements do not affect connections with neighborhood.

[13] Cell intercalation, or convergent extension, leads to coalescence of groups of cells after a coordinated set of interactions controlled by myosin-2 constriction of certain intercellular junctions.

[14] Expansive growth corresponds to a position drift of daughter cells in tumors during cell proliferation.

[15] Emboli in body fluids result from cellular cluster detachment and convection in fluid stream.

[16] Cell streaming is the movement of individual cells behind each other that form chains of cells without tight intercellular junctions.

[17] Cadherin-related neuronal receptors (CNR) that bind Fyn kinase localize to synapses. Protocadherin-7, or NF-protocadherin, is highly expressed in the developing retina and abounds as well as its cytosolic cofactor template-activating factor TAF1 in retinal ganglion cells.

[18] A.k.a. OL-protocadherins. Protocadherins constitute the largest subclass of the cadherin class of calcium-dependent intercellular adhesion molecules.

of contacting cells (but not solitary cells) by recruiting the NAP1–WAVe1 complex at cell junctions. Tissue maintenance and repair also rely on cell motion.

Several mechanisms polarize a migrating cell cohort into leader cells, the so-called tip cells, that guide followers at their rear. In branching ducts, lateral confinement of cell strand results from matrix degradation and deposition of a basement membrane. Cell polarization in the migration direction is caused by chemokines (e.g., stromal cell-derived factor SDF1) and growth factors (e.g., epidermal [EGF], fibroblast [FGF], and vascular endothelial [VEGF] growth factor, as well as transforming growth factor-β, and PDGF- and VEGF-related factors PVF1 and PVF2) produced by stromal cells (paracrine control by fibroblasts, pericytes, and myoepithelial cells of the glandular epithelium) or released from migrating cell cluster (auto- or juxtacrine control). Polarity induction and guidance rely on Notch (Delta-like-4 and Jagged-1) and mitogen-activated protein kinase signaling.

6.4 Diapedesis

Circulating blood cell transmigration through blood vessel walls is achieved by sequential stages that include [649]: (1) capture (or tethering); (2) slow selectin-mediated rolling; (3) chemokine-triggered activation[19] and integrin-dependent arrest from rolling (adhesion stabilization, strengthening, and spreading); (4) intraluminal crawling, seeking preferred sites of transmigration; (5) paracellular and transcellular migration; and (6) migration across the basement membrane.[20]

Leukocytes are recruited to inflammatory sites owing to adhesion molecules (integrins and selectins) and ligands as well as chemokines and their receptors (Table 6.6). Leukocyte crossing of endothelium can be rapid (2–5 mn), but penetrating the endothelial basement membrane takes a much longer time (5–15 mn).

[19] During inflammation, endothelial cells are activated by cytokines to express adhesion molecules and synthesize chemokines and lipid chemoattractants that are afterward located on their luminal surface. Activated endothelial cells also transport chemoattractants from their abluminal surface. Some chemoattractants are generated by proteolytic cleavage in activated mastocytes and platelets and delivered to endothelial cells. Chemokines bind to specific GPCRs and glycosaminoglycans of the endothelial surface. GPCR binding activates PLC that generates IP_3 and DAG. Subsequent calcium influx stimulates VLA4 and arrest of monocytes and neutrophils on inflamed endothelium. GTPases Rap1 and RhoA regulate leukocyte integrin affinity. Guanine nucleotide-exchange factors are activated by Ca^{++} and DAG. Actin-binding talin-1 participates in integrin affinity upregulation.

[20] The endothelial basement membrane is composed of 2 proteic networks composed of vascular laminins (laminin-8 and -10) and collagen-4, which are connected by nidogen-2 and heparin sulfate proteoglycan perlecan.

Table 6.6. Molecules involved in each step of leukocyte extravasation (Source: [649]; ESAM: endothelial cell-selective adhesion molecule; ESL: E-selectin ligand; ICAM: intercellular adhesion molecule; JAM: junctional adhesion molecule; MAdCAM: mucosal vascular addressin cell adhesion molecule; PSGL: P-selectin glycoprotein ligand; PECAM: platelet/endothelial-cell adhesion molecule; VCAM: vascular cell adhesion molecule; CCL: CC-chemokine ligand; CXCL: CXC-chemokine ligand; Vav: RhoGEF). Leukocytes marginate from the blood stream to follow chemokine gradients. Chemokines such as CXCL8 travel to vessel lumen after exocytosis from endothelial cells. Some chemokines activate PI3K isoforms.

Molecule	Extravasation stage (expression and ligands)
ESAM	Paracellular migration of neutrophils
ICAM1	Rolling; arrest; crawling; transcellular migration ($\alpha_L\beta_2$)
ICAM2	Paracellular migration ($\alpha_L\beta_2$)
JAM	Paracellular migration
MAdCAM1	Rolling
PECAM1	Paracellular migration
PSGL1	Leukocyte capture and rolling (selectins)
VCAM1	Rolling, arrest
Selectins	Leukocyte capture and starting rolling; slow rolling; chemokine-activated arrest
E-selectin	Rolling (activated endothelial cells) (PSGL1, glycosylated epican [CD44], ESL1)
L-selectin	Leukocyte rolling (PSGL1)
P-selectin	Rolling (PSGL1) (activated endothelial cells, activated platelets)
VE-cadherin	Paracellular migration
$\alpha_L\beta_2$	Neutrophil rolling; leukocyte arrest (ICAM1)
$\alpha_M\beta_2$	Rolling; arrest; crawling
$\alpha_4\beta_1$	Lymphocyte and monocyte capture and rolling; Leukocyte arrest (VCAM1)
$\alpha_4\beta_7$	Leukocyte rolling (MAdCAM1)
$\alpha_6\beta_1$	Migration across endothelial basement membrane (PECAM1-mediated increased expression)
CD99	Paracellular migration
CCL5	Monocyte activation and arrest
CXCL4/5	Monocyte activation and arrest
Src	Adhesion strengthening, leukocyte spreading
PI3K	Adhesion strengthening, leukocyte spreading
Vav1/2/3	Adhesion strengthening, leukocyte spreading

In endothelial cells, chemokine transport can be carried out by Duffy antigen receptor for chemokines (DARC; or glycoprotein-D; Sect. 6.8.3.1). Chemokine receptors CCR9,[21] CCRL1,[22] and DARC constitute a family of *atypical chemokine binders*. These *chemokine decoy receptors* can act as negative regulators of tumor growth and metastasis. They are interceptors, which are also called decoys and scavengers, that cycle from the plasma membrane to endosomes and internalize their cognate chemokines, thereby participating in the clearance of their chemokine ligands in inflammatory sites [650].

Monomeric chemokines bind to their cognate monomeric G-protein-coupled receptors (Vol. 3 – Chap. 7. G-Protein-Coupled Receptors).[23] However, chemokines, such as CCL2, CCL4, CCL5 and CXCL10, can form oligomers to enhance leukocyte recruitment. Chemokine GPCRs can also constitute dimers. Scaffolds such as arrestins modulate GPCR signaling.

6.5 Wound Healing

Cells involved in wound healing (Vol. 5 – Chap. 11. Tissue Growth, Repair, and Remodeling) have a displacement speed from 10^{-4} to 10^{-3} mm/h. Fibroblasts move at a speed from 120 nm/mn up to 1 µm/mn. Keratinocytes attain a speed of 10 to 40 µm/mn. The velocity of clotting and immune cells is greater. Neutrophils (10^{11} produced per day) look after possible microorganisms throughout the body at a speed up to 2 mm/h. Macrophages migrate to engulf possible invading bodies. Platelets deform and quickly spread out to plug vessel wall breaches.

During wound healing, 2 main coexisting mechanisms occur: (1) "purse-string" closure (especially in small wounds) due to the contraction of the pluricellular actin belt owing to myosin nanomotors and (2) heterogeneous *collective cell migration* in the direction perpendicular to the free edge. Leader cells have strongly polarized shape and well-developed lamellipodia. They do not dissociate from their neighbors, but move faster. Leaders destabilize the cell layer, which acquires motility. Cells remain assembled with a pluricellular actin belt. Leaders in fact drag a column of adjoining follower cells by means of cytoskeletal scaffolds that maintain their motion in the direction of leader cells and strong intercellular cadherin contacts that maintain contacts between leaders and followers [651]. Behind the moving front, followers move collectively in complicated patterns. Autocrine stimulation by hepatocyte growth factor raises the average migration velocity without a leader.

---

[21] A.k.a. chemokine receptor or chemokine-binding protein CCBP2 or D6. Protein CCBP2 is also called CCR9 and CCR10. However, chemokine receptors CCR9 and CCR10 correspond to GPR28 and GPR2, respectively.

[22] Chemokine receptor CCRL1 is also called CCR10 and CCR11; CCRL2 on mastocytes CCR6 and CCR11. Receptor CCRL1 can be regulated by cytokines, such as interleukin-1β, tumor-necrosis factor-α, and interferon-γ. It can bind and scavenge their cognate ligands CCL19, CCL21, CCL25, and CXCL13.

[23] CXCR4- and CCR7-dependent chemotaxis of dendritic cells (but not naive lymphocytes) follows activation of Gi and Gq subunits.

6.6 Metastasis

Malignant tumors develop when cells undergo contact-independent growth, become motile, and cross the basement membrane to form metastases. Tumor cells polarize and prime a reaction set for assembly, maintenance, and turnover of invasive podosomes and invadopodia. *Invadopodia* are rich in actin, cholesterol, and matrix metallopeptidases. These actin-based protrusions form at contact sites between invasive tumor cells and the extracellular matrix.

Normal and transformed cells share similar abilities to degrade the extracellular matrix, adhere, and migrate. However, differences exist between actin-containing structures generated by cells during their displacements.[24] Metallopeptidase content (i.e., membrane type 1-matrix metallopeptidase) of invadopodia degrades the extracellular matrix to allow cell motion and dissemination. Exocyst complex[25] is required for matrix proteolysis.[26]

Most tumors have an epithelial origin but display mesenchymal traits. Epithelial cells at the primary tumor site undergo an epithelial–mesenchymal transition under the influence of combinations of signaling factors that activate various transcription factors (e.g., forkhead box FoxC2, Snail, Twist, and zinc finger E-box binding homeobox Zeb). They lose intercellular contacts and apicobasal polarity. They acquire mesenchymal gene expression. Caveolin-1 that allows internalization of receptors and cadherins contributes to the epithelial–mesenchymal transition. The cytoskeleton remodels and invadopodia can form. Carcinoma cells then migrate.

Cancer stem cells, but not progenitor cells, initiate metastasis. Notch signaling leads to hypoxia-induced epithelial–mesenchymal transition and cell motility [654]. Notch controls Snail-1 (and Snail-2 or Slug) expression as well

---

[24] Podosomes and invadopodia are sites of degradation of the extracellular matrix. Maturation of invadopodia is slower than that of podosomes. Integrins that form clusters owing to talin exist in podosomes and invadopodia. On the other hand, vinculin is a component of podosomes, but not invadopodia. Palladin that participates in stress fibers and focal adhesions is involved in invadopodium formation. Cortactin in invadopodia controls the secretion of matrix metallopeptidases.

[25] Exocyst complex is an octamer (Exoc1–Exoc8, also known as Sec3, Sec5–6, Sec8, Sec10, Sec15, Exo70, and Exo84, respectively) that is involved in vesicular transport, as it tethers vesicle and targets post-Golgi vesicles toward the plasma membrane.

[26] Exocyst subunits Exoc1 (Sec3) and Exoc8 (Exo84) interact with polarity IQ motif-containing GTPase-activating protein IQGAP1 [652]. This interaction is triggered by active CDC42 and RhoA. Exocyst and IQGAP1 are required for accumulation of membrane-type metallopeptidases MT1-MMP at invadopodia. In addition, mt1MMP-dependent matrix degradation at invadopodia is regulated by vSNARE protein VAMP7 (a.k.a. synaptobrevin-like gene-1 protein and tetanus neurotoxin-insensitive vesicle-associated membrane protein [TI-VAMP]), as VAMP7 delivers MT1-MMP to degradation sites. A positive feedback loop exists in which mt1MMP as a cargo of VAMP7-mediated transport vesicles regulates maturation of invadopodia [653].

as its stabilization by lysyl oxidase that is upregulated by hypoxia-inducible factor-1α. Runt-related transcription factor Runx2 that regulates mesenchymal lineage cell differentiation activates expression of adhesion proteins, matrix metallopeptidases, and angiogenic factors in tumor cells and promotes metastasis [655].

Certain tumor cells invade surrounding tissues by taking a rounded shape. Rounded cell motility is controlled by contraction of cortical actin–myosin filaments. Phosphoinositide-dependent kinase PDK1 promotes cancer cell invasion by binding to Rho-associated coiled-coil-containing kinase RoCK1, thereby impeding inhibitory RhoE–RoCK1 interaction [656].[27] Activity of PDK1 that is mainly detected on rounded cells is restricted to the cortical actomyosin network.

Mutations of mitochondrial DNA (i.e., gene that encodes reduced form of nicotinamide adenine dinucleotide dehydrogenase subunit-6) that cause deficiency in respiratory complex-1 (NADH dehydrogenase) and overproduction of reactive oxygen species favor metastasis [657].

Plasmalemmal glycoprotein L1-cell adhesion molecule (L1CAM) that leads to a sustained activation of extracellular signal-regulated kinases is able to promote metastasis [658,659]. Wnt–β-catenin signaling overactivation and intracellular β-catenin accumulation predispose to cancer.

Mesenchymal stem cells can integrate tumor-associated stroma. Mesenchymal chemokine CC-motif ligand CCL5 augments the metastatic potential [660].

Overexpressed adaptor neural precursor cell expressed developmentally downregulated NEDD9 that activates focal adhesion kinases, thereby elicits the formation of focal contacts and activates GTPase Rac via the FAK–CRK–DOCK1 complex, and also promotes tumor cell invasion [661].

Single cells or cell clusters migrate through corridors in the remodeled extracellular matrix. Carcinoma-associated fibroblasts and/or macrophages can lead invasive tumor cells. Tumor cells then escape through the lymph or blood stream and extravasate to form secondary tumors. Small GTPase RhoC allows tumor cells to invade leaky tumor blood vessels [662]. Metastatic RhoC promotes the initial steps of metastasis because it induces the formation of dynamic membrane blebs and protrusions that penetrate blood vessel wall at sites of VEGF-mediated vascular remodeling.

Metastasis needs actin polymerization for cell adhesion, spreading, and motility, and thus RhoA and RhoC. Cell migration depends on $G\alpha_{12/13}$-dependent Rho–Dia signaling. Reorganization of the microtubule-organizing center is influenced by CDC42. Small GTPase RhoC and its effector kinase RoCK regulate activity of myocardin-related transcription factors MRTFa (or MAL) and MRTFb (or MKL2) that are transcriptional coactivators of

---

[27] Small GTPase RhoA also activates RoCK that phosphorylates myosin light chain at the cell cortex, but PDK1 does not affect RhoA-stimulated RoCK phosphorylation of myosin light chain.

transcription factor serum response factor (SRF; Vol. 1 – Chap. 6. Cell Cytoskeleton). Cytoskeletal gene expression is controlled by MRTFs and SRFs, such as non-muscle myosin heavy chain (MYH9 or NMHC2a) and its regulatory light chain (MYL9 or MLC2) that are required for invasion and lung colonization [663]. Furthermore, MRTF–SRF activity allows efficient colonization of the lung from the blood circulation. Factor SRF is activated not only by Rho, but also CDC42 and Rac.

Metastases from primary tumors are often restricted to specific organs. Regional specificity of metastasis could be related to local blood flow patterns, i.e., residence time and regulation mediated by flow-induced stress of cell extravasation and division. Transforming growth factor-β secreted by the tumor environment under hypoxic conditions induces the expression of cytokine angiopoietin-like protein-4 in moving tumor cells that disrupt endothelial intercellular adhesions in lung capillaries and contributes to metastases within lungs (but not bone) of breast cancers [664].

Coiled-coil domain-containing protein-88A (CCDC88a)[28] serves as a non-receptor GEF for $G\alpha_i$ that activates and sequesters Gα subunit to enhance PKB signaling via the Gβγ–PI3K pathway for actin remodeling and cell migration in cancer cells [665].

6.7 Migrating Cells

Cells sense spatial heterogeneities via internal or external signals that are amplified. Spatial cues are then transmitted to the cytoskeleton and secretory apparatus in the case of required associated secretion of granule content. Small GTPases Rho, Rac, and CDC42 transduce such spatial information. Temporal and spatial activation of the RHO family GTPases is achieved by guanine nucleotide (GDP-to-GTP)-exchange factors (GEFs) and GTPase-activating proteins (GAPs). Small GTPases can form complexes to amplify signals that generate persistent cell polarization.

Recruitment of various proteins at distal and proximal sites of moving cells allows action synergy of both cell surface and actomyosin cytoskeleton (Tables 6.7 and 6.8). Signal transduction pathways convert a gradient input into pseudopod formation. During cell displacement, clusters of receptors, adhesion molecules, and cytoskeletal proteins form at the cell cortex to confer cell polarity according to the motion direction. When cells orient along a gradient of chemoattractant concentration, proteic clusters localize at the cell rear with respect to the migration direction. These clusters allow integration of signal and actions needed to control cell protrusions and retraction in the direction of cell movement.

[28] A.k.a. girdin, Gα-interacting vesicle-associated protein (GIV), girder of actin filament, and PKB (Akt) phosphorylation enhancer (APE).

Table 6.7. Proteins involved in the formation of lamellipodium and filopodium (**Part 1**: lamella and lamellipodium; Source: [666]). Stress fibers are generated by formin-mediated actin polymerization owing to actin-related protein-2 and -3 (ARP2–ARP3 complex)-nucleated lamellipodial actin network to yield contractile forces. Enabled homolog and vasodilator-stimulated phosphoproteins (Ena–VASP) include 3 proteins: Enabled, Enabled homolog (EnaH), and VASP. These actin-binding proteins localize to focal adhesions. They prevent capping of actin filament barbed ends. Anti-capping Ena–VASPs also have antibranching and ^{F}actin-bundling activity. Among the 15 known formins, some induce the formation of unbranched actin filaments. Myosin-10 carries proteins to filopodial tips. It binds to VASP, phosphatidylinositol phosphates (e.g., $PI(3,4,5)P_3$), integrins, netrins, and tubulins. Actin filament-bundling fascin crosslinks ^{F}actins (half-life 6–9 s). Brain-specific angiogenesis inhibitor-1-associated protein BAIAP2 interacts with plasmalemmal lipids, actin-filament capping protein EPS8, WAVe2, EnaH, and CDC42 and Rac1 GTPases. Wiskott-Aldrich syndrome protein (WASP) and suppressor of cAMP receptor (SCAR or WAVe) operate via the ARP2–ARP3 complex. Neural WASP (nWASP) that binds actin monomers and the plasma membrane via Rho GTPases attaches growing filament barbed ends to the plasma membrane. Potassium outflux through K^+ channels, in particular calcium-activated K^+ channels concentrated at the leading edge, modulates cell migration by modifying the membrane potential and favoring calcium influx that can activate migration. Stimulated cofilin impedes cell protrusion by increasing actin polymerization and inducing the spatial reorganization of lamellipodium.

Protein	Effect
ARP2–ARP3	Actin filament nucleator (branched lamellipodial ^{F}actin network)
Cofilin	Creation of barbed ends
Ena–VASP	Actin filament elongation
Fascin	Crosslinker of ^{F}actin
Formins	Actin filament elongation
BAIAP2	Plasma membrane deformation
Myosin-2	Stress fiber contraction (lamella)
Myosin-10	Association of actin filament barbed ends
Profilin	Enhancement of Ena–VASP-induced filament polymerization
Rac	Activation of nWASP, WAVe and the ARP2–ARP3 complex Activation of PAK
WASP, WAVe	Stimulation of ^{F}actin nucleation activity of ARP2–ARP3
K_{In} channel	K^+ efflux

Cell polarity is achieved and maintained owing to RNA localization. RNAs accumulate at cell protrusions due to adenomatous polyposis coli tumor suppressor protein (APC). Ubiquitin ligase APC anchors over 50 mRNAs (e.g., Rab13 and plakophilin-4 mRNAs) to microtubule plus-ends at the pseudopodial tips of migrating cells [667].

Table 6.8. Proteins involved in the formation of lamellipodium and filopodium (**Part 2**: filopodium; Source: [666]). Rho in filopodium (RIF) activates actin polymerization by activating diaphanous-related Dia2 formin. Actin crosslinkers (α-actinin, fimbrin, espin, and filamin) are not specific to filopodium shafts. They target more loosely packed ^{F}actins at the lamellipodium–filopodium interface. Lipid phosphatase-related protein-1 (LPR1) induces filopodium formation independently of the ARP2–ARP3 complex and Ena–VASP proteins.

Protein	Effect
CDC42	Filopodium formation
Dia2	Formation of unbranched actin filaments in filopodia
LPR1	Filopodium formation
Myosin-10	Filopodium formation
RIF	Formation of long filopodia

The leading edge constitutes a barrier for lipid diffusion, as cortical and plasmalemmal proteins have a high concentration. This diffusion barrier maintains the polarity of a migrating cell.

6.8 Main Factors and Regulators of Cell Migration

6.8.1 Calcium

Intracellular calcium acts in sensing for guidance of directional movement, cytoskeleton remodeling, force generation, and focal adhesion relocation during cell migration. *Calcium flickers* that correspond to transient high-calcium nanodomains following opening of stretch-activated cation channels of the superfamily of transient receptor potential ion channel occur at the front (lamella) of migrating human embryonic lung fibroblasts toward the chemoattractant source (platelet-derived growth factor gradient) [668]. This intracellular calcium polarization is the opposite of the global rear-to-front calcium gradient with the highest concentration at the trailing edge. Channel TRPM7 transduces mechanical signals into calcium flickers. Calcium release from the cell's stores, mainly the endoplasmic reticulum, via inositol trisphosphate receptors amplifies calcium influx.

Migration of cultured aortic smooth muscle cells is accelerated by pulsatile pressure with respect to steady pressure that is equal to maximum pulsatile pressure [669]. Increased frequency of pulsation further enhances cell migration. At a given mean pressure, the higher the magnitude of pulsation, the greater the number of migrating cells. Upon cell pressurization, calcium concentration increases within seconds, but significant acceleration of cell migration occurs over 6 h of pressurization. Cell phenotype actually changes before migration starts. Cytosolic calcium level is higher in cells that bear pulsatile pressure than steady pressure. In addition, augmentation in mean pressure primes a Rho–RoCK cascade.

6.8.2 Extracellular Matrix Constituents

Extracellular matrix protein-1 (ECM1) is needed for type-2 helper T lymphocytes (T_{H2}) egress from lymphoid tissues [670]. Glycoprotein ECM1 can bind to interleukin-2 receptor and activate sphingosine 1-phosphate receptor $S1P_1$ expression via the Krüppel-like factor KLF2.

6.8.3 Chemoattractants and Guidance Cues

Chemotaxis represents cell responses to a chemoattractant concentration or guidance cue gradient. In steep gradients, cells directly produce pseudopodia in the gradient direction. In weak gradients, pseudopodia are randomly generated and then cells select dominant pseudopodia aligned with the gradient direction. Cells can also become polarized so that they maintain the same front, even when the direction of the chemical gradient changes.

6.8.3.1 Chemokines

Chemokines that are involved in immune and inflammatory reactions induce chemotactic migration of leukocytes as well as other cells such as endothelial cells in angiogenesis. Chemokines do not target specific cells. A given cell responds to various kinds of chemokines.

Many types of chemokines have been identified and categorized according to the spacing of their first 2 cysteine residues close to the N-terminus: (1) C, or γ chemokines (a single N-terminal CysH; family of 2 described chemokines); (2) CC, or β-chemokines (2 adjacent N-terminal CysH; family of 28 known members); (3) CXC, or α-chemokines (2 N-terminal CysH separated by one amino acid, represented by an X; family of 17 detected chemokines); and (4) CX_3C, or δ chemokines (3 amino acids between the 2 cysteines; family with a single member; Tables 6.9 to 6.11).

Chemokines interact with their cognate G-protein-coupled receptors, CCR (CCR1–CCR11), CXCR (CXCR1–CXCR7), XCR1, and CX_3CR (Tables 6.12 and 6.13). A given receptor can bind different types of chemokines. Conversely, a given chemokine can link to several types of receptors. The receptor expression determines the action spectrum of chemokines, and subsequently the response of leukocytes to chemokines. Cells produce several chemokines with an overlapping action spectrum to yield a robust response.

6.8.3.2 Semaphorins and Plexins

Plexins regulate Rho GTPases that activate RoCK kinase and enhance actomyosin interactions and formation of stress fibers. Plexins act on rRas that,

Table 6.9. Chemokines and their receptors (Sources: [671, 672]). (**Part 1**) α-Chemokines, or CXC chemokines. Receptor CCR11 is an atypical chemokine binder, or chemokine decoy receptor, that binds and scavenges CXCL13, in addition to CCL19, CCL21, and CCL25. An alternatively spliced variant of CXCR3 mediates the control of angiogenesis by CXCL4, CXCL9, CXCL10, and CXCL11.

Type	Receptor
CXCL1	CXCR2 > CXCR1
CXCL2	CXCR2
CXCL3	CXCR2
CXCL4	CXCR3b
CXCL5	CXCR2
CXCL6	CXCR1/2
CXCL7	CXCR2
CXCL8	CXCR1/2
CXCL9	CXCR3/3b
CXCL10	CXCR3/3b
CXCL11	CXCR3/3b/7
CXCL12	CXCR4/7
CXCL13	CXCR5, CCR11
CXCL14	Unknown
CXCL15	Unknown
CXCL16	CXCR6
CXCL17	Unknown

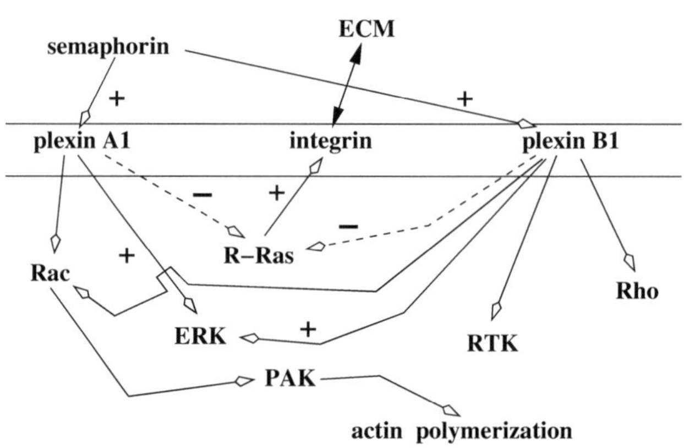

Figure 6.3. Semaphorins and their plexin receptors (Source: [674]). Effects on cell–matrix adhesion and cellular cytoskeleton activity.

in turn, regulate integrins [674] (Fig. 6.3). Plexins also interact with Rac that activates P21-activated kinase to initiate actin polymerization. Rnd1 GTPase binds to plexins.

Table 6.10. Chemokines and their receptors (Sources: [671, 672]). (**Part 2**) β-Chemokines, or CC chemokines (CCL3L1: chemokine [C–C motif] ligand 3-like-1). Chemokine (C–C motif) ligand 9 (CCL9) was previously designated as CCL10. Receptor CCR11 is an atypical chemokine binder, or chemokine decoy receptor, that binds CCL19, CCL21, and CCL25, as well as CXCL13 to direct their endocytosis and clearance. Chemokine CCL16 can bind to and signal via type-4 histamine receptor (H_4). A splice variant of CCL23 connects to and signals via formyl peptide receptor FPR2 (a.k.a. FPRL1 and lipoxin-A4 receptor), a member of a subclass of the G-protein-coupled receptors involved in chemotaxis.

Type	Receptor
CCL1	CCR8
CCL2	CCR2
CCL3	CCR1/5
CCL3L1	CCR1/5
CCL3L2	
CCL3L3	
CCL4	CCR1/5
CCL4L1	
CCL4L2	
CCL5	CCR1/3/5
CCL6	CCR1
CCL7	CCR1/2/3
CCL8	CCR1/2/3/5
CCL9(10)	CCR1
CCL11	CCR2/3/5
CCL12	CCR2
CCL13	CCR1/2/3/5
CCL14	CCR1/5
CCL15	CCR1/3
CCL16	CCR1/2/5/8
CCL17	CCR4
CCL18	CCR1/2
CCL19	CCR3/7, CCR11
CCL20	CCR3/6
CCL21	CCR3/7, CCR11
CCL22	CCR4/7
CCL23	CCR1/5
CCL24	CCR3
CCL25	CCR9/10, CCR11
CCL26	CCR3/6
CCL27	CCR10
CCL28	CCR3/10

Semaphorins,[29] transmembrane or secreted glycoproteins, regulate cell motility and attachment in vascular growth and tumor progression, among

[29] Among the 8 semaphorin subfamilies, subfamilies 3 to 7 are vertebrate semaphorins.

Table 6.11. Chemokines and their receptors (Sources: [671,672]). (**Part 3**) γ and δ chemokines.

Type	Receptor
γ chemokines — C chemokines	
XCL1	XCR1
XCL2	XCR1
δ chemokine — CX_3C chemokines	
CX_3CL1	CX_3CR1

Table 6.12. Examples of receptors of constitutive chemokines (Source: [673]; Bφ: basophil; Eφ: eosinophil; Mφ: macrophage; Mo: monocyte): CCL13 (BCA1: B-cell attracting chemokine); CCL19 (ELC: EBV induced gene-1 ligand chemokine); CCL20 (LARC: liver- and activation-regulated chemokine); CCL22 (MDC: monocyte chemotactic protein); CXCL12 (SDF1: stroma cell-derived factor-1); as well as inducible chemokines: CXCL6 (GCP2: granulocyte chemoattractant protein 2); IL8 (interleukin 8); CCL10 (IP10: interferon γ-inducible protein 10); CX_3CL1 (fractalkine); CCL2/8/7/13 (MCP1/2/3/4: monocyte chemotactic protein; CCL5 (RANTES: regulated on activated normal T-cell expressed and secreted); CXC (NAP2: neutrophil activating peptide 2); CCL3/4 (MIP: macrophage inflammatory protein). Receptors CXCR1 and CXCR2 correspond to interleukin-8 receptors IL8Rα and IL8Rβ, respectively.

Receptors	Ligands	Cells
CCR1–CCR11	CCL2/3/4/5/7/8	Bφ, Eφ, Nφ
	CCL13/17/19	Mo, dendritic cell
	CCL20/22	T and NK cells
CXCR1–CXCR7	CXCL6	Nφ
	CCL10/13	B and NK cells
CX_3CR	CX_3CL1	Mo and T and NK cells

other processes. Semaphorin-bound plexins can act as GAPs for rRas. Although other Ras superfamily members stimulate ERK, rRas slightly affects ERK, but regulates integrin activity. Active rRas increases integrin-based cell adhesion to the extracellular matrix. This effect is diminished by semaphorin binding to plexin.

6.8.4 Polarity Proteins

Directional migration requires polarity proteins, such as partitioning-defective Par6, atypical protein kinase-C (aPKC), and Disc large DLg1. Other

Table 6.13. Chemokine receptors, their main targeted G proteins, and principal ligands (Source: [82]).

Type	Main transducer	Preferential ligands
CCR1	Gi/o	CCL15, CCL23
CCR2	Gi/o	CCL2
CCR3	Gi/o	CCL11, CCL24, CCL26
CCR4	Gi/o	CCL22, CCL17
CCR5	Gi/o	CCL4
CCR6	Gi/o	CCL20
CCR7	Gi/o	CCL19, CCL21
CCR8	Gi/o	CCL1
CCR9	Gi/o	CCL25
CCR10	Gi/o	CCL27, CCL28
CXCR1	Gi/o	CXCL6, CXCL8
CXCR2	Gi/o	CXCL1, CXCL2, CXCL3, CXCL5, CXCL7, CXCL8
CXCR3	Gi/o	CXCL9, CXCL10, CXCL11
CXCR4	Gi/o	CXCL12α/β
CXCR5	Gi/o	CXCL13
CXCR6	Gi/o	CXCL16
CX_3CR1	Gi/o	CX_3CL1 (fractalkine)
XCR1	Gi/o	XCL1α/β (lymphotactins)

polarity proteins, Crumbs-3, palmitoylated membrane protein-5 (MPP5),[30] and protein associated with tight junctions (PATJ),[31] form a complex that regulates the localization of aPKC and Par3 at the leading edge during migration of epithelial cells for wound healing [675].

Cell polarization is the first step in cell migration, during which cells establish protrusive and contractile actin structures in the cell front and rear, respectively. Actin filaments in leading and trailing edges are in fact organized into actin-rich, but distinct, lattices. In the leading edge, actin filaments are organized in dendritic lamellipodial or parallel filopodial arrays. In the trailing edge, actin is organized into contractile structures rich in myosin-2 and often coated with tropomyosins, which influence actin–myosin-2 interaction and could facilitate actin polymerization via formins.

[30] A.k.a. MAGUK P55 subfamily member-5, protein associated with Lin7 (seven) PALS1, and Stardust.

[31] A.k.a. protein associated with Lin7 (seven)-1 (PALS1)-associated tight junction protein (PATJ) and inactivation no afterpotential-D (INAD)-like protein (INADL).

6.8.5 Components of Cell Cytoskeleton and Adhesion

6.8.5.1 Intermediate Filaments

Intermediate filaments intervene in cell motility, especially vimentin. Vimentin phosphorylation affects both its assembly into polymers [676], as well as (like keratin [677]) connections with intermediate filament-associated proteins.

Protein kinase-C phosphorylates vimentin.[32] Protein kinase-C links to transport vesicles and hence controls the endocytosis of integrins to the plasma membrane via the connection of vesicles with intermediate filaments, thereby regulating cell motility [678].

6.8.5.2 Microtubules

Microtubules regulate the turnover of focal adhesions at the cell front and participate in cell detachment at the cell back and establishment and maintenance of cell orientation [679]. Small GTPases of the RHO family capture and stabilize a polarized microtubule mesh via their effectors at the cell cortex. In turn, microtubules modulate the activities of these GTPases.

In migrating cells, 2 sources of asymmetry arise from microtubule organization [680]: centrosome orientation and selective stabilization of a subset of microtubules. Cell centrosome localizes to a position between the nucleus and the leading edge under the control of CDC42, Par6, aPKC, and dynein, as well as CDC42 effectors, such as IQ motif-containing GTPase-activating protein IQGAP1 (RasGAP), myotonic dystrophy kinase-related CDC42-binding kinase (MRCK), and diaphanous Dia1 formin.

Microtubules interact with actin filaments. Actin contributes to the initial polarization of microtubule lattices. Polarized microtubule arrays reinforce cell polarization initiated by the actin cytoskeleton. Besides, microtubules bind to RhoGEF2[33] to dampen its activity.

However, in fibroblasts, the nucleus moves away from the leading edge rather than the centrosome. The restricted formation of stabilized microtubules near the leading edge due to integrin engagement and focal adhesion kinase activity limit Rho GTPase to activate Dia1 only near the cell front. Microtubule plus-end-tracking proteins, such as plus-end-binding protein EB1 and adenomatous polyposis coli protein, as well as cytoplasmic linker protein-associating proteins (CLAsP)[34] in fibroblasts and actin crosslinking family

[32] Activity of PKC is correlated with increased haptotaxis.

[33] A.k.a. RhoGEF H1.

[34] Agent CLAsP is regulated by Rac, GSK3β, and PI3K, rather than by Rho GTPase. Like end-binding protein EB1, cytoplasmic linker protein CLiP1 (or CLiP170), and adenomatous polyposis coli protein (APC), cytoplasmic linker-associated proteins bind to polymerizing microtubules.

protein ACF7 (or microtubule–actin crosslinking factor MACF1) in endodermal cells, are involved in Rho–Dia1-mediated microtubule stabilization.[35] The spectraplakin ACF7 (Sect. 6.8.5.7) localizes to the plasma membrane and ruffles.

Receptor Tyr kinase HER2 regulates microtubule outgrowth to the cell cortex via a complex with mediator of ErbB2 (HER2)-driven motility MEMo, RhoA, and Dia1 formin [681]. Receptor HER2 impedes the activity of glycogen synthase kinase GSK3 via Memo and Dia1 to stabilize microtubules. Agent MEMo enables the localization of microtubule-associated proteins — adenomatous polyposis coli protein and cytoplasmic linker-associated protein CLAsP2 — to the plasma membrane and ruffles. It controls the localization of these +TIP molecules via the regulation of glycogen synthase kinase GSK3. Microtubule extension in the cell periphery also requires EB1 and its interactor, the ACF7 crosslinker. Membrane-bound APC allows the localization of ACF7 to the plasma membrane. In migrating cells, ACF7 is necessary and sufficient for the capture of microtubule in HER2 signaling. The MEMo–RhoA–Dia1 axis also controls the formation of nascent cell adhesions and recruitment of actin-binding proteins such as α-actinin.

6.8.5.3 Microfilaments

Cell locomotion requires molecular motor-driven motions, and assembly and disassembly of actin filament and cell adhesion sites. Cell migration in fact depends on several actin assemblies. Fast-growing barbed ends of actin filaments are orientated toward the plasma membrane in both lamellipodia and filopodia. The elongation of actin filaments pushes the leading edge forward. In addition, stress fibers that are composed of actin and myosin bundles are generated by formin-mediated actin polymerization and from ARP2–ARP3 complex-nucleated lamellipodial actin network to yield contractile forces for cell migration.

Lamellipodia evolve in different stages: protrusion, pause, and retraction. Cell protrusions are achieved by lamellipodia that are composed of diagonal networks of actin filaments (angles 15–90 degrees with respect to the front) [682]. During the transition from protrusion to pause, filaments reorientate (some angles <15 degrees with respect to the front) and filament number at the front decays. Withdrawal is often associated with the formation of actin bundles in a direction more parallel to the cell edge. Filopodial bundles of actin filaments also exhibit a wide angular distribution and variable bilateral polymerization rates along the cell front. The higher the polymerization rate the smaller the angle with the cell edge.

---

35 Plus-end tracking proteins (+TIP) interact with plasma membrane-associated proteins such as Rho GTPases via intermediate proteins, such as IQGAP1 and Dia formin. They can also link directly to cortical actin and components such as microtubule–actin crosslinker ACF7.

Two basic biological models of filopodium formation exist. The convergent elongation model states that filopodial actin filaments are derived from ARP2–ARP3-nucleated lamellipodial actin network. In the de novo nucleation model, filopodial actin filaments are nucleated at filopodial tips by formins. Formins indeed are actin-assembly factors that act downstream from $RhoA^{GTP}$. Unlike ARP2 and ARP3 that create branched actin networks, formins generate elongated actin filaments. However, several different mechanisms regulate the elongation and organization of actin filaments during filopodium formation. An additional model of filopodium formation states that initiation is done by the convergence of uncapped or formin-nucleated actin filament barbed ends at the plasma membrane by *myosin-10* [666]. The lengthening of actin filament barbed ends toward the plasma membrane yields membrane deformation occurring during filopodial elongation that is enhanced by brain-specific angiogenesis inhibitor-1-associated protein BAIAP2.[36] Elongating actin filaments are subsequently crosslinked by fascin and possibly Ena–VASP proteins.

6.8.5.4 Myosin

Migrating cells use actin and myosin filaments to move. Actin filaments polymerize and push to promote cell protrusions, as well as slide with *myosin-2* to retract. Myosin-2 indeed triggers stress fiber contraction. Myosin-2 not only causes retraction of cell protrusions, but also promotes maturation of focal cellular adhesions. Resulting tension can uncover binding sites of adhesion constituents and assist in their recruitment. These constituents then assemble, thereby leading to adhesion maturation. In addition, phosphorylation of myosin-2 regulatory light chain increases myosin-2 assembly into bipolar myosin filaments that assist actin bundling [634]. Myosin-2 is required for ^{F}actin anterograde flow in the cell body and retrograde flow in the lamella. It is absent in the lamellipodium.

Three isoforms of myosin-2 encoded by different genes exist: myosin-2A that associates with RoCK kinases for the organization of actin stress fibers and focal adhesions; myosin-2B that is involved in stabilization of cell front–rear polarity; and myosin-2C.

Spatial segregation of myosin-2A and -2B in cell front and rear is required for cell motility. Creation of a stable trailing region with thick actomyosin bundles and large adhesions impedes lamellipodium appearance in the cell rear and promotes tail extension. Myosin-2A clusters that occur in cell protrusions (but not at the leading edge) participate in the formation of actin cables in cell protrusions during directional migration, whereas myosin-2B is confined to the moving cell center and rear, where phosphorylated myosin light chain that activates myosin-2B intervenes in generation of an extended cell tail [683]. Phosphorylated MLC accumulation in the cell center and rear elicits local actin bundles and large, stable adhesions. On the other hand, activated myosin-2A prevents the formation of an extended cell tail. As the cell

[36] A.k.a. insulin receptor substrate P53 (IRSP53).

moves, the myosin-2A-rich region incorporates myosin-2B and becomes the cell rear. The formation of central and rear, large actin bundles is initiated by myosin-2A from the front of migrating cells, whereas the subsequent formation of thick actin bundles and large adhesions that impede cell protrusion is promoted by myosin-2B.

Both myosin-2 and microtubules operate to establish cell polarity. Myosin-2A isoform operates in cell contractility, focal adhesions, actin stress fiber organization, and tail retraction during cell motion. Nonetheless, myosin-2A can hamper membrane ruffling [684]. In the absence of myosin-2A, augmented ruffling is caused by GTPase Rac1, guanine nucleotide-exchange factor (RacGEF) T-lymphoma invasion and metastasis-inducing protein TIAM1, and KIF11 kinesin family member. A balance between actomyosin and microtubule networks is maintained for cell contractility, polarization, and migration. ^{F}Actin, myosin-2, tubulins, and focal adhesion proteins interact during cell migration. However, myosin-2A favors polymerization of actin rather than that of tubulin. Microtubule disruption enables colocalization of myosin-2B with actin stress fibers. When myosin-2A is absent and microtubule polymerization suppressed, myosin-2B partially compensates for cell contractility, but not for cell migration.

The migratory transmembrane receptor C-type lectin CLec13e[37] that is highly expressed in motile cells is recruited into clathrin-coated pits, then into endosomes. The CLec13e–Rho–RoCK–MLC2 signaling generates contractile forces via phosphorylation of myosin light chain-2, thereby promoting focal adhesion disassembly at the rear of the cell [685]. The activation of the Rho–RoCK pathway requires CLec13e-containing endosomes. C-Type lectin reduces the phosphorylation of 3 RoCK substrates: LIM kinases LIMK1 and LIMK2 and myosin phosphatase-1.

Myosin-10, also at the tips of filopodia along actin bundles (that build the core of filapodia), is a nanomotor required in filopodial formation downstream from CDC42 GTPases [686]. Myosin-10 with tip-complex proteins, such as vasodilator-stimulated phosphoproteins (VASP) and formins, form clusters of actin barbed ends during filopodium formation.[38]

---

[37] A.k.a. urokinase-type plasminogen activator receptor-associated protein (uPARAP), endocytic receptor Endo180, CD280, and mannose receptor-C2 (MRC2). This endocytic receptor that, together with the mannose receptor, M-type phospholipase-A2 receptor, and CD205, constitutes the mannose receptor family of C-type lectins (Vol. 1 – Chap. 7. Plasma Membrane). It regulates collagen remodeling and chemotactic cell migration, as it cooperates with membrane type-1 matrix metallopeptidase (mt1MMP), urokinase-type plasminogen activator (uPA) and UPA receptor (uPAR).

[38] Several molecules are involved in the formation of filopodia, such as formins and the vasodilator-stimulated phosphoprotein (VASP). GTPase CDC42 regulates filopodium formation. It can interact with Wiskott-Aldrich syndrome proteins (nWASP) to activate actin nucleators, such as ARP2 and ARP3 and formins.

6.8.5.5 Actin Cytoskeleton-Associated Proteins

The actin assembly that determines cell shape and operates in intracellular transport is required for cell locomotion. ^{F}Actin cycles through its nucleating and depolymerizing states.

Cofilin

Cofilin increases the number of polymerizing actin filament ends by its severing activity. Therefore, cofilin is able to regulate the interaction of actin networks in the lamellipodium and lamella. High cofilin activity increases spatial overlap of the lamellipodial and lamellar networks and reduces protrusion efficiency. Increased cofilin activity alters the coordination between cell edge protrusion and retraction.

Cofilin activity within the lamellipodium is regulated by a signaling cascade that involves Rac1 GTPase and the sequential downstream activation of PAK1 and LIMK1 kinases [687]. Inactivation of cofilin by active Rac1 stabilizes actin filaments and widens the lamellipodium owing to PAK1 phosphorylation. Conversely, cofilin phosphatase enhances the turnover rate of ^{F}actin and retrograde flow in the lamellipodium. Moreover, PAK1 inhibition decreases actin retrograde flow of the lamella independently of cofilin.

Paxillin and Its Associated Molecules

Paxillin (Pax) is a cytoskeletal and focal adhesion docking protein that regulates cell adhesion and migration. Paxillin acts downstream from focal adhesion integrins and other plasmalemmal receptors [688]. It is implicated in the regulation of integrin and growth factor signaling [689]. It has binding sites for signaling molecules and structural proteins, such as vinculin (Vinc), ArfGAP with PAK-interacting exchange factor (PIX)- and paxillin-binding domains APAP2[39] and focal adhesion kinase (FAK). Paxillin also interacts

---

Protein VASP at tips of filopodia can stimulate filopodium formation. Fascin is an actin-bundling protein in filopodia.

[39] A.k.a. G-protein-coupled receptor kinase (GRK)-interacting ARF GTPase-activating protein (ArfGAP) GIt2, paxillin kinase linker (PKL), and Cool (or RhoGEF6/7)-interacting Tyr-phosphorylated protein CaT2. Members of the GIT family interact with G-protein-coupled receptor kinases and act as adpribosylation factor (ARF) GTPase-activating protein (GAP). Interactor GIT2 links to GIT1 (APAP1). The NCK–PAK–RhoGEF6/7–APAP2 complex is recruited to focal adhesions by paxillin upon integrin engagement and Rac activation [690]. Protein APAP2 is phosphorylated on Tyr residues by both kinases Src and focal adhesion kinase (FAK) that then enable APAP2 recruitment at focal adhesions and paxillin binding. The activation of cytosolic protein Tyr kinases Src and FAK is one of the earliest steps in extracellular signal transduction primed by integrins toward the cytoskeleton. Upon integrin stimulation, integrin-binding

with Rho GTPases,[40] with its downstream target Lin1, Isl1, and Mec3 kinase (LIMK), as well as with the Rac pathway.

Phosphorylated paxillin binds to its effectors and transduces external signals into cellular responses via CRK adaptor (CRK: chicken tumor virus regulator of kinase) in association with CRK-associated substrate (CAS or BCAR1) and mitogen-activated protein kinase modules [691]. In particular, paxillin is phosphorylated following integrin stimulation by FAK and becomes a docking site for CRK [692]. Moreover, inhibitors of these pathways, such as C-terminal Src kinase (CSK), which impedes Src activity, bind directly to paxillin. Vinculin and parvin-α (or actopaxin bind actin directly to regulators of the actin cytoskeletal dynamics, such as RhoGEF6 and -7, APAP2, and P21-activated kinase. These proteins serve as effectors of the ARF and RHO families.

Paxillin connects to the APAP2–PAK–RhoGEF6/7 complex that regulates Rac and Rho GTPases. Activation of Rac at the leading edge of migrating cells is required for stable lamellipodia. Binding of Pax to α_4 integrin prevents Rac activation and thus inhibits lamellipodium formation [693]. Paxillin that recruits the ADP-ribosylation factor GTPase-activating protein (ArfGAP) reduces ARF activity, thereby hampering Rac GTPase.

Coronins

Coronins form a family of actin regulators that include 7 members. Coronins not only bind ^{F}actin, but also the ARP2–ARP3 complex to inhibit actin nucleation.

Peripheral T-lymphocyte deficiency is characterized by a migration defect, especially in thymic egress and lymph node travel, caused by mutation in the actin regulator coronin-1A that induces its mislocation in cell and enhances its inhibition of the ARP2–ARP3 complex [694]. Coronin-1A is mainly expressed by hematopoietic cells. After development in the thymus, mature thymocytes exit into the blood circulation to populate naive T-cell compartments.[41]

Coronin-1B induces the dissociation of the ARP2–ARP3 complex from actin branches in lamellipodia, in addition to its inhibition of actin nucleation

FAK autophosphorylates and is further phosphorylated by Src. Both enzymes then phosphorylate many focal adhesion substrates, such as adaptor BCAR1 (or CAS) and paxillin. Phosphorylated paxillin binds to adaptor CRK to permit Rac activation and cell motility. Small GTPases CDC42, Rac1, and RhoA allow the coordination of the temporospatial progression of events during the signaling that emanates from integrins, from cell polarization to migration. These GTPases are controlled by activators, the guanine nucleotide-exchange factors (CDC42/Rac) RhoGEF6 and -7, and inhibitors, the GTPase-activating proteins APAP1 and -2.

[40] Small GTPases of the RHO family regulate integrin-mediated adhesion via their effects on both the actin filaments and microtubules.

[41] Thymic export of T lymphocytes requires sphingosine 1-phosphate and its $S1P_1$ receptor.

catalyzed by ARP2 and ARP3 [695]. Coronin-1B thus antagonizes ARP2/3-containing branch stabilizer cortactin. Coronin-1B and cortactin reside in different regions, i.e., deeply and superficially in the lamellipodium, respectively.

Actin Nucleators ARP2 and ARP3 and Cofactors WASP and WAVe

Multiple actin nucleation factors collaborate to construct cytoskeletal structures, such as the ARP2–ARP3 complex and members of the formin subclass. Neuronal Wiskott-Aldrich syndrome proteins (nWASP) generate actin elongation for vesicular transport, but hinder spontaneous actin assembly.

Scaffold Abelson interactor AbI1 regulates both WASP-family verprolin homolog (WAVe) and nWASP actin-dependent functions [696]. Protein WAVe that acts downstream from Rac^{GTP} activates the ARP2–ARP3 complex. Among the 3 WAVe isoforms (WAVe1–WAVe3), lymphoid tissue highly expresses ubiquitous WAVe2.

Actin Nucleator Spire

Actin nucleation generates not only branched actin networks, but also nascent unbranched filament. Actin-nucleating factors contribute to both processes. The ARP2–ARP3 heptamer has low intrinsic activity that requires activating protein cofactors, such as Wiskott–Aldrich syndrome protein family members of nucleation-promoting factors regulated by small GTPase Rho as well as suppressor of cAMP receptor (SCAR; i.e., WAVe), among others related to actin branching. Actin nucleator Spire assembles 4 actin monomers that are lined up end to end, but does not form branched actin filaments.

Actin Nucleator SCAR (WAVe)

Both receptor Tyr kinases and G-protein-coupled receptors are able to trigger the reorganization of the actin cytoskeleton in many cell types to modify cell shape and attachment as well as trigger cell motility.

Calcium ion and cyclic nucleotides cAMP and cGMP are ubiquitous, intracellular, second messengers (Vol. 4 – Chap. 10. Signaling Pathways). Intracellular cAMP is produced via adenylyl cyclases upon agonist activation of G-protein-coupled receptors. Intracellular cGMP is synthesized by guanylyl cyclases upon stimulation by natriuretic peptides or nitric oxide (Vol. 3 – Chap. 6. Receptors). Influx Ca^{++} can result from inositol trisphosphate production that liberates stored Ca^{++} through IP_3 receptors. Confined Ca^{++}, cAMP, and cGMP signals to given subcellular regions allow selective activation of subsets of signaling effectors for precise signaling.

In addition, following agonist stimulation, the outflux of these signaling molecules across the plasma membrane through ATP-binding cassette transporters of cyclic nucleotides, particularly ABCc carriers of the MRP family members, solute carrier superclass member SLC22a6, and plasma membrane Ca^{++} ATPases, enables their activity as extracellular messengers that target cognate cell-surface receptors.

On the other hand, extracellular cAMP can be sequentially processed by ectophosphodiesterase to adenosine monophosphate and then by ecto-5′-nucleotidase to adenosine.[42] Adenosine can then act as an auto- or paracrine messenger via adenosine receptors (A_1, A_{2A}, A_{2B}, and A_3). Emptying of calcium stores triggers Ca^{++} influx via store-operated Ca^{++} channels. In addition, extracellular Ca^{++} influences the activity of numerous plasmalemmal Ca^{++} sensors, such as Ca^{++}-dependent ion channels, and extracellular Ca^{++}-sensitive G-protein-coupled receptors, especially calcium-sensing receptor that controls the Ca^{++} and Mg^{++} balance, and gap junction hemichannels that can open in response to a moderate decrease in external Ca^{++} concentration [697].

Suppressor of cAMP receptor (SCAR), or WASP-family verprolin homolog (WAVe), which is related to Wiskott-Aldrich syndrome proteins, acts as an inhibitor of G-protein-coupled signaling and a regulator of the actin cytoskeleton [698].

Actin Nucleator JMY

Like β-catenin, which is a cytoskeletal linker to cell adhesion loci as well as a component of the Wnt signaling pathway, P53 cofactor *junction-mediating and regulatory protein* (JMY) binds to transcriptional coactivators P300 and CREB-binding protein (CREBBP or CBP) that also serve as adaptors, polyubiquitin ligases, and histone acetyltransferases to activate P53-dependent transcription. It shuttles between the nucleus (primary location) and cytoplasm when the cell becomes motile to accumulate at the leading edge.

Actin nucleator JMY then promotes actin nucleation [699].[43] Increased JMY concentration enhances cell migration. It activates ARP2/3-induced actin branch formation. Furthermore, it is able to catalyze actin polymerization in the absence of ARP2 and ARP3. It can actually, like Spire, participate in new unbranched filament formation. It increases the speed at which new filaments are formed.

---

42 In Dictyostelium discoideum (amoebozoa – mycetozoa), the extracellular messenger cAMP can serve as a chemoattractant and morphogenetic signal that is transduced via a family of G-protein-coupled receptors, the cAMP chemoattractant receptors (cAR1–cAR4; not the Ca^{++}-sensing receptor [CaR]). Receptor cAR1 stimulates actin polymerization and crosslinking. In multicellular morphogenesis, receptor cAR2 is expressed by prestalk cells that form the presumptive tip.

43 Protein JMY possesses a WWWCA sequence that is composed of an actin monomer-binding domain, a central domain that tethers to actin and ARP2 and ARP3, and an ARP2/3-binding acidic domain.

6.8.5.6 Adhesion Molecules

Three major forms of adhesion contact can be defined: (1) focal complexes, (2) focal adhesions, and (3) fibrillar adhesions. These types of cellular contacts correspond to different stages of the interaction of cells with the extracellular matrix.

During cell migration, fibrillar adhesions form initially and immediately behind the leading edge of a cell to anchor filopodial struts and lamellipodial meshes of actomyosin filaments that mediate membrane protrusion. When protrusion ceases or retracts, fibrillar adhesions transform into larger focal adhesions to provide robust anchorage via contractile actomyosin stress fibers. Afterward, focal adhesions evolve into centrally located fibrillar adhesions, which are sites of fibronectin matrix deposition. The composition of adhesion contacts varies according to their type. Focal complexes lack zyxin; fibrillar adhesions lack α_V integrins [700].

Confluent, cohesive epithelial sheet cells can coordinate and migrate. Crosstalk between E-cadherin-mediated, cell–cell and integrin-based, cell–matrix adhesion complexes exist. E-Cadherin regulates lamellipodium activity and cell migration directionality, but not the migration rate [701]. Lamellipodia enables local exploration, before any motion.

Integrins

Coordinated assembly and disassembly of focal adhesions at the leading edge of the cell is a feature of migrating cells. Integrin-mediated cell motility can be decomposed into 3 stages: (1) formation of an integrin-dependent protrusion and new adhesion sites at the leading edge; (2) contraction of the actin–myosin cytoskeleton; and (3) detachment of the trailing edge after degradation of integrins at the cell back.

Integrins are heterodimeric, transmembrane, glycoproteic receptors that do or do not directly bind to components of the extracellular matrix and cytoskeleton. They are involved in cellular adhesion dynamics during cell migration. Clustering of activated integrins of the leading edge of a cell forms initial adhesion sites via the recruitment various proteins, such as actin stress fibers and other focal adhesion complex components (focal adhesion kinase, paxillin, and talin) to form mature adhesions that anchor to actin stress fibers.

Focal adhesion formation requires inhibition of myosin-2-mediated cell contractility behind leading membrane extensions during cell migration.

Talin activates integrin binding to the extracellular matrix and connects matrix-bound integrins (especially fibronectin-bound integrins) to the actin cytoskeleton. The initial cell-spreading reaction (primary adhesion and early cell-edge extension) is unaffected by talin-1 and 2 as actin rearrangements induced by integrin commitment and early activation of Src kinases are not mediated by talin [702]. Nevertheless, further reorganization of actin filaments and formation of focal adhesions require talin that hinders persistent, rapid, myosin-2-dependent rearward flow of actin.[44] Talin actually favors full activation of focal adhesion kinases via tension generated between matrix-bound integrins and actomyosin cytoskeleton. Moreover, adequate location of talin-binding focal adhesion proteins vinculin and paxillin in addition to activated focal adhesion kinases is required for stress fiber and focal adhesion formation.

Released integrins are recycled to the cell front, using actin-binding motor myosin-10 [703]. Protein phosphorylation by protein kinase-C and other Ser/Thr kinases regulates integrin transport. Protein kinase-B acts on the endosomal transport and recycling of integrins, by phosphorylating (inactivating) glycogen synthase kinase GSK3 [704].

Integrins and syndecans (membrane-intercalated proteoglycans)[45] cooperate to control adhesion-dependent directional cell migration [700]. Focal adhesion formation during cell spreading and migration on fibronectin depends on the commitment of $\alpha_5\beta_1$ integrin and syndecan-4. Integrins $\alpha_V\beta_3$ and $\alpha_V\beta_5$ cooperate with syndecan-1 for adhesion to vitronectin. Integrins $\alpha_2\beta_1$ and $\alpha_6\beta_4$ interact with syndecans for adhesion to laminin. Syndecans serve as receptors not only for ECM proteins (e.g., fibronectin), but also for growth factors (e.g., fibroblast growth factor-2 and vascular endothelial growth factor). Syndecan-4 activates protein kinase-Cα to regulate Rac1. Syndecan-4 is required in wound healing with simultaneous engagement of $\alpha_5\beta_1$ integrin. Syndecan-1 is also involved in re-epithelialization during wound healing, whereas syndecan-4 acts mainly in fibroblast migration and wound contraction. Syndecan-1 and -4 regulate angiogenesis owing to their synergistic activity with $\alpha_V\beta_3$, $\alpha_V\beta_5$, and $\alpha_5\beta_1$ integrins, as well as heparan sulfate-dependent regulation of growth-factor location. Syndecan-3 and various β_1 integrins participate in neuronal migration.

Adaptor Numb binds to several endocytic proteins. It also belongs to the machinery for directional integrin transfer in migrating cells [705]. In endo- and epithelial cell cultures, Numb and β integrin colocalize at focal adhesions. Numb at clathrin-coated structures binds to partitioning-defective polarization complex Par3 at the leading edge of migrating cells. Atypical protein kinase-C that belongs to the Par3 complex phosphorylates Numb to disrupt Numb–integrin association and release Numb from clathrin-coated structures. Protein Numb binds to free integrins and recruits them to clathrin-coated structures to initiate integrin recycling.

Integrin-mediated cell migration is due to integrin regulation of the activity of the RHO family GTPases. Localized activation of Rac for lamellipodial extension is promoted at the leading edge of cells by phosphorylation

[44] Following initial spreading, myosin-2-mediated contractile force generates a rearward flux of actin that is constrained to the cell periphery by focal adhesions and stress-fiber formation. Early cell spreading extensions depend on Src kinases, but not on talin. As talin-1 and -2 activate β_1 and β_3 integrins, initial cell spreading is mediated by heterodimers that do not contain these integrin subunits.

[45] Whereas syndecan-4 is ubiquitous, syndecan-1, -2, and -3 have a restricted tissue distribution.

of $\alpha_4\beta_1$ integrin and subsequent unbinding of adaptor paxillin, whereas it is impeded at the trailing edge by dephosphorylation of $\alpha_4\beta_1$ integrin, binding of paxillin, and inactivation of Rac GTPase.

Integrins $\alpha_9\beta_1$ engaged with specific ligands in the extracellular matrix modulate polyamine metabolism and activation of inward rectifier $K_{IR}4.2$ to increase cell migration [706]. Channels $K_{IR}4.2$ are concentrated at the leading edge of migrating cells, where they colocalize with focal adhesions. Inward rectifier potassium channels are blocked by intracellular polyamines[46] and Mg^{++}. Spermidine–spermine N1-acetyltransferase (SSAT) that binds to $\alpha_9\beta_1$ integrin acetylates high-order polyamines spermine and spermidine to relieve the blocking of $K_{IR}4.2$.

Cadherins and Catenins

Catenin-δ1 binds to the juxtamembrane domain of cadherins in intercellular adherens junctions. Overexpression of catenin-δ1 modulates the activity of Rho GTPases and augments cell migration. Catenin-δ1 colocated with cortactin-containing actin structures in both junctional and extrajunctional sites regulates lamellipodial dynamics [707]. At the leading edge of migrating cells, catenin-δ1 cooperates with cortactin and its partner ARP3 for lamellipodial extension and focal adhesion formation.

6.8.5.7 Spectraplakins

Adequate focal adhesion dynamics (formation, maturation, and disassembly) are required during cell migration. Small Rho GTPases and focal adhesion kinases control focal adhesion turnover. However, assembly and disassembly rates as well as focal adhesion features (size and strength) depend on spectraplakins.

Spectraplakins constitute a set of giant, cytoskeletal linker proteins that are able to directly bind both microtubules and actin filaments and thus crosslink microtubule and ^{F}actin networks. Coordination and interaction between microtubules and filamentous actin regulates cell migration, as microtubule linkage to focal adhesions allows focal adhesion turnover. Spectraplakins are involved in the maintenance of both cell–cell (via cadherins) and cell–matrix (via integrins) adhesions.

Spectraplakins share a similarity with members of the spectrin and plakin superfamilies. All isoforms contain a microtubule-binding motif at the C-terminus, and some isoforms contain an actin-binding site at the N-terminus.[47]

[46] High-order polyamines yield strong channel block compared with that by low-order polyamines. Spermidine is a polyamine that inhibits nitric oxide synthase NOS1, but stimulates T4 polynucleotide kinase and T7 RNA polymerase. Spermine is a polyamine formed from spermidine that serves as growth factor in some bacteria.

[47] These 2 cytoskeletal-connecting domains can be separated by up to 8,000 amino acids that form spectrin repeats or both plakin and spectrin repeats.

Spectraplakin gene transcripts are alternatively spliced to generate a wide diversity of isoforms. The spectraplakin family includes dystonin[48] and microtubule–actin crosslinking factor MACF1.[49]

Microtubule–actin crosslinker MACF1 guides microtubule growth along filamentous actin toward focal adhesions [708]. Spectraplakin MACF1 binds not only to microtubule and ^{F}actin, but also microtubule plus-end proteins, such as end-binding protein EB1 and cytoplasmic linker protein-associating proteins (CLAsP), and coordinates their location with respect to ^{F}actin and focal adhesions. Both crosslinking and ATPase activities of MACF1 are needed for synergistic motion of microtubules and actin cables and actin-guided tracking of microtubules to peripheral focal adhesions.

6.8.6 Peptidases

6.8.6.1 Calpains

Cell migration involves the successive formation and degradation of cell–matrix adhesions. Members of the calpain family (calpain-1–calpain-3, calpain-5–calpain-15) of ubiquitous, calcium-dependent, non-lysosomal, cysteine peptidases regulate focal adhesion dynamics.[50]

Calcium-activated neutral peptidase (calpain) small regulatory subunit CAPNS1 (formerly called CAPN4) operates in the production of traction forces and mediation of mechanical stress detection during fibroblast migration [709]. Stress fibers are aberrant and also contain fewer colocalized vinculin-containing adhesions in CAPNS1 gene-deficient cells than CAPN1- and CAPN2-knockdown cells.

Focal adhesions rely on integrins that operate as mechanical sensors, chemical receptors, and force transmitters. Calpain cleaves many focal adhesion constituents and cytoskeletal linkages, such as integrins, in particular β_1 and β_3 integrins, focal adhesion kinase, and talin [710]. Activation of calpain-1 and -2 requires calcium concentrations in the micromolar and millimolar range, respectively.

---

[48] A.k.a. bullous pemphigoid antigen BPAg1 and microtubule-actin crosslinking factor MACF2.

[49] A.k.a. macrophin 1, trabeculin-α, 620-kDa actin-binding protein ABP620, and actin crosslinking family protein ACF7.

[50] The calpain proteolytic system comprises: (1) calcium-activated neutral peptidases (CANP or CAPN), or calpain peptidases and (2) calpain-specific inhibitor calpastatin (CaSt). The best characterized members of the calpain family — calpain-1 and -2 — are heterodimers composed of large subunits CAPN1 and CAPN2, respectively, that share the small subunit CAPNS1 (formerly CAPN4). The genes of the calpain family also encodes small subunit CAPNS2 as well as CAPN3, CAPN5 to CAPN14, and small optic lobe homolog SOLH, i.e., calpain-3, calpain-5 to calpain-15.

6.8.6.2 Caspases

The pathway that includes caspase-11 and actin-interacting protein-1 (AIP1),[51] which cooperates with cofilin to optimize actin depolymerization, modulates the migration of inflammatory cells [711].

6.8.7 Secreted Transcription Factors

Nuclease-sensitive element-binding protein NSEP1[52] of the cold shock protein family[53] functions not only in gene transcription, but also in RNA processing. It is an miR216a target.[54] Once secreted by renal mesangial cells and monocytes upon inflammation using vesicles and ATP-binding cassette transporters, NSEP1 may operate as a mitogen in different cell types and as a promigratory agent [714].

6.8.8 Receptors

Adenosine triphosphate release from the leading edge of neutrophils amplifies chemotaxis via $P2Y_2$ receptors [715]. Neutrophils rapidly hydrolyze released ATP to adenosine, and A_3 adenosine receptors are recruited to the leading edge. Adenosine via A_3 receptors thus promotes cell migration.[55]

---

[51] A.k.a. WD repeat domain-containing protein WDR1.

[52] A.k.a. Y-box-binding transcription factor YBx1 (or YB1), CCAAT-binding transcription factor-1 subunit-A, DNA-binding protein-B (DBPb), and enhancer factor-1 subunit-A (EF1a). Overproduction of NSEP1 can reduce the synthesis of epithelial proteins, such as E-cadherin, ZO1, mucin-1, and cytokeratin-18, but can upregulate that of mesenchymal proteins, such as N-cadherin, vimentin, fibronectin, and smooth muscle actin-α, thereby favoring epithelial–mesenchymal transition [712]. It may also provoke a cell cycle arrest.

[53] Survival of organisms requires adaptation to environmental stress, such as cold, heat, and acid shocks, as well as pressure and osmotic stresses. In some bacteria, proteins produced at low temperatures constitute the group of cold-induced (CIP) or cold shock (CSP) proteins. Antifreeze glycoproteins (AFGP) circulate in blood of some animal species.

[54] In diabetic nephropathy, transforming growth factor-β1 causes accumulation of extracellular matrix proteins such as collagen-1α2 and hypertrophy of renal mesangial cells. Factor TGFβ also enhances the production of TGFβ-stimulated clone TSC22 as well as microRNA-216a, thereby downregulating Ybx1 [713]. In normal conditions, NSEP1 forms a ribonucleoproteic complex with Tsc22 mRNA. In diabetic nephropathy, TSC22, the concentration of which rises, interacts with E-box regulator transcription factor binding to immunoglobulin heavy chain (IGHM) enhancer TFE3.

[55] Cyclic adenosine monophosphate is detected by G-protein-coupled cAMP receptor-1 in some living species. In Dictyostelium discoideum, activated receptor stimulates RasC and RasG that prime different pathways mediated by phosphatidylinositol 3-kinases, target of rapamycin complex-2, phospholipase-A2, and

Neutrophils respond to both endogenous interleukin-8 and bacterial fMLP (^{N}formyl methionyl-leucyl-phenylalanine) chemoattractant. Neutrophils that experience these 2 conflicting gradients follow PI3K-independent fMLP chemotaxis rather than PI3K-dependent IL8 chemotaxis [716].

Urokinase-type plasminogen activator receptor (uPAR) localizes to the leading edge of migrating cells, where it can foster motility by activating Rac [717]. Glycosyl-phosphatidylinositol-anchored uPAR may rely on transmembrane coreceptors, such as G-protein-coupled receptors, tetraspanins, low-density lipoprotein receptor-related protein, associated protein uPARAP (or CLec13e), and integrins, for intracellular signaling. Receptor uPAR may interact with integrins via vitronectin, even in the absence of urokinase-type plasminogen activator [717]. Receptor uPAR cooperates with β_3 integrin and can then trigger the formation of the CAS-CRK complex that recruits the GEF DOCK1 to the plasma membrane (CAS–CRK–DOCK1 complex) to activate Rac GTPase (vitronectine–uPAR–integrin–FAK–CAS–CRK–DOCK1–Rac axis).

6.8.9 Small Guanosine Triphosphatases

The activity of the actin cytoskeleton (cellular transport, cell motility, phagocytosis, etc.) is regulated by RHO family GTPases, which are activated by Rho GEFs, inactivated by Rho GAPs, and sequestered by Rho GDIs. The 3 main members of the RHO family — CDC42, Rac1, and RhoA — stimulate the formation of filopodia, lamellipodia, and contractile actin–myosin filaments, respectively (Fig. 6.4).

Small GTPases Rho, Rac, and CDC42 regulate the actin cytoskeletal dynamics, especially for shape changes and cell motility [718] (Table 6.14 and Fig. 6.5). Small GTPases foster the formation and polarization of actin filaments and microtubules as well as the polarized transfer of cargos and distribution of adhesion molecules. Mutual antagonism between Rac and RhoA GTPases contributes to their cell-polarizing effects.

In neutrophils, Rac mediates the protrusion of the pseudopod by stimulating ^{F}actin assembly. RhoA controls the retraction of the cell back via actin–myosin contractility. At the front, CDC42 guides the response. The chemoattractant binds to its receptor and activates: (1) RoCK enzyme via RhoGEF and RhoA; (2) PI3K that transforms phosphatidylinositol (4,5)-bisphosphate into phosphatidylinositol (3,4,5)-trisphosphate; and (3) CDC42 via the CDC42–PAK1–RhoGEF6/7 complex. Activated PI3K increases the concentration of PIP_3 at the cell front. Phosphatidylinositol trisphosphate and activated CDC42 cooperate for ^{F}actin polymerization via Rac GTPase. Activated RoCK binds and stimulates PTen phosphatase that localizes with

guanylyl cyclases [716]. Enzyme PLA2 stimulated by cAMP during chemotaxis releases free fatty acids and lysophospholipids from phospholipids. Guanylyl cyclases produce cGMP that is required for myosin 2 filament formation at the back of the cell.

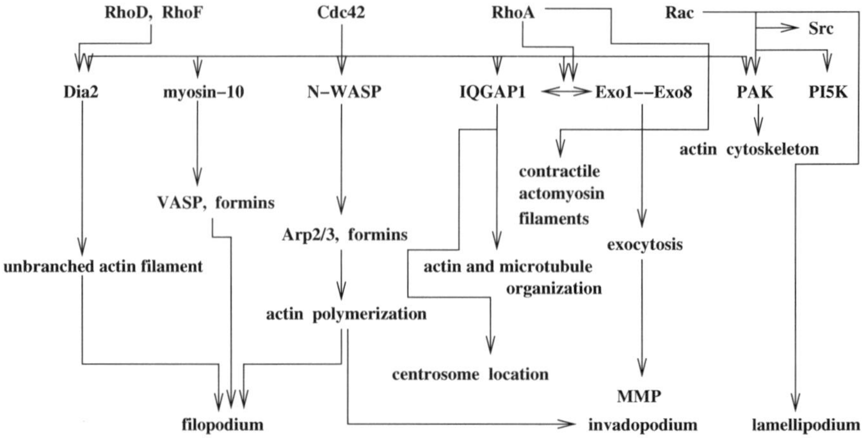

Figure 6.4. Small GTPases and cellular protrusions.

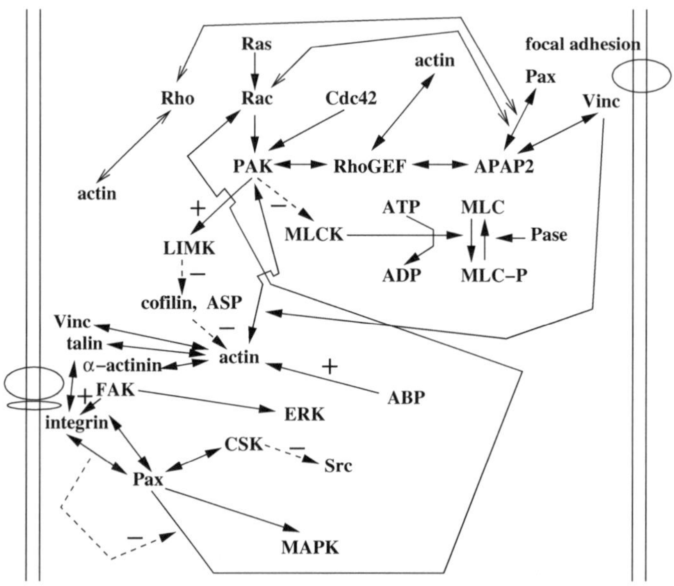

Figure 6.5. Rho GTPases, focal adhesions, and the cytoskeleton.

RhoA and RoCK to the back and sides of the motile cell [719] (Fig. 6.6). Small GTPase RhoA acts synergistically with PTen to reduce the levels of the phosphorylated protein kinase-B.

A set of positive feedback loops exists at the cell front that enables the maintenance of cellular protrusions: (1) localized activation of Rac GTPase initiates and maintains cellular protrusions, as it promotes the recruitment and activation of PI3K that, in turn, supports Rac via PIP_3-sensitive

Table 6.14. RHO family GTPases and cell activity (cell motility, smooth muscle contraction–relaxation, endothelium permeability, platelet activation, and leukocyte migration; ARP: actin-related protein; PAK: P21-activated kinase; RoCK: Rho-associated, coiled-coil-containing protein kinase; nWASP: neuronal Wiskott-Aldrich syndrome protein; WAVe: WASP-family verprolin homolog). Rac GTPases inhibit RhoA GTPases and conversely. Activity of RHO family GTPases depends on the balance between Rho guanine nucleotide (GDP–GTP)-exchange factors (GEFs) and Rho GTPase-activating proteins (GAPs). Rac GTPases can inhibit Rho via the activation of P190RhoGAP, whereas Rho represses Rac via RhoGAP2. Among other regulators, APAP ArfGAPs bind to paxillin, RhoGEF6/7, and FAK and GRK kinases. On the other hand, involved GEFs include RhoGEF6 and -7 (a.k.a. α- and β PIX as well as Cool2 and Cool1, respectively), DOCK1 (for Rac1), RapGEF1, and P190RhoGEF (or RasGRF1; for RhoA). Whereas RhoA is activated near the cell edge during leading-edge progression, CDC42 and Rac1 are activated slightly downstream from the leading edge with a delay of about 40 s. In the lamellum, localized activation of Rho fosters adhesion maturation. Stimulation by Rho activates myosin-2 via RoCK kinase. Myosin-2-generated tension elicits both adhesion maturation (local action) and disassembly (cell rear action).

GTPase	Function
	Cell leading region
CDC42	Filopodium, focal complex Stimulation of nWASP, WAVe, ARP2–ARP3 Activation of PAK
Rac	Lamellipodium, focal complex Activation of nWASP, WAVe, ARP2–ARP3 Activation of PAK
Rho	Actin polymerization via interaction with formins Stimulation of Diaphanous
	Cell trailing region
Rho	Stress fiber, focal adhesion Activation of RoCK

RacGEFs [720]; (2) microtubule polymerization activates Rac and Rac stabilizes microtubules; (3) stimulated integrins recruit and activate Rac that initiates new adhesion foci at the leading edge that recruit and support the clustering of activated integrins to the edge of lamellipodia, and (4) RalA activated by guanine nucleotide-exchange factors (GEF; RGL1–RGL4) of RalA recruited by Ras promotes the formation of lamellipodia that contribute to RalA activation [721].

Spatially and temporally restricted changes in the concentration of calcium — Ca^{++} flickers — happen near the leading edge of migrating cells. Nanodomains of calcium ion support cell migration, as they control the location

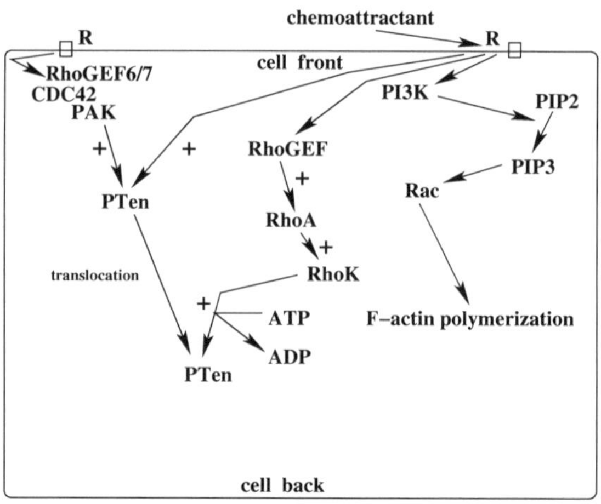

Figure 6.6. Chemotaxis. Chemoattractants activate RoCK via RhoGEF and RhoA. Activated RoCK binds and phosphorylates phosphatase and tensin homolog deleted on chromosome 10 (ten), or phosphatidylinositol 3-phosphatase (PTen). PTen localizes with RhoA and RoCK to the back and sides of the motile cell. Activation of the PI3K and CDC42–PAK1–RhoGEF6 complex at the cell front (leading edge) further increases PTen activity.

of Rho GTPases. Receptor-activated transient receptor potential canonical channels TRPC5 and TRPC6 (Vol. 3 – Chap. 2. Membrane Ion Carriers) regulate actin remodeling and cell motility [722]. The former complexes with Rac1; the latter with RhoA GTPase. Calcium influx through TRPC5 activates Rac1, thereby fostering the generation of lamellipodia; Ca^{++} import increases RhoA activity, hence supporting the formation of stress fibers and focal contacts.

6.8.9.1 Ras GTPase

During the initial stage of chemotaxis (directional sensing stage), small GTPases Ras localize rapidly (<3 s) to the plasma membrane, first uniformly, then at the cell front. Small Ras GTPases activate cRaf and the extracellular signal-regulated kinase cascade. Activation of cRaf requires prohibitin [723]. In addition, X-linked (xIAP) and cellular (cIAP) inhibitors of apoptosis bind directly to and destabilize cRaf, hence impeding cell migration [724].[56] Small GTPases of the Ras superfamily also activate the Rac pathway. Activity of Rac^{GTP} depends on the mode of membrane anchoring. Cyclic AMP-based chemotactic signaling targets Ras GTPases that activate phosphatidylinositol 3-kinases and target of rapamycin TORC2 complex.

[56] Inhibitors of apoptosis are caspase inhibitors that also act in various cellular processes, such as cell division and signaling. Agent XIAP promotes cRaf ubiquitination via HSP90.

Table 6.15. Small GTPases Rho and Rac and cell migration.

GTPase	Effect
Rho–formin	Actin polymerization, stress fiber formation
Rho–RoCK–MLC	Myosin activity, stress fiber contraction
Rho–RoCK$\ominus \longrightarrow$ MLCP	Myosin activity, stress fiber contraction
Rac–WAVe–ARP2/3	Actin polymerization, filipodium formation

6.8.9.2 Rho GTPase

Activation of RhoA GTPase and, subsequently, of its effector, the RoCK kinase, promotes retraction, suppressing lamella formation. RhoA GTPase promotes the formation of contractile stress fibers in the cell body and at the rear (Table 6.15).

Both RhoD and RhoF directly activate Diaphanous-related formin Dia2 that promotes the directed growth of actin filaments. Transcription factors Mesoderm posterior protein MesP and forkhead box FoxF as well as fibroblast growth factor upregulate small GTPases RhoD and RhoF to control this process [725]. Monomeric GTPases RhoD and -F contribute to the formation of protrusions in migrating cells.[57]

6.8.9.3 Rac GTPase

Activation of Rac GTPase supports actin polymerization and the protrusion of leading lamellae and lamellipodia. At the leading edge, Rac1 and CDC42 promote cell motility through the formation of lamellipodia and filopodia, respectively. Temporal and spatial control of these competing signalings allows the coordination of elongation and shortening of actin filaments, thereby the cycling between cellular protrusions and retractions for an effective cell displacement.

Rac GTPases and Src Tyr kinase that undergo a mutual regulation differ in subcellular distribution within the same signaling feedback loop [726]. During angiogenesis, vascular endothelial growth factor transiently activates Src to disrupt adherens junctions. Platelet-derived growth factor targets both Src and Rac to regulate cell migration. In response to PDGF, Rac activity is highly polarized at the leading edge, whereas Src activity is relatively uniform. PDGF activates Rac in coordination with integrins that colocalize with PDGF receptors at the leading edge of migrating cellss. Activated PDGFRs then recruit PI3K that synthesizes PIP_3 to enroll RacGEFs, such as Vav2,

[57] Transcription factor Mesoderm posterior homolog in response to T-box factor TBx6 and FGF activates transcription factors of the E26 (ETS) family ETS1 and ETS2 and forkhead box FoxF to target heart precursor trunk ventral cells and cause their migration.

TIAM1, and DOCK180. Simultaneously, at the leading edge, integrins can gather Rac to membrane rafts. Activated Rac forms positive feedback loops to maintain a persistent polarization that is mediated by Src kinases. Enzymes Src phosphorylate BCAR1 scaffold[58] that recruits CRK adaptor and DOCK1 guanine nucleotide-exchange factor.[59] Kinase Src also hastens PI3K enzyme. The RacGEFs Vav2, TIAM1, and DOCK1 are recruited to the plasma membrane by PIP_3 or directly phosphorylated by Src for Rac activation. On the other hand, Rac mediates suitable Src location. PDGF-induced Src activation is facilitated by Src translocation from perinuclear regions to the plasma membrane that is mediated by Rac and actin filaments. In addition, Src hastens P190RhoGAP and subsequently inhibits RhoA at focal adhesion sites, whereas Src activates RhoA at podosomes. RhoA activity is also characterized by a subcellular compartmentation, as it targets kinase RoCK at a contractile region behind lamellipodium and the cell rear, but colocalizes with Dia at the leading edge of a migrating cell.

Signal transduction pathways involved in actin polymerization require small Rac GTPase and its Rac-associated phosphatidylinositol 5-kinase. Numerous effectors have been implicated in Rho-mediated signaling. P21-activated kinase (PAK) is implicated in both CDC42- and Rac-mediated organization of the actin cytoskeleton [727]. Activation of Rac can occur via RhoGEF6 and -7.[60] Binding of PAK to RhoGEF6(7) coordinates the activation of CDC42 and Rac associated with phosphatidylinositol 3-kinase. Proteins RhoGEF6 and -7 can tether to the ArfGAP APAP1.[61] Protein APAP1 links the PAK–RhoGEF6(7) complex and focal contacts. Actin-associated synaptopodin regulates RhoA signaling [728]. Synaptopodin competes with SMURF1 Ub ligase for RhoA binding, and thus prevents SMURF1-mediated ubiquitination of RhoA for proteasomal degradation.

GTPase-activating protein RhoGAP24 that specifically targets Rac GTPase binds to *filamin-A* [729], resides at both the leading and trailing edges of polarized cells, and manages the coordination of antagonist reactions. Activation of Rac leads to actin-filament elongation. Small GTPase Rac is antagonized by filamin-A in membrane protrusions. Phosphorylation of RhoGAP24 by RoCK kinase stimulates its RacGAP activity, which requires filamin-A. Agent RhoGAP24 suppresses leading-edge protrusions.

The phosphatidylinositol (3,4,5)-trisphosphate-dependent Rac exchanger PREx1 is a guanine nucleotide-exchange factor for Rac downstream from receptor Tyr kinases such as HERs (HER1–HER4) bound to epidermal growth factor and heregulin as well as G-protein-coupled receptors such as chemokine receptor CXCR4 that trigger cell migration [730]. Protein PREx1 that is

---

[58] Breast cancer anti-estrogen resistance protein BCAR1 is a.k.a. CAS or P130CAS.

[59] Dedicator of cytokinesis DOCK1 is a.k.a. 180-kDa protein downstream of CRK (DOCK180).

[60] A.k.a. PAK-interacting exchange factors α- and β-PIX.

[61] A.k.a. G-protein-coupled receptor kinase-interacting protein GIT1.

Table 6.16. Signaling pathways controlled by CDC42 (Source: [720]). Small GTPase CDC42 localizes the microtubule-organizing center (MTOC) and Golgi body in front of the nucleus toward the leading edge. Stathmin is a regulator of microtubule dynamics that sequesters tubulin, but also fosters rapid microtubule remodeling upon phosphorylation. CDC42 activates RalA GTPase that interacts directly with components of the exocyst, an octamer involved in vesicular tranport. Microtubule tip protein Cap-Gly domain-containing linker protein CLiP1 interacts with the CDC42–Rac effector IQGAP and mediates transient capture of microtubules at the cell cortex of the leading edge of polarized cells.

Pathway	Effect
CDC42–Par3/6–aPKC	Polarity
CDC42–WASP–ARP2/3	Actin polymerization, filipodium formation
CDC42–PAK–LIMK	Actin polymerization
CDC42–PAK–stathmin	Tubulin polymerization
CDC42–RalA–exocyst	Polarized transfer of cargos
CDC42–IQGAP–CLiP1	Microtubule polarization
CDC42–MRCK–myosin	Nucleus displacement
CDC42–Ral–exocyst	Filipodium formation

activated by the $G\beta\gamma$ subunit of heterotrimeric G proteins and phosphatidylinositol (3,4,5)-trisphosphate can thus integrate signals from both RTKs and transactivated GPCRs to activate Rac and cause cell migration. In response to heregulin, PREx1 is phosphorylated (Ser605, Ser606, and Ser1169; promoted activity) and dephosphorylated (Ser313 and Ser319; restricted activity) to control the activation of Rac1 [731].

Small GTPase Rac1, especially ^{GTP}Rac, can be sumoylated. Small ubiquitin-like modifier (SUMO) ligase PIAS3 interacts with Rac1 for optimal cell migration in response to hepatocyte growth factor stimulation [732].

Collective cell migration happens during morphogenesis and tumor metastasis. Small GTPase Rac controls collective cell migration in response to guidance signals via epidermal (EGFR), platelet-derived (PDGFR), and vascular endothelial (VEGFR) growth factor receptors. Localized activation of Rac in a single cell suffices for collective migration of cells [733]. Signaling downstream from Rac includes Jun N-terminal kinases that maintain communication between cells, but not for the control of directional protrusions [733].

6.8.9.4 CDC42 GTPase

CDC42 interacts with multiple proteins (Table 6.16). Small GTPase CDC42 contributes to the cell polarization as it localizes to the microtubule-organizing center and Golgi body in front of the nucleus toward the leading edge, thereby promoting the microtubule growth into the lamella and microtubule-mediated delivery of Golgi-derived vesicles to the leading edge.

Table 6.17. Activators of CDC42 for cell migration (Source: [720]). In neurons, cadherins (Cdh) can stimulate CDC42 via the GEF Vav1 or inactivate via β-catenin and RhoGAP32 (or RICS). Activated T-cell receptors (TCR) recruit CDC42 and actin polymerizers and activate CDC42 via interleukin-2-inducible T-cell kinase (ITK). In lymphoid and myeloid cells, another GEF, interferon regulatory factor (IRF4)-binding protein (IBP), or differentially expressed in FDCP6 homolog (DEF6), is recruited to the immunological synapse and stimulates CDC42. Chemokines connect to their cognate G-protein-coupled receptors (GPCR). The Gβγ subunit of the trimeric G protein binds to the CDC42 effector PAK1 that interacts with RhoGEF6 (CDC42 GEF). Epidermal (EGF) and platelet-derived (PDGF) growth factor bind to their receptor Tyr kinase (RTK) that activate CDC42 via Src and Vav2.

Initiating mechanism	Pathway
Intercellular contact	TCR–Vav2–CDC42
	TCR–IBP (DEF6)–CDC42
	Cdh–Vav1–CDC42
Soluble factors	Chemokine–GPCR–Gβγ–RhoGEF6–CDC42
	Growth factor–RTK–Src–Vav2–CDC42

Cell polarization by CDC42 can result from [720]: (1) stimulated intercellular contact, especially between immune cells; (2) soluble factors, such as chemoattractants and growth factors, particularly epidermal (EGF) and platelet-derived (PDGF) growth factor, that target their cognate receptor Tyr kinases; and (3) physical constraints (Table 6.17).

Small GTPase CDC42 activates the WASP (nWASP)–ARP2/3 pathway. Association of CDC42 with WASP or nWASP and binding to phosphatidylinositol (4,5)-bisphosphate relieves the auto-inhibited conformation of WASPs to activate the ARP2–ARP3 complex and induce ARP2–ARP3 complex-dependent actin filament nucleation.

Alternatively, CDC42 acts on brain-specific angiogenesis inhibitor-1-associated protein BAIAP2[62] that is able to bind GTPase CDC42, its effector WAVe2, and Enabled homolog (EnaH) protein of the Ena–VASP family. Therefore, CDC42 can regulate filopodium formation by activating actin filament nucleation via WASP or nWASP and membrane deformation via BAIAP2 [666]. Scaffold BAIAP2 may recruit EnaH to filopodial tips downstream from CDC42 and activate WAVe2 downstream from Rac1 to induce ARP2/3-promoted actin filament nucleation.

A positive feedback loop exists between CDC42 and proton efflux by Na^+–H^+ exchanger NHE1 (or SLC9a1) at the leading edge of migrating fibroblasts [734]. Small GTPase CDC42 acts upstream from NHE1 to promote H^+ efflux. Conversely, NHE1 leads to CDC42 activation by guanine nucleotide-exchange factor and operates for the location of $CDC42^{GTP}$ in the leading edge

[62] A.k.a. 53-kDa insulin receptor substrate protein (IRSP53).

of migrating cells. Binding of GEF to phosphotidylinositol (4,5)-bisphosphate indeed depends on pH.

GTPase-activating protein RhoGAP10 for CDC42 regulates the ARP2–ARP3 complex and ^{F}actin polymerization on Golgi membranes in coordination with ARF1 [735]. RhoGAP10 bridges ARF1 and CDC42 signaling pathways, which require actin dynamics.

Small GTPase CDC42 activates RalA GTPase that interacts directly with components of the exocyst, an octamer involved in vesicular tranport [721]. Microtubule tip protein Cap-Gly domain-containing linker protein CLiP1 interacts with the CDC42–Rac effector IQGAP and mediates transient capture of microtubules at the cell cortex of the leading edge of polarized cells [736].

6.8.9.5 Other Monomeric GTPases

In addition to CDC42, small GTPase RIF also induces filopodium formation in cells. RIF activates actin polymerization by activating diaphanous-related formin Dia2. Other Rho GTPases of the RHO family (RhoQ, RhoT, and RhoU) also elicit filopodial formation [666].

Rnd proteins[63] that regulate the organization of the actin cytoskeleton, modulate contractility in smooth muscle cells, which involves RhoA (reciprocal inhibition). GTPase RhoE (Rnd3) regulates the activity of actin filaments by binding and inhibiting RoCK1,[64] which induces actin–myosin contractility. Phosphorylated (by RoCK1) and unphosphorylated RhoE are located in the cytosol and plasma membrane, respectively. Kinase RoCK, stimulated by PDGF, phosphorylates RhoE. Phosphorylation that controls the RhoE stability requires the protein kinase-C pathway [737]. Phosphorylation of RhoE induces stress fiber disruption.

Small Rnd GTPases interact with RhoGAP [738]. In addition, Rnd1 acts on adaptor GRB7 involved in cell migration. Small GTPase Rnd1 inhibits calcium sensitization in vascular smooth muscle cells, particularly in varicose veins, with subsequent decreased contractility.

6.8.10 Protein Kinases and Phosphatases

The rapid, reversible phosphorylation of mediators enables at least partly the control of localization, timing, and duration of migratory processes as well as their coordination. Mediator phosphorylation state is regulated by kinases and phosphatases (Table 6.18).

---

[63] Rnd3 (RhoE) is also involved in the regulation of cell cycle.

[64] Conversely, Rho effector RoCK1 (but not RoCK2, also activated by RhoA) phosphorylates Rnd3.

Table 6.18. Regulators of cell migration and pathways (Source: [639]; $\ominus \longrightarrow$: inhibition; $\downarrow$: decrease; PPM1e, PPM1f: PP2c domain-containing protein phosphatase-1E and -1F [a.k.a. partner of PAK-interacting exchange factor (PIX, or RhoGEF6/7) POPX1 and POPX2]).

Pathway	Effect
PP2–Rac1	Maintenance of FAK–Src–paxillin complexes
PP2–eCadh–βCtn–Rac–IQGAP1	Maintenance of adherens junction, control of interactions of microfilaments with focal adhesions
PP2–PKCζ	Inhibition tight junction assembly
RhoA–RoCK–MLCP	Contraction of stress fibers
PTPn12$\ominus \longrightarrow$ Rac1	Membrane ruffling $\downarrow$
PTPn20$\ominus \longrightarrow$ Rho	Via P190RhoGAP
PTen$\ominus \longrightarrow$ Rac1, CDC42	Migration $\downarrow$
PPM1e/f$\ominus \longrightarrow$ PAK1	
Dynamin–FAK	Focal adhesion disassembly (microtubules)
Rac/CDC42–PAK1	Lamellipodium formation, membrane ruffling
PAK1–LIMK, PAK1–MLCK	Actin polymerization
Ca^{++}–calpain	Adhesion plaque disassembly
Ca^{++}–PP3	Adhesion plaque disassembly

Table 6.19. Protein kinases and cell migration. (CRMP: collapsin response mediator protein [a family of molecules involved in repulsive guidance of growth cones]; MARCKS: myristoylated alanine-rich C-kinase substrate).

Kinase	Activators, partners, and substrates
FAK	PLCγ, PI3KR1, PTen, Src, PTPn12 GRB2/7, CAS, SHC ASAP1, RhoGAP2 Paxillin, talin
PAK	CDC42 LIMK
RoCK	Rho MLCP, LIMK MLC, adducin, calponin, ERM, NHE1, CRMP2, Vimentin

6.8.10.1 Protein Kinases

Protein kinases regulate cell migration by phosphorylating adaptors and enzymes involved in the cytoskeleton organization and cell–matrix adhesions (Table 6.19).

Focal Adhesion Kinase

Focal adhesion kinase controls the focal adhesion turnover. In particular, FAK remodels cell–matrix adhesions during cell chemotaxis, haptotaxis, and durotaxis [739]. Its phosphorylation by Src kinase and dephosphorylation by many phosphatases regulate its activity. Phosphatase PTPn11 limits the phosphorylation degree of FAK and hence the number of focal adhesions to foster the migration. It modulates Rac and Rho activity as well as that of the Src–ERK axis. Elevated ERK activity fosters cell contractility.

The FAK–Src complex phosphorylates 2 scaffold proteins, paxillin and Crk-associated substrate (CAS or BCAR1). Paxillin and CAS recruit other molecules to adhesion sites and regulate the assembly and organization of the cortical actin cytoskeleton [740]. Paxillin localizes in dynamical adhesion sites with relatively high turnover rate, i.e., that assemble and rapidly disassemble, such as at the front of migrating cells. Focal adhesion kinase resides with paxillin in adhesion loci near the leading edge.

The FAK–Src may a priori regulate the rate of cell adhesion formation, maturation, and/or disassembly. In fact, the kinases FAK and Src, the adaptors CAS and paxillin, and the effector kinases ERK and MLCK are required for adhesion disassembly [740]. In addition, FAK may activate Rac or inhibit Rho.

P21-Activated Kinase

P21-activated kinase is activated by RHO family GTPases, as the Rho–PAK binding causes its conformational change and autophosphorylation. Antagonists CDC42, Rac, and RhoA control the cell contraction that relies on actomyosin stress fibers. RhoA activates RoCK that phosphorylates (inactivates) MLCP; CDC42 targets MRCK that phosphorylates myosin. On the other hand, Rac activates PAK that phosphorylates (inactivates) MLCK (reduced contractility) as well as MLC (elevated contractility). The predominant effect depends on the localization and regulator types.

In the leading edge of migrating cells, microtubule growth depends on Rac1 and PAK. Kinase PAK phosphorylates (inhibits) the microtubule remodeler stathmin [741]. Cell adhesions either mature or turn over. The complex constituted by FAK, Src, paxillin, CRK, CAS, PAK, and APAP operates as a signaling module that controls Rho GTPases.

G-protein subunit $G\beta\gamma$ released after the activation of G-protein-coupled receptors signals via RhoGEF6, CDC42, and PAK upon chemoattractant signaling to regulate directional sensing and migration of myeloid cells [743]. Protein PAK1 functions not only as a CDC42 effector, but also as a scaffold for CDC42 activation.

Leucine-rich repeat (LRR) and PDZ domain (LAP)-containing proteins constitute a family that comprises Scrib1,[65] Erbin,[66] densin-180,[67] and LRR and no PDZ-containing protein LRRC1.[68] Among these members, the cytoplasmic multimodular protein Scrib1 that localizes to epithelial adherens junctions (in which it promotes E-cadherin-mediated cell adhesion) and neuronal presynaptic regions complexes with RhoGEF7 (CDC42–Rac GEF) and APAP1 (ARF6 GAP) and links to PAK at the leading edge of migrating cells [742]. It also supports the polarized distribution of active Rac at the leading edge.

RoCK Kinases

Effectors of Rho GTPase, the kinases RoCK1 and RoCK2, phosphorylate various substrates, such as myosin light-chain phosphatase, myosin light chain, ezrin–radixin–moesin proteins, and LIM kinases [744]. They foster the formation of actin stress fibers and focal adhesions. They are required for cell rear retraction.

Src Kinase

Three ubiquitous SRC superfamily kinases — Src, Yes, and Fyn — are involved in cell spreading and motility. Kinase Src with its partners, either kinases, such as focal adhesion (FAK), extracellular signal-regulated (ERK), and myosin light-chain kinase (MLCK), and adaptors, such as CAS and paxillin, participate in adhesion turnover at the cell front (e.g., Src–CAS–CRK–DOCK1–Rac and Src–PI3K–RacGEF–Rac pathways).

Mitogen-Activated Protein Kinases

The cell adaptation to environmental osmolarity changes requires the activation of the P38MAPK pathways. Scaffold Disc large homolog DLg1 is phosphorylated by P38γ[69] and dissociated from Disc large-associated protein DLgAP1,[70] and is therefore released from the cytoskeleton [745]. This process might regulate the integrity of intercellular junctional complexes and cell shape in response to osmotic stress.

Activation of the ERK pathway at the cell periphery is required for adhesion disassembly before cell spreading and migration. Moreover, ERK effectors, such as calcium-dependent calpain-2 and myosin light-chain kinase,

[65] A.k.a. Scribble homolog and LAP4.
[66] A.k.a. LAP2 and ErbB2-interacting protein (ErbB2IP).
[67] A.k.a. LAP1.
[68] A.k.a. LANo.
[69] A.k.a. SAPK3 and ERK6.
[70] A.k.a. guanylyl kinase-associated protein (GKAP) and SAP90 (DLg4)-associated protein SAPAP1.

which cleaves focal adhesion kinases and other components of focal adhesions and phosphorylates myosin light chain, respectively, are involved in focal adhesion turnover and disassembly. Adaptor sorbin and SH3 domain-containing protein SorbS1,[71] which binds to cytoskeletal proteins paxillin, vinculin, and actin, localizes to adhesion sites between the cell and the extracellular matrix. Adaptor SorbS1 precludes cell spreading via the PAK–ERK pathway and focal adhesion turnover [746].

Collective cell migration is regulated by the Jun N-terminal kinase pathway. Signaling via JNK modulates intercellular and cell–matrix contacts to sustain collective migration by regulating several effectors, such as polarity factor ParD3 (partitioning-defective-3 homolog, or Bazooka) and cytoskeletal adaptor paxillin [747]. Whereas transcription factor Ras-responsive element-binding protein RREB1 regulates the cohesion of border cell clusters via the JNK cascade, it controls the collective migration by signal transducers and activators of transcription (STAT) pathway [748].

Scaffold proteins for MAPKs allow specific and efficient signal transduction of the relevant MAPK cascades. Scaffold MAPK8IP4[72] regulates cell migration via a $G\alpha_{13}$ pathway [749].

6.8.10.2 Protein and Lipid Phosphatases

Protein phosphatases involved in the regulation of cell migration comprise [639]: (1) dual protein Tyr and lipid phosphatase and tensin homolog (PTen); (2) dual-specificity phosphatases (protein Tyr–Ser–Thr phosphatases); (3) receptor, soluble, and low-molecular-weight protein Tyr phosphatases, such as PTPn11, PTPn12; (4) protein Ser/Thr holophosphatases, such as protein phosphatases PP2 and PP3, among others; and lipid phosphatases SHIPs.

Phosphatases regulate cell–matrix and cell–cell adhesions, particularly adherens junctions. They contribute to the control of the formation and maintenance of the actin cytoskeleton, especially lamellipodial protrusion and rear retraction during cell migration via their inhibitory or stimulatory effects [639].

Protein Tyr Phosphatase Receptors

Some protein Tyr phosphatase receptor participate in cell migration (Tables 6.3, 6.4, and 6.18). Protein Tyr phosphatase receptor PTPRu

[71] Adaptor SorbS1, also termed CBL-associated protein (CAP) and ponsin, constitutes with SorbS2, also called Arg/Abl-interacting protein ArgBP2, and SorbS3, also named vinexin, the sorbin homology (SoHo) family of adaptors. They are involved in the regulation of glucose transport, transcriptional activation, and ubiquitination.

[72] A.k.a. Jun N-terminal kinase (JNK)-associated leucine zipper protein (JLP), JNK-interacting protein JIP4, and sperm-associated antigen SpAg9.

(or PTPRo) stimulates cell migration via adaptor paxillin, whereas PTPRh inhibits this process via CAS [639]. On the other hand, PTPRf impedes E-cadherin phosphorylation and favors epithelial integrity.

Cytosolic Protein Tyr Phosphatases

Cytosolic protein Tyr phosphatases (non-receptor) PTPn1b is implicated in the regulation of integrin signaling cell migration via Src and CAS [750]. Phosphatase PTPn11 is required for the propagation of the signal via the ERK and PI3K–PKB pathways downstream from numerous protein Tyr kinase receptors [750]. Phosphatases PTPn11 and PTPn12 control the cytoskeletal rearrangement and number of adhesions for proper cell migration [639]. Yet, PTPn11 together with EPHa2 impedes cell migration. Phosphatase PTPn12 (or PTPpest) plays a dual role in cell cytoskeleton organization, as it promotes the turnover of focal adhesions needed for cell migration and regulates the Pro–Ser–Thr phosphatase (PEST)-interacting protein (PSTPIP) phosphorylation level [751]. The latter interacts with PTPn18 and WASP, among others.

Protein Ser/Thr Phosphatases

Some holoenzymes of the subclass of protein Ser/Thr phosphatases are involved in cell migration. Protein Ser/Thr phosphatases can associate with and maintain E-cadherin–β-catenin complexes. Phosphatase PP2 is an inhibitor of protein Tyr kinases of the SRC superfamily. Phosphatase PP3 is an inhibitor of PP2.

Heterotrimeric holoenzyme PP2 is composed of 3 subunits among a set encoded by distinct genes (a structural A [α or β], regulatory B [α, β, or γ), and B″α], and catalytic C [α or β] subunit, as well as regulatory subunit 4, a PP2-, PP4-, and PP6-associated protein. The latter modifies PP2 enzymatic activity and substrate specificity and participates in the regulation of cell migration, as it activates Rac1.[73]

Phosphatase PP2 is an important regulator of the FAK–Src–paxillin complex. In addition, PP2 dephosphorylates RalA GTPase. Calmodulin-dependent kinase CamK2 dissociates the PP2-IQGAP1 complex from the integrin-Rac complex. Phosphorylated FAK interacts with SRC superfamily kinases that phosphorylate FAK partners CRK-associated substrate (CAS or BCAR1) and paxillin.

Phosphatase PP3 is responsible for the calcium-dependent release of adhesion plaques in the rear of neutrophils that migrate on vitronectin, but not

[73] Phosphatase PP2 can form a cytoplasmic complex with β-catenin, adenomatous polyposis coli protein, glycogen-synthase kinase-3β, and axin. Phosphatase PTen promotes GSK3β activation. When the kinase activity predominates, β-catenin is phosphorylated, then ubiquitinated, and degraded. When the PP2 activity predominates, β-catenin is stabilized and transported into the nucleus.

on fibronectin [710]. Calcineurin preferentially targets vitronectin associated with $\alpha_V\beta_3$ integrin [710].

Magnesium-dependent members of the PP2C family are inhibitors of the stress response. Phosphatase PPM1d (or P53-induced PP2cδ) prevents P38MAPK activity. In the presence of nitric oxide produced by inducible nitric oxide synthase (NOS2), insulin stimulates the cGMP production to inhibit via the protein phosphatase PPM1f [752] (at least in fibroblasts) Ca^{++}–calmodulin-dependent protein kinase CamK2 and the migration of vascular smooth muscle cells [753]. Insulin precludes Rho signaling, as it impedes RhoA translocation via the NO–cGMP axis and myosin phosphorylation (activation) [754].

Phosphatase and Tensin Homolog

Phosphatase and tensin homolog deleted on chromosome 10 (ten) protein (PTen) acts on phosphatidylinositol (3,4,5)-triphosphate produced by phosphatidylinositol 3-kinase. It hinders cell migration, especially that of colorectal cancer cells [755]. Phosphatase PTen interacts with focal adhesion kinase to reduce its phosphorylation level.

INPP5 Family Phosphatase

The termination of PI3K signaling by degradation of $PI(3,4,5)P_3$ not only results from PTen, but also SH2-containing inositol 5-phosphatase (SHIP). Members of the inositol polyphosphate 5-phosphatase (INPP5) family hydrolyze the 5′phosphate from phosphatidylinositol (3,4,5)-trisphosphate and inositol (1,3,4,5)-tetrakisphosphate. Protein SHIP2 can interact with SHC and CAS adaptors.

Slingshot Phosphatase

The leading edge of migrating cells is characterized by cell protrusions. Actin remodeling at the leading edge determines the migration direction and initiates lamellipodium growth. Actin remodeling is controlled by a temporospatial coordination of RHO family GTPases, kinases, and phosphatases.[74] Upon small GTPase RhoA activation,[75] protein kinase-D1[76]

[74] Rapid actin polymerization–depolymerization cycle involves cofilin activation and inactivation cycle.
[75] Small GTPase RhoA inhibits actin depolymerization.
[76] Cleavage of $PI(4,5)P_2$ by PLC releases cofilin from plasmalemmal cofilin–$PI(4,5)P_2$ complexes and generates diacylglycerol. The latter activates novel protein kinase-C that excites PKD1.

binds to ^{F}actin at lamellipodium edges and precludes cofilin dephosphorylation (activation)[77] by slingshot phosphatase SSh1[78] at the cell front [756].[79] Protein kinase-D1 thus hinders the formation of free ^{F}actin barbed ends (preferred targets for dendritic nucleation by ARP2–ARP3 complex) by depolymerizing factor cofilin, hence impeding cell migration.[80]

6.8.11 Signaling Lipids and Phosphatidylinositol 3-Kinases

Phosphatidylinositols mediate random[81] and chemoattractant-induced directional migration of cells. Chemotactic gradients cause cells such as neutrophils to produce aligned gradients of phosphatidylinositol (3,4,5)-trisphosphate in their plasma membrane. Cell reorganization achieves a quick reversible asymmetry (chemotactic switch). In addition, PIP_2 and PIP_3 control actin polymerization.

Phosphatidylinositol 3-kinases recruited to the plasma membrane of the leading edge during chemotaxis generate accumulation of phosphatidylinositol (3,4,5)-trisphosphate, whereas phosphoinositide 3-phosphatases PTen and SHIP1 recruited at the back and sides of the moving cell dephosphorylate PIP_3. Phosphatidylinositol (3,4,5)-trisphosphate produced at the cell front ahead of the actin pseudopods accelerates the motion. Kinases and phosphatases that produce PIP_3 gradients are important for chemotaxis driven by weak gradients, but not strong gradients.

In endothelial cells, phosphoinositide 3-kinases are involved in angiogenesis. Activity of PI3K is required for endothelial cell proliferation, migration, and survival that depend on vascular endothelial growth factor-A. The ubiquitous isoform PI3KC1Cα (p110α encoded by the Pik3ca gene) is targeted by VEGFa to regulate endothelial cell migration via activated RhoA [758].

Gangliosides and Tetraspanins

Glycosphingolipids such as gangliosides interact with plasmalemmal proteins, such as growth factor receptors, integrins, tetraspanins, and cytoplasmic enzymes (e.g., SRC superfamily kinases and small GTPases), to form nanodomains that control glycosphingolipid-dependent cell adhesion, growth, and motility.

[77] Cofilin phosphorylation by Lin11, Isl1, and Mec3 kinases LIMK1 or LIMK2 and testicular protein kinase (TESK) prevents cofilin binding to actin.

[78] Slingshot phosphatase activity is enhanced by its binding to filamentous actin.

[79] Activation by RhoA of PKD1 leads to SSh1 phosphorylation. Phosphorylated SSh1 is then sequestered by 14-3-3 protein in the cytoplasm.

[80] Both cell motion velocity and directionality decay.

[81] In Dictyostelium, in the absence of a chemoattractant and any other external stimulus, self-organized waves of phosphatidylinositol (3,4,5)-trisphosphate are generated owing to phosphatase and tensin homolog (PTen) and phosphoinositide 3-kinase [757].

Ganglioside GM3 tethers to tetraspanin-29 and the GM3–Tspan29 complex interacts with $\alpha_3\beta_1$ or $\alpha_5\beta_1$ integrin to hinder the motility of tumor cells [759]. The GM3–Tspan28–Tspan29 complex inhibits fibroblast growth factor receptor and prevents integrin–FGFR interactions.

Gangliosides GM2 and GM3 as well as the GM2–GM3 complex (in the presence of Ca^{++}) that can complex with tetraspanin-27[82] impedes HGFR kinase activity, thereby hampering cell motility and growth [759]. The GM2–Tspan27 complex also precludes interactions between laminin-5 and HGFR kinase.

6.8.12 Growth Factors

Platelet-derived growth factor-B (PDGFb) stimulates cytoskeletal remodeling and chemotaxis via NCK adaptors [760].[83] Adaptors NCKs transduce the signal from ligand-stimulated PDGFβ receptors using BCAR1 docker. Platelet-derived growth factor uses a pathway that implicates Src, the transcription and translation regulator heterogeneous nuclear ribonucleoprotein hnRNPk and the mRNA encoding myosin regulatory light chain (MRLC)-interacting protein MyLIP, a ubiquitin ligase. The latter leads to the degradation of the myosin regulatory light chain required for reorganization of the actin cytoskeleton to prepare fibroblast locomotion [761].[84]

Hepatocyte growth factor promotes scattering and motility as well as collective migration of endothelial or epithelial cells. HGF is able to accelerate healing in some circumstances. During collective cell migration, HGF suppresses the differentiation between leader and follower cells [651].

Epidermal growth factor promotes normal cell migration owing to the ERK cascade. EGFR regulates cell migration by transient activation of peptidases, kinases, phospholipases, small GTPases, and gene expression by transcription factor complex AP1. Furthermore, it switches the expression of 2 members of the adaptor tensin family[85] from tensin-3 to C-terminal tensin-like protein (CTen or Tns4) that causes dissociation of integrin–tensin–actin complexes and disintegration of actin stress fibers [762].

Sphingosine 1-phosphate signals via S1P receptor $S1P_1$ to promote the exit of lymphocytes from lymphoid organs (spleen and thymus) into lymph and blood. Separate sources provide S1P to plasma and lymph. Erythrocytes are major contributors of plasma S1P [763]. Sphingosine kinases SphK1 and

[82] A.k.a. CD82, metastasis suppressor Kangai-1, and suppressor of tumorigenicity ST6.
[83] Two types of NCKs exist: NCKα, or NCK1, and NCKβ, or NCK2.
[84] Cell motility needs: (1) reduction in stress fibers and focal adhesions and, then, (2) formation of lamellipodia.
[85] Members of the family of tensins (Tns1–Tns4) regulate focal adhesions, thereby cell migration. Tensins bind to both integrins and actin. However, C-terminal tensin-like protein (CTen, or Tns4) lacks the actin-binding domain.

SphK2 control S1P production. Agent S1P is possibly supplied to lymph by lymphatic endothelial cells.

6.8.13 Morphogen Wnt

The non-canonical Wnt pathway that targets the cell cytoskeleton controls cell orientation, polarity, and directional movement by the redistribution of adhesion receptors.

Morphogen Wnt5a confers the polarization of target molecules and actomyosin cytoskeleton [764]. It allows the recruitment of actin, myosin-2B, Frizzled-3, and melanoma cell adhesion molecule (MCAM) into Wnt-mediated receptor–actin–myosin polarity structure (WRAMP) within 30 mn. The latter also requires Dvl2 and PKC. Peripheral, asymmetrically distributed (polarized) WRAMP initiates membrane contractility and motion in the direction of membrane retraction owing to endosome trafficking. Morphogen Wnt5a operates via several small and large GTPases: (1) RhoB that controls the motion of endosomes; (2) Rab4 that stimulates the formation of multivesicular bodies; and (3) dynamin that regulates the creation of endosomes at the cell surface. Morphogen Wnt5a polarizes migrating cells by promoting the recycling of membrane components.

Intercellular signaling occurs between cells in the leading and trailing zones of a migrating cell cluster to maintain its polarity during bulk migration. Interaction between the Wnt–β-catenin and fibroblast growth factor pathways causes mutually exclusive activation for polarized distribution of chemokine receptors that coordinate collective cell migration [765]. Wnt–β-catenin signaling induces FGF as well as FGF inhibitor expression. Conversely, the FGF pathway blocks Wnt–β-catenin signaling by priming generation of a Wnt–β-catenin inhibitor. Localized activation of the Wnt–β-catenin pathway in the leading cells restricts FGF signaling to trailing cells and conversely.

6.8.14 Endo- and Exocytic Molecules

The exocyst is an octamer that tethers secretory vesicles to specific domains of the plasma membrane for exocytosis and cell surface expansion. It is also involved in actin-based membrane protrusions and cell migration controlled by Ral GTPase. Exocyst component Exo70 that induces actin-based membrane protrusions interacts with the ARP2–ARP3 complex [766]. The RalA–exocyst–ARP2/3 association that is regulated by epidermal growth factor can coordinate actin dynamics and membrane transport to promote efficient membrane protrusions at the leading edge of migrating cells.

The endocytic cycle recycles the back of a moving cell to the front. It could also particpate in adjustment of the surface area of moving cells to match shape change [666]. Additional surface components can be supplied by exocytosis of cell membrane. ^{N}Ethylmaleimide-sensitive fusion protein (NSF) intervenes with SNARE complexes that are formed when a vesicle docks to a membrane. Clathrin and NSF are involved in cell motion.

Caveolin-1 has a polarized location in migrating endothelial cells. Caveolin-1 accumulates in the front of migrating cells or in the rear, whether endothelial cells cross a pore or move on a plane. The locomotion mode and the polarization of caveolin-1 depend on the substrate topology, but not on chemoattractants [767].

6.9 An Example: Migration of Airway Smooth Muscle Cells

Migration of smooth muscle cells occurs during formation of airways as well as blood vessels and gastrointestinal tract. During lung development, migration and differentiation of airway smooth muscle precursor cells is directed by auto- and paracrine factors. In the mature lung, inflammatory mediators trigger airway wall remodeling that involves both smooth muscle hypertrophy and hyperplasia. Increased cell number results from migration of remote precursors and proliferation of cells within airway wall smooth muscle bundles. In addition, myofibroblasts of the lamina propria can differentiate into smooth muscle-like cells that are characterized by an increased expression of smooth muscle-restricted contractile proteins [768].

Numerous pro- and antimigratory molecules influence displacements of airway and vascular smooth muscle cells (Vol. 5 – Chap. 8. Smooth Muscle Cells): hormones, biogenic amines, growth factors (e.g., PDGF), cytokines, and extracellular matrix components (collagens, fibronectin, and laminins, as well as matrix metallopeptidases and tissue inhibitors of metallopeptidases). Each agent activates signaling cascades to (Table 6.20): (1) remodel the cytoskeleton, as both actin polymerization and depolymerization are required for cell migration (lamellipodial extension and cell rear retraction); (2) stimulate nanomotors for cell contraction; and (3) change adhesiveness between the moving cell and the extracellular matrix, as leading cell region attaches, whereas trailing area detaches, and conversely during the next displacement stage. Cell migration is initiated by activated receptors, such as G-protein-coupled receptors, receptor Tyr kinases, and cell adhesion contributors and receptors, particularly integrins that prime cytoskeleton remodeling and reorganization of cell organelles.

Intracellular migratory substances include Ca^{++}, calmodulin, actin-binding proteins, components of focal adhesion complexes, and enzymes, such as phosphatidylinositol 3-kinase, phospholipase-C, phosphoinositide-dependent kinase PDK1, small GTPases of the RAS hyperfamily, especially those of the RHO family (CDC42, Rac, and Rho), as well as focal adhesion, SRC superfamily, Rho-activated, P21-activated, and mitogen-activated protein kinases [768]. After receptor activation, Ca^{++}, phosphatidylinositol (4,5)-bisphosphate, and

Table 6.20. Cell migration relies on actin polymerization and activity of myosin nanomotors (Source: [768]). Signaling pathways involved in migration of smooth muscle cells thus control actin polymerization and cell contraction generated by nanomotor myosin-2 associated with actin filaments. Smooth muscle myosin-2 is regulated by: (1) Ca^{++}–calmodulin activation of myosin light-chain kinase (MLCK) and (2) Ca^{++}-independent small GTPase stimulation of Rho-activated protein kinase (RoCK) that phosphorylates (inhibits) myosin light-chain phosphatase (MLCP), which dephosphorylates myosin-2 regulatory light chains. Kinase LIMK reduces cofilin-mediated actin depolymerization. Kinases PAKs (P21-activated kinase) target other protein kinases (e.g., MLCK), actin-binding proteins (cofilin, cortactin, and caldesmon), myosin-2, and intermediate filament components (vimentin) to favor actin polymerization and actomyosin contraction. Kinases of the SRC superfamily (Src, Fyn, Lyn, and Yes) phosphorylate focal adhesion proteins. Phosphatidylinositol 3-kinase (PI3K) targets PKB, S6K, extracellular signal-regulated protein kinase (ERK), and P38MAPK. Kinases ERKs can phosphorylate caldesmon and MLCK. Substrates of P38MAPK include mitogen-activated protein kinase (MAPK)-activated protein kinase MAPKAPK2 and heat shock protein HSP27 that regulates actin remodeling.

Molecular cascade	Effect
Ca^{++}–calmodulin–MLCK–myosin-2	Actomyosin fiber contraction
CDC42–WASP–ARP2/3	Actin polymerization
MAP3K3/6–P38MAPK–MAPKAPK–HSP27	Actin polymerization
Rac–WAVe–ARP2/3	Actin polymerization
Rac–PAK1–MLCK–myosin-2	Actomyosin fiber contraction
Ras–Raf–ERK–MLCK–myosin-2	Actomyosin fiber contraction
RhoA–RoCK–MLCP–myosin-2	Actomyosin fiber contraction
RhoA–RoCK–LIMK–cofilin	Actin polymerization
RhoA–Dia1	Actin polymerization

small GTPases mediate intracellular reaction cascades.[86] Focal adhesion complexes contain integrins, adaptors (paxillin, talin, tensin, and vinculin), regulators, i.e., kinases (e.g., Src and FAK) and their adaptors such as BCAR1, and controllers of myosin-2 activity, such as kinases MLCK, PAK, and RoCK and phosphatase MLCP.

In the leading region of a moving cell, actin polymerizes and focal adhesions assemble.[87] Myosin-2 nanomotors linked to actin filaments in the cell body

[86] Monomeric GTPases activate formins (Dia1 and Dia2), Wiskott-Aldrich syndrome protein (WASP), and WASP–verprolin homolog (WAVe) that regulate actin polymerization, as well as P38MAPK and ERK, RoCK kinases, and P21-activated protein kinases.

[87] Actin-nucleating proteins (formins Dia1 and Dia2 as well as VASP) that promote filament formation at barbed (plus) end and their extension act together with profilin that adds ^{G}actin monomers. Actin branching is controlled by small GTPases that activate Wiskott-Aldrich syndrome protein (WASP) and WASP–verprolin homolog (WAVe). Effectors WAVe and WASP, in turn, stimulate the

generate force that pulls the cell forward.[88] Cell organelles that are tethered by adaptors and nanomotors move along the cytoskeleton. In the cell rear, the cytoskeleton depolymerizes and focal adhesions disassemble.[89]

6.10 Chemotaxis Modeling

During chemotaxis, migrating cells are able to sense slight chemical gradients between the back and the front regions that guide their migration. Chemoattractant concentration can differ by only a few percent from one side of the cell to the other. Cells may be sensitive to the amplitude of the concentration gradient of chemoattractant as well as local magnitude in concentration of chemoattractant [769].[90] Chemotaxis can then result from establishment of a spatial distribution of plasmalemmal, liganded receptors and the output signal field within the cell that generates cellular extension and retraction to support cell displacement.

In particular, chemotaxis enables the recruitment of immune cells to sites of inflammation. Macrophages move according to a gradient of complement component C5a. In addition, chemotaxis relies on ATP release and an autocrine regulation that involves receptors for ATP, $P2Y_2$, for ADP, $P2Y_{12}$, and for adenosine, A_{2A}, A_{2B}, and A_3 [770].

G-protein-coupled chemoattractant receptors initiate chemotaxis. Chemotaxis relies on 3 main processes: (1) cell alignment along the chemoattractant gradient; (2) cell polarization; and (3) protrusion at the leading edge (cell front) and retraction mainly at the trailing edge that both implicate small GTPases [771].

The time gradient of the cell number depends on cell diffusion flux, then on cell diffusivity, chemotactic flux, cell proliferation, and death rates. The chemotactic flux depends on: (1) a chemotactic response function of the available cell number and of the chemoattractant concentration; and (2) chemoattractant concentration gradient [297]. The time gradient of the chemoattractant concentration depends on the production and destruction rates as well

---

ARP2–ARP3 complex. Gelsolin, a potent actin-severing member of the gelsolin–villin superfamily that is activated by increased cytosolic Ca^{++} concentration, promotes actin nucleation when PIP_2 binds to gelsolin and releases gelsolin from ^{F}actin.

[88] Myosin-2 is activated by kinases (MLCK, PAK, RoCK) as well as inhibition of myosin light-chain phosphatase.

[89] Gelsolin and cofilin depolymerize actin filaments. Focal adhesion disassembly is achieved by proteic complexes that are associated with microtubules that emanate from the microtubule-organizing center.

[90] Minimal concentration gradient in chemoattractant required for a directional response depends on the local concentration. Cells exposed to a 1.25% gradient between the back and the front respond to large enough concentration (10–30 nmol) [769]. For small local concentrations (< 10 nmol) the intracellular information loss is insignificant. Information processing is suboptimal with steeper gradients and higher local concentrations.

as its diffusion flux. The Keller-Segel model is widely used for the chemical control of cell movement [772]:

$$\begin{aligned} \frac{\partial n}{\partial t} &= \mathcal{D}_c \boldsymbol{\nabla}^2 n - \boldsymbol{\nabla} \cdot (n \kappa_s \boldsymbol{\nabla} c_{ca}) + \mathsf{P} n, \\ \frac{\partial c_{ca}}{\partial t} &= \mathcal{D}_{ca} \boldsymbol{\nabla}^2 c_{ca} + \mathsf{P}_{ca} n - \tau^{-1} c_{ca}, \end{aligned} \tag{6.1}$$

where n is the cell density, c_{ca} the chemoattractant concentration, $\mathcal{D}_c$ and $\mathcal{D}_{ca}$ the cell and chemoattractant diffusivity, P the cell division rate, P_{ca} the chemoattractant production rate, κ_s the cell sensitivity to chemoattractant, and τ the chemoattractant half-life.

Numerous models were developed afterward, e.g., such as the following one that takes into account needed nutrients (concentration c_n) to be consumed [297]:

$$\begin{aligned} \frac{\partial}{\partial t} n &= \mathcal{D}_c \nabla^2 n - \nabla \Big(\frac{\kappa_1 n}{(\kappa_2 + c_{ca})^2} \nabla c_{ca} \Big) + \kappa_3 n \Big(\frac{\kappa_4 c_n^2}{\kappa_9 + c_n^2} - n \Big), \\ \frac{\partial}{\partial t} c_{ca} &= \mathcal{D}_{ca} \nabla^2 c_{ca} + \kappa_5 \frac{n^2}{\kappa_6 + n^2} c_n - \kappa_7 n c_n, \\ \frac{\partial}{\partial t} c_n &= \mathcal{D}_n \nabla^2 c_n - \kappa_8 \frac{c_n^2}{\kappa_9 + c_n^2} n. \end{aligned} \tag{6.2}$$

A new formulation of the system of partial differential equations has been obtained by the introduction of a new variable and is approximated via a mixed finite element technique [773]. More recently, a computational model for cell migration in 3D matrices has been developed [774]. Effects due to matrix sterics and mechanics arising in 3D situations are exhibited. Current trends in taxis modeling rely on hyperbolic equations [775] instead of the classical Keller-Segel model and parabolic equations [776].

A transport model for cell movement (with velocity $\mathbf{v}$) in a tissue (cell and fiber density ρ_c and ρ_f, cell and fiber distribution density function $p = p(t, \mathbf{x}, \mathbf{v})$ and $q = q(t, \mathbf{x}, \mathbf{a})$ [$\mathbf{a}$: angle of fibers], the orientation of the fiber network being described by a symmetrical, positive, definite tensor $\mathbf{O}$) primed by a chemotactic force $\mathbf{f}(c)$ that depends on a given chemical distribution $c(t, \mathbf{x})$ (e.g., $f(c) = \kappa \boldsymbol{\nabla}_x c$) can be written as [777]:

$$\partial_t p + \mathbf{v} \cdot \boldsymbol{\nabla}_x p + \boldsymbol{\nabla}_v \cdot [\mathbf{f}(c) p] = \mathsf{I}_{cc} + \mathsf{I}_{cm}, \tag{6.3}$$

where I_{cc} and I_{cm} are the cell–cell and cell–matrix interaction operators, assuming mass conservation. Moment equations for the cell population density (ρ_c) and momentum ($\rho_c V$) at the macroscopic scale are then expressed with encounter rates with a cell that has a given velocity (c_c) and fibers that has a given orientation (c_f) per unit volume and unit time.[91]

[91] The encounter rate c_c does not depend on the particular incoming velocities. The encounter rate c_f depends neither on the particular incoming velocity nor on the fiber orientation.

$$\begin{aligned}
\partial_t \rho_c &+ \boldsymbol{\nabla}_x \cdot \Big(\frac{\mathsf{K} \rho_c \boldsymbol{\nabla}_x c}{\mathsf{c}_c \rho_c + \mathsf{c}_f \rho_f} \Big) \\
&= \boldsymbol{\nabla}_x \cdot \Big(\Big(\frac{1}{\mathsf{c}_c \rho_c + \mathsf{c}_f \rho_f} \Big) \boldsymbol{\nabla}_x (\sigma \rho_c) \Big) \\
&\quad + \boldsymbol{\nabla}_x \cdot \Big(\frac{1}{\mathsf{c}_c \rho_c + \mathsf{c}_f \rho_f} \boldsymbol{\nabla}_x \cdot \Big[\frac{\mathsf{c}_f \sigma \rho_c \rho_f}{\mathsf{c}_c \rho_c + \mathsf{c}_f \rho_f} (\mathbf{O} - \mathbf{I}) \Big] \Big),
\end{aligned} \tag{6.4}$$

where $\mathbf{I}$ is the identity matrix and σ the variance of the transition probability densities assumed to be equal for both cell–cell and cell–matrix interactions. The first and second term on the right-hand side are related to an isotropic transfer and a fiber-caused anisotropy.

Concluding Remarks

A science — so the Savants say,
"Comparative Anatomy"
By which a single bone —
Is made a secret to unfold
Of some rare tenant of the mold,
Else perished in the stone —

(Emily Dickinson [1830–1886])

The present volume of the set of textbooks devoted to Circulatory and Ventilatory Systems in the framework of Biomathematical and Biomechanical Modeling gives the basic information on cell behavior, especially when cells react to various stimuli, particularly mechanical stresses. Numerous processes can be modeled to quickly assess effects of parameters, all other agents remaining constant, once the mathematical model has been validated. The advantage of mathematical models is their ability to yield the complete quantity fields, whereas measurements are made in some points or correspond to averages of exploration windows of the investigated variable field.

In physiological systems associated with flows, the magnitude and direction of mechanical stresses applied by the flowing fluid on the wetted surface of conduit wall (i.e., vascular endothelium with its glycocalyx or respiratory epithelium with the mucus layer and periciliary fluid) as well as within the vessel wall vary during the cardiac and ventilatory cycles. The heart generates unsteady flows with a given frequency spectrum in a network of blood vessels characterized by complicated architecture and variable geometry both in space and time. Breathing results from the deformation of the thoracic cage imposed by respiratory muscles with a poorer frequency content, but complete flow reversal in the same duct network from inspiration to expiration, and vice versa. The thoracic muscular cage that cyclically inflates and deflates lowers and heightens the intrathoracic pressure, hence dilating and collapsing lung alveoli and airways to inhale and exhale air.

In the vasculature as well as the respiratory tract, the vessel geometry varies over short distances. The vascular network of curved blood vessels is composed of successive geometrical singularities, mainly branchings. The respiratory tract is characterized by a large wetted surface inside a small volume, especially in the nose and thorax. Therefore, the bronchial tree is a network of successive branchings (at inspiration, junctions at expiration) between short, more or less curved pipes of corrugated walls in large bronchi due to the presence of partial or complete cartilaginous rings. The narrower the airway caliber, the shorter the length. In the nose, turbinates allows heat and water exchange, but render air current highly bent with change in curvature over a tiny distance. In addition, the laryngeal constriction, the aperture of which varies during the ventilatory cycle, provokes air jet.

Moreover, blood and air streams correspond to time-dependent, three-dimensional, developing flow, as they are conveyed in conduit entrance length, where the boundary layer develops (Vol. 7).

Furthermore, blood vessels and airways are deformable. Changes in transmural pressure (pressure difference between the pressure at the wetted surface of the lumen applied by the moving fluid on the conduit wall and the pressure at the external wall side that depends on the activity on the neighboring organs) can also influence the shape of the vessel cross-section, especially when it becomes negative. In addition, in the arterial compartment, especially the aorta, the change in cross-section shape is due to taper. More generally, possible prints of adjacent organs with more or less progressive constriction and enlargment, and adaptation to branching (transition zone) also give rise to 3-dimensional flows. These flows are commonly displayed by virtual transverse currents, even if the vessel is considered straight. Furthermore, geometrical singularities influence flow pattern both upstream and downstream.

Local changes in the direction of stress components can also be caused by flow separation and flow reversal during the cardiac and respiratory cycles. Flow separation is set by an adverse pressure gradient when inertia forces and fluid vorticity are high enough, especially in branching segments. Due to its time-dependent feature, flow separation regions spread over a variable length during the flow cycle and can move. The variable location and size of the flow separation region in unsteady flows depends on the flow distribution between branches that can vary during the flow cycle.

Flow reversal occurs during the diastole of the left ventricle in elastic arteries, such as the aorta, and most of the muscular arteries, such as brachial and femoral arteries (but not in the carotid arteries), as well as during alternations from inspiratory decelerating flow phase and expiratory accelerating flow phase and conversely. In arteries, flow reversal can be observed either in a region near the wall, more or less wide with respect to the position of the local center of vessel curvature, or in the entire lumen.

Consequently, the stress field exerted by the flowing fluid and experienced by the biological tissues at various depths of the wall are strongly variable both in time and space. Cellular sensors then process mechanical signals by

Table 6.21. Mechanotransduction types and compartments of the arterial network.

Arterial compartment	Function	Goal
Large arteries	Body' territory irrigation	Regulation of cardiac postload
Small resistive arteries	Organ perfusion	Maintenance of a constant flow rate
Arterioles	Tissue perfusion	Vasodilation caused by increased local metabolism

ensemble averaging not only to raise the signal-to-noise ratio, but also to adequately adapt the local size of the conduit lumen, i.e., the local flow resistance to maintain either flow rate or pressure, only in the case of sustained, abnormal stress field.

Applied mechanical stresses influence cell fate. Angiogenesis that relies on cell proliferation, migration, and death to produce and model new blood vessels from existing segments of the vasculature depends, at least partly, on experienced stress and strain fields.

Mechanotransduction is the process that starts with the detection of mechanical stresses and its translation into chemical signals at the plasma membrane (Vols. 3 and 4). This translation comprises a chain of chemical reactions that are controlled both in time and space. Tranduction begins at the cell cortex, where signaling effectors are recruited. Certain signaling mediators are transported by the cell cytoskeleton from the cytosol to the nucleoplasm to trigger adequate gene expression and protein synthesis. The cytoskeleton itself transmits mechanical constraints and reacts to the stress field to adapt the cell shape. Regulators are newly synthesized or stored in vesicles and released in the extracellular matrix to carry out their auto-, juxta-, para-, and eventually endocrine functions. Three types of mechanotransductions exist according to the compartment of the arterial bed: large and small resistive arteries and arterioles (Vol. 5 and Table 6.21).

References

Introduction

1. Lucretius (1997) De rerum natura (Book I-615) [On the Nature of Things]. Garnier Flammarion, Paris

Chap. 1. Remote Control Cells

2. Nury H, Poitevin F, Van Renterghem C, Changeux JP, Corringer PJ, Delarue M, Baaden M (2010) One-microsecond molecular dynamics simulation of channel gating in a nicotinic receptor homologue. Proceedings of the National Academy of Sciences of the United States of America 107:6275-6280
3. Yu YC, Bultje RS, Wang X, Shi SH (2009) Specific synapses develop preferentially among sister excitatory neurons in the neocortex. Nature 458:501–504
4. Bissiere S, Zelikowsky M, Ponnusamy R, Jacobs NS, Blair HT, Fanselow MS (2011) Electrical synapses control hippocampal contributions to fear learning and memory. Science 331:87–91
5. Guyenet PG (2006) The sympathetic control of blood pressure. Nature Reviews – Neuroscience 7:335–346
6. Nakamura K, Morrison SF (2010) A thermosensory pathway mediating heat-defense responses. Proceedings of the National Academy of Sciences of the United States of America 107:8848–8853
7. Gourine AV, Kasymov V, Marina N, Tang F, Figueiredo MF, Lane S, Teschemacher AG, Spyer KM, Deisseroth K, Kasparov S (2010) Astrocytes control breathing through pH-dependent release of ATP. Science 329:571–575
8. Schultz HD, Li YL, Ding Y (2007) Arterial chemoreceptors and sympathetic nerve activity: implications for hypertension and heart failure. Hypertension 50:6–13
9. Schultz HD, Li YL (2007) Carotid body function in heart failure. Respiratory Physiology and Neurobiology 157:171–185

10. Li YL, Xia XH, Zheng H, Gao L, Li YF, Liu D, Patel KP, Wang W, Schultz HD (2006) Angiotensin II enhances carotid body chemoreflex control of sympathetic outflow in chronic heart failure rabbits. Cardiovascular Research 71:129–138

11. Cotter L, Ozçelik M, Jacob C, Pereira JA, Locher V, Baumann R, Relvas JB, Suter U, Tricaud N (2010) Dlg1–PTEN Interaction regulates myelin thickness to prevent damaging peripheral nerve overmyelination. Science 328:1415–1418

12. Charles P, Tait S, Faivre-Sarrailh C, Barbin G, Gunn-Moore F, Denisenko-Nehrbass N, Guennoc AM, Girault JA, Brophy PJ, Lubetzki C (2002) Neurofascin is a glial receptor for the paranodin/Caspr-contactin axonal complex at the axoglial junction. Current Biology 12:217–220

13. Denisenko-Nehrbass N, Oguievetskaia K, Goutebroze L, Galvez T, Yamakawa H, Ohara O, Carnaud M, Girault JA (2003) Protein 4.1B associates with both Caspr/paranodin and Caspr2 at paranodes and juxtaparanodes of myelinated fibres. European Journal of Neuroscience 17:411–416

14. Traka M, Goutebroze L, Denisenko N, Bessa M, Nifli A, Havaki S, Iwakura Y, Fukamauchi F, Watanabe K, Soliven B, Girault JA, Karagogeos D (2003) Association of TAG-1 with Caspr2 is essential for the molecular organization of juxtaparanodal regions of myelinated fibers. Journal of Cell Biology 162:1161–1172

15. Chao DL, Ma L, Shen K (2009) Transient cell–cell interactions in neural circuit formation. Nature Reviews – Neuroscience 10:262–271

16. Epsztein J, Lee AK, Chorev E, Brecht M (2010) Impact of spikelets on hippocampal CA1 pyramidal cell activity during spatial exploration. Science 327:474–477

17. Ko J, Fuccillo MV, Malenka RC, Südhof TC (2009) LRRTM2 functions as a neurexin ligand in promoting excitatory synapse formation. Neuron 64:791–798

18. de Wit J, Sylwestrak E, O'Sullivan ML, Otto S, Tiglio K, Savas JN, Yates JR 3rd, Comoletti D, Taylor P, Ghosh A (2009) LRRTM2 interacts with Neurexin1 and regulates excitatory synapse formation. Neuron 64:799–806

19. Feng W, Zhang M (2009) Organization and dynamics of PDZ-domain-related supramodules in the postsynaptic density. Nature Reviews – Neuroscience 10:87–99

20. Schratt G (2009) MicroRNAs at the synapse. Nature Reviews – Neuroscience 10:842–849

21. Chandrasekar V, Dreyer JL (2009) MicroRNAs miR-124, let-7d and miR-181a regulate cocaine-induced plasticity. Molecular and Cellular Neuroscience 42:350–362

22. Gao FB (2010) Context-dependent functions of specific microRNAs in neuronal development. Neural Development 5:25

23. Sensi SL, Paoletti P, Bush AI, Sekler I (2009) Zinc in the physiology and pathology of the CNS. Nature Reviews – Neuroscience 10:780–791

24. Brocard JB, Rajdev S, Reynolds IJ (1993) Glutamate-induced increases in intracellular free Mg^{2+} in cultured cortical neurons. Neuron 11:751–757

25. Günzel D, Zimmermann F, Durry S, Schlue WR (2001) Apparent intracellular Mg^{2+} buffering in neurons of the leech Hirudo medicinalis. Biophysical Journal 80:1298–1310
26. Wingard JN, Chan J, Bosanac I, Haeseleer F, Palczewski K, Ikura M, Ames JB (2005) Structural analysis of Mg^{2+} and Ca^{2+} binding to CaBP1, a neuron-specific regulator of calcium channels. Journal of Biological Chemistry 280:37461–37470
27. Johansen J, Taft WC, Yang J, Kleinhaus AL, DeLorenzo RJ (1985) Inhibition of Ca^{2+} conductance in identified leech neurons by benzodiazepines. Proceedings of the National Academy of Sciences of the United States of America 82:3935–3939
28. Siechen S, Yang S, Chiba A, Saif T (2009) Mechanical tension contributes to clustering of neurotransmitter vesicles at presynaptic terminals. Proceedings of the National Academy of Sciences of the United States of America 106:12611–12616
29. Sawada K, Echigo N, Juge N, Miyaji T, Otsuka M, Omote H, Yamamoto A, Moriyama Y (2008) Identification of a vesicular nucleotide transporter. Proceedings of the National Academy of Sciences of the United States of America 105:5683–5686
30. Prado VF, Prado MA, Parsons SM (2008) Vesicular acetylcholine transporter (VAChT). Signaling Gateway (www.signaling-gateway.org)
31. Stein BE, Stanford TR (2008) Multisensory integration: current issues from the perspective of the single neuron. Nature Reviews – Neuroscience 9:255–266
32. Silver RA (2010) Neuronal arithmetic. Nature Reviews – Neuroscience 11:474–489
33. Groffen AJ, Martens S, Díez Arazola R, Cornelisse LN, Lozovaya N, de Jong AP, Goriounova NA, Habets RL, Takai Y, Borst JG, Brose N, McMahon HT, Verhage M (2010) Doc2b is a high-affinity Ca^{2+} sensor for spontaneous neurotransmitter release. Science 327:1614–1618
34. Sun J, Pang ZP, Qin D, Fahim AT, Adachi R, Südhof TC (2007) A dual-Ca^{2+}-sensor model for neurotransmitter release in a central synapse. Nature 450:676–682
35. Fasano C, Tercé F, Niel JP, Nguyen TT, Hiol A, Bertrand-Michel J, Mallet N, Collet X, Miolan JP (2007) Neuronal conduction of excitation without action potentials based on ceramide production. PLoS ONE 2:e612
36. Abraham WC (2008) Metaplasticity: tuning synapses and networks for plasticity. Nature Reviews – Neuroscience 9:387
37. Zhang C, Wu B, Beglopoulos V, Wines-Samuelson M, Zhang D, Dragatsis I, Südhof TC, Shen J (2009) Presenilins are essential for regulating neurotransmitter release. Nature 460:632–636
38. Rust MB, Gurniak CB, Renner M, Vara H, Morando L, Görlich A, Sassoè-Pognetto M, Banchaabouchi MA, Giustetto M, Triller A, Choquet D, Witke W (2010) Learning, AMPA receptor mobility and synaptic plasticity depend on n-cofilin-mediated actin dynamics. EMBO Journal 29:1889–1902

39. Attwell D, Buchan AM, Charpak S, Lauritzen M, Macvicar BA, Newman EA (2010) Glial and neuronal control of brain blood flow. Nature 468:232–243
40. Cui K, Yuan XB (2007) TRP channels and axon pathfinding (Chap. 4). In Liedtke WB, Heller S (eds.) TRP Ion Channel Function in Sensory Transduction and Cellular Signaling Cascades, CRC Press (Taylor & Francis Group), Boca Raton, FL, USA
41. Prozorovski T, Schulze-Topphoff U, Glumm R, Baumgart J, Schröter F, Ninnemann O, Siegert E, Bendix I, Brüstle O, Nitsch R, Zipp F, Aktas O (2008) Sirt1 contributes critically to the redox-dependent fate of neural progenitors. Nature – Cell Biology 10:385–394
42. Shelly M, Lim BK, Cancedda L, Heilshorn SC, Gao H, Poo MM (2010) Local and long-range reciprocal regulation of cAMP and cGMP in axon/dendrite formation. Science 327:547–552
43. Giaume C, Koulakoff A, Roux L, Holcman D, Rouach N (2010) Astroglial networks: a step further in neuroglial and gliovascular interactions. Nature Reviews – Neuroscience 11:87–99
44. Schummers J, Yu H, Sur M (2008) Tuned responses of astrocytes and their influence on hemodynamic signals in the visual cortex. Science 320:1638–1643
45. Gordon GRJ, Choi HB, Rungta RL, Ellis-Davies GCR, MacVicar BA (2008) Brain metabolism dictates the polarity of astrocyte control over arterioles. Nature 456:745–749
46. Hughes EG, Elmariah SB, Balice-Gordon RJ (2010) Astrocyte secreted proteins selectively increase hippocampal GABAergic axon length, branching, and synaptogenesis. Molecular and Cellular Neurosciences 43:136–145
47. Hamilton NB, Attwell D (2010) Do astrocytes really exocytose neurotransmitters? Nature Reviews – Neuroscience 11:227–238
48. Agulhon C, Fiacco TA, McCarthy KD (2010) Hippocampal short- and long-term plasticity are not modulated by astrocyte Ca^{2+} signaling. Science 327:1250–1254
49. De Young GW, Keizer J (1992) A single-pool inositol 1,4,5-trisphosphate-receptor-based model for agonist-stimulated oscillations in Ca^{2+} concentration. Proceedings of the National Academy of Sciences of the United States of America 89:9895–9899
50. Li YX, Rinzel J (1994) Equations for InsP3 receptor-mediated $[Ca^{2+}]_i$ oscillations derived from a detailed kinetic model: a Hodgkin-Huxley like formalism. Journal of Theoretical Biology 166:461–473
51. Shuai JW, Jung P (2003) Langevin modeling of intracellular calcium dynamics. In Falcke M, Malchow E (eds) Understanding Calcium Dynamics, Lecture Notes in Physics, Springer-Verlag, Berlin
52. Behm DJ, Ogbonna A, Wu C, Burns-Kurtis CL, Douglas SA (2009) Epoxyeicosatrienoic acids function as selective, endogenous antagonists of native thromboxane receptors: identification of a novel mechanism of vasodilation. Journal of Pharmacology and Experimental Therapeutics 328:231–239
53. Filosa JA, Bonev AD, Straub SV, Meredith AL, Wilkerson MK, Aldrich RW, Nelson MT (2006) Local potassium signaling couples neuronal activity to vasodilation in the brain. Nature – Neuroscience 9:1397-1403

54. Boison D, Chen JF, Fredholm BB (2010) Adenosine signaling and function in glial cells. Cell Death and Differentiation 17:1071–1082
55. Kurbel S, Dodig K, Radic R (2002) The osmotic gradient in kidney medulla: a retold story. Advances in Physiology Education 26:278–281
56. Satlin LM, Carattino MD, Liu W, Kleyman TR (2006) Regulation of cation transport in the distal nephron by mechanical forces. American Journal of Physiology – Renal Physiology 291:F923-F931
57. Frontini A, Cinti S (2010) Distribution and development of brown adipocytes in the murine and human adipose organ. Cell Metabolism 11:253-256
58. Wozniak SE, Gee LL, Wachtel MS, Frezza EE (2009) Adipose tissue: the new endocrine organ? A review article. Digestive Diseases and Sciences 54:1847–1856
59. Ye ZW, Wu XM, Jiang JG (2009) Expression changes of angiotensin II pathways and bioactive mediators during human preadipocytes-visceral differentiation. Metabolism 58:1288–1296
60. Isakson P, Hammarstedt A, Gustafson B, Smith U (2009) Impaired preadipocyte differentiation in human abdominal obesity: role of Wnt, tumor necrosis factor-alpha, and inflammation. Diabetes 58:1550–1557
61. Hajer GR, van Haeften TW, Visseren FLJ (2008) Adipose tissue dysfunction in obesity, diabetes, and vascular diseases. European Heart Journal 29:2959–2971
62. Tang W, Zeve D, Suh JM, Bosnakovski D, Kyba M, Hammer RE, Tallquist MD, Graff JM (2008) White fat progenitor cells reside in the adipose vasculature. Science 322:583–586
63. Tseng YH, Kokkotou E, Schulz TJ, Huang TL, Winnay JN, Taniguchi CM, Tran TT, Suzuki R, Espinoza DO, Yamamoto Y, Ahrens MJ, Dudley AT, Norris AW, Kulkarni RN, Kahn CR (2008) New role of bone morphogenetic protein 7 in brown adipogenesis and energy expenditure. Nature. 454:1000–1004
64. Casteilla L, Dani C (2006) Adipose tissue-derived cells: from physiology to regenerative medicine. Diabetes and Metabolism 32:393–401
65. Tang W, Zeve D, Suh JM, Bosnakovski D, Kyba M, Hammer RE, Tallquist MD, Graff JM (2008) White fat progenitor cells reside in the adipose vasculature. Science 322:583–586
66. Miranville A, Heeschen C, Sengenes C, Curat CA, Busse R, Bouloumie A (2004) Improvement of postnatal neovascularization by human adipose tissue-derived stem cells. Circulation 110:349–355
67. Planat-Benard V, Silvestre JS, Cousin B, André M, Nibbelink M, Tamarat R, Clergue M, Manneville C, Saillan-Barreau C, Duriez M, Tedgui A, Levy B, Pénicaud L, Casteilla L (2004) Plasticity of human adipose lineage cells toward endothelial cells: physiological and therapeutic perspectives. Circulation 109:656–663
68. Jumabay M, Zhang R, Yao Y, Goldhaber JI, Boström KI (2010) Spontaneously beating cardiomyocytes derived from white mature adipocytes. Cardiovascular Research 85:17–27

69. Berry DC, Jin H, Majumdar A, Noy N (2011) Signaling by vitamin A and retinol-binding protein regulates gene expression to inhibit insulin responses. Proceedings of the National Academy of Sciences of the United States of America 15;108(11):4340–4345
70. Vegiopoulos A, Müller-Decker K, Strzoda D, Schmitt I, Chichelnitskiy E, Ostertag A, Diaz MB, Rozman J, Hrabe de Angelis M, Nüsing RM, Meyer CW, Wahli W, Klingenspor M, Herzig S (2010) Cyclooxygenase-2 controls energy homeostasis in mice by de novo recruitment of brown adipocytes. Science 328:1158–1161
71. Djouder N, Tuerk RD, Suter M, Salvioni P, Thali RF, Scholz R, Vaahtomeri K, Auchli Y, Rechsteiner H, Brunisholz RA, Viollet B, Mäkelä TP, Wallimann T, Neumann D, Krek W (2010) PKA phosphorylates and inactivates AMPKα to promote efficient lipolysis. EMBO Journal 29:469–481
72. Bastelica D, Morange P, Berthet B, Borghi H, Lacroix O, Grino M, Juhan-Vague I, Alessi MC (2002) Stromal cells are the main plasminogen activator inhibitor-1-producing cells in human fat: evidence of differences between visceral and subcutaneous deposits. Arteriosclerosis, Thrombosis, and Vascular Biology 22:173–178
73. Skurk T, van Harmelen V, Blum WF, Hauner H (2005) Angiotensin II promotes leptin production in cultured human fat cells by an ERK1/2-dependent pathway. Obesity Research 13:969–973
74. Jayasooriya AP, Mathai ML, Walker LL, Begg DP, Denton DA, Cameron-Smith D, Egan GF, McKinley MJ, Rodger PD, Sinclair AJ, Wark JD, Weisinger HS, Jois M, Weisinger RS (2008) Mice lacking angiotensin-converting enzyme have increased energy expenditure, with reduced fat mass and improved glucose clearance. Proceedings of the National Academy of Sciences of the United States of America 105:6531–6536
75. Wölfing B, Buechler C, Weigert J, Neumeier M, Aslanidis C, Schöelmerich J, Schäffler A (2008) Effects of the new C1q/TNF-related protein (CTRP-3) "cartonectin" on the adipocytic secretion of adipokines. Obesity 16:1481–1486
76. Kishore U, Gaboriaud C, Waters P, Shrive AK, Greenhough TJ, Reid KB, Sim RB, Arlaud GJ (2004) C1q and tumor necrosis factor superfamily: modularity and versatility. Trends in Immunology 25:551–561
77. Yamauchi T, Kamon J, Minokoshi Y, Ito Y, Waki H, Uchida S, Yamashita S, Noda M, Kita S, Ueki K, Eto K, Akanuma Y, Froguel P, Foufelle F, Ferre P, Carling D, Kimura S, Nagai R, Kahn BB, Kadowaki T (2002) Adiponectin stimulates glucose utilization and fatty-acid oxidation by activating AMP-activated protein kinase. Nature – Medicine 8:1288–1295
78. Van Gaal LF, Mertens IL, De Block CE (2006) Mechanisms linking obesity with cardiovascular disease. Nature 444:875–880
79. Kim HS, Yumkham S, Choi JH, Son GH, Kim K, Ryu SH, Suh PG (2006) Serotonin stimulates GnRH secretion through the c-Src-PLCγ1 pathway in GT1-7 hypothalamic cells. Journal of Endocrinology 190:581–591
80. Fichna J, Janecka A, Costentin J, Do Rego JC (2007) The endomorphin system and its evolving neurophysiological role. Pharmacological Reviews 59:88–123

81. Sakurai T (2007) The neural circuit of orexin (hypocretin): maintaining sleep and wakefulness. Nature Reviews – Neuroscience 8:171–181
82. Alexander SPH, Mathie A, Peters JA (2007) Guide to receptors and channels. British Journal of Pharmacology 150:S1–S168
83. Nakamura T, Kawabe K, Sapru HN (2009) Cardiovascular responses to microinjections of urocortin 3 into the nucleus tractus solitarius of the rat. American Journal of Physiology – Heart and Circulatory Physiology 296:H325–H332
84. Huang Y, Yao XQ, Lau CW, Chan YC, Tsang SY, Chan FL (2004) Urocortin and cardiovascular protection. Acta pharmacologica Sinica 25:257–265
85. Wiley KE, Davenport AP (2004) CRF2 receptors are highly expressed in the human cardiovascular system and their cognate ligands urocortins 2 and 3 are potent vasodilators. British Journal of Pharmacology 143:508–514
86. Rowlinson SW, Yoshizato H, Barclay JL, Brooks AJ, Behncken SN, Kerr LM, Millard K, Palethorpe K, Nielsen K, Clyde-Smith J, Hancock JF, Waters MJ (2008) An agonist-induced conformational change in the growth hormone receptor determines the choice of signalling pathway. Nature – Cell Biology 10:740–747
87. Chu JYS, Lee LTO, Lai CH, Vaudry H, Chan YS, Yung WH, Chow BKC (2009) Secretin as a neurohypophysial factor regulating body water homeostasis. Proceedings of the National Academy of Sciences of the United States of America 106:15961–15966
88. Carrillo-Sepúlveda MA, Ceravolo GS, Fortes ZB, Carvalho MH, Tostes RC, Laurindo FR, Webb RC, Barreto-Chaves MLM (2010) Thyroid hormone stimulates NO production via activation of the PI3K/Akt pathway in vascular myocytes. Cardiovascular Research 85:560–570
89. Cueille C, Garel JM (2004) RAMP et récepteurs couplés aux protéines G [RAMPs and G protein-coupled receptors]. Médecine Sciences 20:773–778
90. Tastan I, Schreckenberg R, Mufti S, Abdallah Y, Piper HM, Schlüter KD (2009) Parathyroid hormone improves contractile performance of adult rat ventricular cardiomyocytes at low concentrations in a non-acute way. Cardiovascular Research 82:77-83
91. Atfi A, Baron R (2010) PTH battles TGF-β in bone. Nature – Cell Biology 12:205–207
92. Hay DL, Smith DM (2001) Adrenomedullin receptors: molecular identity and function. Peptides 22:1753–1763
93. Hay DL, Howitt SG, Conner AC, Schindler M, Smith DM, Poyner DR (2003) CL/RAMP2 and CL/RAMP3 produce pharmacologically distinct adrenomedullin receptors: a comparison of effects of adrenomedullin22-52, CGRP8-37 and BIBN4096BS. British Journal of Pharmacology 140:477–486
94. Yamashiro T, Kuge H, Zhang J, Honke K (2010) Calcineurin mediates the angiotensin II-induced aldosterone synthesis in the adrenal glands by upregulation of transcription of the CYP11B2 gene. Journal of Biochemistry 148:115–123
95. Mahapatra NR (2008) Catestatin is a novel endogenous peptide that regulates cardiac function and blood pressure. Cardiovascular Research 80:330–338

96. Perusquía M, Stallone JN (2010) Do androgens play a beneficial role in the regulation of vascular tone? Nongenomic vascular effects of testosterone metabolites. American Journal of Physiology – Heart and Circulatory Physiology 298:H1301–H1307
97. Cooper GJ (1994) Amylin compared with calcitonin gene-related peptide: structure, biology, and relevance to metabolic disease. Endocrine Reviews 15:163–201
98. Hasegawa K, Kar S, Carr BI (1994) Stimulation of hepatocyte DNA synthesis by neurotensin. Journal of Cell Biology 158:215–222
99. Zhang D, Shooshtarizadeh P, Laventie BJ, Colin DA, Chich JF, Vidic J, de Barry J, Chasserot-Golaz S, Delalande F, Van Dorsselaer A, Schneider F, Helle K, Aunis D, Prévost G, Metz-Boutigue MH (2009) Two chromogranin A-derived peptides induce calcium entry in human neutrophils by calmodulin-regulated calcium independent phospholipase A2. PLoS One 4:e4501
100. Ibelgaufts H (2010) Cytokines and Cells Online Pathfinder Encyclopaedia (www.copewithcytokines.de/cope.cgi)
101. Helle KB (2010) The chromogranin A-derived peptides vasostatin-I and catestatin as regulatory peptides for cardiovascular functions. Cardiovascular Research 85:9–16
102. Gomez RA, Pentz ES, Jin X, Cordaillat M, Sequeira Lopez ML (2009) CBP and p300 are essential for renin cell identity and morphological integrity of the kidney. American Journal of Physiology – Heart and Circulatory Physiology 296:H1255-H1262
103. Prosser HC, Forster ME, Richards AM, Pemberton CJ (2009) Cardiac chymase converts rat proAngiotensin-12 (PA12) to angiotensin II: effects of PA12 upon cardiac haemodynamics. Cardiovascular Research 82:40-50
104. Jessup JA, Trask AJ, Chappell MC, Nagata S, Kato J, Kitamura K, Ferrario CM (2008) Localization of the novel angiotensin peptide, angiotensin-(1-12), in heart and kidney of hypertensive and normotensive rats. American Journal of Physiology – Heart and Circulatory Physiology 294:H2614–H2618
105. San-Cristobal P, Pacheco-Alvarez D, Richardson C, Ring AM, Vazquez N, Rafiqi FH, Chari D, Kahle KT, Leng Q, Bobadilla NA, Hebert SC, Alessi DR, Lifton RP, Gamba G (2009) Angiotensin II signaling increases activity of the renal Na-Cl cotransporter through a WNK4-SPAK-dependent pathway. Proceedings of the National Academy of Sciences of the United States of America 106:4384–4389
106. Jeong Y, Chaupin DF, Matsushita K, Yamakuchi M, Cameron SJ, Morrell CN, Lowenstein CJ (2009) Aldosterone activates endothelial exocytosis. Proceedings of the National Academy of Sciences of the United States of America 106:3782–3787
107. Das BB, Solinger R (2009) Role of natriuretic peptide family in cardiovascular medicine. Cardiovascular and Hematological Agents in Medicinal Chemistry 7:29–42
108. Lanfear DE (2008) Genetic variation in the natriuretic peptide system and heart failure. Heart Failure Reviews (http://dx.doi.org/10.1007/s10741-008-9113-y)

109. Vesely DL (2007) Urodilatin: a better natriuretic peptide? Current Heart Failure Report 4:147–152
110. Koller KJ, Goeddel DV (1992) Molecular biology of the natriuretic peptides and their receptors. Circulation 86:1081–1088
111. Citarella MR, Choi MR, Gironacci MM, Medici C, Correa AH, Fernández BE (2009) Urodilatin and dopamine: a new interaction in the kidney. Regulatory Peptides 153:19–24
112. Lee CY, Burnett JC (2007) Natriuretic peptides and therapeutic applications. Heart Failure Reviews 12:131–142
113. Piechota M, Banach M, Jacon A, Rysz J (2008) Natriuretic peptides in cardiovascular diseases. Cellular and Molecular Biology Letters 13:155–181
114. Mendonça MC, Koles N, Doi SQ, Sellitti DF (2010) Transforming growth factor-β1 regulation of C-type natriuretic peptide expression in human vascular smooth muscle cells: dependence on TSC22D1. American Journal of Physiology – Heart and Circulatory Physiology 299:H2018–H2027

Chap. 2. Cell Growth and Proliferation

115. Ikenoue T, Inoki K, Yang Q, Zhou X, Guan KL (2008) Essential function of TORC2 in PKC and Akt turn motif phosphorylation, maturation and signalling. EMBO Journal 27:1919–1931
116. Facchinetti V, Ouyang W, Wei H, Soto N, Lazorchak A, Gould C, Lowry C, Newton AC, Mao Y, Miao RQ, Sessa WC, Qin J, Zhang P, Su B, Jacinto E (2008) The mammalian target of rapamycin complex 2 controls folding and stability of Akt and protein kinase C. EMBO Journal 27: 1932–1943
117. Hotamisligil GS, Erbay E (2008) Nutrient sensing and inflammation in metabolic diseases. Nature Reviews – Immunology 8:923–934
118. Buettner C, Muse ED, Cheng A, Chen L, Scherer T, Pocai A, Su K, Cheng B, Li X, Harvey-White J, Schwartz GJ, Kunos G, Rossetti L, Buettner C (2008) Leptin controls adipose tissue lipogenesis via central, STAT3-independent mechanisms. Nature – Medicine 14:667–675
119. Wada T, Stepniak E, Hui L, Leibbrandt A, Katada T, Nishina H, Wagner EF, Penninger JM (2008) Antagonistic control of cell fates by JNK and p38-MAPK signaling. Cell Death and Differentiation 15:89–93
120. Finck BN, Kelly DP (2006) PGC-1 coactivators: inducible regulators of energy metabolism in health and disease. Journal of Clinical Investigation 116:615–622
121. Rohas LM, St Pierre J, Uldry M, Jager S, Handschin C, Spiegelman BM (2007) A fundamental system of cellular energy homeostasis regulated by PGC-1α. Proceedings of the National Academy of Sciences of the United States of America 104:7933–7938
122. St-Pierre J, Drori S, Uldry M, Silvaggi JM, Rhee J, Jager S, Handschin C, Zheng K, Lin J, Yang W, Simon DK, Bachoo R, Spiegelman BM (2006) Suppression of reactive oxygen species and neurodegeneration by the PGC-1 transcriptional coactivators. Cell 127:397–408

123. Oishi Y, Manabe I, Tobe K, Ohsugi M, Kubota T, Fujiu K, Maemura K, Kubota N, Kadowaki T, Nagai R (2008) SUMOylation of Krüppel-like transcription factor 5 acts as a molecular switch in transcriptional programs of lipid metabolism involving PPAR-δ. Nature – Medicine 14:656–666

124. Finkel T, Deng CX, Mostoslavsky R (2009) Recent progress in the biology and physiology of sirtuins. Nature 460:587–591

125. Guarente L (2006) Sirtuins as potential targets for metabolic syndrome. Nature 444:868–874

126. Ahn BH, Kim HS, Song S, Lee IH, Liu J, Vassilopoulos A, Deng CX, Finke T (2008) A role for the mitochondrial deacetylase Sirt3 in regulating energy homeostasis. Proceedings of the National Academy of Sciences of the United States of America 105:14447–14452

127. Hay N, Sonenberg N (2004) Upstream and downstream of mTOR. Genes and Development 18:1926–1945

128. Sarbassov DD, Guertin DA, Ali SM, Sabatini DM (2005) Phosphorylation and regulation of Akt/PKB by the rictor-mTOR complex. Science 307:1098–1101

129. Sancak Y, Peterson TR, Shaul YD, Lindquist RA, Thoreen CC, Bar-Peled L, Sabatini DM (2008) The Rag GTPases bind Raptor and mediate amino acid signaling to mTORC1. Science 320:1496–1501

130. Sun Y, Fang Y, Yoon MS, Zhang C, Roccio M, Zwartkruis FJ, Armstrong M, Brown HA, Chen J (2008) Phospholipase D1 is an effector of Rheb in the mTOR pathway. Proceedings of the National Academy of Sciences of the United States of America 105:8286–8291

131. Düvel K, Yecies JL, Menon S, Raman P, Lipovsky AI, Souza AL, Triantafellow E, Ma Q, Gorski R, Cleaver S, Vander Heiden MG, MacKeigan JP, Finan PM, Clish CB, Murphy LO, Manning BD (2010) Activation of a metabolic gene regulatory network downstream of mTOR complex 1. Molecular Cell 39:171–183

132. Sarbassov DD, Sabatini DM (2005) Redox regulation of the nutrient-sensitive raptor-mTOR pathway and complex. Journal of Biological Chemistry 280:39505–39509

133. Ma XM, Blenis J (2009) Molecular mechanisms of mTOR-mediated translational control. Nature Reviews – Molecular Cell Biology 10:307–318

134. Kantidakis T, Ramsbottom BA, Birch JL, Dowding SN, White RJ (2010) mTOR associates with TFIIIC, is found at tRNA and 5S rRNA genes, and targets their repressor Maf1. Proceedings of the National Academy of Sciences of the United States of America 107:11823–11828

135. Oficjalska-Pham D, Harismendy O, Smagowicz WJ, Gonzalez de Peredo A, Boguta M, Sentenac A, Lefebvre O (2006) General repression of RNA polymerase III transcription is triggered by protein phosphatase type 2A-mediated dephosphorylation of Maf1. Molecular Cell 22:623–632

136. Moir RD, Lee J, Haeusler RA, Desai N, Engelke DR, Willis IM (2006) Protein kinase A regulates RNA polymerase III transcription through the nuclear localization of Maf1. Proceedings of the National Academy of Sciences of the United States of America 103:15044–15049

137. Towpik J, Graczyk D, Gajda A, Lefebvre O, Boguta M (2008) Derepression of RNA polymerase III transcription by phosphorylation and nuclear export of its negative regulator, Maf1. Journal of Biological Chemistry 283:17168–17174
138. Bai X, Ma D, Liu A, Shen X, Wang QJ, Liu Y, Jiang Y (2007) Rheb activates mTOR by antagonizing its endogenous inhibitor, FKBP38. Science 318:977–980
139. Sancak Y, Bar-Peled L, Zoncu R, Markhard AL, Nada S, Sabatini DM (2010) Ragulator-Rag complex targets mTORC1 to the lysosomal surface and is necessary for its activation by amino acids. Cell 141:290–303
140. Nada S, Hondo A, Kasai A, Koike M, Saito K, Uchiyama Y, Okada M (2009) The novel lipid raft adaptor p18 controls endosome dynamics by anchoring the MEK-ERK pathway to late endosomes. EMBO Journal 28:477–489
141. Karni R, Hippo Y, Lowe SW, Kraine AR (2008) The splicing-factor oncoprotein SF2/ASF activates mTORC1. Proceedings of the National Academy of Sciences of the United States of America 105:15323–15327
142. Costa-Mattioli M, Sonenberg N (2008) RAPping production of type I interferon in pDCs through mTOR. Nature – Immunology 9:1097–1099
143. Cao W, Manicassamy S, Tang H, Kasturi SP, Pirani A, Murthy N, Pulendran N (2008) Toll-like receptor-mediated induction of type I interferon in plasmacytoid dendritic cells requires the rapamycin-sensitive PI(3)K-mTOR-p70S6K pathway. Nature – Immunology 9:1157–1164
144. Weichhart T, Costantino G, Poglitsch M, Rosner M, Zeyda M, Stuhlmeier KM, Kolbe T, Stulnig TM, Hörl WH, Hengstschläger M, Müller M, Säemann MD (2008) The TSC-mTOR signaling pathway regulates the innate inflammatory response. Immunity 29:565–577
145. Matsuda T, Zhai P, Maejima Y, Hong C, Gao S, Tian B, Goto K, Takagi H, Tamamori-Adachi M, Kitajima S, Sadoshima J (2008) Distinct roles of GSK-3α and GSK-3β phosphorylation in the heart under pressure overload. Proceedings of the National Academy of Sciences of the United States of America 105:20900–20905
146. Beharry Z, Mahajan S, Zemskova M, Lin YW, Tholanikunnel BG, Xia Z, Smith CD, Kraft AS (2011) The Pim protein kinases regulate energy metabolism and cell growth. Proceedings of the National Academy of Sciences of the United States of America 108:528–533
147. Sang L, Coller HA, Roberts JM (2008) Control of the reversibility of cellular quiescence by the transcriptional repressor HES1. Science 321:1095–1100
148. Meaburn KJ, Misteli T (2007) Cell biology: chromosome territories. Nature 445:379–381
149. Hutchins JR, Toyoda Y, Hegemann B, Poser I, Hériché JK, Sykora MM, Augsburg M, Hudecz O, Buschhorn BA, Bulkescher J, Conrad C, Comartin D, Schleiffer A, Sarov M, Pozniakovsky A, Slabicki MM, Schloissnig S, Steinmacher I, Leuschner M, Ssykor A, Lawo S, Pelletier L, Stark H, Nasmyth K, Ellenberg J, Durbin R, Buchholz F, Mechtler K, Hyman AA, Peters JM (2010) Systematic analysis of human protein complexes identifies chromosome segregation proteins. Science 328:593–599 (MitoCheck consortium, www.mitocheck.org)

150. Shimamoto Y, Kapoor TM (2010) Mechanoregulation: cellular seat belts. Nature 468:518–519

151. Akiyoshi B, Sarangapani KK, Powers AF, Nelson CR, Reichow SL, Arellano-Santoyo H, Gonen T, Ranish JA, Asbury CL, Biggins S (2010) Tension directly stabilizes reconstituted kinetochore–microtubule attachments. Nature 468:576–579

152. Folco HD, Pidoux AL, Urano T, Allshire RC (2008) Heterochromatin and RNAi are required to establish CENP-A chromatin at centromeres. Science 319:94–97

153. Anderson DJ, Hetzer MW (2008) Reshaping of the endoplasmic reticulum limits the rate for nuclear envelope formation. Journal of Cell Biology 182:911–924

154. Théry M, Racine V, Pépin A, Piel M, Chen Y, Sibarita JB, Bornens M (2005) The extracellular matrix guides the orientation of the cell division axis. Nature – Cell Biology 7:947–953

155. Théry M, Jiménez-Dalmaroni A, Racine V, Bornens M, Jülicher F (2007) Experimental and theoretical study of mitotic spindle orientation. Nature 447:493–496

156. Amaro AC, Samora CP, Holtackers R, Wang E, Kingston IJ, Alonso M, Lampson M, McAinsh AD, Meraldi P (2010) Molecular control of kinetochore-microtubule dynamics and chromosome oscillations. Nature – Cell Biology 12:319–329

157. Vavylonis D, Wu JQ, Hao S, O'Shaughnessy B, Pollard TD (2008) Assembly mechanism of the contractile ring for cytokinesis by fission yeast. Science 319:97–100

158. Lee HH, Elia N, Ghirlando R, Lippincott-Schwartz J, Hurley JH (2008) Midbody targeting of the ESCRT machinery by a noncanonical coiled coil in CEP55. Science 322:576–580

159. Montagnac G, Chavrier P (2010) Abscission accomplished by PtdIns(3)P. Nature – Cell Biology 12:308–310

160. Sagona AP, Nezis IP, Pedersen NM, Liestol K, Poulton J, Rusten TE, Skotheim RI, Raiborg C, Stenmark H (2010) PtdIns(3)P controls cytokinesis through KIF13A-mediated recruitment of FYVE-CENT to the midbody. Nature – Cell Biology 12:362–371

161. Güttinger S, Laurell E, Kutay U (2009) Orchestrating nuclear envelope disassembly and reassembly during mitosis. Nature Reviews – Molecular Cell Biology 10:178–191

162. Coller HA (2007) What's taking so long: S-phase entry from quiescence versus proliferation. Nature Reviews – Molecular Cell Biology 8:667–670

163. Sang L, Coller HA, Roberts JM (2008) Control of the reversibility of cellular quiescence by the transcriptional repressor HES1. Science 321:1095–1100

164. Kawashima SA, Yamagishi Y, Honda T, Ishiguro KI, Watanabe Y (2010) Phosphorylation of H2A by Bub1 prevents chromosomal instability through localizing shugoshin. Science 327:172–177

165. Javerzat JP (2010) Directing the centromere guardian. Science 327:150–151

166. Yoon YM, Baek KH, Jeong SJ, Shin HJ, Ha GH, Jeon AH, Hwang SG, Chun JS, Lee CW (2004) WD repeat-containing mitotic checkpoint proteins act as transcriptional repressors during interphase. FEBS Letters 575:23–29

167. Nilsson J, Yekezare M, Minshull J, Pines J (2008) The APC/C maintains the spindle assembly checkpoint by targeting Cdc20 for destruction. Nature – Cell Biology 10:1411–1420

168. Malureanu LA, Jeganathan KB, Hamada M, Wasilewski L, Davenport J, van Deursen JM (2009) BubR1 N terminus acts as a soluble inhibitor of cyclin B degradation by APC/C(Cdc20) in interphase. Developmental Cell 16:118–131

169. Kulukian A, Han JS, Cleveland DW (2009) Unattached kinetochores catalyze production of an anaphase inhibitor that requires a Mad2 template to prime Cdc20 for BubR1 binding. Developmental Cell 16:105–117

170. Kimata Y, Baxter JE, Fry AM, Yamano H (2008) A role for the Fizzy/Cdc20 family of proteins in activation of the APC/C distinct from substrate recruitment. Molecular Cell 32:576–583

171. Dephoure N, Zhou C, Villén J, Beausoleil SA, Bakalarski CE, Elledge SJ, Gygi SP (2008) A quantitative atlas of mitotic phosphorylation. Proceedings of the National Academy of Sciences of the United States of America 105:10762–10767

172. Morgan DO (2007) The Cell Cycle; Principles of Control. Chap. 3 The Cell-Cycle Control System. New Science Press, London, UK

173. Berndt N (1999) Protein dephosphorylation and the intracellular control of the cell number. Frontiers in Bioscience 4:d22–d42

174. Hochegger H, Takeda S, Hunt T (2008) Cyclin-dependent kinases and cell-cycle transitions: does one fit all? Nature Reviews – Molecular Cell Biology 9:910–916

175. Huang S, Chen CS, Ingber DE (1998) Control of cyclin D1, p27(Kip1), and cell cycle progression in human capillary endothelial cells by cell shape and cytoskeletal tension. Molecular Biology of the Cell 9:3179–3193

176. Vodermaier HC (2004) APC/C and SCF: controlling each other and the cell cycle. Current Biology 14:R787-R796

177. Orlando DA, Lin CY, Bernard A, Wang JY, Socolar JES, Iversen ES, Hartemink AJ, Haase SB (2008) Global control of cell-cycle transcription by coupled CDK and network oscillators. Nature 453:944–947

178. Fung TK, Poon RY (2005) A roller coaster ride with the mitotic cyclins. Seminars in Cell and Developmental Biology 16:335–342

179. Fung TK, Poon RY (2006) Cyclin A1. UCSD-Nature Molecule Pages. UCSD-Nature Signaling Gateway (www.signaling-gateway.org)

180. Poon RY, Fung TK (2007) Cyclin A2. UCSD-Nature Molecule Pages. UCSD-Nature Signaling Gateway (www.signaling-gateway.org)

181. Katona RL, Lahti JM (2006) Cyclin C. UCSD-Nature Molecule Pages. UCSD-Nature Signaling Gateway (www.signaling-gateway.org)

182. Martin SG, Berthelot-Grosjean M (2009) Polar gradients of the DYRK-family kinase Pom1 couple cell length with the cell cycle. Nature 459:852–856

183. Moseley JB, Mayeux A, Paoletti A, Nurse P (2009) A spatial gradient coordinates cell size and mitotic entry in fission yeast. Nature 459:857–860
184. Merrick KA, Larochelle S, Zhang C, Allen JJ, Shokat KM, Fisher RP (2008) Distinct activation pathways confer cyclin-binding specificity on Cdk1 and Cdk2 in human cells. Molecular Cell 32:662–672
185. Akoulitchev S, Chuikov S, Reinberg D (2000) TFIIH is negatively regulated by cdk8-containing mediator complexes. Nature 407:102–106
186. Katona RL, Lahti JM (2006) Cdk8. UCSD-Nature Molecule Pages. UCSD-Nature Signaling Gateway (www.signaling-gateway.org)
187. Trembley JH, Lahti JM (2006) Cdk11. UCSD-Nature Molecule Pages. UCSD-Nature Signaling Gateway (www.signaling-gateway.org)
188. Ullah Z, Lee CY, Depamphilis ML (2009) Cip/Kip cyclin-dependent protein kinase inhibitors and the road to polyploidy. Cell Division 4:10
189. Ray A, James MK, Larochelle S, Fisher RP, Blain SW (2009) p27Kip1 inhibits cyclin D-cyclin-dependent kinase 4 by two independent modes. Molecular and Cellular Biology 29:986–999
190. Sinuani I, Weissgarten J, Beberashvili I, Rapoport MJ, Sandbank J, Feldman L, Albeck M, Averbukh Z, Sredni B (2009) The cyclin kinase inhibitor p57(kip2) regulates TGF-β-induced compensatory tubular hypertrophy: effect of the immunomodulator AS101. Nephrology, Dialysis, Transplantation 24:2328–2338
191. Mitra P, Ghule PN, van der Deen M, Medina R, Xie RL, Holmes WF, Ye X, Nakayama KI, Harper JW, Stein JL, Stein GS, van Wijnen AJ (2009) CDK inhibitors selectively diminish cell cycle controlled activation of the histone H4 gene promoter by p220NPAT and HiNF-P. Journal of Cellular Physiology 219:438–448
192. Nakano N, Urasawa K, Takagi Y, Saito T, Kaneta S, Ishikawa S, Higashi H, Tsutsui H, Hatakeyama M, Kitabatake A (2005) Downregulation of cyclin-dependent kinase inhibitor; p57(kip2), is involved in the cell cycle progression of vascular smooth muscle cells. Biochemical and Biophysical Research Communications 338:1661–1667
193. Yu L, Quinn DA, Garg HG, Hales CA (2006) Gene expression of cyclin-dependent kinase inhibitors and effect of heparin on their expression in mice with hypoxia-induced pulmonary hypertension. Biochemical and Biophysical Research Communications 345:1565–1572
194. Parker LL, Walter SA, Young PG, Piwnica-Worms H (1993) Phosphorylation and inactivation of the mitotic inhibitor Weel by the nim1/cdr1 kinase. Nature 363:736–738
195. Watanabe N (2008) Wee1. UCSD-Nature Molecule Pages. UCSD-Nature Signaling Gateway (www.signaling-gateway.org)
196. Olsen BB, Kreutzer JN, Watanabe N, Holm T, Guerra B (2010) Mapping of the interaction sites between Wee1 kinase and the regulatory β-subunit of protein kinase CK2. International Journal of Oncology 36:1175–1182
197. Kelly AE, Ghenoiu C, Xue JZ, Zierhut C, Kimura H, Funabiki H (2010) Survivin reads phosphorylated histone h3 threonine 3 to activate the mitotic kinase aurora B. Science 330:235–239

198. Wang F, Dai J, Daum JR, Niedzialkowska E, Banerjee B, Stukenberg PT, Gorbsky GJ, Higgins JM (2010) Histone h3 thr-3 phosphorylation by haspin positions aurora B at centromeres in mitosis. Science 330:231–235
199. Yamagishi Y, Honda T, Tanno Y, Watanabe Y (2010) Two histone marks establish the inner centromere and chromosome bi-orientation. Science 330:239–243
200. Tsukahara T, Tanno Y, Watanabe Y (2010) Phosphorylation of the CPC by Cdk1 promotes chromosome bi-orientation. Nature 467:719–723
201. Archambault V, Glover DM (2009) Polo-like kinases: conservation and divergence in their functions and regulation. Nature Reviews – Molecular Cell Biology 10:265–275
202. Seki A, Coppinger JA, Jang CY, Yates JR, Fang G (2008) Bora and the kinase Aurora A cooperatively activate the kinase Plk1 and control mitotic entry. Science 320:1655–1658
203. Yim H, Erikson RL (2010) Cell division cycle 6, a mitotic substrate of polo-like kinase 1, regulates chromosomal segregation mediated by cyclin-dependent kinase 1 and separase. Proceedings of the National Academy of Sciences of the United States of America 107:19742–19747
204. Rosario CO, Ko MA, Haffani YZ, Gladdy RA, Paderova J, Pollett A, Squire JA, Dennis JW, Swallow CJ (2010) Plk4 is required for cytokinesis and maintenance of chromosomal stability. Proceedings of the National Academy of Sciences of the United States of America 107:6888–6893
205. Barré B, Perkins ND (2007) A cell cycle regulatory network controlling NF-κB subunit activity and function. EMBO Journal 26:4841–4855
206. Hershko DD (2008) Oncogenic properties and prognostic implications of the ubiquitin ligase Skp2 in cancer. Cancer 112:1415–1424
207. Jonason JH, Gavrilova N, Wu M, Zhang H, Sun H (2007) Cell Cycle 6:951–961
208. Gao D, Inuzuka H, Tseng A, Chin RY, Toker A, Wei W (2009) Phosphorylation by Akt1 promotes cytoplasmic localization of Skp2 and impairs APCCdh1-mediated Skp2 destruction. Nature – Cell Biology 11:397–408
209. Lin HK, Wang G, Chen Z, Teruya-Feldstein J, Liu Y, Chan CH, Yang WL, Erdjument-Bromage H, Nakayama KI, Nimer S, Tempst P, Pandolfi PP (2009) Phosphorylation-dependent regulation of cytosolic localization and oncogenic function of Skp2 by Akt/PKB. Nature – Cell Biology 11:420–432
210. Saurin AT, Durgan J, Cameron AJ, Faisal A, Marber MS, Parker PJ (2008) The regulated assembly of a PKCε complex controls the completion of cytokinesis. Nature – Cell Biology 10:891–901
211. Tian B, Yang Q, Mao Z (2009) Phosphorylation of ATM by Cdk5 mediates DNA damage signalling and regulates neuronal death. Nature – Cell Biology 11:211–218
212. Katsuno Y, Suzuki A, Sugimura K, Okumura K, Zineldeen DH, Shimada M, Niida H, Mizuno T, Hanaoka F, Nakanishi M (2009) Cyclin A-Cdk1 regulates the origin firing program in mammalian cells. Proceedings of the National Academy of Sciences of the United States of America 106:3184–3189
213. Chen Y, Pereira EA, Sanchez Y (2007) Chk1. UCSD-Nature Molecule Pages. UCSD-Nature Signaling Gateway (www.signaling-gateway.org)

214. Mills GB, Schmandt R, McGill M, Amendola A, Hill M, Jacobs K, May C, Rodricks AM, Campbell S, Hogg D (1992) Expression of TTK, a novel human protein kinase, is associated with cell proliferation. Journal of Biological Chemistry 267:16000–16006

215. Wei JH, Chou YF, Ou YH, Yeh YH, Tyan SW, Sun TP, Shen CY, Shieh SY (2005) TTK/hMps1 participates in the regulation of DNA damage checkpoint response by phosphorylating CHK2 on threonine 68. Journal of Biological Chemistry 280:7748–7757

216. BioGRID: General Repository for Interaction Datasets; database of physical and genetic interactions for model organisms (www.thebiogrid.org)

217. Huang YF, Chang MD, Shieh SY (2009) TTK/hMps1 mediates the p53-dependent postmitotic checkpoint by phosphorylating p53 at Thr18. Molecular and Cellular Biology 29:2935–2944

218. Nihira K, Taira N, Miki Y, Yoshida K (2008) TTK/Mps1 controls nuclear targeting of c-Abl by 14-3-3-coupled phosphorylation in response to oxidative stress. Oncogene 27:7285–7295

219. Burgess A, Vigneron S, Brioudes E, Labbé JC, Lorca T, Castro A (2010) Loss of human Greatwall results in G2 arrest and multiple mitotic defects due to deregulation of the cyclin B-Cdc2/PP2A balance. Proceedings of the National Academy of Sciences of the United States of America 107:12564-12569

220. McCormick JA, Yang CL, Ellison DH (2008) WNK Kinases and renal sodium transport in health and disease: an integrated view. Hypertension 51:588–596

221. Jiang ZY, Zhou QL, Holik J, Patel S, Leszyk J, Coleman K, Chouinard M, Czech MP (2005) Identification of WNK1 as a substrate of Akt/protein kinase B and a negative regulator of insulin-stimulated mitogenesis in 3T3-L1 cells. Journal of Biological Chemistry 280:21622–21628

222. Sun X, Gao L, Yu RK, Zeng G (2006) Down-regulation of WNK1 protein kinase in neural progenitor cells suppresses cell proliferation and migration. Journal of Neurochemistry 99:1114–1121

223. Jun P, Hong C, Lal A, Wong JM, McDermott MW, Bollen AW, Plass C, Held WA, Smiraglia DJ, Costello JF (2009) Epigenetic silencing of the kinase tumor suppressor WNK2 is tumor-type and tumor-grade specific. Neuro-Oncology 11:414–422

224. Tu SW, Bugde A, Luby-Phelps K, Cobb MH (2011) WNK1 is required for mitosis and abscission. Proceedings of the National Academy of Sciences of the United States of America 108:1385–1390

225. Moniz S, Matos P, Jordan P (2008) WNK2 modulates MEK1 activity through the Rho GTPase pathway. Cellular Signalling 20:1762–1768

226. Veríssimo F, Silva E, Morris JD, Pepperkok R, Jordan P (2006) Protein kinase WNK3 increases cell survival in a caspase-3-dependent pathway. Oncogene 25:4172–4182

227. Morishita D, Katayama R, Sekimizu K, Tsuruo T, Fujita N (2008) Pim kinases promote cell cycle progression by phosphorylating and down-regulating p27Kip1 at the transcriptional and posttranscriptional levels. Cancer Research 68:5076–5085

228. Patterson KI, Brummer T, O'Brien PM, Daly RJ (2009) Dual-specificity phosphatases: critical regulators with diverse cellular targets. Biochemical Journal 418:475–489

229. Esteban V, Vázquez-Novelle MD, Calvo E, Bueno A, Sacristán MP (2006) Human Cdc14A reverses CDK1 phosphorylation of Cdc25A on serines 115 and 320. Cell Cycle 5:2894–2898

230. Ferguson AM, White LS, Donovan PJ, Piwnica-Worms H (2005) Normal cell cycle and checkpoint responses in mice and cells lacking Cdc25B and Cdc25C protein phosphatases. Molecular and Cellular Biology 25:2853–2860

231. Mochida S, Maslen SL, Skehel M, Hunt T (2010) Greatwall phosphorylates an inhibitor of protein phosphatase 2A that is essential for mitosis. Science 330:1670-1673

232. Gharbi-Ayachi A, Labbé JC, Burgess A, Vigneron S, Strub JM, Brioudes E, Van-Dorsselaer A, Castro A, Lorca T (2010) The substrate of Greatwall kinase, Arpp19, controls mitosis by inhibiting protein phosphatase 2A. Science 330:1673-1677

233. Ianari A, Natale T, Calo E, Ferretti E, Alesse E, Screpanti I, Haigis K, Gulino A, Lees JA (2009) Proapoptotic function of the retinoblastoma tumor suppressor protein. Cancer Cell 15:184–194

234. Liu Y, Clem B, Zuba-Surma EK, El-Naggar S, Telang S, Jenson AB, Wang Y, Shao H, Ratajczak MZ, Chesney J, Dean DC (2009) Mouse fibroblasts lacking RB1 function form spheres and undergo reprogramming to a cancer stem cell phenotype. Cell Stem Cell 4:336–347

235. Shamma A, Takegami Y, Miki T, Kitajima S, Noda M, Obara T, Okamoto T, Takahashi C (2009) Rb Regulates DNA damage response and cellular senescence through E2F-dependent suppression of N-Ras isoprenylation. Cancer Cell 15:255–269

236. Moshe Y, Boulaire J, Pagano M, Hershko A (2004) Role of Polo-like kinase in the degradation of early mitotic inhibitor 1, a regulator of the anaphase promoting complex/cyclosome. Proceedings of the National Academy of Sciences of the United States of America 101:7937–7942

237. Canman JC, Lewellyn L, Laband K, Smerdon SJ, Desai A, Bowerman B, Oegema K (2008) Inhibition of Rac by the GAP activity of centralspindlin is essential for cytokinesis. Science 322:1543–1546

238. Miller AL, Bement WM (2009) Regulation of cytokinesis by Rho GTPase flux. Nature – Cell Biology 11:71–77

239. Lin DI, Aggarwal P, Diehl JA (2008) Phosphorylation of MCM3 on Ser-112 regulates its incorporation into the MCM2–7 complex. Proceedings of the National Academy of Sciences of the United States of America 105:8079–8084

240. Schwenen HD, Monteiro LJ, Wilson MSC, Lam EW (2009) FOXM1. Transcription Factor Encyclopedia (http://www.cisreg.ca/cgi-bin/tfe/articles.pl?tfid=437)

241. Baugh LR, Sternberg PW (2006) FOXO regulates transcription of cki-1/Cip/Kip and repression of lin-4 during C. elegans L1 arrest. Current Biology 16:780–785

242. Giangrande PH, Hallstrom TC, Tunyaplin C, Calame K, Nevins JR (2003) Identification of E-box factor TFE3 as a functional partner for the E2F3 transcription factor. Molecular and Cellular Biology 23:3707–3720

243. Bretscher A, Edwards K, Fehon RG (2002) ERM proteins and merlin: integrators at the cell cortex. Nature Reviews – Molecular Cell Biology 3:586–599

244. McClatchey AI, Fehon RG (2009) Merlin and the ERM proteins–regulators of receptor distribution and signaling at the cell cortex. Trends in Cell Biology 19:198–206

245. Huang J, Chen J (2008) VprBP targets Merlin to the Roc1-Cul4A-DDB1 E3 ligase complex for degradation. Oncogene 27:4056–4064

246. Li W, You L, Cooper J, Schiavon G, Pepe-Caprio A, Zhou L, Ishii R, Giovannini M, Hanemann CO, Long SB, Erdjument-Bromage H, Zhou P, Tempst P, Giancotti FG (2010) Merlin/NF2 suppresses tumorigenesis by inhibiting the E3 ubiquitin ligase $CRL4^{DCAF1}$ in the nucleus. Cell 140:477–490

247. Ahn J, Novince Z, Concel J, Byeon CH, Makhov AM, Byeon IJ, Zhang P, Gronenborn AM (2011) The Cullin-RING E3 ubiquitin ligase CRL4–DCAF1 complex dimerizes via a short helical region in DCAF1. Biochemistry (DOI: 10.1021/bi101749s)

248. Kaldis P, Pagano M (2009) Wnt signaling in mitosis. Developmental Cell 17:749–750

249. Davidson G, Shen J, Huang YL, Su Y, Karaulanov E, Bartscherer K, Hassler C, Stannek P, Boutros M, Niehrs C (2009) Cell cycle control of Wnt receptor activation. Developmental Cell 17:788–799

250. Wilker EW, van Vugt MATM, Artim SA, Huang PH, Petersen CP, Reinhardt HC, Feng Y, Sharp PA, Sonenberg N, White FM, Yaffe MB (2007) 14-3-3σ controls mitotic translation to facilitate cytokinesis. Nature 446:329–332

251. Lin HK, Wang G, Chen C, Teruya-Feldstein J, Liu Y, Chan CH, Yang WL, Erdjument-Bromage H, Nakayama KI, Nimer S, Tempst P, Pandolfi PP (2009) Phosphorylation-dependent regulation of cytosolic localization and oncogenic function of Skp2 by Akt/PKB. Nature – Cell Biology 11:420–432

252. Gao D, Inuzuka H, Tseng A, Chin RY, Toker A, Wei W (2009) Phosphorylation by Akt1 promotes cytoplasmic localization of Skp2 and impairs APC–Cdh1-mediated Skp2 destruction. Nature – Cell Biology 11:397-408

253. Gutierrez GJ, Tsuji T, Chen M, Jiang W, Ronai ZA (2010) Interplay between Cdh1 and JNK activity during the cell cycle. Nature – Cell Biology 12:686–695

254. Skaar JR, Pagano M (2008) Cdh1: a master G0/G1 regulator. Nature – Cell Biology 10:755–757

255. Toettcher JE, Loewer A, Ostheimer GJ, Yaffe MB, Tidor B, Lahav G (2009) Distinct mechanisms act in concert to mediate cell cycle arrest. Proceedings of the National Academy of Sciences of the United States of America 106:785–790

256. Lavi O, Louzoun Y (2009) What cycles the cell? – Robust autonomous cell cycle models. Mathematical Medicine and Biology 26:337–359

257. Csikász-Nagy A, Battogtokh D, Chen KC, Novák B, Tyson JJ (2006) Analysis of a generic model of eukaryotic cell-cycle regulation. Biophysical Journal 90:61–4379

258. Tyson JJ (1991) Modeling the cell division cycle: cdc2 and cyclin interactions. Proceedings of the National Academy of Sciences of the United States of America 88:7328–7332

259. Tyson JJ, Novak B, Chen K, Val J (1995) Checkpoints in the cell cycle from a modeler's perspective. Progress in Cell Cycle Research 1:1-8

260. Goldbeter A (1991) A minimal cascade model for the mitotic oscillator involving cyclin and cdc2 kinase. Proceedings of the National Academy of Sciences of the United States of America 88:9107-9111

261. Coudreuse D, Nurse P (2010) Driving the cell cycle with a minimal CDK control network. Nature 468:1074–1079

262. Gérard C, Goldbeter A (2011) A skeleton model for the network of cyclin-dependent kinases driving the mammalian cell cycle. Interface Focus 1:24–35

263. Altinok A, Gonze D, Lévi F, Goldbeter A (2011) An automaton model for the cell cycle. Interface Focus 1:36–47

264. Bekkal Brikci F, Clairambault J, Ribba B, Perthame B (2008) An age-and-cyclin-structured cell population model for healthy and tumoral tissues. Journal of Mathematical Biology 57:91–110

265. Clairambault J. Personal communication

266. Lim KH, Ancrile BB, Kashatus DF, Counter CM (2008) Tumour maintenance is mediated by eNOS. Nature 452:646–649

267. Chien Y, Kim S, Bumeister R, Loo YM, Kwon SW, Johnson CL, Balakireva MG, Romeo Y, Kopelovich L, Gale M Jr, Yeaman C, Camonis JH, Zhao Y, White MA (2006) RalB GTPase-mediated activation of the IκB family kinase TBK1 couples innate immune signaling to tumor cell survival. Cell 127:157–170

268. Silva JM, Marran K, Parker JS, Silva J, Golding M, Schlabach MR, Elledge SJ, Hannon GJ, Chang K (2008) Profiling essential genes in human mammary cells by multiplex RNAi screening. Science 319:617–620

269. Schlabach MR, Luo J, Solimini NL, Hu G, Xu Q, Li MZ, Zhao Z, Smogorzewska A, Sowa ME, Ang XL, Westbrook TF, Liang AC, Chang K, Hackett JA, Harper JW, Hannon GJ, Elledge SJ (2008) Cancer proliferation gene discovery through functional genomics. Science 319:620–624

270. Pelham RJ, Rodgers L, Hall I, Lucito R, Nguyen KC, Navin N, Hicks J, Mu D, Powers S, Wigler M, Botstein D (2006) Identification of alterations in DNA copy number in host stromal cells during tumor progression. Proceedings of the National Academy of Sciences of the United States of America 103:19848–19853.

271. Zeisberg EM, Potenta S, Xie L, Zeisberg M, Kalluri R (2007) Discovery of endothelial to mesenchymal transition as a source for carcinoma-associated fibroblasts. Cancer Research 67:10123–10128

272. Zitvogel L, Tesniere A, Kroeme G (2006) Cancer despite immunosurveillance: immunoselection and immunosubversion. Nature Reviews – Immunology 6:715–727

273. Stangl S, Gehrmann M, Riegger J, Kuhs K, Riederer I, Sievert W, Hube K, Mocikat R, Dressel R, Kremmer E, Pockley AG, Friedrich L, Vigh L, Skerra

A, Multhoff G (2011) Targeting membrane heat-shock protein 70 (Hsp70) on tumors by cmHsp70.1 antibody. Proceedings of the National Academy of Sciences of the United States of America 108:733–738

274. Axelrod R, Axelrod DE, Pienta KJ (2006) Evolution of cooperation among tumor cells. Proceedings of the National Academy of Sciences of the United States of America 103:13474–13479
275. Zou W (2006) Regulatory T cells, tumour immunity and immunotherapy. Nature Reviews – Immunology 6:295–307
276. Ohta A, Gorelik E, Prasad SJ, Ronchese F, Lukashev D, Wong MK, Huang X, Caldwell S, Liu K, Smith P, Chen JF, Jackson EK, Apasov S, Abrams S, Sitkovsky M (2006) A2A adenosine receptor protects tumors from antitumor T cells. Proceedings of the National Academy of Sciences of the United States of America 103:13132–13137
277. Watts RN, Hawkins C, Ponka P Richardson DR (2006) Nitrogen monoxide-mediated iron release from cells is linked to NO-induced glutathione efflux via multidrug resistance-associated protein 1. Proceedings of the National Academy of Sciences of the United States of America 103:7670–7675
278. Feng H, Shuda M, Chang Y, Moore PS (2008) Clonal integration of a polyomavirus in human Merkel cell carcinoma. Science 319:1096–1100
279. Karin M, Greten FR (2005) NF-κB: linking inflammation and immunity to cancer development and progression. Nature Reviews – Immunology 5:749–759
280. Zakrzewski JL, Suh D, Markley JC, Smith OM, King C, Goldberg GL, Jenq R, Holland AM, Grubin J, Cabrera-Perez J, Brentjens RJ, Lu SX, Rizzuto G, Sant'Angelo DB, Riviere I, Sadelain M, Heller G, Zúñiga-Pflücker JC, Lu C, van den Brink MRM (2008) Tumor immunotherapy across MHC barriers using allogeneic T-cell precursors. Nature – Biotechnology 26:453–461
281. Clarke C, Smyth MJ (2007) Calreticulin exposure increases cancer immunogenicity. Nature – Biotechnology 25:192–193
282. Obeid M, Tesniere A, Ghiringhelli F, Fimia GM, Apetoh L, Perfettini JL, Castedo M, Mignot G, Panaretakis T, Casares N, Metivier D, Larochette N, van Endert P, Ciccosanti F, Piacentini M, Zitvogel L, Kroemer G (2007) Calreticulin exposure dictates the immunogenicity of cancer cell death. Nature – Medicine 13:54–61
283. Pardo O, Wellbrock C, Khanzada UK, Aubert M, Arozarena I, Davidson S, Bowen F, Parker PJ, Filonenko VV, Gout IT, Sebire N, Marais R, Downward J, Seckl MJ (2006) FGF-2 protects small cell lung cancer cells from apoptosis through a complex involving PKCϵ, B-Raf and S6K2. EMBO Journal 25:3078–3088
284. Timmerbeul I, Garrett-Engele CM, Kossatz U, Chen X, Firpo E, Grünwald V, Kamino K, Wilkens L, Lehmann U, Buer J, Geffers R, Kubicka S, Manns MP, Porter PL, Roberts JM, Malek NP (2006) Testing the importance of p27 degradation by the SCFskp2 pathway in murine models of lung and colon cancer. Proceedings of the National Academy of Sciences of the United States of America 103:14009–14014

285. Noren NK, Foos G, Hauser CA, Pasquale EB (2006) The EphB4 receptor suppresses breast cancer cell tumorigenicity through an Abl–Crk pathway. Nature – Cell Biology 8:815–825

286. Gery S, Komatsu N, Baldjyan L, Yu A, Koo D, Koeffler HP (2006) The circadian gene Per1 plays an important role in cell growth and DNA damage control in human cancer cells. Molecular Cell 2:375–382

287. Pouysségur J, Dayan F, Mazure NM (2006) Hypoxia signalling in cancer and approaches to enforce tumour regression. Nature 441:437–443

288. Zhang Q, Wang SY, Nottke AC, Rocheleau JV, Piston DW, Goodman RH (2006) Redox sensor CtBP mediates hypoxia-induced tumor cell migration. Proceedings of the National Academy of Sciences of the United States of America 103:9029-9033

289. Podsypanina K, Du YC, Jechlinger M, Beverly LJ, Hambardzumyan D, Varmus H (2008) Seeding and propagation of untransformed mouse mammary cells in the lung. Science 321:1841–1844

290. Rosenfeld N, Aharonov R, Meiri E, Rosenwald S, Spector Y, Zepeniuk M, Benjamin H, Shabes N, Tabak S, Levy A, Lebanony D, Goren Y, Silberschein E, Targan N, Ben-Ari A, Gilad S, Sion-Vardy N, Tobar A, Feinmesser M, Kharenko O, Nativ O, Nass D, Perelman M, Yosepovich A, Shalmon B, Polak-Charcon S, Fridman E, Avniel A, Bentwich I, Bentwich Z, Cohen D, Chajut A, Barshack I (2008) MicroRNAs accurately identify cancer tissue origin. Nature – Biotechnology 26:462–469

291. Bandyopadhyay S, Zhan R, Chaudhuri A, Watabe M, Pai SK, Hirota S, Hosobe S, Tsukada T, Miura K, Takano Y, Saito K, Pauza ME, Hayashi S, Wang Y, Mohinta S, Mashimo T, Iiizumi M, Furuta E, Watabe K (2006) Interaction of KAI1 on tumor cells with DARC on vascular endothelium leads to metastasis suppression. Nature – Medicine 12:933–938

292. Kowanetz M, Wu X, Lee J, Tan M, Hagenbeek T, Qu X, Yu L, Ross J, Korsisaari N, Cao T, Bou-Reslan H, Kallop D, Weimer R, Ludlam MJ, Kaminker JS, Modrusan Z, van Bruggen N, Peale FV, Carano R, Meng YG, Ferrara N (2010) Granulocyte-colony stimulating factor promotes lung metastasis through mobilization of $Ly6G^{+}Ly6C^{+}$ granulocytes. Proceedings of the National Academy of Sciences of the United States of America 107:21248–21255

293. Duda DG, Duyverman AM, Kohno M, Snuderl M, Steller EJ, Fukumura D, Jain RK (2010) Malignant cells facilitate lung metastasis by bringing their own soil. Proceedings of the National Academy of Sciences of the United States of America 107:21677–21682

294. Medvedev G S, Ono K, Holmes PJ (2003) Travelling wave solutions of the degenerate Kolmogorov-Petrovski-Piskunov equation. European Journal of Applied Mathematics 14:343–367

295. Bages M, Martinez P, Roquejoffre JM (2008) Large-time dynamics for a class of KPP-type equations in periodic media. Comptes Rendus de l'Académie des sciences, Paris, Ser. I (Mathematics) 346:1051-1056

296. Kolmogorov AN, Petrovsky IG, Piscounov NS (1937) Etude de l'équation de la diffusion avec croissance de la quantité de matière et son application à un problème biologique. Bulletin de l'Université d'Etat de Moscou (Bjul.

Moskowskogo Gos. Univ.), Série Internationale, Section A1, 1–26 [A study of the equation of diffusion with increase in the quantity of matter, and its application to a biological problem. In Pelce P. (ed), 1988, Dynamics of curved fronts, Academic Press, Boston]

297. Murray JD (2002) Mathematical Biology. I: Introduction and II: Spatial Models and Biomedical Applications. Springer, New York
298. Clairambault J, Michel P, Perthame B (2006) Circadian rhythm and tumour growth. Comptes Rendus de l'Académie des sciences, Paris, Ser. I (Mathematics) 342:17–22
299. Basdevant C, Clairambault J, Lévi F (2005) Optimisation of time-scheduled regimen for anti-cancer drug infusion. Mathematical Modelling and Numerical Analysis 39:1069–1086
300. Buchakjian MR, Kornbluth S (2010) The engine driving the ship: metabolic steering of cell proliferation and death. Nature Reviews – Molecular Cell Biology 11:715–727

Chap. 3. Growth Factors

301. Lorenz HP, Longaker MT (2004) Wounds: biology, pathology, and management. http://recon.stanford.edu/Articles/LorenzWH.pdf
302. Woundbiotech (www.bu.edu/woundbiotech/growthfactors/gfdevtest.html)
303. Borel J-P, Maquart F-X (1991) La cicatrisation[The healing]. La Recherche 236:1174–1181
304. Tedgui A, Lévy B (1994) Biologie de la paroi artérielle [Biology of the arterial wall]. Masson, Paris
305. Edinger AL, Thompson CB (2002) Akt maintains cell size and survival by increasing mTOR-dependent nutrient uptake. Molecular Biology of the Cell 13:2276–2288
306. http://www.sigmaaldrich.com/
307. de Jong JS, van Diest PJ, van der Valk P, Baak JP (1998) Expression of growth factors, growth-inhibiting factors, and their receptors in invasive breast cancer II: correlations with proliferation and angiogenesis. Journal of Pathology 184:53–57
308. Brigstock DR (2003) The CCN family: a new stimulus package. Journal of Endocrinology 178:169–175
309. Brigstock DR, Goldschmeding R, Katsube KI, Lam SC, Lau LF, Lyons K, Naus C, Perbal B, Riser B, Takigawa M, Yeger H (2003) Proposal for a unified CCN nomenclature. Molecular Pathology – Journal of Clinical Pathology 56:127–128
310. Brigstock DR (1999) The connective tissue growth factor/cysteine-rich 61/nephroblastoma overexpressed (CCN) family. Endocrine Reviews 20:189–206
311. Moussad EE, Brigstock DR (2000) Connective tissue growth factor: what's in a name? Molecular Genetics and Metabolism 71:276–292

312. Li MH, Sanchez T, Pappalardo A, Lynch KR, Hla T, Ferrer F (2008) Induction of antiproliferative connective tissue growth factor expression in Wilms' tumor cells by sphingosine-1-phosphate receptor 2. Molecular Cancer Research 6:1649–1656

313. Zhang X, Gureasko J, Shen K, Cole PA, Kuriyan J (2006) An allosteric mechanism for activation of the kinase domain of epidermal growth factor receptor. Cell 125:1137–1149

314. Avraham R, Sas-Chen A, Manor O, Steinfeld I, Shalgi R, Tarcic G, Bossel N, Zeisel A, Amit I, Zwang Y, Enerly E, Russnes HG, Biagioni F, Mottolese M, Strano S, Blandino G, Børresen-Dale AL, Pilpel Y, Yakhini Z, Segal E., Yarden Y (2010) EGF decreases the abundance of microRNAs that restrain oncogenic transcription factors. Science Signaling 3:ra43

315. Amit I, Citri A, Shay T, Lu Y, Katz M, Zhang F, Tarcic G, Siwak D, Lahad J, Jacob-Hirsch J, Amariglio N, Vaisman N, Segal E, Rechavi G, Alon U, Mills GB, Domany E, Yarden Y (2007) A module of negative feedback regulators defines growth factor signaling. Nature – Genetics 39:503–512

316. Nanba D, Higashiyama S. (2004) Dual intracellular signaling by proteolytic cleavage of membrane-anchored heparin-binding EGF-like growth factor. Cytokine and Growth Factor Reviews 15:13–19

317. Kim HS, Shin HS, Kwak HJ, Cho CH, Lee CO, Koh GY (2003) Betacellulin induces angiogenesis through activation of mitogen-activated protein kinase and phosphatidylinositol 3'-kinase in endothelial cell. FASEB Journal 17:318–320

318. Schaefer G, Fitzpatrick VD, Sliwkowski MX (1997) Gamma-heregulin: a novel heregulin isoform that is an autocrine growth factor for the human breast cancer cell line, MDA-MB-175. Oncogene 15:1385–1394

319. Falls DL (2003) Neuregulins: functions, forms, and signaling strategies. Experimental Cell Research 284:14–30

320. Ky B, Kimmel SE, Safa RN, Putt ME, Sweitzer NK, Fang JC, Sawyer DB, Cappola TP (2009) Neuregulin-1 beta is associated with disease severity and adverse outcomes in chronic heart failure. Circulation 120:310–317

321. Eckert JM, Byer SJ, Clodfelder-Miller BJ, Carroll SL (2009) Neuregulin-1 β and neuregulin-1 α differentially affect the migration and invasion of malignant peripheral nerve sheath tumor cells. Glia 57:1501–1520

322. Ponomareva ON, Ma H, Dakour R, Raabe TD, Lai C, Rimer M (2005) Stimulation of acetylcholine receptor transcription by neuregulin-2 requires an N-box response element and is regulated by alternative splicing. Neuroscience 134:495–503

323. Carteron C, Ferrer-Montiel A, Cabedo H (2006) Characterization of a neural-specific splicing form of the human neuregulin 3 gene involved in oligodendrocyte survival. Journal of Cell Science 119:898–909

324. Hayes NV, Blackburn E, Smart LV, Boyle MM, Russell GA, Frost TM, Morgan BJ, Baines AJ, Gullick WJ (2007) Identification and characterization of novel spliced variants of neuregulin 4 in prostate cancer. Clinical Cancer Research 13:3147–3155

325. Itoh N, Ornitz DM (2008) Functional evolutionary history of the mouse Fgf gene family. Developmental Dynamics 237:18–27

326. BioGRID: General Repository for Interaction Datasets; database of physical and genetic interactions for model organisms (www.thebiogrid.org)

327. FGF Pathway. Sigma-Aldrich company (www.sigmaaldrich.com/life-science/your-favorite-gene-search/pathway-overviews/fgf-pathway.html)

328. Duan DS, Werner S, Williams LT (1992) A naturally occurring secreted form of fibroblast growth factor (FGF) receptor 1 binds basic FGF in preference over acidic FGF. Journal of Biological Chemistry 267:16076–16080

329. Zhang X, Ibrahimi OA, Olsen SK, Umemori H, Mohammadi M, Ornitz DM (2006) Receptor specificity of the fibroblast growth factor family. The complete mammalian FGF family. Journal of Biological Chemistry 281:15694–15700

330. Kalinina J, Byron SA, Makarenkova HP, Olsen SK, Eliseenkova AV, Larochelle WJ, Dhanabal M, Blais S, Ornitz DM, Day LA, Neubert TA, Pollock PM, Mohammadi M (2009) Homodimerization controls the fibroblast growth factor 9 subfamily's receptor binding and heparan sulfate-dependent diffusion in the extracellular matrix. Molecular and Cellular Biology 29:4663–4678

331. Mohammadi M, Olsen SK, Ibrahimi OA (2005) Structural basis for fibroblast growth factor receptor activation. Cytokine and Growth Factor Reviews 16:107–137

332. Wu X, Ge H, Lemon B, Vonderfecht S, Baribault H, Weiszmann J, Gupte J, Gardner J, Lindberg R, Wang Z, Li Y (2010) Separating mitogenic and metabolic activities of fibroblast growth factor 19 (FGF19). Proceedings of the National Academy of Sciences of the United States of America 107:14158–14163

333. Ogawa Y, Kurosu H, YamamotoM, Nandi A, Rosenblatt KP, Goetz R, Eliseenkova AV, Mohammadi M, Kuro-o M (2007) βKlotho is required for metabolic activity of fibroblast growth factor 21. Proceedings of the National Academy of Sciences of the United States of America 104:7432–7437

334. Tomiyama KI, Maeda R, Urakawa I, Yamazaki Y, Tanaka T, Ito S, Nabeshima Y, Tomita T, Odori S, Hosoda K, Nakao K, Imura A, Nabeshima YI (2010) Relevant use of Klotho in FGF19 subfamily signaling system in vivo. Proceedings of the National Academy of Sciences of the United States of America 107:1666–1671

335. Terauchi A, Johnson-Venkatesh EM, Toth AB, Javed D, Sutton MA, Umemori H (2010) Distinct FGFs promote differentiation of excitatory and inhibitory synapses. Nature 465:783–787

336. Kharitonenkov A, Chen Z, Sures I, Wang H, Schilling J, Ullrich A (1997) A family of proteins that inhibit signalling through tyrosine kinase receptors. Nature 386:181–186

337. Neugebauer JM, Amack JD, Peterson AG, Bisgrove BW, Yost HJ (2009) FGF signalling during embryo development regulates cilia length in diverse epithelia. Nature 458:651–654

338. Tolbert WD, Daugherty-Holtrop J, Gherardi E, Vande Woude G, Xu HE (2010) Structural basis for agonism and antagonism of hepatocyte growth factor. Proceedings of the National Academy of Sciences of the United States of America 107:13264–13269

339. Kajiya K, Hirakawa S, Ma B, Drinnenberg I, Detmar M (2005) Hepatocyte growth factor promotes lymphatic vessel formation and function. EMBO Journal 24:2885–2895

340. Naïmi M, Van Obberghen E (2008) Facteur de transcription FoxO1. Une passerelle nucléaire entre métabolisme et mitogenèse [Nuclear FoxO1 as a bridge between metabolism and mitogenesis]. Médecine – Sciences (Paris) 24:635-639

341. Saito M, Lessard SJ, Rivas DA, Reeder DW, Hawley JA, Yaspelkis BB (2008) Activation of atypical protein kinase Czeta toward TC10 is regulated by high-fat diet and aerobic exercise in skeletal muscle. Metabolism 57:1173–1180

342. Luan B, Zhao J, Wu H, Duan B, Shu G, Wang W, Li D, Jia W, Kang J, Pei G (2009) Deficiency of a β-arrestin-2 signal complex contributes to insulin resistance. Nature 457:1146–1149

343. Trajkovski M, Hausser J, Soutschek J, Bhat B, Akin A, Zavolan M, Heim MH, Stoffel M (2011) MicroRNAs 103 and 107 regulate insulin sensitivity. Nature 474:649–653

344. Neri Serneri GG, Boddi M, Modesti PA, Cecioni I, Coppo M, Padeletti L, Michelucci A, Colella A, Galanti G (2001) Increased cardiac sympathetic activity and insulin-like growth factor-I formation are associated with physiological hypertrophy in athletes. Circulation Research 89:977–982

345. Butler AA, LeRoith D (2001) Tissue-specific versus generalized gene targeting of the igf1 and igf1r genes and their roles in insulin-like growth factor physiology. Endocrinology 142:1685–1688

346. Kojima I, Nishimoto I, Iiri T, Ogata E, Rosenfeld R (1988) Evidence that type II insulin-like growth factor receptor is coupled to calcium gating system. Biochemical and Biophysical Research Communications 154:9–19

347. Chen DY, Stern SA, Garcia-Osta A, Saunier-Rebori B, Pollonini G, Bambah-Mukku D, Blitzer RD, Alberini CM (2011) A critical role for IGF-II in memory consolidation and enhancement. Nature 469:491–497

348. Zhu W, Shiojima I, Ito Y, Li Z, Ikeda H, Yoshida M, Naito AT, Nishi JI, Ueno H, Umezawa A, Minamino T, Nagai T, Kikuchi A, Asashima M, Komuro I (2008) IGFBP-4 is an inhibitor of canonical Wnt signalling required for cardiogenesis. Nature 454:345–349

349. Lu H, Xu X, Zhang M, Cao R, Bråkenhielm E, Li C, Lin H, Yao G, Sun H, Qi L, Tang M, Dai H, Zhang Y, Su R, Bi Y, Zhang Y, Cao Y (2007) Combinatorial protein therapy of angiogenic and arteriogenic factors remarkably improves collaterogenesis and cardiac function in pigs. Proceedings of the National Academy of Sciences of the United States of America 104:12140–12145

350. Yuasa S, Itabashi Y, Koshimizu U, Tanaka T, Sugimura K, Kinoshita M, Hattori F, Fukami S, Shimazaki T, Okano H, Ogawa S, Fukuda K (2005) Transient inhibition of BMP signaling by Noggin induces cardiomyocyte differentiation of mouse embryonic stem cells. Nature – Biotechnology 23:607–611

351. Shi M, Zhu J, Wang R, Chen X, Mi L, Walz T, Springer TA (2011) Latent TGF-β structure and activation. Nature 474:343–349
352. Kuninger D, Kuns-Hashimoto R, Nili M, Rotwein P (2008) Pro-protein convertases control the maturation and processing of the iron-regulatory protein, RGMc/hemojuvelin. BMC Biochemistry 9:9
353. Schmierer B, Hill CS (2007) TGFβ–SMAD signal transduction: molecular specificity and functional flexibility. Nature Reviews – Molecular Cell Biology 8:970-982
354. ten Dijke P, Arthur HM (2007) Extracellular control of TGFβ signalling in vascular development and disease. Nature Reviews – Molecular Cell Biology 8:857–869
355. Kane R, Godson C, O'Brien C (2008) Chordin-like 1, a bone morphogenetic protein-4 antagonist, is upregulated by hypoxia in human retinal pericytes and plays a role in regulating angiogenesis. Molecular Vision 14:1138–1148
356. Nakayama N, Han CY, Cam L, Lee JI, Pretorius J, Fisher S, Rosenfeld R, Scully S, Nishinakamura R, Duryea D, Van G, Bolon B, Yokota T, Zhang K (2004) A novel chordin-like BMP inhibitor, CHL2, expressed preferentially in chondrocytes of developing cartilage and osteoarthritic joint cartilage. Development 131:229–240
357. ten Dijke P, Krause C, de Gorter DJ, Löwik CW, van Bezooijen RL (2008) Osteocyte-derived sclerostin inhibits bone formation: its role in bone morphogenetic protein and Wnt signaling. Journal of Bone and Joint Surgery. American Volume 1:31–35
358. Wrighton KH, Lin X, Feng XH (2008) Critical regulation of TGFβ signaling by Hsp90. Proceedings of the National Academy of Sciences of the United States of America 105:9244–9249
359. Rodríguez-Vita J, Sánchez-López E, Esteban V, Rupérez M, Egido J, Ruiz-Ortega M (2005) Angiotensin II activates the Smad pathway in vascular smooth muscle cells by a transforming growth factor-β-independent mechanism. Circulation 111:2509–2517
360. Kandias NG, Chasapis CT, Bentrop D, Episkopou V, Spyroulias GA (2009) High yield expression and NMR characterization of Arkadia E3 ubiquitin ligase RING-H2 finger domain. Biochemical and Biophysical Research Communications 378:498–502
361. Mizutani A, Saitoh M, Imamura T, Miyazawa K, Miyazono K (2010) Arkadia complexes with clathrin adaptor AP2 and regulates EGF signalling. Journal of Biochemistry 148:733–741
362. Groppe J, Hinck CS, Samavarchi-Tehrani P, Zubieta C, Schuermann JP, Taylor AB, Schwarz PM, Wrana JL, Hinck AP (2008) Cooperative assembly of TGF-β superfamily signaling complexes is mediated by two disparate mechanisms and distinct modes of receptor binding. Molecular Cell 29:157–168
363. Kang JS, Saunier EF, Akhurst RJ, Derynck R (2008) The type I TGF-β receptor is covalently modified and regulated by sumoylation. Nature – Cell Biology 10:654–664
364. Dahle O, Kumar A, Kuehn MR (2010) Nodal signaling recruits the histone demethylase Jmjd3 to counteract Polycomb-mediated repression at target genes. Science Signaling 3:ra48

365. Ho J, Cocolakis E, Dumas VM, Posner BI, Laporte SA, Lebrun J-J (2005) The G protein-coupled receptor kinase-2 is a TGF-inducible antagonist of TGF signal transduction. EMBO Journal 24:3247–3258

366. Sorrentino A, (2008) The type I TGF-β receptor engages TRAF6 to activate TAK1 in a receptor kinase-independent manner. Nature – Cell Biology 10:1199–1207

367. Shukla A, Malik M, Cataisson C, Ho Y, Friesen T, Suh KS, Yuspa SH (2009) TGF-β signalling is regulated by Schnurri-2-dependent nuclear translocation of CLIC4 and consequent stabilization of phospho-Smad2 and 3. Nature – Cell Biology 11:777–784

368. Kato M, Putta S, Wang M, Yuan H, Lanting L, Nair I, Gunn A, Nakagawa Y, Shimano H, Todorov I, Rossi JJ, Natarajan R (2009) TGF-β activates Akt kinase through a microRNA-dependent amplifying circuit targeting PTEN. Nature – Cell Biology 11:881–889

369. Watanabe Y, Itoh S, Goto T, Ohnishi E, Inamitsu M, Itoh F, Satoh K, Wiercinska E, Yang W, Shi L, Tanaka A, Nakano N, Mommaas AM, Shibuya H, Ten Dijke P, Kato M (2010) TMEPAI, a transmembrane TGF-β-inducible protein, sequesters Smad proteins from active participation in TGF-β signaling. Molecular Cell 37:123–134

370. Yoshimura A, Wakabayashi Y, Mori T (2010) Cellular and molecular basis for the regulation of inflammation by TGF-β. Journal of Biochemistry 147:781–792

371. Chan MC, Hilyard AC, Wu C, Davis BN, Hill NS, Lal A, Lieberman J, Lagna G, Hata A (2010) Molecular basis for antagonism between PDGF and the TGF family of signalling pathways by control of miR-24 expression. EMBO Journal 29:559–573

372. Helbing T, Volkmar F, Goebel U, Heinke J, Diehl P, Pahl HL, Bode C, Patterson C, Moser M (2010) Krüppel-like factor 15 regulates BMPER in endothelial cells. Cardiovascular Research 85:551–559

373. Kamba T, Tam BY, Hashizume H, Haskell A, Sennino B, Mancuso MR, Norberg SM, O'Brien SM, Davis RB, Gowen LC, Anderson KD, Thurston G, Joho S, Springer ML, Kuo CJ, McDonald DM (2006) VEGF-dependent plasticity of fenestrated capillaries in the normal adult microvasculature. American Journal of Physiology – Heart Circulation Physiology 290:H560–H576

374. Lieb W, Safa R, Benjamin EJ, Xanthakis V, Yin X, Sullivan LM, Larson MG, Smith HM, Vita JA, Mitchell GF, Sawyer DB, Vasan RS (2009) Vascular endothelial growth factor, its soluble receptor, and hepatocyte growth factor: clinical and genetic correlates and association with vascular function. European Heart Journal 30:1121–1127

375. Adams RH, Alitalo K (2007) Molecular regulation of angiogenesis and lymphangiogenesis. Nature Reviews – Molecular Cell Biology 8:464–478

376. Turunen MP, Ylä-Herttuala S (2011) Epigenetic regulation of key vascular genes and growth factors. Cardiovascular Research 90:441–446

377. Zhang F, Tang Z, Hou X, Lennartsson J, Li Y, Koch AW, Scotney P, Lee C, Arjunan P, Dong L, Kumar A, Rissanen TT, Wang B, Nagai N, Fons P, Fariss R, Zhang Y, Wawrousek E, Tansey G, Raber J, Fong GH, Ding

H, Greenberg DA, Becker KG, Herbert JM, Nash A, Yla-Herttuala S, Cao Y, Watts RJ, Li X (2009) VEGF-B is dispensable for blood vessel growth but critical for their survival, and VEGF-B targeting inhibits pathological angiogenesis. Proceedings of the National Academy of Sciences of the United States of America 106:6152–6157

378. Hagberg CE, Falkevall A, Wang X, Larsson E, Huusko J, Nilsson I, van Meeteren LA, Samen E, Lu L, Vanwildemeersch M, Klar J, Genove G, Pietras K, Stone-Elander S, Claesson-Welsh L, Ylä-Herttuala S, Lindahl P, Eriksson U (2010) Vascular endothelial growth factor B controls endothelial fatty acid uptake. Nature 464:917–921
379. McColl BK, Baldwin ME, Roufail S, Freeman C, Moritz RL, Simpson RJ, Alitalo K, Stacker SA, Achen MG (2003) Plasmin activates the lymphangiogenic growth factors VEGF-C and VEGF-D. Journal of Experimental Medicine 198:863–868
380. Schoppmann SF, Birner P, Stöckl J, Kalt R, Ullrich R, Caucig C, Kriehuber E, Nagy K, Alitalo K, Kerjaschki D (2002) Tumor-associated macrophages express lymphatic endothelial growth factors and are related to peritumoral lymphangiogenesis. American Journal of Pathology 161:947–956
381. Vlahakis NE, Young BA, Atakilit A, Sheppard D (2005) The lymphangiogenic vascular endothelial growth factors VEGF-C and -D are ligands for the integrin $\alpha_9\beta_1$. Journal of Biological Chemistry 280:4544–4552
382. Sainz J, Sata M (2010) Is PlGF a plaque growth factor? Cardiovascular Research 86:4–5
383. Roncal C, Buysschaert I, Gerdes N, Georgiadou M, Ovchinnikova O, Fischer C, Stassen JM, Moons L, Collen D, De Bock K, Hansson GK, Carmeliet P (2010) Short-term delivery of anti-PlGF antibody delays progression of atherosclerotic plaques to vulnerable lesions. Cardiovascular Research 86:29–36
384. Joukov V, Pajusola K, Kaipainen A, Chilov D, Lahtinen I, Kukk E, Saksela O, Kalkkinen N, Alitalo K (1996) A novel vascular endothelial growth factor, VEGF-C, is a ligand for the Flt4 (VEGFR-3) and KDR (VEGFR-2) receptor tyrosine kinases. EMBO Journal 15:290–298
385. Stacker SA, Caesar C, Baldwin ME, Thornton GE, Williams RA, Prevo R, Jackson DG, Nishikawa S, Kubo H, Achen MG (2001) VEGF-D promotes the metastatic spread of tumor cells via the lymphatics. Nature – Medecine 7:186–191
386. Leblanc AJ, Shipley RD, Kang LS, Muller-Delp JM (2008) Age impairs Flk-1 signaling and NO-mediated vasodilation in coronary arterioles. American Journal of Physiology – Heart and Circulatory Physiology 295:H2280–H2288
387. Wang Y, Chang J, Chen KD, Li S, Li JYS, Wu C, Chien S (2007) Selective adapter recruitment and differential signaling networks by VEGF vs. shear stress. Proceedings of the National Academy of Sciences of the United States of America 104:8875–8879
388. Greenberg JI, Shields DJ, Barillas SG, Acevedo LM, Murphy E, Huang J, Scheppke L, Stockmann C, Johnson RS, Angle N, Cheresh DA (2008) A role for VEGF as a negative regulator of pericyte function and vessel maturation. Nature 456:809–813

389. Advani A, Kelly DJ, Advani SL, Cox AJ, Thai K, Zhang Y, White KE, Gow RM, Marshall SM, Steer BM, Marsden PA, Rakoczy E, Gilbert RE (2007) Role of VEGF in maintaining renal structure and function under normotensive and hypertensive conditions. Proceedings of the National Academy of Sciences of the United States of America 104:14448-14453

390. Partanen TA, Makinen T, Arola J, Suda T, Weich HA, Alitalo K (1999) Endothelial growth factor receptors in human fetal heart. Circulation 100:583–586

391. Zhou Y, Bourcy K, Kang YJ (2009) Copper-induced regression of cardiomyocyte hypertrophy is associated with enhanced vascular endothelial growth factor receptor-1 signalling pathway. Cardiovascular Research 84:54–63

392. Huusko J, Merentie M, Dijkstra MH, Ryhänen MM, Karvinen H, Rissanen TT, Vanwildemeersch M, Hedman M, Lipponen J, Heinonen SE, Eriksson U, Shibuya M, Ylä-Herttuala S (2010) The effects of VEGF-R1 and VEGF-R2 ligands on angiogenic responses and left ventricular function in mice. Cardiovascular Research 86:122–130

393. Muramatsu T (2002) Midkine and pleiotrophin: two related proteins involved in development, survival, inflammation and tumorigenesis. Journal of Biochemistry 132:359–371

394. Stoica GE, Kuo A, Powers C, Bowden ET, Sale EB, Riegel AT, Wellstein A (2002) Midkine binds to anaplastic lymphoma kinase (ALK) and acts as a growth factor for different cell types. Journal of Biological Chemistry 277:35990–35998

395. Sumi Y, Muramatsu H, Takei Y, Hata K, Ueda M, Muramatsu T (2002) Midkine, a heparin-binding growth factor, promotes growth and glycosaminoglycan synthesis of endothelial cells through its action on smooth muscle cells in an artificial blood vessel model. Journal of Cell Science 115:2659–2667

396. Shintani Y, Takashima S, Asano Y, Kato H, Liao Y, Yamazaki S, Tsukamoto O, Seguchi O, Yamamoto H, Fukushima T, Sugahara K, Kitakaze M, Hori M (2006) Glycosaminoglycan modification of neuropilin-1 modulates VEGFR2 signaling. EMBO Journal 25:3045–3055

397. Takamatsu H, Takegahara N, Nakagawa Y, Tomura M, Taniguchi M, Friedel RH, Rayburn H, Tessier-Lavigne M, Yoshida Y, Okuno T, Mizui M, Kang S, Nojima S, Tsujimura T, Nakatsuji Y, Katayama I, Toyofuku T, Kikutani H, Kumanogoh A (2010) Semaphorins guide the entry of dendritic cells into the lymphatics by activating myosin II. Nature – Immunology 11:594–600

398. Zacchigna S, Lambrechts D, Carmeliet P (2008) Neurovascular signalling defects in neurodegeneration. Nature Reviews – Neuroscience 9:169–181

399. Kitamura K, Kangawa K, Kawamoto M, Ichiki Y, Nakamura S, Matsuo H, Eto T (1993) Adrenomedullin: a novel hypotensive peptide isolated from human pheochromocytoma. Biochemical and Biophysical Research Communications 192:553–560

400. Thomson LM, Kapas S, Hinson JP (2003) Paracrine effects of PAMP and adrenomedullin on the human adrenal H295R cell line: PAMP but not adrenomedullin stimulates DHEA secretion. Regulatory Peptides 112:3–7

401. Masada K, Nagayama T, Hosokawa A, Yoshida M, Suzuki-Kusaba M, Hisa H, Kimura T, Satoh S (1999) Effects of adrenomedullin and PAMP on adrenal catecholamine release in dogs. American Journal of Physiology – Regulatory, Integrative and Comparative Physiology 276:R1118–R1124

402. Kato J, Tsuruda T, Kita T, Kitamura K, Eto T (2005) Adrenomedullin: a protective factor for blood vessels. Arteriosclerosis, Thrombosis, and Vascular Biology 25:2480–2487

403. Ikeda U, Kanbe T, Shimada K (1996) Adrenomedullin increases inducible nitric oxide synthase in rat vascular smooth muscle cells stimulated with interleukin-1. Hypertension 27:1240–1244

404. Horio T, Nishikimi T, Yoshihara F, Nagaya N, Matsuo H, Takishita S, Kangawa K (1998) Production and secretion of adrenomedullin in cultured rat cardiac myocytes and nonmyocytes: stimulation by interleukin-1β and tumor necrosis factor-α. Endocrinology 139:4576–4580

405. Charles CJ, Lainchbury JG, Nicholls MG, Rademaker MT, Richards AM, Troughton RW (2003) Adrenomedullin and the renin-angiotensin-aldosterone system. Regulatory Peptides 112:41–49

406. Hippenstiel S, Witzenrath M, Schmeck B, Hocke A, Krisp M, Krüll M, Seybold J, Seeger W, Rascher W, Schütte H, Suttorp N (2002) Adrenomedullin reduces endothelial hyperpermeability. Circulation Research 91:618–625

407. Iwase T, Nagaya N, Fujii T, Itoh T, Ishibashi-Ueda H, Yamagishi M, Miyatake K, Matsumoto T, Kitamura S, Kangawa K (2005) Adrenomedullin enhances angiogenic potency of bone marrow transplantation in a rat model of hindlimb ischemia. Circulation 111:356–362

408. Cantarella G, Martinez G, Cutulia VM, Loreto C, D'Alcamo M, Prato A, Amico-Roxas M, Bernardini R, Clementi G (2005) Adrenomedullin modulates COX-2 and HGF expression in reserpine-injuried gastric mucosa in the rat. European Journal of Pharmacology 518:221–226

409. Jin DH, Harada K, Ohnishi S, Yamahara K, Kangawa K, Nagaya N (2008) Adrenomedullin induces lymphangiogenesis and ameliorates secondary lymphoedema. Cardiovascular Research 80:339–345

410. Tang W, Lu Y, Tian QY, Zhang Y, Guo FJ, Liu GY, Syed NM, Lai Y, Lin EA, Kong L, Su J, Yin F, Ding AH, Zanin-Zhorov A, Dustin ML, Tao J, Craft J, Yin Z, Feng JQ, Abramson SB, Yu XP, Liu CJ (2011) The growth factor progranulin binds to TNF receptors and is therapeutic against inflammatory arthritis in mice. Science 332:478–484

411. Ahmed Z, Mackenzie IR, Hutton ML, Dickson DW (2007) Progranulin in frontotemporal lobar degeneration and neuroinflammation. Journal of Neuroinflammation 4:7

412. He Z, Bateman A (2003) Progranulin (granulin-epithelin precursor, PC-cell-derived growth factor, acrogranin) mediates tissue repair and tumorigenesis. Journal of Molecular Medicine (Berlin, Germany) 81:600–612

413. Daniel R, He Z, Carmichael KP, Halper J, Bateman A (2000) Cellular localization of gene expression for progranulin. Journal of Histochemistry and Cytochemistry 48:999-1009

414. Bateman A, Bennett HPJ (2009) The granulin gene family: from cancer to dementia. BioEssays 31:1245–1254

415. Ngan ES, Tam PK (2008) Prokineticin-signaling pathway. International Journal of Biochemistry and Cell Biology 40:1679–1684

416. Urayama K, Guilini C, Messaddeq N, Hu K, Steenman M, Kurose H, Ertl G, Nebigil CG (2007) The prokineticin receptor-1 (GPR73) promotes cardiomyocyte survival and angiogenesis. FASEB Journal 21:2980–2993

417. Monnier J, Samson M (2008) Cytokine properties of prokineticins. FEBS Journal 275:4014–4021

418. Attramadal H (2009) Prokineticins and the heart: diverging actions elicited by signalling through prokineticin receptor-1 or -2. Cardiovascular Research 81:3–4

419. Urayama K, Guilini C, Turkeri G, Takir S, Kurose H, Messaddeq N, Dierich A, Nebigil CG (2008) Prokineticin receptor-1 induces neovascularization and epicardial-derived progenitor cell differentiation. Arteriosclerosis, Thrombosis, and Vascular Biology 28:841–849

420. Urayama K, Dedeoglu DB, Guilini C, Frantz S, Ertl G, Messaddeq N, Nebigil CG (2009) Transgenic myocardial overexpression of prokineticin receptor-2 (GPR73b) induces hypertrophy and capillary leakage. Cardiovascular Research 81:28-37

421. Cheng MY, Bullock CM, Li C, Lee AG, Bermak JC, Belluzzi J, Weaver DR, Leslie FM, Zhou QY (2002) Prokineticin 2 transmits the behavioural circadian rhythm of the suprachiasmatic nucleus. Nature 417:405–410

422. Ng KL, Li JD, Cheng MY, Lee AG, Zhou Q-Y (2005) Dependence of olfactory bulb neurogenesis on prokineticin 2 signaling. Science 308:1923–1927

423. Cook JS, Sauder CL, Ray CA (2011) Melatonin differentially affects vascular blood flow in humans. American Journal of Physiology – Heart and Circulatory Physiology 300:H670–H674

424. Alvarez SE, Harikumar KB, Hait NC, Allegood J, Strub GM, Kim EY, Maceyka M, Jiang H, Luo C, Kordula T, Milstien S, Spiegel S (2010) Sphingosine-1-phosphate is a missing cofactor for the E3 ubiquitin ligase TRAF2. Nature 465:1084–1088

425. Natarajan V, Jayaram HN, Scribner WM, Garcia JG (1994) Activation of endothelial cell phospholipase D by sphingosine and sphingosine-1-phosphate. American Journal of Respiratory Cell and Molecular Biology 11:221–229

426. Rosen H, Goetz EJ (2005) Sphingosine 1-phosphate and its receptors: an autocrine and paracrine network. Nature Reviews – Immunology 5:560–570

427. Wojciak-Stothard B, Ridley AJ (2003) Shear stress-induced endothelial cell polarization is mediated by Rho and Rac but not Cdc42 or PI 3-kinases. Journal of Cell Biology 161:429–439

428. McVerry BJ, Garcia JGN (2005) In vitro and in vivo modulation of vascular barrier integrity by sphingosine 1-phosphate: mechanitic insights. Cellular Signalling 17:131–139

429. Wojciak-Stothard, B, Ridley, AJ (2002) RhoGTPases and the regulation of endothelial permeability. Vascular Pharmacology 39:187–199

430. Lee MJ, Thangada S, Claffey KP, Ancellin N, Liu CH, Kluk M, Volpi M, Sha'afi RI, Hla T (1999) Vascular endothelial cell adherens junction assembly and morphogenesis induced by sphingosine-1-phosphate. Cell 99:301–312

431. Rivera J, Proia RL, Olivera A (2008) The alliance of sphingosine-1-phosphate and its receptors in immunity. Nature Reviews – Immunology 8:753–763

432. Zachariah MA, Cyster JG (2010) Neural crest-derived pericytes promote egress of mature thymocytes at the corticomedullary junction. Science 328:1129–1135

433. Wei SH, Rosen H, Matheu MP, Sanna MG, Wang SK, Jo E, Wong CH, Parker I, Cahalan MD (2005) Sphingosine 1-phosphate type 1 receptor agonism inhibits transendothelial migration of medullary T cells to lymphatic sinuses. Nature – Immunology 6:1228–1235

434. Gualillo O, González-Juanatey JR, Lago F (2007) The emerging role of adipokines as mediators of cardiovascular function: physiologic and clinical perspectives. Trends in Cardiovascular Medicine 17:275–283

435. Okamoto Y, Arita Y, Nishida M, Muraguchi M, Ouchi N, Takahashi M, Igura T, Inui Y, Kihara S, Nakamura T, Yamashita S, Miyagawa J, Funahashi T, Matsuzawa Y (2000) An adipocyte-derived plasma protein, adiponectin, adheres to injured vascular walls. Hormone and Metabolic Research 32:47–50

436. Goldstein BJ, Scalia RG, Ma XL (2008) Protective vascular and myocardial effects of adiponectin. Nature – Clinical Practice – Cardiovascular Medicine 6:27–35

437. Holland WL, Miller RA, Wang ZV, Sun K, Barth BM, Bui HH, Davis KE, Bikman BT, Halberg N, Rutkowski JM, Wade MR, Tenorio VM, Kuo MS, Brozinick JT, Zhang BB, Birnbaum MJ, Summers SA, Scherer PE (2011) Receptor-mediated activation of ceramidase activity initiates the pleiotropic actions of adiponectin. Nature – Medicine 17:55–63

438. Date H, Imamura T, Ideguchi T, Kawagoe J, Sumi T, Masuyama H, Onitsuka H, Ishikawa T, Nagoshi T, Eto T (2006) Adiponectin produced in coronary circulation regulates coronary flow reserve in nondiabetic patients with angiographically normal coronary arteries. Clinical Cardiology 29:211–214

439. Kobayashi H, Ouchi N, Kihara S, Walsh K, Kumada M, Abe Y, Funahashi T, Matsuzawa Y (2004) Selective suppression of endothelial cell apoptosis by the high molecular weight form of adiponectin. Circulation Research 94:e27-31

440. Ouchi N, Kobayashi H, Kihara S, Kumada M, Sato K, Inoue T, Funahashi T, Walsh K (2004) Adiponectin stimulates angiogenesis by promoting cross-talk between AMP-activated protein kinase and Akt signaling in endothelial cells. Journal of Biological Chemistry 279:1304–1309

441. Motoshima H, Wu X, Mahadev K, Goldstein BJ (2004) Adiponectin suppresses proliferation and superoxide generation and enhances eNOS activity in endothelial cells treated with oxidized LDL. Biochemical and Biophysical Research Communications 315:264–271

442. Mahadev K, Wu X, Donnelly S, Ouedraogo R, Eckhart AD, Goldstein BJ (2008) Adiponectin inhibits vascular endothelial growth factor-induced migration of human coronary artery endothelial cells. Cardiovascular Research 78:376-84

443. Ouchi N, Kihara S, Arita Y, Okamoto Y, Maeda K, Kuriyama H, Hotta K, Nishida M, Takahashi M, Muraguchi M, Ohmoto Y, Nakamura T, Yamashita S, Funahashi T, Matsuzawa Y (2000) Adiponectin, an adipocyte-derived plasma protein, inhibits endothelial NF-κB signaling through a cAMP-dependent pathway. Circulation 102:1296–1301

444. Wozniak SE, Gee LL, Wachtel MS, Frezza EE (2009) Adipose tissue: the new endocrine organ? A review article. Digestive Diseases and Sciences 54:1847–1856

445. Tilg H, Moschen AR (2006) Adipocytokines: mediators linking adipose tissue, inflammation and immunity. Nature Reviews – Immunology 6:772–783

446. Kumada M, Kihara S, Ouchi N, Kobayashi H, Okamoto Y, Ohashi K, Maeda K, Nagaretani H, Kishida K, Maeda N, Nagasawa A, Funahashi T, Matsuzawa Y (2004) Adiponectin specifically increased tissue inhibitor of metalloproteinase-1 through interleukin-10 expression in human macrophages. Circulation 109:2046–2049

447. Masri B, Knibiehler B, Audigier Y (2005) Apelin signalling: a promising pathway from cloning to pharmacology. Cellular Signalling 17:415–426

448. Kleinz MJ, Davenport AP (2005) Emerging roles of apelin in biology and medicine. Pharmacology and Therapeutics 107:198–211

449. Leeper NJ, Tedesco MM, Kojima Y, Schultz GM, Kundu RK, Ashley EA, Tsao PS, Dalman RL, Quertermous T (2009) Apelin prevents aortic aneurysm formation by inhibiting macrophage inflammation. American Journal of Physiology – Heart and Circulatory Physiology 296:H1329–H1335

450. Ozcan L, Ergin AS, Lu A, Chung J, Sarkar S, Nie D, Myers MG Jr, Ozcan U (2009) Endoplasmic reticulum stress plays a central role in development of leptin resistance. Cell Metabolism 9:35–51

451. Ahima RS (2006) Adipose tissue as an endocrine organ. Obesity 14:242S–249S

452. Rosen ED, Spiegelman BM (2006) Adipocytes as regulators of energy balance and glucose homeostasis. Nature 444:847–853

453. Satoh N, Ogawa Y, Katsuura G, Numata Y, Tsuji T, Hayase M, Ebihara K, Masuzaki H, Hosoda K, Yoshimasa Y, Nakao K (1999) Sympathetic activation of leptin via the ventromedial hypothalamus: leptin-induced increase in catecholamine secretion. Diabetes 48:1787–1793

454. Jiang L, You J, Yu X, Gonzalez L, Yu Y, Wang Q, Yang G, Li W, Li C, Liu Y (2008) Tyrosine-dependent and -independent actions of leptin receptor in control of energy balance and glucose homeostasis. Proceedings of the National Academy of Sciences of the United States of America 105:18619–18624

455. Moult PR, Cross A, Santos SD, Carvalho AL, Lindsay Y, Connolly CN, Irving AJ, Leslie NR, Harvey J. (2010) Leptin regulates AMPA receptor trafficking via PTEN inhibition. Journal of Neuroscience 30:4088–4101

456. Oda A, Taniguchi T, Yokoyama M (2001) Leptin stimulates rat aortic smooth muscle cell proliferation and migration. Kobe Journal of Medical Sciences 47:141–150

457. Sierra-Honigmann MR, Nath AK, Murakami C, García-Cardena G, Papapetropoulos A, Sessa WC, Madge LA, Schechner JS, Schwabb MB, Polverini

PJ, Flores-Riveros JR (1998) Biological action of leptin as an angiogenic factor. Science 281:1683–1686

458. Polak J (2008) Multiple roles of leptin in blood pressure regulation, from central to peripheral mechanisms. American Journal of Hypertension 21:1172
459. Haynes WG, Morgan DA, Walsh SA, Mark AL, Sivitz WI (1997) Receptor-mediated regional sympathetic nerve activation by leptin. Journal of Clinical Investigation 100:270–278
460. Beltowski J, Wojcicka G, Górny D, Marciniak A (2002) Human leptin administered intraperitoneally stimulates natriuresis and decreases renal medullary Na^+, K^+-ATPase activity in the rat – impaired effect in dietary-induced obesity. Medical Science Monitor 8:BR221–BR229
461. Ortiz PA, Garvin JL (2002) Role of nitric oxide in the regulation of nephron transport. American Journal of Physiology – Renal Physiology 282:F777–F784
462. Vecchione C, Maffei A, Colella S, Aretini A, Poulet R, Frati G, Gentile MT, Fratta L, Trimarco V, Trimarco B, Lembo G (2002) Leptin effect on endothelial nitric oxide is mediated through Akt-endothelial nitric oxide synthase phosphorylation pathway. Diabetes 51:168–173
463. Mattson DL, Lu S, Nakanishi K, Papanek PE, Cowley AW (1994) Effect of chronic renal medullary nitric oxide inhibition on blood pressure. American Journal of Physiology – Heart and Circulatory Physiology 266:H1918–H1926
464. Torday JS, Rehan VK (2006) Up-regulation of fetal rat lung parathyroid hormone-related protein gene regulatory network down-regulates the Sonic Hedgehog/Wnt/β-catenin gene regulatory network. Pediatric Research 60:382–388
465. Torday JS, Sun H, Wang L, Torres E, Sunday ME, Rubin LP (2002) Leptin mediates the parathyroid hormone-related protein paracrine stimulation of fetal lung maturation. American Journal of Physiology – Lung Cellular and Molecular Physiology 283:L405–L410
466. Torday JS, Rehan VK (2002) Stretch-stimulated surfactant synthesis is coordinated by the paracrine actions of PTHrP and leptin. American Journal of Physiology – Lung Cellular and Molecular Physiology 283:L130–L135
467. Shi Y, Yadav VK, Suda N, Liu XS, Guo XE, Myers MG Jr, Karsenty G (2008) Dissociation of the neuronal regulation of bone mass and energy metabolism by leptin in vivo. Proceedings of the National Academy of Sciences of the United States of America 105:20529–20533
468. Chen C, Jiang J, Lü JM, Chai H, Wang X, Lin PH, Yao Q (2010) Resistin decreases expression of endothelial nitric oxide synthase through oxidative stress in human coronary artery endothelial cells. American Journal of Physiology – Heart and Circulatory Physiology 299:H193–H201
469. Fukuhara A, Matsuda M, Nishizawa M, Segawa K, Tanaka M, Kishimoto K, Matsuki Y, Murakami M, Ichisaka T, Murakami H, Watanabe E, Takagi T, Akiyoshi M, Ohtsubo T, Kihara S, Yamashita S, Makishima M, Funahashi T, Yamanaka S, Hiramatsu R, Matsuzawa Y, Shimomura I (2005) Visfatin: a protein secreted by visceral fat that mimics the effects of insulin. Science 307:426–430

470. Yamawaki H, Hara N, Okada M, Hara Y (2009) Visfatin causes endothelium-dependent relaxation in isolated blood vessels. Biochemical and Biophysical Research Communications 383:503–508

471. Bae YH, Bae MK, Kim SR, Lee JH, Wee HJ, Bae SK (2009) Upregulation of fibroblast growth factor-2 by visfatin that promotes endothelial angiogenesis. Biochemical and Biophysical Research Communications 379:206–211

472. Adya R, Tan BK, Chen J, Randeva HS (2009) Pre-B cell colony enhancing factor (PBEF)/visfatin induces secretion of MCP-1 in human endothelial cells: role in visfatin-induced angiogenesis. Atherosclerosis 205:113–119

473. Wang P, Xu TY, Guan YF, Su DF, Fan GR, Miao CY (2009) Perivascular adipose tissue-derived visfatin is a vascular smooth muscle cell growth factor: role of nicotinamide mononucleotide. Cardiovascular Research 81:370–380

474. Tan BK, Chen J, Brown J, Adya R, Ramanjaneya M, Menon V, Bailey CJ, Lehnert H, Randeva HS (2009) In vivo and ex vivo regulation of visfatin production by leptin in human and murine adipose tissue: role of mitogen-activated protein kinase and phosphatidylinositol 3-kinase signalling pathways. Endocrinology 150:3530–3539

475. Song HK, Lee MH, Kim BK, Park YG, Ko GJ, Kang YS, Han JY, Han SY, Han KH, Kim HK, Cha DR (2008) Visfatin: a new player in mesangial cell physiology and diabetic nephropathy. American Journal of Physiology – Renal Physiology 295:F1485–F1494

476. Storka A, Vojtassakova E, Mueller M, Kapiotis S, Haider DG, Jungbauer A, Wolzt M (2008) Angiotensin inhibition stimulates PPARγ and the release of visfatin. European Journal of Clinical Investigation 38:820–826

477. Hintze KJ, Nabor D, Timbimboo H, Snow D (2009) Investigations of iron, hypoxia and adipocyte hypoxia on hepcidin promoter activity. FASEB Journal 23:105.5

478. Vokurka M, Lacinova Z, Kremen J, Kopecky P, Blaha J, Pelinkova K, Haluzik M, Necas E (2009) Hepcidin expression in adipose tissue increases during cardiac surgery. Physiological Research (www.biomed.cas.cz/physiolres/pdf/prepress/1759.pdf)

479. Cazes A, Galaup A, Chomel C, Bignon M, Bréchot N, Le Jan S, Weber H, Corvol P, Muller L, Germain S, Monnot C (2006) Extracellular matrix-bound angiopoietin-like 4 inhibits endothelial cell adhesion, migration, and sprouting and alters actin cytoskeleton. Circulation Research 99:1207–1215

480. Hamilton JA (2008) Colony-stimulating factors in inflammation and autoimmunity. Nature Reviews – Immunology 8:533–544

481. Croft M (2009) The role of TNF superfamily members in T-cell function and diseases. Nature Reviews – Immunology 9:271–285

482. Kishore U, Gaboriaud C, Waters P, Shrive AK, Greenhough TJ, Reid KB, Sim RB, Arlaud GJ (2004) C1q and tumor necrosis factor superfamily: modularity and versatility. Trends in Immunology 25:551–561

483. Sims JE, Smith DE (2010) The IL-1 family: regulators of immunity. Nature Reviews – Immunology 10:89–102

484. Kurzrock R (2003) Hematopoietic growth factors. In Bast RC, Kufe DW, Pollock RE, Weichselbaum RR, Holland JF, Frei E (eds) Cancer Medicine. BC Decker, Hamilton, Ontario, Canada

485. Weber A, Wasiliew P, Kracht M (2010) Interleukin-1 (IL-1) pathway. Science Signaling 3:cm1

486. Dinarello CA (1998)Interleukin-1b, interleukin-18, and the interleukin-1b converting enzyme. Annals of the New York Academy of Sciences 856:1–11

487. Weinberg EO, Shimpo M, De Keulenaer GW, MacGillivray C, Tominaga S, Solomon SD, Rouleau JL, Lee RT (2002) Expression and regulation of ST2, an interleukin-1 receptor family member, in cardiomyocytes and myocardial infarction. Circulation 106:2961–2966

488. Akashi K, Kondo M, Weissman IL (1998) Role of interleukin-7 in T-cell development from hematopoietic stem cells. Immunological Reviews 165:13–28

489. Renauld J-C (1995) Interleukin-9: structural characteristics and biologic properties. In Kurzrock R, Talpaz M, Norwell MA (eds) Cytokines: Interleukins and Their Receptors. Kluwer Academic Publishers, Norwell, MA

490. Veldhoen M, Uyttenhove C, van Snick J, Helmby H, Westendorf A, Buer J, Martin B, Wilhelm C, Stockinger B (2008) Transforming growth factor-β "reprograms" the differentiation of T helper 2 cells and promotes an interleukin 9-producing subset. Nature – Immunology 9:1341–1346

491. Chang HC, Sehra S, Goswami R, Yao W, Yu Q, Stritesky GL, Jabeen R, McKinley C, Ahyi AN, Han L, Nguyen ET, Robertson MJ, Perumal NB, Tepper RS, Nutt SL, Kaplan MH (2010) The transcription factor PU.1 is required for the development of IL-9-producing T cells and allergic inflammation. Nature – Immunology 11:527–534

492. Noelle RJ, Nowak EC (2010) Cellular sources and immune functions of interleukin-9. Nature Reviews – Immunology 10:683–687

493. Ziegler SF, Artis D (2010) Sensing the outside world: TSLP regulates barrier immunity. Nature – Immunology 11:289–293

494. Muraguchi A, Hirano T, Tang B, Matsuda T, Horii Y, Nakajima K, Kishimoto T (1988) The essential role of B cell stimulatory factor 2 (BSF-2/IL-6) for the terminal differentiation of B cells. Journal of Experimental Medicine 167:332–344

495. Petersen EW, Carey AL, Sacchetti M, Steinberg GR, Macaulay SL, Febbraio MA, Pedersen BK (2005) Acute IL-6 treatment increases fatty acid turnover in elderly humans in vivo and in tissue culture in vitro. American Journal of Physiology – Endocrinology and Metabolism 288:E155–E162

496. Senn JJ, Klover PJ, Nowak IA, Zimmers TA, Koniaris LG, Furlanetto RW, Mooney RA (2003) Suppressor of cytokine signaling-3 (SOCS-3), a potential mediator of interleukin-6-dependent insulin resistance in hepatocytes. Journal of Biological Chemistry 278:13740–13746

497. Pedersen BK (2007) IL-6 signalling in exercise and disease. Biochemical Society Transactions 35:1295–1297

498. Du X, Williams DA (1997) Interleukin-11: review of molecular, cell biology, and clinical use. Blood 89:3897–3908

499. González A, López B, Ravassa S, Beaumont J, Arias T, Hermida N, Zudaire A, Díez J (2009) Biochemical markers of myocardial remodelling in hypertensive heart disease. Cardiovascular Research 81:509–518
500. Asadullah K, Sterry W, Volk HD (2003) Interleukin-10 therapy – review of a new approach. Pharmacological Reviews 55:241–269
501. Saraiva M, O'Garra A (2010) The regulation of IL-10 production by immune cells. Nature Reviews – Immunology 10:170–181
502. Rousset F, Garcia E, Defrance T, Péronne C, Vezzio N, Hsu DH, Kastelein R, Moore KW, Banchereau J (1992) IL-10 is a potent growth and differentiation factor for activated human B lymphocytes. Proceedings of the National Academy of Sciences of the United States of America 89:1890–1893
503. Jacobsen SE (1995) IL12, a direct stimulator and indirect inhibitor of haematopoiesis. Research in Immunology 146:506–514
504. Cua DJ, Tato CM (2010) Innate IL-17-producing cells: the sentinels of the immune system. Nature Reviews – Immunology 10:479–489
505. Fossiez F, Djossou O, Chomarat P, Flores-Romo L, Ait-Yahia S, Maat C, Pin JJ, Garrone P, Garcia E, Saeland S, Blanchard D, Gaillard C, Das Mahapatra B, Rouvier E, Golstein P, Banchereau J, Lebecque S (1996) T cell interleukin-17 induces stromal cells to produce proinflammatory and hematopoietic cytokines. Journal of Experimental Medicine 183:2593–2603
506. Aarvak T, Chabaud M, Miossec P, Natvig JB (1999) IL-17 is produced by some proinflammatory Th1/Th0 cells but not by Th2 cells. Journal of Immunology 162:1246–1251
507. Yao Z, Painter SL, Fanslow WC (1995) Human IL-17: a novel cytokine derived from T cells. Journal of Immunology 155:5483–5486
508. Lin H, Lee E, Hestir K, Leo C, Huang M, Bosch E, Halenbeck R, Wu G, Zhou A, Behrens D, Hollenbaugh D, Linnemann T, Qin M, Wong J, Chu K, Doberstein SK, Williams LT (2008) Discovery of a cytokine and its receptor by functional screening of the extracellular proteome. Science 320:807–811
509. Baggiolini M, Clark-Lewis I (1992) Interleukin-8, a chemotactic and inflammatory cytokine. FEBS Letters 307:97–101
510. Platanias LC (2005) Mechanisms of type-I and type-II-interferon-mediated signalling. Nature Reviews Immunology 5:375–386
511. Dale M, Nicklin MJ (1999) Interleukin-1 receptor cluster: gene organization of IL1R2, IL1R1, IL1RL2 (IL-1Rrp2), IL1RL1 (T1/ST2), and IL18R1 (IL-1Rrp) on human chromosome 2q. Genomics 57:177–179.

Chap. 4. Cell Survival and Death

512. Mehlen P (2010) Dependence receptors: the trophic theory revisited. Science Signaling 3:pe47
513. Guenebeaud C, Goldschneider D, Castets M, Guix C, Chazot G, Delloye-Bourgeois C, Eisenberg-Lerner A, Shohat G, Zhang M, Laudet V, Kimchi A, Bernet A, Mehlen P (2010) The dependence receptor UNC5H2/B triggers apoptosis via PP2A-mediated dephosphorylation of DAP kinase. Molecular Cell 40:863–876

514. Boucher J, Macotela Y, Bezy O, Mori MA, Kriauciunas K, Kahn CR (2010) Kinase-independent role for unoccupied insulin and IGF-1 receptors in the control of apoptosis. Science Signaling 3:ra87

515. Nikoletopoulou V, Lickert H, Frade JM, Rencurel C, Giallonardo P, Zhang L, Bibel M, Barde YA (2010) Neurotrophin receptors TrkA and TrkC cause neuronal death whereas TrkB does not. Nature 467:59–63

516. Kraft C, Deplazes A, Sohrmann M, Peter M (2008) Mature ribosomes are selectively degraded upon starvation by an autophagy pathway requiring the Ubp3p/Bre5p ubiquitin protease. Nature – Cell Biology 10:602–610

517. LockshinRA, Williams CM (1964) Programmed cell death – II. Endocrine potentiation of the breakdown of the intersegmental muscles of silkmoths. Journal of Insect Physiology 10:643–649

518. Kerr JF, Wyllie AH, Currie AR (1972) Apoptosis: a basic biological phenomenon with wide-ranging implications in tissue kinetics. British Journal of Cancer 26:239–257

519. Youle RJ, Strasser A (2008) The BCL-2 protein family: opposing activities that mediate cell death. Nature Reviews – Molecular Cell Biology 9:47–59

520. Eisenberg-Lerner A, Bialik S, Simon HU, Kimchi A (2009) Life and death partners: apoptosis, autophagy and the cross-talk between them. Cell Death and Differentiation 16:966–975

521. Kim H, Rafiuddin-Shah M, Tu HC, Jeffers JR, Zambetti GP, Hsieh JJD, Cheng EHY (2006) Hierarchical regulation of mitochondrion-dependent apoptosis by BCL-2 subfamilies. Nature – Cell Biology 8:1348–1358

522. Fox JL, Ismail F, Azad A, Ternette N, Leverrier S, Edelmann MJ, Kessler BM, Leigh IM, Jackson S, Storey A (2010) Tyrosine dephosphorylation is required for Bak activation in apoptosis. EMBO Journal 29:3853–3868

523. Castello PR, Woo DK, Ball K, Wojcik J, Liu L, Poyton RO (2008) Oxygen-regulated isoforms of cytochrome c oxidase have differential effects on its nitric oxide production and on hypoxic signaling. Proceedings of the National Academy of Sciences of the United States of America 105:8203–8208

524. Berdichevsky A, Guarente L, Bose A (2010) Acute oxidative stress can reverse insulin resistance by inactivation of cytoplasmic JNK. Journal of Biological Chemistry 285:21581–21589

525. Kumar S (2007) Caspase function in programmed cell death. Cell Death and Differentiation 14:32–43

526. Santoro MM, Samuel T, Mitchell T, Reed JC, Stainier DYR (2007) Birc2 (cIap1) regulates endothelial cell integrity and blood vessel homeostasis. Nature – Genetics 39:1397–1402

527. Martinon F, Tschopp J (2007) Inflammatory caspases and inflammasomes: master switches of inflammation. Cell Death and Differentiation 14:10–22

528. Lamkanfi M, Festjens N, Declercq W, Vanden Berghe T, Vandenabeele P (2007) Caspases in cell survival, proliferation and differentiation. Cell Death and Differentiation 14:44–55

529. Benhar M, Forrester MT, Hess DT, Stamler JS (2008) Reduced thioredoxin denitrosylates caspase-3, a key protein involved in cell death. Science 320:1019–1020

530. Festjens N, Vanden Berghe T, Cornelis S, Vandenabeele P (2007) RIP1, a kinase on the crossroads of a cell's decision to live or die. Cell Death and Differentiation 14:400–410

531. Thornton TM, Pedraza-Alva G, Deng B, Wood CD, Aronshtam A, Clements JL, Sabio G, Davis RJ, Matthews DE, Doble B, Rincon M (2008) Phosphorylation by p38 MAPK as an alternative pathway for GSK3β inactivation. Science 320:667–670

532. Kroemer G, Galluzzi L, Vandenabeele P, Abrams J, Alnemri ES, Baehrecke EH, Blagosklonny MV, El-Deiry WS, Golstein P, Green DR, Hengartner M, Knight RA, Kumar S, Lipton SA, Malorni W, Nuñez G, Peter ME, Tschopp J, Yuan J, Piacentini M, Zhivotovsky B, Melino G (2009) Classification of cell death: recommendations of the Nomenclature Committee on Cell Death 2009. Cell Death and Differentiation 16:3–11

533. Galluzzi L, Maiuri MC, Vitale I, Zischka H, Castedo M, Zitvogel L, Kroemer G (2007) Cell death modalities: classification and pathophysiological implications. Cell Death and Differentiation 14:1237–1243

534. Degterev A, Yuan J (2008) Expansion and evolution of cell death programmes. Nature Reviews – Molecular Cell Biology 9:378–390

535. Mehrpour M, Esclatine A, Beau I, Codogno P (2010) Overview of macroautophagy regulation in mammalian cells. Cell Research 20:748–762

536. Mariño G, Kroemer G (2010) Ammonia: a diffusible factor released by proliferating cells that induces autophagy. Science Signaling 3:pe19

537. Liang J, Shao SH, Xu ZX, Hennessy B, Ding Z, Larrea M, Kondo S, Dumont DJ, Gutterman JU, Walker CL, Slingerland JM, Mills GB (2007) The energy sensing LKB1–AMPK pathway regulates p27kip1 phosphorylation mediating the decision to enter autophagy or apoptosis. Nature – Cell Biology 9:218–224

538. Egan DF, Shackelford DB, Mihaylova MM, Gelino S, Kohnz RA, Mair W, Vasquez DS, Joshi A, Gwinn DM, Taylor R, Asara JM, Fitzpatrick J, Dillin A, Viollet B, Kundu M, Hansen M, Shaw RJ (2011) Phosphorylation of ULK1 (hATG1) by AMP-activated protein kinase connects energy sensing to mitophagy. Science 331:456–461

539. Behrends C, Sowa ME, Gygi SP, Harper JW (2010) Network organization of the human autophagy system. Nature 466:68–76

540. Lee SB, Kim S, Lee J, Park J, Lee G, Kim Y, Kim JM, Chung J (2007) ATG1, an autophagy regulator, inhibits cell growth by negatively regulating S6 kinase. EMBO Reports 8:360–365

541. Scherz-Shouval R, Shvets E, Fass E, Shorer H, Gil L, Elazar Z (2007) Reactive oxygen species are essential for autophagy and specifically regulate the activity of Atg4. EMBO Journal 26:1749–1760

542. Simonsen A, Cumming RC, Brech A, Isakson P, Schubert DR, Finley KD (2008) Promoting basal levels of autophagy in the nervous system enhances longevity and oxidant resistance in adult Drosophila. Autophagy 4:176–184

543. Nakai A, Yamaguchi O, Takeda T, Higuchi Y, Hikoso S, Taniike M, Omiya S, Mizote I, Matsumura Y, Asahi M, Nishida K, Hori M, Mizushima N, Otsu K (2007) The role of autophagy in cardiomyocytes in the basal state and in response to hemodynamic stress. Nature – Medicine 13:619–624

544. Furuya T, Kim M, Lipinski M, Li J, Kim D, Lu T, Shen Y, Rameh L, Yankner B, Tsai LH, Yuan J (2010) Negative regulation of Vps34 by Cdk mediated phosphorylation. Molecular Cell 38:500–511

545. Maiuri MC, Criollo A, Kroemer G (2010) Crosstalk between apoptosis and autophagy within the Beclin 1 interactome. EMBO Journal 29:515–516

546. Lee JY, Koga H, Kawaguchi Y, Tang W, Wong E, Gao YS, Pandey UB, Kaushik S, Tresse E, Lu J, Taylor JP, Cuervo AM, Yao TP (2010) HDAC6 controls autophagosome maturation essential for ubiquitin-selective quality-control autophagy. EMBO Journal 29:969–980

547. Williams A, Sarkar S, Cuddon P, Ttofi EK, Saiki S, Siddiqi FH, Jahreiss L, Fleming A, Pask D, Goldsmith P, O'Kane CJ, Floto RA, Rubinsztein DC (2008) Novel targets for Huntington's disease in an mTOR-independent autophagy pathway. Nature – Chemical Biology 4:295–305

548. Nishida K, Kyoi S, Yamaguchi O, Sadoshima J, Otsu K (2009) The role of autophagy in the heart. Cell Death and Differentiation 16:31–38

549. Bruno P, Brinkmann CR, Boulanger MC, Flinterman M, Klanrit P, Landry MC, Portsmouth D, Borst J, Tavassoli M, Noteborn M, Backendorf C, Zimmerman RME (2009) Family at last: highlights of the first international meeting on proteins killing tumour cells. Cell Death and Differentiation 16:184–186

550. Maiuri M, Tazdemir E, Criollo A, Morselli E, Vicencio J, Carnuccio R, Kroemer G (2009) Control of autophagy by oncogenes and tumor suppressor genes. Cell Death and Differentiation 16:87–93

551. Campisi J (2005) Senescent cells, tumor suppression, and organismal aging: good citizens, bad neighbors. Cell 120:513–522

552. Campisi J, d'Adda di Fagagna F (2007) Cellular senescence: when bad things happen to good cells. Nature Reviews – Molecular Cell Biology 8:729–740

553. Hayashi T, Matsui-Hirai H, Miyazaki-Akita A, Fukatsu A, Funami J, Ding QF, Kamalanathan S, Hattori Y, Ignarro LJ, Iguchi A (2006) Endothelial cellular senescence is inhibited by nitric oxide: implications in atherosclerosis associated with menopause and diabetes. Proceedings of the National Academy of Sciences of the United States of America 103:17018–17023

554. Passos JF, Nelson G, Wang C, Richter T, Simillion C, Proctor CJ, Miwa S, Olijslagers S, Hallinan J, Wipat A, Saretzki G, Rudolph KL, Kirkwood TBL, von Zglinicki T (2010) Feedback between p21 and reactive oxygen production is necessary for cell senescence. Molecular Systems Biology 6:347

555. Zahn JM, Poosala S, Owen AB, Ingram DK, Lustig A, Carter A, Weeraratna AT, Taub DD, Gorospe M, Mazan-Mamczarz K, Lakatta EG, Boheler KR, Xu X, Mattson MP, Falco G, Ko MS, Schlessinger D, Firman J, Kummerfeld SK, Wood WH 3rd, Zonderman AB, Kim SK, Becker KG (2007) AGEMAP: a gene expression database for aging in mice. PLoS Genetics 3:e201

556. Adler AS, Sinha S, Kawahara TL, Zhang JY, Segal E, Chang HY (2007) Motif module map reveals enforcement of aging by continual NF-κB activity. Genes and Development 21:3244–3257

557. Das M, Jiang F, Sluss HK, Zhang C, Shokat KM, Flavell RA, Davis RJ (2007) Suppression of p53-dependent senescence by the JNK signal transduction pathway. Proceedings of the National Academy of Sciences of the United States of America 104:15759–15764

558. Kurosu H, Yamamoto M, Clark JD, Pastor JV, Nandi A, Gurnani P, McGuinness OP, Chikuda H, Yamaguchi M, Kawaguchi H, Shimomura I, Takayama Y, Herz J, Kahn CR, Rosenblatt KP, Kuro-o M (2005) Suppression of aging in mice by the hormone Klotho. Science 309:1829–1833

559. Imura A, Tsuji Y, Murata M, Maeda R, Kubota K, Iwano A, Obuse C, Togashi K, Tominaga M, Kita N, Tomiyama K, Iijima J, Nabeshima Y, Fujioka M, Asato R, Tanaka S, Kojima K, Ito J, Nozaki K, Hashimoto N, Ito T, Nishio T, Uchiyama T, Fujimori T, Nabeshima Y (2007) α-Klotho as a regulator of calcium homeostasis. Science 316:1615–1618

560. Augert A, Payré C, de Launoit Y, Gil J, Lambeau G, Bernard D (2009) The M-type receptor PLA2R regulates senescence through the p53 pathway. EMBO Reports 10:271–277

561. Killilea DW, Ames BN (2008) Magnesium deficiency accelerates cellular senescence in cultured human fibroblasts. Proceedings of the National Academy of Sciences of the United States of America 105:5768–5773

562. Ow YLP, Green DR, Hao Z, Mak TW (2008) Cytochrome c: functions beyond respiration. Nature Reviews – Molecular Cell Biology 9:532–542

563. Riedl SJ, Salvesen GS (2007) The apoptosome: signalling platform of cell death. Nature Reviews – Molecular Cell Biology 8:405–413

564. Taylor RC, Cullen SP, Martin SJ (2008) Apoptosis: controlled demolition at the cellular level. Nature Reviews – Molecular Cell Biology 9:231–241

565. Willis SN, Fletcher JI, Kaufmann T, van Delft MF, Chen L, Czabotar PE, Ierino H, Lee EF, Fairlie WD, Bouillet P, Strasser A, Kluck RM, Adams JM, Huang DCS (2007) Apoptosis initiated when BH3 ligands engage multiple Bcl-2 homologs, not Bax or Bak. Science 315:856–859

566. Gavathiotis E, Suzuki M, Davis ML, Pitter K, Bird GH, Katz SG, Tu HC, Kim H, Cheng EHY, Tjandra N, Walensky LD (2008) Bax activation is initiated at a novel interaction site. Nature 455:1076–1081

567. Sen N, Hara MR, Kornberg MD, Cascio MB, Bae BI, Shahani N, Thomas B, Dawson TM, Dawson VL, Snyder SH, Sawa A (2008) Nitric oxide-induced nuclear GAPDH activates p300/CBP and mediates apoptosis. Nature – Cell Biology 10:866–873

568. Liu L, Sakai T, Sano N, Fukui K (2004) Nucling mediates apoptosis by inhibiting expression of galectin-3 through interference with nuclear factor κB signalling. Biochemical Journal 380:31–41

569. Tran NH, Sakai T, Kim SM, Fukui K (2010) NF-κB regulates the expression of Nucling, a novel apoptosis regulator, with involvement of proteasome and caspase for its degradation. Journal of Biochemistry 148:573–580

570. Giannoni E, Buricchi F, Grimaldi G, Parri M, Cialdai F, Taddei ML, Raugei G, Ramponi G, Chiarugi P (2008) Redox regulation of anoikis: reactive oxygen species as essential mediators of cell survival. Cell Death and Differentiation 15:867–878

571. Moro L, Dolce L, Cabodi S, Bergatto E, Boeri Erba E, Smeriglio M, Turco E, Retta SF, Giuffrida MG, Venturino M, Godovac-Zimmermann J, Conti A, Schaefer E, Beguinot L, Tacchetti C, Gaggini P, Silengo L, Tarone G, Defilippi P (2002) Integrin-induced epidermal growth factor (EGF) receptor

activation requires c-Src and p130Cas and leads to phosphorylation of specific EGF receptor tyrosines. Journal of Biological Chemistry 277:9405–9414

572. Maiuri MC, Zalckvar E, Kimchi A, Kroemer G (2007) Self-eating and self-killing: crosstalk between autophagy and apoptosis. Nature Reviews – Molecular Cell Biology 8:741–752
573. Vandenabeele P, Galluzzi L, Vanden Berghe T, Kroemer G (2010) Molecular mechanisms of necroptosis: an ordered cellular explosion. Nature Reviews – Molecular Cell Biology 11:700–714
574. Lim SY, Davidson SM, Mocanu MM, Yellon DM, Smith CC (2007) The cardioprotective effect of necrostatin requires the cyclophilin-D component of the mitochondrial permeability transition pore. Cardiovascular Drugs and Therapy 21:467–469
575. Ditsworth D, Zong WX, Thompson CB (2007) Activation of poly(ADP)-ribose polymerase (PARP-1) induces release of the pro-inflammatory mediator HMGB1 from the nucleus. Journal of Biological Chemistry 82:17845–17854
576. Yu SW, Andrabi SA, Wang H, Kim NS, Poirier GG, Dawson TM, Dawson VL (2006) Apoptosis-inducing factor mediates poly(ADP-ribose) (PAR) polymer-induced cell death. Proceedings of the National Academy of Sciences of the United States of America 103:18314–18319

Chap. 5. Circadian Clock

577. O'Neill JS, Reddy AB (2011) Circadian clocks in human red blood cells. Nature 469:498–503
578. Tsai TYC, Choi YS, Ma W, Pomerening JR, Tang C, Ferrell JE (2008) Robust, tunable biological oscillations from interlinked positive and negative feedback loops. Science 321:126–129
579. Zhang EE, Kay SA (2010) Clocks not winding down: unravelling circadian networks. Nature Reviews – Molecular Cell Biology 11:764–776
580. Duong HA, Robles MS, Knutti D, Weitz CJ (2011) A molecular mechanism for circadian clock negative feedback. Science 332:1436–1439
581. Kuzmichev A, Zhang Y, Erdjument-Bromage H, Tempst P, Reinberg D (2002) Role of the Sin3-histone deacetylase complex in growth regulation by the candidate tumor suppressor p33(ING1). Molecular and Cellular Biology 22:835–848
582. Bass J, Takahashi JS (2011) Circadian rhythms: redox redux. Nature 469:476–478
583. O'Neill JS, Reddy AB (2011) Circadian clocks in human red blood cells. Nature 469:498–503
584. Meredith AL, Wiler SW, Miller BH, Takahashi JS, Fodor AA, Ruby NF, Aldrich RW (2006) BK calcium-activated potassium channels regulate circadian behavioral rhythms and pacemaker output. Nature – Neuroscience 9: 1041–1049

585. Takahashi JS, Hong HK, Ko CH, McDearmon EL (2008) The genetics of mammalian circadian order and disorder: implications for physiology and disease. Nature Reviews – Genetics 9:764–775

586. Doyle SE, Yoshikawa T, Hillson H, Menaker M (2008) Retinal pathways influence temporal niche. Proceedings of the National Academy of Sciences of the United States of America 105:13133–13138

587. Herzog ED (2007) Neurons and networks in daily rhythms. Nature Reviews – Neuroscience 8:790–802

588. Buhr ED, Yoo SH, Takahashi JS (2010) Temperature as a universal resetting cue for mammalian circadian oscillators. Science 330:379–385

589. Pennartz CM, de Jeu MT, Bos NP, Schaap J, Geurtsen AM (2002) Diurnal modulation of pacemaker potentials and calcium current in the mammalian circadian clock. Nature 416:286–290

590. Danovi SA (2010) Circadian rhythms: in the heat of the night. Nature Reviews – Neuroscience 11:788–789

591. Reinke H, Saini C, Fleury-Olela F, Dibner C, Benjamin IJ, Schibler U (2008) Differential display of DNA-binding proteins reveals heat-shock factor 1 as a circadian transcription factor. Genes and Development 22:331–345

592. Ukai-Tadenuma M, Kasukawa T, Ueda HR (2008) Proof-by-synthesis of the transcriptional logic of mammalian circadian clocks. Nature – Cell Biology 10:1154–1163

593. Cai W, Rambaud J, Teboul M, Masse I, Benoit G, Gustafsson JA, Delaunay F, Laudet V, Pongratz I (2008) Expression levels of estrogen receptor β are modulated by components of the molecular clock. Molecular and Cellular Biology 28:784–793

594. DeBruyne JD, Weaver DR, Reppert SM (2007) CLOCK and NPAS2 have overlapping roles in the suprachiasmatic circadian clock. Nature – Neuroscience 10:543–545

595. Dibner C, Sage D, Unser M, Bauer C, d'Esmond T, Naef F, Schibler U (2009) Circadian gene expression is resilient to large fluctuations in overall transcription rates. EMBO Journal 28:123–134

596. O'Neill JS, Maywood ES, Chesham JE, Takahashi JS, Hastings MH (2008) cAMP-dependent signaling as a core component of the mammalian circadian pacemaker. Science 320:949–953

597. Cha J, Chang SS, Huang G, Cheng P, Liu Y (2008) Control of WHITE COLLAR localization by phosphorylation is a critical step in the circadian negative feedback process. EMBO Journal 27:3246–3255

598. Tataro O, Schafmeier T (2010) Of switches and hourglasses: regulation of subcellular traffic in circadian clocks by phosphorylation. EMBO Reports 11:927–935

599. Sanada K, Okano T, Fukada Y (2002) Mitogen-activated protein kinase phosphorylates and negatively regulates basic helix-loop-helix PAS transcription factor BMAL1. Journal of Biological Chemistry 277:267–271

600. Busino L, Bassermann F, Maiolica A, Lee C, Nolan PM, Godinho SI, Draetta GF, Pagano M (2007) SCFFbxl3 controls the oscillation of the circadian clock by directing the degradation of cryptochrome proteins. Science 316:900–904

601. Godinho SI, Maywood ES, Shaw L, Tucci V, Barnard AR, Busino L, Pagano M, Kendall R, Quwailid MM, Romero MR, O'Neill J, Chesham JE, Brooker D, Lalanne Z, Hastings MH, Nolan PM (2007) The after-hours mutant reveals a role for Fbxl3 in determining mammalian circadian period. Science 316:897–900

602. Isojima Y, Nakajima M, Ukai H, Fujishima H, Yamada RG, Masumoto KH, Kiuchi R, Ishida M, Ukai-Tadenuma M, Minami Y, Kito R, Nakao K, Kishimoto W, Yoo SH, Shimomura K, Takao T, Takano A, Kojima T, Nagai K, Sakaki Y, Takahashi JS, Ueda HR (2009) CKIϵ/δ-dependent phosphorylation is a temperature-insensitive, period-determining process in the mammalian circadian clock. Proceedings of the National Academy of Sciences of the United States of America 106:15744–15749

603. Tamaru T, Hirayama J, Isojima Y, Nagai K, Norioka S, Takamatsu K, Sassone-Corsi P (2009) CK2α phosphorylates BMAL1 to regulate the mammalian clock. Nature – Structural and Molecular Biology 16:446–448

604. Robles MS, Boyault C, Knutti D, Padmanabhan K, Weitz CJ (2010) Identification of RACK1 and protein kinase Cα as integral components of the mammalian circadian clock. Science 327:463–466

605. Ohsaki K, Oishi K, Kozono Y, Nakayama K, Nakayama KI, Ishida N (2008) The role of β-TrCP1 and β-TrCP2 in circadian rhythm generation by mediating degradation of clock protein PER2. Journal of Biochemistry 144:609–618

606. Nakahata Y, Kaluzova M, Grimaldi B, Sahar S, Hirayama J, Chen D, Guarente LP, Sassone-Corsi P (2008) The NAD^+-dependent deacetylase SIRT1 modulates CLOCK-mediated chromatin remodeling and circadian control. Cell 134:329–340

607. Asher G, Gatfield D, Stratmann M, Reinke H, Dibner C, Kreppel F, Mostoslavsky R, Alt FW, Schibler U (2008) SIRT1 regulates circadian clock gene expression through PER2 deacetylation. Cell 134:212–214

608. Katada S, Sassone-Corsi P (2010) The histone methyltransferase MLL1 permits the oscillation of circadian gene expression. Nature – Structural and Molecular Biology 17:1414–1421

609. So AY, Bernal TU, Pillsbury ML, Yamamoto KR, Feldman BJ (2009) Glucocorticoid regulation of the circadian clock modulates glucose homeostasis. Proceedings of the National Academy of Sciences of the United States of America 106:17582–17587

610. Lamia KA, Sachdeva UM, DiTacchio L, Williams EC, Alvarez JG, Egan DF, Vasquez DS, Juguilon H, Panda S, Shaw RJ, Thompson CB, Evans RM (2009) AMPK regulates the circadian clock by cryptochrome phosphorylation and degradation. Science 326:437–440

611. Wang X, Tang J, Xing L, Shi G, Ruan H, Gu X, Liu Z, Wu X, Gao X, Xu Y (2010) Interaction of MAGED1 with nuclear receptors affects circadian clock function. EMBO Journal 29:1389–1400

612. Delaunay F, Laudet V (2002) Circadian clock and microarrays: mammalian genome gets rhythm. Trends in Genetics 18:595–597

613. Green CB, Douris N, Kojima S, Strayer CA, Fogerty J, Lourim D, Keller SR, Besharse JC (2007) Loss of Nocturnin, a circadian deadenylase, confers

resistance to hepatic steatosis and diet-induced obesity. Proceedings of the National Academy of Sciences of the United States of America 104:9888–9893.

614. Liu C, Li S, Liu T, Borjigin J, Lin J (2007) Transcriptional coactivator PGC-1alpha integrates the mammalian clock and energy metabolism. Nature 447:477–481

615. Yin L, Wu N, Curtin JC, Qatanani M, Szwergold NR, Reid RA, Waitt GM, Parks DJ, Pearce KH, Wisely GB, Lazar MA (2007) Rev-erbα, a heme sensor that coordinates metabolic and circadian pathways. Science 318:1786–1789

616. Lamia KA, Storch KF, Weitz CJ (2008) Physiological significance of a peripheral tissue circadian clock. Proceedings of the National Academy of Sciences of the United States of America 105:15172–15177

617. Vollmers C, Gill S, DiTacchio L, Pulivarthy SR, Le HD, Panda S (2009) Time of feeding and the intrinsic circadian clock drive rhythms in hepatic gene expression. Proceedings of the National Academy of Sciences of the United States of America 106:21453–21458

618. Zhang EE, Liu Y, Dentin R, Pongsawakul PY, Liu AC, Hirota T, Nusinow DA, Sun X, Landais S, Kodama Y, Brenner DA, Montminy M, Kay SA (2010) Cryptochrome mediates circadian regulation of cAMP signaling and hepatic gluconeogenesis. Nature Medicine 16:1152–1156

619. Eckel-Mahan K, Sassone-Corsi P (2009) Metabolism control by the circadian clock and vice versa. Nature – Structural and Molecular Biology 16:462–467

620. Bass J, Takahashi JS (2010) Circadian integration of metabolism and energetics. Science 330:1349–1354

621. Zuber AM, Centeno G, Pradervand S, Nikolaeva S, Maquelin L, Cardinaux L, Bonny O, Firsov D (2009) Molecular clock is involved in predictive circadian adjustment of renal function. Proceedings of the National Academy of Sciences of the United States of America 106:16523–16528

622. Son GH, Chung S, Choe HK, Kim HD, Baik SM, Lee H, Lee HW, Choi S, Sun W, Kim H, Cho S, Lee KH, Kim K (2008) Adrenal peripheral clock controls the autonomous circadian rhythm of glucocorticoid by causing rhythmic steroid production. Proceedings of the National Academy of Sciences of the United States of America 105:20970–20975

623. Kon N, Hirota T, Kawamoto T, Kato Y, Tsubota T, Fukada Y (2008) Activation of TGF-β/activin signalling resets the circadian clock through rapid induction of Dec1 transcripts. Nature – Cell Biology 10:1463–1469

624. Méndez-Ferrer S, Lucas D, Battista M, Frenette PS (2008) Haematopoietic stem cell release is regulated by circadian oscillations. Nature 452:442–447

625. Keller M, Mazuch J, Abraham U, Eom GD, Herzog ED, Volk HD, Kramer A, Maier B (2009) A circadian clock in macrophages controls inflammatory immune responses. Proceedings of the National Academy of Sciences of the United States of America 106:21407–21412

626. Ibelgaufts H (2010) Cytokines and Cells Online Pathfinder Encyclopaedia (www.copewithcytokines.de/cope.cgi)

627. Mihrshahi R, Barclay AN, Brown MH (2009) Essential roles for Dok2 and RasGAP in CD200 receptor-mediated regulation of human myeloid cells. Journal of Immunology 183:4879–4886

628. Leloup JC, Goldbeter A (2004) Modeling the mammalian circadian clock: sensitivity analysis and multiplicity of oscillatory mechanisms. Journal of Theoretical Biology 230:541–562

629. Lepoutre T, Clairambault J, Gaubert S, Perthame B (2009) personal communication

630. Forger DB (2011) Signal processing in cellular clocks. Proceedings of the National Academy of Sciences of the United States of America 108:4281–4285

Chap. 6. Cell Motility

631. Sato MJ, Kuwayama H, van Egmond WN, Takayama AL, Takagi H, van Haastert PJ, Yanagida T, Ueda M (2009) Switching direction in electric-signal-induced cell migration by cyclic guanosine monophosphate and phosphatidylinositol signaling. Proceedings of the National Academy of Sciences of the United States of America 106:6667–6672

632. Brock A, Chang E, Ho CC, LeDuc P, Jiang X, Whitesides GM, Ingber DE (2003) Geometric determinants of directional cell motility revealed using microcontact printing. Langmuir 19:1611–1617

633. Saez A, Ghibaudo M, Buguin A, Silberzan P, Ladoux B (2007) Rigidity-driven growth and migration of epithelial cells on microstructured anisotropic substrates. Proceedings of the National Academy of Sciences of the United States of America 104:8281–8286

634. Parsons JT, Horwitz AR, Schwartz MA (2010) Cell adhesion: integrating cytoskeletal dynamics and cellular tension. Nature Reviews – Molecular Cell Biology 11:633–643

635. Wang Y, Ding SJ, Wang W, Jacobs JM, Qian WJ, Moore RJ, Yang F, Camp DG, Smith RD, Klemke RL (2007) Profiling signaling polarity in chemotactic cells. Proceedings of the National Academy of Sciences of the United States of America 104:8328–8333

636. Gomes ER, Jani S, Gundersen GG (2005) Nuclear movement regulated by Cdc42, MRCK, myosin, and actin flow establishes MTOC polarization in migrating cells. Cell 121:451–463

637. Reif K, Cyster J (2002) The CDM protein DOCK2 in lymphocyte migration. Trends in Cell Biology 12:368–373

638. Webb DJ, Donais K, Whitmore LA, Thomas SM, Turner CE, Parsons JT, Horwitz AF (2004) FAK-Src signalling through paxillin, ERK and MLCK regulates adhesion disassembly. Nature – Cell Biology 6:154–161

639. Larsen M, Tremblay ML, Yamada KM (2003) Phosphatases in cell–matrix adhesion and migration. Nature Reviews – Molecular Cell Biology 4:700–711

640. Tountas NA, Brautigan DL (2004) Migration and retraction of endothelial and epithelial cells require PHI-1, a specific protein-phosphatase-1 inhibitor protein. Journal of Cell Science 117:5905–5912

641. Ezratty EJ, Partridge MA, Gundersen GG (2005) Microtubule-induced focal adhesion disassembly is mediated by dynamin and focal adhesion kinase. Nature – Cell Biology 7:581–590

642. Sontag JM, Sontag E (2006) Regulation of cell adhesion by PP2A and SV40 small tumor antigen: an important link to cell transformation. Cellular and Molecular Life Sciences 63:2979–2991

643. Dormann D, Weijer CJ (2006) Imaging of cell migration. EMBO Journal 25:3480–3493

644. Bretscher MS (1987) How animal cells move. Scientific American 257:89-90

645. Harms BD, Bassi GM, Horwitz AR, Lauffenburger DA (2005) Directional persistence of EGF-induced cell migration is associated with stabilization of lamellipodial protrusions. Biophysical Journal 88:1479–1488

646. Shroff H, Galbraith CG, Galbraith JA, Betzig E (2008) Live-cell photoactivated localization microscopy of nanoscale adhesion dynamics. Nature – Methods 5:417–423

647. Friedl P, Gilmour D (2009) Collective cell migration in morphogenesis, regeneration and cancer. Nature Reviews – Molecular Cell Biology 10:445–457

648. Nakao S, Platek A, Hirano S, Takeichi M (2008) Contact-dependent promotion of cell migration by the OL-protocadherin-Nap1 interaction. Journal of Cell Biology 182:395–410

649. Ley K, Laudanna C, Cybulsky MI, Nourshargh S (2007) Getting to the site of inflammation: the leukocyte adhesion cascade updated. Nature Reviews – Immunology 7:678–689

650. Thelen M, Stein JV (2008) How chemokines invite leukocytes to dance. Nature – Immunology 9:953–959

651. Poujade M, Grasland-Mongrain E, Hertzog A, Jouanneau J, Chavrier P, Ladoux B, Buguin A, Silberzan P (2007) Collective migration of an epithelial monolayer in response to a model wound. Proceedings of the National Academy of Sciences of the United States of America 104:15988–15993

652. Sakurai-Yageta M, Recchi C, Le Dez G, Sibarita JB, Daviet L, Camonis J, D'Souza-Schorey C, Chavrier P (2008) The interaction of IQGAP1 with the exocyst complex is required for tumor cell invasion downstream of Cdc42 and RhoA. Journal of Cell Biology 181:985–998

653. Steffen A, Le Dez G, Poincloux R, Recchi C, Nassoy P, Rottner K, Galli T, Chavrier P (2008) MT1-MMP-dependent invasion is regulated by TI-VAMP/VAMP7. Current Biology 18:926–931

654. Sahlgren C, Gustafsson MV, Jin S, Poellinger L, Lendahl U (2008) Notch signaling mediates hypoxia-induced tumor cell migration and invasion. Proceedings of the National Academy of Sciences of the United States of America 105:6392–6397

655. Pratap J, Javed A, Languino LR, van Wijnen AJ, Stein JL, Stein GS, Lian JB (2005) The Runx2 osteogenic transcription factor regulates matrix metalloproteinase 9 in bone metastatic cancer cells and controls cell invasion. Molecular and Cellular Biology 25:8581–8591

656. Pinner S, Sahai E (2008) PDK1 regulates cancer cell motility by antagonising inhibition of ROCK1 by RhoE. Nature – Cell Biology 10:127–137

657. Ishikawa K, Takenaga K, Akimoto M, Koshikawa N, Yamaguchi A, Imanishi H, Nakada K, Honma Y, Hayashi JI (2008) ROS-Generating mitochondrial DNA mutations can regulate tumor cell metastasis. Science 320:661–664

658. Silletti S, Yebra M, Perez B, Cirulli V, McMahon M, Montgomery AM (2004) Extracellular signal-regulated kinase (ERK)-dependent gene expression contributes to L1 cell adhesion molecule-dependent motility and invasion. Journal of Biological Chemistry 279:28880–28888

659. Gavert N, Conacci-Sorrell M, Gast D, Schneider A, Altevogt P, Brabletz T, Ben-Ze'ev A (2005) L1, a novel target of β-catenin signaling, transforms cells and is expressed at the invasive front of colon cancers. Journal of Cell Biology 168:633–642

660. Karnoub AE, Dash AB, Vo AP, Sullivan A, Brooks MW, Bell GW, Richardson AL, Polyak K, Tubo R, Weinberg RA (2007) Mesenchymal stem cells within tumour stroma promote breast cancer metastasis. Nature 449:557–563

661. Kim M, Gans JD, Nogueira C, Wang A, Paik JH, Feng B, Brennan C, Hahn WC, Cordon-Cardo C, Wagner SN, Flotte TJ, Duncan LM, Granter SR, Chin L (2006) Comparative oncogenomics identifies NEDD9 as a melanoma metastasis gene. Cell 125:1269–1281

662. Stoletov K, Montel V, Lester RD, Gonias SL, Klemke R (2007) High-resolution imaging of the dynamic tumor cell vascular interface in transparent zebrafish. Proceedings of the National Academy of Sciences of the United States of America 104:17406–17411

663. Medjkane S, Perez-Sanchez C, Gaggioli C, Sahai E, Treisman R (2009) Myocardin-related transcription factors and SRF are required for cytoskeletal dynamics and experimental metastasis. Nature – Cell Biology 11:257–268

664. Padua D, Zhang XH, Wang Q, Nadal C, Gerald WL, Gomis RR, Massagué J (2008) TGFβ primes breast tumors for lung metastasis seeding through angiopoietin-like 4. Cell 133:66–77

665. Garcia-Marcos M, Ghosh P, Farquhar MG (2009) GIV is a nonreceptor GEF for Galphai with a unique motif that regulates Akt signaling. Proceedings of the National Academy of Sciences of the United States of America 106:3178–3183

666. Mattila PK, Lappalainen P (2008) Filopodia: molecular architecture and cellular functions. Nature Reviews – Molecular Cell Biology 9:446–454

667. Mili S, Moissoglu K, Macara IG (2008) Genome-wide screen reveals APC-associated RNAs enriched in cell protrusions. Nature 453:115–119

668. Wei C, Wang X, Chen M, Ouyang K, Song LS, Cheng H (2009) Calcium flickers steer cell migration. Nature 457:901–905

669. Tada T, Nawata J, Wang H, Onoue N, Zhulanqiqige D, Ito K, Sugimura K, Fukumoto Y, Shimokawa H (2008) Enhanced pulsatile pressure accelerates vascular smooth muscle migration: implications for atherogenesis of hypertension. Cardiovascular Research 80:346–353

670. Li Z, Zhang Y, Liu Z, Wu X, Zheng Y, Tao Z, Mao K, Wang J, Lin G, Ji Y, Qin M, Sun S, Zhu X, Sun B (2011) ECM1 controls TH2 cell egress from lymph nodes through re-expression of $S1P_1$. Nature – Immunology 12:178–185

671. Bacon K, Baggiolini M, Broxmeyer H, Horuk R, Lindley I, Mantovani A, Matsushima K, Murphy P, Nomiyama H, Oppenheim J, Rot A, Schall T, Tsang M, Thorpe R, Van Damme J, Wadhwa M, Yoshie O, Zlotnik A, Zoon K (2001)

Chemokine/Chemokine Receptor Nomenclature. International Union of Immunological Societies/World Health Organization Subcommittee on Chemokine Nomenclature. Journal of Leukocyte Biology 70:465–466

672. Zlotnik A, Yoshie O, Nomiyama H (2006) The chemokine and chemokine receptor superfamilies and their molecular evolution. Genome Biology 7:243
673. Mantovani A (1999) The chemokine system: redundancy for robust outputs. Immunology Today 20:254–257
674. Kruger RP, Aurandt J, Guan KL (2005) Semaphorins command cells to move. Nature Reviews – Molecular Cell Biology 6:789–800
675. Shin K, Wang Q, Margolis B (2007) PATJ regulates directional migration of mammalian epithelial cells. EMBO Reports 8:158–164
676. Inagaki M, Inagaki N, Takahashi T, Takai Y (1997) Phosphorylation-dependent control of structures of intermediate filaments: a novel approach using site- and phosphorylation state-specific antibodies. Journal of Biochemistry (Tokyo) 121:407–414
677. Ku NO, Liao J, Omary MB (1998) Phosphorylation of human keratin 18 serine 33 regulates binding to 14-3-3 proteins. EMBO Journal 17:1892–1906
678. Ivaska J, Vuoriluoto K, Huovinen T, Izawa I, Inagaki M, Parker PJ (2005) PKC-mediated phosphorylation of vimentin controls integrin recycling and motility. EMBO Journal 24:3834–3845
679. Watanabe T, Noritake J, Kaibuchi K (2005) Regulation of microtubules in cell migration. Trends in Cell Biology 15:76–83
680. Li R, Gundersen GG (2008) Beyond polymer polarity: how the cytoskeleton builds a polarized cell. Nature Reviews – Molecular Cell Biology 9:860–873
681. Zaoui K, Benseddik K, Daou P, Salaün D, Badache A (2010) ErbB2 receptor controls microtubule capture by recruiting ACF7 to the plasma membrane of migrating cells. Proceedings of the National Academy of Sciences of the United States of America 107:18517–18522
682. Koestler SA, Auinger S, Vinzenz M, Rottner K, Small JV (2008) Differentially oriented populations of actin filaments generated in lamellipodia collaborate in pushing and pausing at the cell front. Nature – Cell Biology 10:306–313
683. Vicente-Manzanares M, Koach MA, Whitmore L, Lamers ML, Horwitz AF (2008) Segregation and activation of myosin IIB creates a rear in migrating cells. Journal of Cell Biology 183:543–554
684. Even-Ram S, Doyle AD, Conti AM, Matsumoto K, Adelstein RS, Yamada KM (2007) Myosin IIA regulates cell motility and actomyosin–microtubule crosstalk. Nature – Cell Biology 9:299–309
685. Sturge J, Wienke D, Isacke CM (2006) Endosomes generate localized Rho-ROCK-MLC2-based contractile signals via Endo180 to promote adhesion disassembly. Journal of Cell Biology 175:337–347
686. Bohil AB, Robertson BW, Cheney RE (2006) Myosin-X is a molecular motor that functions in filopodia formation. Proceedings of the National Academy of Sciences of the United States of America 103:12411–12416

687. Delorme V, Machacek M, DerMardirossian C, Anderson KL, Wittmann T, Hanein D, Waterman-Storer C, Danuser G, Bokoch GM (2007) Cofilin activity downstream of Pak1 regulates cell protrusion efficiency by organizing lamellipodium and lamella actin networks. Developmental Cell 13:646–662

688. Turner CE (2000) Paxillin and focal adhesion signalling. Nature – Cell Biology 2:E231–E236

689. Turner CE (2000) Paxillin interactions. Journal of Cell Science 113:4139–4140

690. Brown MC, Cary LA, Jamieson JS, Cooper JA, Turner CE (2005) Src and FAK kinases cooperate to phosphorylate paxillin kinase linker, stimulate its focal adhesion localization, and regulate cell spreading and protrusiveness. Molecular Biology of the Cell 16:4316–4328

691. Cary LA, Guan JL (1999) Focal adhesion kinase in integrin-mediated signaling. Frontiers in Bioscience 4:D102–D113

692. Schaller MD, Parsons JT (1995) pp125FAK-dependent tyrosine phosphorylation of paxillin creates a high-affinity binding site for Crk. Molecular and Cellular Biology 15:2635–2645

693. Nishiya N, Kiosses WB, Han J, Ginsberg MH (2005) An α_4 integrin-paxillin-Arf-GAP complex restricts Rac activation to the leading edge of migrating cells. Nature – Cell Biology 7:343–352

694. Shiow LR, Roadcap DW, Paris K, Watson SR, Grigorova IL, Lebet T, An J, Xu Y, Jenne CN, Föger N, Sorensen RU, Goodnow CC, Bear JE, Puck JM, Cyster JG (2008) The actin regulator coronin 1A is mutant in a thymic egress-deficient mouse strain and in a patient with severe combined immunodeficiency. Nature – Immunology 9:1307–1315

695. Cai L, Makhov AM, Schafer DA, Bear JE (2008) Coronin 1B antagonizes cortactin and remodels Arp2/3-containing actin branches in lamellipodia. Cell 134:828–842.

696. Innocenti M, Gerboth S, Rottner K, Lai FPL, Hertzog M, Stradal TEB, Frittoli E, Didry D, Polo S, Disanza A, Benesch S, Di Fiore PP, Carlier MF, Scita G (2005) Abi1 regulates the activity of N-WASP and WAVE in distinct actin-based processes. Nature – Cell Biology 7:969–976

697. Hofer AM, Lefkimmiatis K (2007) Extracellular calcium and cAMP: second messengers as "third messengers"? Physiology 22:320–327

698. Bear JE, Rawls JF, Saxe CL (1998) SCAR, a WASP-related protein, isolated as a suppressor of receptor defects in late Dictyostelium development. Journal of Cell Biology 142:1325–1335

699. Zuchero JB, Coutts AS, Quinlan ME, La Thangue NB, Mullins RD (2009) p53-Cofactor JMY is a multifunctional actin nucleation factor. Nature – Cell Biology 11:451–459

700. Morgan MR, Humphries MJ, Bass MD (2007) Synergistic control of cell adhesion by integrins and syndecans. Nature Reviews – Molecular Cell Biology 8:957–969

701. Borghi N, Lowndes M, Maruthamuthu V, Gardel ML, Nelson WJ (2010) Regulation of cell motile behavior by crosstalk between cadherin- and integrin-mediated adhesions. Proceedings of the National Academy of Sciences of the United States of America 107:13324–13329

702. Zhang X, Jiang G, Cai Y, Monkley SJ, Critchley DR, Sheetz MP (2008) Talin depletion reveals independence of initial cell spreading from integrin activation and traction. Nature – Cell Biology 10:1062–1068

703. Zhang H, Berg JS, Li Z, Wang Y, Lang P, Sousa AD, Bhaskar A, Cheney RE, Stromblad S (2004) Myosin-X provides a motor-based link between integrins and the cytoskeleton. Nature – Cell Biology 6:523–531

704. Roberts MS, Woods AJ, Dale TC, van der Sluijs P, Norman JC (2004) Protein kinase B/Akt acts via glycogen synthase kinase 3 to regulate recycling of alpha v beta 3 and alpha 5 beta 1 integrins. Molecular and Cellular Biology 24:1505–1515

705. Nishimura T, Kaibuchi K (2007) Numb controls integrin endocytosis for directional cell migration with aPKC and PAR-3. Developmental Cell 13:15-28

706. deHart GW, Jin T, McCloskey DE, Pegg AE, Sheppard D (2008) The $\alpha_9\beta_1$ integrin enhances cell migration by polyamine-mediated modulation of an inward-rectifier potassium channel. Proceedings of the National Academy of Sciences of the United States of America 105:7188–7193

707. Boguslavsky S, Grosheva I, Landau E, Shtutman M, Cohen M, Arnold K, Feinstein E, Geiger B, Bershadsky A (2007) p120 catenin regulates lamellipodial dynamics and cell adhesion in cooperation with cortactin. Proceedings of the National Academy of Sciences of the United States of America 104:10882–10887

708. Wu X, Kodama A, Fuchs E (2008) ACF7 regulates cytoskeletal-focal adhesion dynamics and migration and has ATPase activity. Cell 135:137–148

709. Undyala VV, Dembo M, Cembrola K, Perrin BJ, Huttenlocher A, Elce JS, Greer PA, Wang YL, Beningo KA (2008) The calpain small subunit regulates cell-substrate mechanical interactions during fibroblast migration. Journal of Cell Science 121:3581–3588

710. Huttenlocher A, Palecek SP, Lu Q, Zhang W, Mellgren RL, Lauffenburger DA, Ginsberg MH, Horwitz AF (2007) Regulation of cell migration by the calcium-dependent protease calpain. Journal of Biological Chemistry 272:32719–32722

711. Li J, Brieher WM, Scimone ML, Kang SJ, Zhu H, Yin H, von Andrian UH, Mitchison T, Yuan J (2007) Caspase-11 regulates cell migration by promoting Aip1–Cofilin-mediated actin depolymerization. Nature – Cell Biology 9:276–286

712. Evdokimov V, Tognon C, Ng T, Sorensen PHB (2009) Reduced proliferation and enhanced migration: two sides of the same coin? Molecular mechanisms of metastatic progression by YB 1. Cell Cycle 8:2901–2906

713. Kato M, Wang L, Putta S, Wang M, Yuan H, Sun G, Lanting L, Todorov I, Rossi JJ, Natarajan R (2010) Post-transcriptional up-regulation of Tsc-22 by Ybx1, a target of miR-216a, mediates TGF-β-induced collagen expression in kidney cells. Journal of Biological Chemistry 285:34004–34015

714. Frye BC, Halfter S, Djudjaj S, Muehlenberg P, Weber S, Raffetseder U, En Nia A, Knott H, Baron JM, Dooley S, Bernhagen J, Mertens PR (2009) Y-box protein-1 is actively secreted through a non-classical pathway and acts as an extracellular mitogen. EMBO Reports 10:783–789

715. Chen Y, Corriden R, Inoue Y, Yip L, Hashiguchi N, Zinkernage A, Nizet V, Insel PA, Junger WG (2006) ATP release guides neutrophil chemotaxis via P2Y2 and A3 receptors. Science 314:1792–1795

716. Kay RR, Langridge L, Traynor D, Hoeller O (2008) Changing directions in the study of chemotaxis. Nature Reviews – Molecular Cell Biology 9:455–463

717. Smith HW, Marra P, Marshall CJ (2008) uPAR promotes formation of the p130Cas-Crk complex to activate Rac through DOCK180. Journal of Cell Biology 182:777–790

718. Nobes CD, Hall A (1995) Rho, rac, and cdc42 GTPases regulate the assembly of multimolecular focal complexes associated with actin stress fibers, lamellipodia, and filopodia. Cell 81:53–62

719. Li Z, Dong X, Wang Z, Liu W, Deng N, Ding Y, L Tang L, Hla T, Zeng R, Li L, Wu D (2005) Regulation of PTEN by Rho small GTPases. Nature – Cell Biology 7:399–404

720. Etienne-Manneville S (2004) Cdc42 – the centre of polarity. Journal of Cell Science 117:1291–1300

721. Takaya A, Ohba Y, Kurokawa K, Matsuda M (2004) RalA activation at nascent lamellipodia of epidermal growth factor-stimulated Cos7 cells and migrating Madin-Darby canine kidney cells. Molecular Biology of the Cell 15:2549–2557

722. Tian D, Jacobo SMP, Billing D, Rozkalne A, Gage SD, Anagnostou T, Pavenstaedt H, Hsu HH, Schlondorff J, Ramos A, Greka A (2010) Antagonistic regulation of actin dynamics and cell motility by TRPC5 and TRPC6 channels. Science Signaling 3:ra77

723. Rajalingam K, Wunder C, Brinkmann V, Churin Y, Hekman M, Sievers C, Rapp UR, Rudel T (2005) Prohibitin is required for Ras-induced Raf-MEK-ERK activation and epithelial cell migration. Nature – Cell Biology 7:837–843

724. Dogan T, Harms GS, Hekman M, Karreman K, Oberoi TK, Alnemri ES, Rapp UR, Rajalingam K (2008) X-linked and cellular IAPs modulate the stability of C-RAF kinase and cell motility. Nature – Cell Biology 10:1447–1455

725. Christiaen L, Davidson B, Kawashima T, Powell W, Nolla H, Vranizan K, Levine M (2008) The transcription/migration interface in Heart precursors of Ciona intestinalis. Science 320:1349–1352

726. Ouyang M, Sun J, Chien S, Wang Y (2008) Determination of hierarchical relationship of Src and Rac at subcellular locations with FRET biosensors. Proceedings of the National Academy of Sciences of the United States of America 105:14353–14358

727. Kiosses WB, Daniels RH, Otey C, Bokoch, GM, Schwartz, MA (1999) A role for p21-activated kinase in endothelial cell migration. Journal of Cell Biology 147:831-844

728. Asanuma K, Yanagida-Asanuma E, Faul C, Tomino Y, Kim K, Mundel P (2006) Synaptopodin orchestrates actin organization and cell motility via regulation of RhoA signalling. Nature – Cell Biology 8:485–491

729. Ohta Y, Hartwig JH, Stossel TP (2006) FilGAP, a Rho- and ROCK-regulated GAP for Rac binds filamin A to control actin remodelling. Nature – Cell Biology 8:803–814

730. Sosa MS, Lopez-Haber C, Yang C, Wang H, Lemmon MA, Busillo JM, Luo J, Benovic JL, Klein-Szanto A, Yagi H, Gutkind JS, Parsons RE, Kazanietz MG (2010) Identification of the Rac-GEF P-Rex1 as an essential mediator of ErbB signaling in breast cancer. Molecular Cell 40:877–892

731. Montero JC, Seoane S, Ocaña A, Pandiella A (2011) P-Rex1 participates in Neuregulin-ErbB signal transduction and its expression correlates with patient outcome in breast cancer. Oncogene 30:1059–1071

732. Castillo-Lluva S, Tatham MH, Jones RC, Jaffray EG, Edmondson RD, Hay RT, Malliri A (2010) SUMOylation of the GTPase Rac1 is required for optimal cell migration. Nature Cell Biology 12:1078–1085

733. Wang X, He L, Wu YI, Hahn KM, Montell DJ (2010) Light-mediated activation reveals a key role for Rac in collective guidance of cell movement in vivo. Nature – Cell Biology 12:591–597

734. Frantz C, Karydis A, Nalbant P, Hahn KM, Barber DL (2007) Positive feedback between Cdc42 activity and H+ efflux by the Na-H exchanger NHE1 for polarity of migrating cells. Journal of Cell Biology 179:403–410.

735. Dubois T, Paléotti O, Mironov AA, Fraisier V, Stradal TEB, De Matteis MA, Franco M, Chavrier P (2005) Golgi-localized GAP for Cdc42 functions downstream of ARF1 to control Arp2/3 complex and F-actin dynamics. Nature – Cell Biology 7:353–364

736. Gundersen GG (2002) Microtubule capture: IQGAP and CLIP-170 expand the repertoire. Current Biology 12:R645–R647

737. Riento K, Totty N, Villalonga P, Garg R, Guasch R, Ridley AJ (2005) RhoE function is regulated by ROCK I-mediated phosphorylation. EMBO Journal 24:1170–1180

738. Chardin P (2006) Function and regulation of Rnd proteins. Nature Reviews – Molecular Cell Biology 7:54–62

739. Hauck CR, Hsia DA, Schlaepfer DD (2002) The focal adhesion kinase–a regulator of cell migration and invasion. IUBMB Life 53:115–119

740. Webb DJ, Donais K, Whitmore LA, Thomas SM, Turner CE, Parsons JT, Horwitz AF (2004) FAK-Src signalling through paxillin, ERK and MLCK regulates adhesion disassembly. Nature – Cell Biology 6:154–161

741. Wittmann T, Bokoch GM, Waterman-Storer CM (2004) Regulation of microtubule destabilizing activity of Op18/stathmin downstream of Rac1. Journal of Biological Chemistry 279:6196–6203

742. Nola S, Sebbagh M, Marchetto S, Osmani N, Nourry C, Audebert S, Navarro C, Rachel R, Montcouquiol M, Sans N, Etienne-Manneville S, Borg JP, Santoni MJ (2008) Scrib regulates PAK activity during the cell migration process. Human Molecular Genetics 17:3552–3565

743. Li Z, Hannigan M, Mo Z, Liu B, Lu W, Wu Y, Smrcka AV, Wu G, Li L, Liu M, Huang CK, Wu D (2003) Directional sensing requires Gβγ-mediated PAK1 and PIXα-dependent activation of Cdc42. Cell 114:215–227

744. Riento K, Ridley AJ (2003) ROCKs: multifunctional kinases in cell behaviour. Nature Reviews – Molecular Cell Biology 4, 446-456

745. Sabio G, Arthur JSC, Kuma Y, Peggie M, Carr J, Murray-Tait V, Centeno F, Goedert M, Morrice NA, Cuenda A (2005) p38 regulates the localisation of SAP97 in the cytoskeleton by modulating its interaction with GKAP. EMBO Journal 24:1134–1145

746. Zhang M, Liu J, Cheng A, DeYoung SM, Chen X, Dold LH, Saltiel AR (2006) CAP interacts with cytoskeletal proteins and regulates adhesion-mediated ERK activation and motility. EMBO Journal 25:5284–5293

747. Llense F, Martín-Blanco E (2008) JNK signaling controls border cell cluster integrity and collective cell migration. Current Biology 18:538–544

748. Melani M, Simpson KJ, Brugge JS, Montell D (2008) Regulation of cell adhesion and collective cell migration by Hindsight and its human homolog RREB1. Current Biology 18:532–537

749. Gantulga D, Tuvshintugs B, Endo Y, Takino T, Sato H, Murakami S, Yoshioka K (2008) The scaffold protein c-Jun NH2-terminal kinase-associated leucine zipper protein regulates cell migration through interaction with the G protein $G\alpha_{13}$. Journal of Biochemistry 144:693–700

750. Tiganis T, Bennett AM (2007) Protein tyrosine phosphatase function: the substrate perspective. Biochemical Journal 402:1–15

751. Angers-Loustau A, Côté JF, Charest A, Dowbenko D, Spencer S, Lasky LA, Tremblay ML (1999) Protein tyrosine phosphatase-PEST regulates focal adhesion disassembly, migration, and cytokinesis in fibroblasts. Journal of Cell Biology 144:1019–1031

752. Harvey BP, Banga SS, Ozer HL (2004) Regulation of the multifunctional Ca^{2+}/calmodulin-dependent protein kinase II by the PP2C phosphatase PPM1F in fibroblasts. Journal of Biological Chemistry 279:24889–24898

753. Zhang S, Yang Y, Kone BC, Allen JC, Kahn AM (2003) Insulin-stimulated cyclic guanosine monophosphate inhibits vascular smooth muscle cell migration by inhibiting Ca^{2+}/calmodulin-dependent protein kinase II. Circulation 107:1539–1544

754. Sandu OA, Ito M, Begum N (2001) Selected contribution: insulin utilizes NO/cGMP pathway to activate myosin phosphatase via Rho inhibition in vascular smooth muscle. Journal of Applied Physiology 91:1475–1482

755. Langlois M-J, Bergeron S, Bernatchez G, Boudreau F, Saucier C, Perreault N, Carrier JC, Rivard N (2010) The PTEN phosphatase controls intestinal epithelial cell polarity and barrier function: role in colorectal cancer progression. PLoS One 5:e15742

756. Eiseler T, Döppler H, Yan IK, Kitatani K, Mizuno K, Storz P (2009) Protein kinase D1 regulates cofilin mediated F-actin reorganization and cell motility via slingshot. Nature – Cell Biology 11:545–556

757. Arai Y, Shibata T, Matsuoka S, Sato MJ, Yanagida T, Ueda M (2010) Self-organization of the phosphatidylinositol lipids signaling system for random cell migration. Proceedings of the National Academy of Sciences of the United States of America 107:12399–12404

758. Graupera M, Guillermet-Guibert J, Foukas LC, Phng LK, Cain RJ, Salpekar A, Pearce W, Meek S, Millan J, Cutillas PR, Smith AJH, Ridley AJ, Ruhrberg C, Gerhardt H, Vanhaesebroeck B (2008) Angiogenesis selectively requires the

p110α isoform of PI3K to control endothelial cell migration. Nature 453:662–666

759. Todeschini AR, Dos Santos JN, Handa K, Hakomori SI (2008) Ganglioside GM2/GM3 complex affixed on silica nanospheres strongly inhibits cell motility through CD82/cMet-mediated pathway. Proceedings of the National Academy of Sciences of the United States of America 105:1925–1930
760. Rivera GM, Antoku S, Gelkop S, Shin NY, Hanks SK, Pawson T, Mayer BJ (2006) Requirement of Nck adaptors for actin dynamics and cell migration stimulated by platelet-derived growth factor B. Proceedings of the National Academy of Sciences of the United States of America 103:9536–9541
761. Nagano K, Bornhauser BC, Warnasuriya G, Entwistle A, Cramer R, Lindholm D, Naaby-Hansen S (2006) PDGF regulates the actin cytoskeleton through hnRNP-K-mediated activation of the ubiquitin E3-ligase MIR. EMBO Journal 25:1871–1882
762. Katz M, Amit I, Citri A, Shay T, Carvalho S, Lavi S, Milanezi F, Lyass L, Amariglio N, Jacob-Hirsch J, Ben-Chetrit N, Tarcic G, Lindzen M, Avraham R, Liao YC, Trusk P, Lyass A, Rechavi G, Spector NL, Lo SH, Schmitt F, Bacus SS, Yarden Y (2007) A reciprocal tensin-3–cten switch mediates EGF-driven mammary cell migration. Nature – Cell Biology 9:961–969
763. Pappu R, Schwab SR, Cornelissen I, Pereira JP, Regard JB, Xu Y, Camerer E, Zheng YW, Huang Y, Cyster JG, Coughlin SR (2007) Promotion of lymphocyte egress into blood and lymph by different sources of sphingosine-1-phosphate. Science 316:295–298
764. Witze ES, Litman ES, Argast GM, Moon RT, Ahn NG (2008) Wnt5a control of cell polarity and directional movement by polarized redistribution of adhesion receptors. Science 320:365–369
765. Aman A, Piotrowski T (2008) Wnt/β-catenin and FGF signaling control collective cell migration by restricting chemokine receptor expression. Developmental Cell 15:749–761
766. Zuo X, Zhang J, Zhang Y, Hsu SC, Zhou D, Guo W (2006) Exo70 interacts with the Arp2/3 complex and regulates cell migration. Nature – Cell Biology 8:1383–1388
767. Santilman V, Baran J, Anand-Apte B, Fox PL, Parat MO (2006) Caveolin-1 polarization in migrating endothelial cells is directed by substrate topology not chemoattractant gradient. Cell Motility and the Cytoskeleton 63:673–680
768. Gerthoffer WT (2008) Migration of airway smooth muscle cells. Proceedings of the American Thoracic Society 5:97–105
769. Fuller D, Chen W, Adler M, Groisman A, Levine H, Rappel WJ, Loomis WF (2010) External and internal constraints on eukaryotic chemotaxis. Proceedings of the National Academy of Sciences of the United States of America 107:9656–9659
770. Kronlage M, Song J, Sorokin L, Isfort K, Schwerdtle T, Leipziger J, Robaye B, Conley PB, Kim HC, Sargin S, Schön P, Schwab A, Hanley PJ (2010) Autocrine purinergic receptor signaling is essential for macrophage chemotaxis. Science Signaling 3:ra55

771. Meili R, Sasaki AT, Firtel RA (2005) Rho rocks PTEN. Nature – Cell Biology 7:334–335

772. Keller EF, Segel LA (1971) Model for chemotaxis. Journal of Theoretical Biology 30:225–234

773. Marrocco A (2003) Numerical simulation of chemotactic bacteria aggregation via mixed finite elements. Mathematical Modelling and Numerical Analysis 37:617–630

774. Zaman MH, Kamm RD, Matsudaira P, Lauffenburger DA (2005) Computational model for cell migration in three-dimensional matrices. Biophysical Journal 89:1389–1397

775. Perthame B (2004) PDE models for chemotactic movements. Parabolic, hyperbolic and kinetic. Applied Mathematics 49:539–564

776. Mantzaris NV, Webb S, Othmer HG (2004) Mathematical modeling of tumor-induced angiogenesis. Journal of Mathematical Biology 49:111–187

777. Chauvière A, Preziosi L, Hillen T (2007) A continuum model for cell motion in network tissues. Discrete and Continuous Dynamical Systems (Supplements) 1–10 (http://calvino.polito.it/ preziosi/pubs/CHP1.pdf)

A

Notation Rules: Aliases and Symbols

A.1 Aliases for Molecules

Aliases include all written variants, i.e., any abbreviation, in particular acronyms.[1] *Acronym* corresponds to a word made from the initial letters or syllables of nouns that are pronounceable as a word. Acronyms are generally written with all letters in upper case. Yet, some acronyms are treated as words and written in lower case (e.g., laser [originally LASER] is an acronym for light amplification by stimulated emission of radiation, sonar [originally SONAR] for sound navigation and ranging). A substance name can derive from its chemical name (e.g., amphetamine: α-methylphenethylamine).

Acronyms can give rise to molecule names by adding a scientific suffix such as "-in", a common ending of molecule nouns (e.g., sirtuin stands for silent information regulator-2 [alias SIRT]). Other scientific prefixes and suffixes can be frequently detected throughout the present text. Their meaning is given in Notations, particularly for readers from Asia. Many prefixes are used to specify position, configuration and behavior, quantity, direction and motion, structure, timing, frequency, and speed.

Initialisms are abbreviations that are formed from initial letters of a single long noun or several nouns and, instead of being pronounced like an ordinary word, are read letter-by-letter (e.g., DNA that stands deoxyribonucleic acid).

Some abbreviations can give rise to alphabetisms that are written as new words (e.g., Rho-associated, coiled coil containing protein kinase [RoCK] that is also called Rho kinase). In biochemistry, multiple-letter abbreviations can also be formed from a single word that can be long (e.g., Cam stands for

[1] In general, abbreviations exclude the initials of short function words, such as "and", "or", "of", or "to". However they are sometimes included in acronyms to make them pronounceable (e.g., radar [originally RADAR] for radio detection and ranging). These letters are often written in lower case. In addition, both cardinal (size, molecular weight, etc.) and ordinal (isoform discovery order) numbers in names are represented by digits.

calmodulin, which is itself a portmanteau word, Trx for thioredoxin, etc.) as well as short (e.g., Ttn for titin, etc.). In addition, single-letter symbols of amino acids are often used to define a molecule alias (e.g., tyrosine can be abbreviated as Tyr or Y, hence SYK stands for spleen tyrosine kinase).

A *portmanteau* is a word that combines initials and some inner letters of at least 2 words (e.g., calmodulin stands for calcium modulated protein; caspase, a cysteine-dependent aspartate-specific protease; chanzyme, an ion channel and enzyme; chemokine, a chemoattractant cytokine;[2] emilin, an elastin microfibril interfacer; porin, a pore-forming protein; restin, a Reed-Steinberg cell-expressed intermediate filament-associated protein, an alias for cytoplasmic linker protein CLiP1 (or CLiP170); serpin, a serine protease inhibitor; siglec, a sialic acid-binding Ig-like lectin; sirtuin, a silent information regulator-2 (two); transceptor, a transporter-related receptor; and Prompt, a promoter upstream transcript).[3]

Aliases use, in general, capital letters and can include hyphens and dots. Yet, as a given protein can represent a proto-oncogene[4] encoded by a gene that can give rise to an oncogene (tumor promoter) after gain- or loss-of-function mutations,[5] the same acronym represents 3 different entities.[6]

[2] Cytokines are peptidic, proteic, or glycoproteic regulators that are secreted by cells of the immune system. These immunomodulating agents serve as auto- or paracrine signals.

[3] The upper case initial P in Prompt is used to avoid confusion with command-line interpreter prompt or prompt book to direct precise timing of actions on theater stage.

[4] In 1911, P. Rous isolated a virus that was capable of generating tumors of connective tissue (sarcomas) in chicken. Proteins were afterward identified, the activity of which, when uncontrolled, can provoke cancer, hence the name oncogene given to genes that encode these proteins. Most of these proteins are enzymes, more precisely kinases. The first oncogene was isolated from the avian Rous virus by D. Stéhelin and called Src (from sarcoma). This investigator demonstrated that the abnormal functioning of the Src protein resulted from mutation of a normal gene, or proto-oncogene, which is involved in cell division.

[5] Loss-of-function mutations cause complete or partial loss of function of gene products that operate as tumor suppressors, whereas gain-of-function mutations generate gene products with new or abnormal function that can then act as oncogenes. Typical tumor-inducing agents are enzymes, mostly regulatory kinases and small guanosine triphosphatases, that favor proliferation of cells, which do normally need to be activated to exert their activities. Once their genes are mutated, these enzymes become constitutively active. Other oncogenes include growth factors (a.k.a. mitogens) and transcription factors. Mutations can also disturb signaling axis regulation, thereby raising protein expression. Last, but not least, chromosomal translocation can also provoke the expression of a constitutively active hybrid protein.

[6] Like Latin-derived shortened expressions – as well as foreign words – that are currently written in italics, genes can be italicized. However, this usage is not required in scientific textbooks published by Springer. Italic characters are then used to highlight words within a text to easily target them. Proteins are currently

Besides, a given abbreviation can designate distinct molecules without necessarily erroneous consequence in a given context (e.g., PAR: polyADPribose or protease-activated receptor and GCK: germinal center kinases or glucokinase; in the latter case, the glucokinase abbreviation should be written as GcK or, better, GK).

In addition, a large number of aliases that designate a single molecule results from the fact that molecules have been discovered independently several times with possibly updated functions. Some biochemists uppercase the name of a given molecule, whereas others lowercase (e.g., cell division cycle guanosine triphosphatase of the Rho family CDC42 or Cdc42, adaptor growth factor receptor-bound protein GRB2 or Grb2, chicken tumor virus regulator of kinase CRK or Crk, guanine nucleotide-exchange factor Son-of-sevenless SOS or Sos, etc.). Acronyms are then not always capitalized. Printing style of aliases should not only avoid confusion, but also help one in remembering alias meaning.

In the present textbook, choice of lower and upper case letters in molecule aliases is dictated by the following criteria. (1) An upper case letter is used for initials of words that constitute molecule nouns (e.g., receptor tyrosine kinase RTK). An alias of any compound takes into account added atoms or molecules (e.g., PI: phosphoinositide and PIP: phosphoinositide phosphate) as well as their number (e.g., PIP2: phosphatidylinositol bisphosphate and DAG: diacylglycerol).

(2) A lower case letter is used when a single letter denotes a subfamily or an isoform when it is preceded by a capital letter (e.g., PTPRe: protein tyrosine phosphatase receptor-like type-E). Nevertheless, an upper case letter is used in an alias after a single or several lower case letters to distinguish the isoform type (e.g., RhoA isoform and DNA-repair protein RecA for recombination protein-A), but OSM stands for oncostatin-M, not osmole Osm[7] to optimize molecule identification.

These criteria enable to use differently written aliases with the same sequence of letters for distinct molecules (e.g., CLIP for corticotropin-like intermediate peptide, CLiP: cytoplasmic CAP-Gly domain-containing linker protein, and iCliP: intramembrane-cleaving protease).

As the exception proves the rule, current aliases, such as PKA and PLA that designate protein kinase-A and phospholipase-A, respectively, have been kept. Preceded by only 2 upper case letters, a lower case letter that should be used to specify an isoform can bring confusion with acronyms of other protein types (e.g., phospholamban alias PLb).

---

romanized (ordinary print), but with a capital initial. Nevertheless, names (not aliases) of chemical species are entirely lowercased in most – if not all – scientific articles, except to avoid confusion with a usual word (e.g., hedgehog animal vs. Hedgehog protein).

[7] Osmole: the amount of osmotically active particles that exerts an osmotic pressure of 1 atm when dissolved in 22.4 l of solvent at 0 C.

Nouns (e.g., hormone-like fibroblast growth factor [hFGF] and urokinase-type plasminogen activator [uPA]) or adjectives (e.g., intracellular FGF isoform [iFGF]) that categorize a subtype of a given molecule correspond to a lower case letter to emphasize the molecule species. Hence, an upper case letter with a commonly used hyphen (e.g., I[R]-SMAD that stands for inhibitory [receptor-regulated] SMAD; V-ATPase for vacuolar adenosine triphosphatase; MT1-MMP for membrane type-1 matrix metalloproteinase; and T[V]-SNARE for target [vesicle-associated] soluble ^{N}ethylmaleimide-sensitive factor-attachment protein receptor) is then replaced by a lower case letter (e.g., i[r]SMAD, vATPase, mt1MMP, and t[v]SNARE), as is usual for RNA subtypes (mRNA, rRNA, snRNA, and tRNA for messenger, ribosomal, small nuclear, and transfer RNA, respectively). Similarly, membrane-bound and secreted forms of receptors and coreceptors that can derive from alternative mRNA splicing are defined by a lower case letter (e.g., sFGFR for secreted extracellular FGFR form and sFRP for soluble Frizzled-related protein), as well as eukaryotic translation elongation (eEF) and initiation (eIF) factors.

(3) Although `l`, `r`, and `t` can stand for molecule-like, -related, and -type, respectively, when a chemical is related to another one, in general, upper case letters are used for the sake of homogenity and to clearly distinguish between the letter L and numeral 1 (e.g., KLF: Krüppel-like factor, CTK: C-terminal Src kinase (CSK)-type kinase, and SLA: Src-like adaptor).

(4) An upper case letter is most often used for initials of adjectives contained in the molecule name (e.g., AIP: actin-interacting protein; BAX: BCL2-associated X protein; HIF: hypoxia-inducible factor; KHC: kinesin heavy chain; LAB: linker of activated B lymphocytes; MAPK: mitogen-activated protein kinase; and SNAP: soluble N-ethylmaleimide-sensitive factor-attachment protein);

(5) Lower case letters are used when alias letters do not correspond to initials (e.g., Fox – not fox –: forkhead box), except for portmanteau words that are entirely written in minuscules (e.g., gadkin: γ1-`adaptin` and `kinesin interactor`).

This rule applies, whether alias letters do correspond to successive noun letters (e.g., Par: partitioning defective protein and Pax: paxillin, as well as BrK: breast tumor kinase and ChK: checkpoint kinase, whereas CHK denotes C-terminal Src kinase [CSK]-homologous kinase) or not (e.g., Fz: Frizzled and HhIP: Hedgehog-interacting protein),[8] except for composite chemical species (e.g., DAG: diacylglycerol). However, some current usages have been kept for

[8] The Hedgehog gene was originally identified in the fruit fly Drosophila melanogaster. It encodes a protein involved in the determination of segmental polarity and intercellular signaling during morphogenesis. Homologous gene and protein exist in various vertebrate species. The name of the mammal hedgehog comes from hecg and hegge (dense row of shrubs or low trees), as it resides in hedgerows, and hogg and hogge, due to its pig-like, long projecting nose (snout). The word Hedgehog hence is considered as a seamless whole.

short aliases of chemical species name (e.g., Rho for Ras homolog rather than RHo).

In any case, molecule (super)family (class) aliases as well as those of their members are written in capital letters, such as the IGSF (IGSF*i*: member *i*; immunoglobulin), KIF (KIF*i*; kinesin), SLC (SLC*i*; solute carrier), TNFSF (TNFSF*i*; tumor-necrosis factor), and TNFRSF (TNFRSF*i*; tumor-necrosis factor receptor) superfamily.

Gene names are also written with majuscules when the corresponding protein name contains at least one minuscule, otherwise only the gene name initial is written with an upper case letter that is then followed by lower case letters.

To highlight its function, substrate aliases (e.g., ARF GTPases) contained in a molecule alias are partly written with lower case letters (e.g., ArfRP, ArfGEF, ArfGAP stand for ARF-related protein, ARF guanine-nucleotide exchange factor, and ARF GTPase-activating protein, respectively).

Last, but not least, heavy and pedantic designation of protein isoforms based on roman numerals has been avoided and replaced by usual arabic numerals (e.g., angiotensin-2 rather than angiotensin-II), except for coagulation (or clotting) factors. Moreover, character I can mean either letter I or number 1 without obvious discrimination at first glance (e.g., GAPI that stands for Ras GTPase-activating protein GAP1, but can be used to designate a growth-associated protein inhibitor).

Unnecessary hyphenation in aliases of substances (between an upper case letter, which can define the molecule function, and the chemical alias, or between it and assigned isotype number) has been avoided. In any case, the Notation section serves not only to define aliases, but also, in some instances, as disambiguation pages.

A.2 Symbols for Physical Variables

Unlike substances aliases, symbols for physical quantities are most often represented by a single letter of the Latin or Greek alphabet (i: current; J: flux; L: length; m: mass; p: pressure; P: power; T: temperature; t: time; u: displacement; v: velocity; x: space; λ: wavelength; μ: dynamic viscosity; ρ: mass density; etc.). These symbols are specified using sub- and superscripts (c_p and c_v: heat capacity at constant pressure and volume, respectively; $\mathcal{D}_T$: thermal diffusivity; G_h: hydraulic conductivity; G_T: thermal conductivity; α_k: kinetic energy coefficient; α_m: momentum coefficient; etc.).

A physical quantity associated with a given point in space at a given time can be: (1) a scalar uniquely defined by its magnitude; (2) a vector characterized by a magnitude, a support, and a direction represented by an oriented line segment defined by a unit vector; and (3) a tensor specified by a magnitude and a few directions. To ensure a straightforward meaning of symbols used for scalar, vectorial, and tensorial quantities, bold face upper (**T**) and lower (**v**) case letters are used to denote a tensor and a vector, respectively, whereas both roman (plain, upright)-style upper and lower case letters designate a scalar.

List of Currently Used Prefixes and Suffixes

Prefixes (localization)

"ab-" (Latin) and "apo-" (Greek: απο): away from or off (abluminal: endothelial edge opposite to wetted surface; apolipoproteins: lipid carriers that cause egress [also ingress] from cells; aponeurosis (απονευρωσις; νευρον: sinew, tendon) muscle sheath that limits radial motion and enhances axial contraction; and apoptosis: separation ["-ptosis": fall (πτωσις): as leaves fall away from a tree], a type of programmed cell death)

"acr-" (variant "acro-" [ακρος]): top or apex

"ad-" (adfecto: to reach; adfio: to blow toward; adfluo: to flow toward): toward (ad- becomes "ac-" before c, k, or q; "af-" before f [afferent]; "ag-" before g [agglutination]; "al-" before l; "ap-" before p [approximation]; "as-" before s; and "at-" before t)

"cis-", "juxta-", and "para-" (παρα): near, beside, or alongside

"contra-": opposite side; "ipsi-" (ipse): same side; "latero-": side;

"ecto-" (εκτος), "exo-" (εξο), and "extra-": outside, outer, external, or beyond (exogenous chemicals produced by an external source, or xenobiotics ["xeno-": foreigner])

"endo-" (ενδον) and "intra-": inside (endogenous substances synthesized by the body's cells; endomembranes at organelle surfaces within the cell)

"ep-" (variant "eph-", or "epi-" [επι]): upon (epigenetics refers to the inheritance ("-genetic": ability to procreate [γεννητικος]) of variations in gene expression beyond ("epi-": on, upon, above, close to, beside, near, toward, against, among, beyond, and also) change in the DNA sequence.

"front-" and "pre-": anterior or in front of

"post-": behind

"infra-" and "sub-": under or below

"super-" and "supra-": above

"inter-": between or among

"peri-" (περι): around

"tele-" (τελε): remote

"trans-": across

Prefixes (composition)

"an-" and "aniso-" (ανισος): unequal, uneven, heterogeneous

"iso-" (ισος): equal, alike (isomer [μερος: part, portion]

"mono-" (μονος) and "uni-" (unicus): single

"oligo-" (ολιγος): few, little, small

"multi-" (multus), "pluri-" (plus, plures), and "poly-" (πολυς): many, much
"ultra-": in excess.

Prefixes (quantity)

"demi-" (dimidius) and "hemi-" (ημι): half
"sesqui-": one and a half (half more)
"di-" or "dis-" (δυο; δις) as well as "bi-" or "bis-": 2, twice
"tri" (τρεις, τρι-; tres, tria): 3
"tetra-" (τετρα), "quadri-" (variant: "quadr-" and "quadru-"): 4
"penta-" (πεντας; pentas), "quinqu-", and "quint-": 5
"hexa-" (εξ) and "sexa-": 6
"hepta-" (επτα): 7
"octa-" (οκτα): 8
"nona-" (εννεα): 9 (ninth part)
"deca-" (δεκα): 10
"quadra-" (quadragenarius): 40 (elements)
"quinqua-" (quinquagenarius): 50
"sexa-" (sexagenarius [sex: 6]: 60
"septua-" (septuagenarius [septem: 7]): 70
"nona-" (nonagenarius): 90

Prefixes (motion and direction)

"af-": toward the center (single master object); e.g., nerve and vascular afferents (ferre: to carry) to brain and heart, respectively, rather than toward any slave, supplied tissue from the set of the body's organs; also affector, i.e., chemical messenger that brings a signal to the cell considered as the object of interest, this exploration focus being virtually excised from the organism with its central command system, except received signals
"ef-" (effero: to take away): from the center (efferent; effector, i.e., chemical transmitter recruited by the previous mediator of a signaling cascade at a given locus to possibly translocate to another subcellular compartment)
"antero-" (anterior): before, in front of, facing, or forward
"retro-": behind or backward
"tropo-" (τροπος): duct direction; (tropa: rotation; celestial revolution); e.g., tropomyosin (μυς, musculus: muscle; μυο-: refers to muscle [μυοτρωτος: injured at a muscle])

Prefixes (structure and size)

"macro-" (μακρος): large, long, or big
"mega-" (μεγας): great, large
"meso-" (μεσος): middle
"micro-" (μικρος): small
"nano-" (νανος): dwarf, tiny
"homo-" (ομο-): same (ομολογος: agreeing, corroborating; variant: "homeo-" [homeostasis])
"hetero-" (ετερο-): other

Prefixes (timing)

"ana-" (ανα): culminating (anaphase of the cell division cycle), up, above (ανοδος: a way up, anode [positive electrode; οδος; way, path, road, track])
"ante-": before
"circa-": approximately, around (circadian: approximately one day)
"infra-": below, shorter (infradian: rhythm with lower frequency than that of circadian rhythm, not smaller period)
"inter-": among, between, during
"meta-" (μετα): after, beyond, behind, later; in the middle of (metaphase of the cell division cycle); as well as connected to, but with a change of state (metabolism) and about (metadata)
"post-": after
"pre-": earlier

"pro-" (προ): preceding, first, before (prophase of the cell division cycle)
"telo-" (τελος): end, completion
"ultra-": beyond, longer (ultradian: period smaller than that of 24–28-hour cycle, i.e., frequency greater than that of the circadian rhythm)

Prefixes (functioning modality)

"auto-" (αυτος): same, self
"brady-" (βραδυς): slow (decelerate)
"tachy-" (ταχος): rapid (accelerate)
"amphi-" (αμφι): both (amphiphilic substances are both hydrophilic and lipophilic; amphisomes are generated by both autophagosomes and endosomes)
"ana-" : upward (anabolism) or against (anaphylaxis)
"cata-" (κατα): downward (catabolism, cathode [negative electrode; οδος; way, path, road, track])
"anti-" (αντι): against
"pro-": favoring
"co-" (coaccedo: add itself to): together
"contra-": adverse, against, beside, next to, opposite
"de-": remove, reduce, separation after association (Latin de; e.g., deoxy-)
"dys-" (δυς): abnormal (δυσαης): ill-blowing)
"equi-" (æque): equal or alike
"hem-" or "hemat-" (αιμα: blood): related to blood
"hypo-" (υπο): under, beneath, and low
"hyper-" (υπερ): above, beyond, and large
"per-": through (e.g., percutaneous) and during (e.g., peroperative)
"pseudo-" (ψευδο): pretended, false
"re-"; again

Scientific suffixes

"-ase": enzyme (synthase, lipase, etc.)
"-ate": salt of a base
"-cyte" (κυτος): cell (erythro- [ερυθρος: red], leuko- [λευκος: light, bright, clear, white], thrombo- [θρομβος: lump, clot], adipo- [adeps: fat; adipalis, adipatus, adipeus, adipinus: fatty], fibro- [fibra: fiber, filament], myo- [μυς: muscle, mouse, mussel], myocardiocyte [κραδια: heart; cardiacus: related to heart, stomach; to have heart trouble, stomach trouble], etc.);
"-crine" (κρινω): to decide, to separate, and to secrete (e.g., endocrine regulator) (ευκρινεω: keep in order)
"-elle": small (organelle in a cell [like an organ in a body])
"-ium", "-ion", "-isk", and "-iscus": little ("-ium": tissue interface and envelope, such as endothelium and pericardium)
"-phil" (φιλια): attracted (αφιλια: want of friends)
"-phob" (φοβια): repulsed (υδροφοβια, hydrophobia [Latin]: horror of water)
"-phore" (φερω): carrier (αμφερω: to bring up)
"-yl" denotes a radical (molecules with unpaired electrons)
"-ploid" (πλοω): double, fold (diploid, twofold; διπλοω: to double; διαπλοω: unfold)
"-emia": in relation to flow (ανεμια: flatulence; ευηνεμια: fair wind), particularly blood condition
"-genesis" (γενεσις): cause, generation, life source, origin, productive force
"-iasis": for diseased condition
"-itis": inflammation
"-lemma" (λεμμα: skin): sheath
"-ole" and "-ule": small (arteriole and venule; variant "-ula" [blastula] and "-ulum")
"-plasma" (πλασμα): anything molded (plasma: creature generated from silt of earth)
"-plasia" (πλασια): formation, molding

"-podium" (ποδος: foot; podium [Latin]: small knoll, small protuberance): protrusion
"-poiesis" (ποιεω): production
"-soma" (σωμα): body
"-sclerosis" (σκλημα): hardness, induration
"-stasis" (στασις): stabilization (αποκαταστασις: restoration; ανυποστασις: migration)
"-stomosis" (στομα: mouth): equipped with an outlet
"-taxy/tactic" (ταχυ: rapid; τακτικος: to maneuver): related to motion (also prefix, i.e., ταχυκινησις: quick motion; ταχυνω: to accelerate; and ταχυπνοια: short breath; not [δια]ταξις: disposition, arrangement)
"-trophy/trophic" (τροφις: well fed): related to growth
"-oma": tumor of
"-pathy" (παθος, παθεια): disease of
"-tomy" (τομια) and "-ectomy": surgical removal (απλοτομια: simple incision; φαûrhουγγοτομια: laryngotomy)

List of Aliases

A

$\mathcal{A}$: Avogadro number
$\mathcal{A}(p)$: area-pressure relation
$\mathbf{A}$: Almansi strain tensor
A: cross-sectional area
`A`: actin-binding site
$\mathbf{a}$: acceleration
a: major semi-axis
AA: arachidonic acid
AAA: ATPase associated with diverse cellular activities
AAA: abdominal aortic aneurysm
AAAP: aneurysm-associated antigenic protein
AAK: adaptin-associated kinase
ABC: ATP-binding cassette transporter (transfer ATPase)
AbI: Abelson kinase interactor
Abl: Abelson leukemia viral proto-oncogene product (NRTK)
ABLIM: actin-binding LIM domain-containing protein
ABP: actin-binding protein
AC: atrial contraction
ACAP: ArfGAP with coiled-coil, ankyrin repeat, PH domains
ACase: adenylyl cyclase
ACi: adenylyl cyclase isoform i
ACAT: acylCoA–cholesterol acyltransferase
ACC: acetyl coenzyme-A carboxylase
ACE: angiotensin-converting enzyme
ACh: acetylcholine
ACK: activated CDC42-associated kinase
ACTH: adrenocorticotropic hormone
Cav-actin: caveolin-associated F actin
^{F}actin: filamentous actin
^{G}actin: monomeric globular actin
AcvR: activin receptor (TGFβ receptor superfamily)
Ad: adrenaline
ADAM: a disintegrin and metallopeptidase (adamalysin)
ADAMTS: a disintegrin and metallopeptidase with thrombospondin
ADAP: adhesion and degranulation-promoting adaptor protein
ADAP: ArfGAP with dual PH domains
ADF: actin depolymerizing factor (cofilin-related destrin)
ADH: antidiuretic hormone (vasopressin)
ADMA: asymmetric dimethylarginine
ADP: adenosine diphosphate
AE: anion exchanger
AEA: N-arachidonoyl ethanolamine (anandamide)
AF: atrial fibrillation
AFAP: ArfGAP with phosphoinositide-binding and PH domains
aFGF: acidic fibroblast growth factor (FGF1)
AGAP: ArfGAP with GTPAse, ankyrin repeat, and PH domains
AGF: autocrine growth factor

AGFG: ArfGAP with FG repeats
Ago: Argonaute protein
AGS: activator of G-protein signaling
AHR: aryl hydrocarbon receptor
AIF: apoptosis-inducing factor
AIP: actin-interacting protein
AIRe: autoimmune regulator
AKAP: A-kinase (PKA)-anchoring protein
ALE: arbitrary Eulerian Lagrangian
ALIX: apoptosis-linked gene-2-interacting protein-X
ALK: anaplastic lymphoma kinase
ALK*i*: type-*i* activin receptor-like kinase (TGFβ receptor superfamily)
ALOx5: arachidonate 5-lipoxygenase
ALOx5AP: arachidonate 5-lipoxygenase activation protein
ALP: actinin-associated LIM protein (PDLIM3)
alsin: amyotrophic lateral sclerosis protein
ALX: adaptor in lymphocytes of unknown function X
AMAP: A multidomain ArfGAP protein
AMBRA: activating molecule in beclin-1-regulated autophagy protein
AMHR: anti-Müllerian hormone receptor (TGFβ receptor superfamily)
AMIS: apical membrane initiation site (lumenogenesis)
AMPAR: α-amino 3-hydroxy 5-methyl 4-isoxazole propionic acid receptor
AMPK: AMP-activated protein kinase
AMSH: associated molecule with SH3 domain (deubiquitinase)
AmyR: amylin receptor
Ang: angiopoietin
AngL: angiopoietin-like molecule
Ank: ankyrin
ANP: atrial natriuretic peptide
ANPR (NP_1): atrial natriuretic peptide receptor (guanylyl cyclase)
ARNT: aryl hydrocarbon nuclear receptor translocator
ANS: autonomic nervous system
ANT: adenine nucleotide transporter
Anx: annexin
AOC: amine oxidase copper-containing protein
AoV: aortic valve
AP: (clathrin-associated) adaptor proteic complex
AP: Activator protein (transcription factor)
AP: activating enhancer-binding protein
AP4A: diadenosine tetraphosphate
APAF: apoptotic peptidase-activating factor
APAP: ArfGAP with PIx- and paxillin-binding domains
APC: antigen-presenting cell
APC: adenomatous polyposis coli protein (Ub ligase)
APC/C: anaphase-promoting complex (or cyclosome; Ub ligase)
APH: anterior pharynx defective phenotype homolog
aPKC: atypical protein kinase C
APl: action potential
Apn: adiponectin
Apo: apolipoprotein
ApoER: apolipoprotein-E receptor
APPL: adaptor containing phospho-Tyr interaction, PH domain, and Leu zipper
APS: adaptor with a PH and SH2 domain
Aqp: aquaporin
AR: adrenergic receptor (adrenoceptor)
AR: androgen receptor (nuclear receptor NR3c4; transcription factor)
AR: area ratio
ARAP: ArfGAP with RhoGAP, ankyrin repeat, PH domains
Areg: amphiregulin (EGF superfamily member)
ARF: ADP-ribosylation factor
ArfRP: ARF-related protein
ARFTS: CKI2A-locus alternate reading frame tumor suppressor (ARF or $p14^{ARF}$)
ARH: autosomal recessive hypercholesterolemia adaptor (low-density lipoprotein receptor adaptor)
ARH: aplysia Ras-related homolog
ArhGEF: RhoGEF

ARL: ADP-ribosylation factor-like protein
ARNO: Arf nucleotide site opener
ARP: absolute refractory period
ARP: actin-related protein
ARPP: cAMP-regulated phosphoprotein
ART: arrestin-related transport adaptor (α-arrestin)
ART: adpribosyltransferase
Artn: artemin
ARVCF: armadillo repeat gene deleted in velocardiofacial syndrome
ARVD: arrythmogenic right ventricular dystrophy
AS: Akt (PKB) substrate
ASAP: artery-specific antigenic protein
ASAP: ArfGAP with SH3, ankyrin repeat, PH domains
ASIC: acid-sensing ion channel
ASK: apoptosis signal-regulating kinase
aSMC: airway smooth muscle cell
ASP: actin-severing protein
AT: antithrombin
ATAA: ascending thoracic aortic aneurysm
ATF: activating transcription factor
AtG: autophagy-related gene product
ATMK: ataxia telangiectasia mutated kinase
ATn: angiotensin
ATng: angiotensinogen
AtOx: anti-oxidant protein (metallochaperone)
ATP: adenosine triphosphate
ATPase: adenosine triphosphatase
ATR ($AT_{1/2}$): angiotensin receptor
ATRK: ataxia telangiectasia and Rad3-related kinase
AVN: atrioventricular node
AVV: atrioventricular valves
AW: analysis window

B

$\mathbf{B}$: Biot–Finger strain tensor
B: bulk modulus
$\mathcal{B}$: bilinear form
b: minor semi-axis
$\mathbf{b}$: body force
$\hat{\mathbf{b}}$: unit binormal
BACE: β-amyloid precursor protein-converting enzyme
BAD: BCL2 antagonist of cell death
BAF: barrier-to-autointegration factor
BAG: BCL2-associated athanogene (chaperone regulator)
BAI: brain-specific angiogenesis inhibitor (adhesion-GPCR)
BAIAP: brain-specific angiogenesis inhibitor-1-associated protein (insulin receptor substrate)
BAK: BCL2-antagonist–killer
(i)BALT: (inducible) bronchus-associated lymphoid tissue
BAnk: B-cell scaffold with ankyrin repeats
Barkor: beclin-1-associated autophagy-related key regulator
BAT: brown adipose tissue
BATF: basic leucine zipper ATF-like transcription factor (B-cell-activating transcription factor)
BAX: BCL2-associated X protein
BBB: blood–brain barrier
BC: boundary condition
bCAM: basal cell adhesion molecule (Lutheran blood group glycoprotein)
BCAP: B-cell adaptor for phosphatidylinositol 3-kinase
BCAR: Breast cancer anti-estrogen resistance docking protein
BCL: B-cell lymphoma (leukemia) protein
BCLxL: B-cell lymphoma extra-large protein
BCR: B-cell receptor
BCR: breakpoint cluster region protein
Bdk: bradykinin
BDNF: brain-derived neurotrophic factor
Becn, beclin: BCL2-interacting protein
BEM: boundary element method
Best: bestrophin
bFGF: basic fibroblast growth factor (FGF2)
BFUe: burst-forming unit erythroid

BFUmeg: burst-forming unit megakaryocyte
BGT: betaine–GABA transporter
BID: BH3-interacting domain death agonist
BIG: brefeldin-A-inhibited GEFs for ARFs
BIK: BCL2-interacting killer
BIM: BH3-containing protein BCL2-like 11 (BCL2L11)
BK: high-conductance, Ca^{++}-activated, voltage-gated K^+ channel
BLK: B-lymphoid tyrosine kinase
Blm: Bloom syndrome, RecQ DNA helicase-like protein
BLnk: B-cell linker protein
BM: basement membrane
BMAL: brain and muscle ARNT-like protein (gene Bmal)
BMAT: bone-marrow adipose tissue
BMF: BCL2 modifying factor
BMP: bone morphogenetic protein (TGFβ superfamily)
BMPR: bone morphogenetic protein receptor
BNIP: BCL2/adenovirus E1B 19-kDa protein-interacting protein
BNP: B-type natriuretic peptide
BMX: bone marrow Tyr kinase gene in chromosome-X product
BOC: brother of CDO
BOK: BCL2-related ovarian killer
BORG: binder of Rho GTPase
BRAG: brefeldin-resistant ArfGEF
BrCa: breast cancer-associated (susceptibility) protein (tumor suppressor; DNA-damage repair; a.k.a. FancD1)
BrD: bromodomain-containing protein
BrK: breast tumor kinase
BrSK: brain-selective kinase
BSEP: bile salt export pump
BTF: basic transcription factor
BTK: Bruton Tyr kinase
BUB: budding uninhibited by benzimidazoles

C

$\mathbf{C}$: stress tensor
$\mathtt{C}$: compliance
C: heat capacity
C: chronotropy
C$\mathbf{x}$: type-$\mathtt{x}$ chemokine C (γ)
C_D: drag coefficient
C_f: friction coefficient
C_L: lift coefficient
C_p: pressure coefficient
$\mathbf{c}$: stress vector
c_τ: shear
c_w: wall shear stress
c: concentration
$c(p)$: wave speed
c_p: isobar heat capacity
c_v: isochor heat capacity
C1P: ceramide 1-phosphate
C-terminus: carboxy (carboxyl group COOH)-terminus
C/EBP: CCAAT/enhancer-binding protein
CA: computed angiography
CAi: carbonic anhydrase isoform i
Ca: calcium
Ca_V: voltage-gated Ca^{++} channel
Ca_V1.$\mathbf{x}$: L-type high-voltage-gated Ca^{++} channel
Ca_V2.$\mathbf{x}$: P/Q/R-type Ca^{++} channel
Ca_V3.$\mathbf{x}$: T-type low-voltage-gated Ca^{++} channel
CAAT: cationic amino acid transporter
CABG: coronary artery bypass grafting
Cables: CDK5 and Abl enzyme substrate
CAK: CDK-activating kinase (pseudokinase)
Cam: calmodulin (calcium-modulated protein)
CamK: calmodulin-dependent kinase
cAMP: cyclic adenosine monophosphate
CAP: adenylyl cyclase-associated protein
CAP: carboxyalkylpyrrole protein adduct
CAP: chromosome-associated protein (BrD4)
CAPN: calpain gene
CaPON: carboxy-terminal PDZ ligand of NOS1

CAR: constitutive androstane receptor (NR1i3)
CaR: calcium-sensing receptor
CARP: cell division cycle and apoptosis regulatory protein
CAS: cellular apoptosis susceptibility protein
CAS: CRK-associated substrate (or P130CAS and BCAR1)
CAs: cadherin-associated protein
CASK: calcium–calmodulin-dependent serine kinase (pseudokinase)
CASL: CRK-associated substrate-related protein (CAS2)
CASP: cytohesin-associated scaffold protein
caspase: cysteine-dependent aspartate-specific peptidase
Cav: caveolin
CBF: coronary blood flow
CBF: core-binding factor
CBL: Casitas B-lineage lymphoma adaptor and Ub ligase
CBLb: CBL-related adaptor
CBP: cap-binding protein
CBP: CREB-binding protein
CBP: C-terminal Src kinase-binding protein
CBS: cystathionine β-synthase (H_2S production)
CCDC: coiled-coil domain-containing protein
CCICR: calcium channel-induced Ca^{++} release
CCK4: colon carcinoma kinase 4 (PTK7)
CCL: chemokine CC-motif ligand
CCN: CyR61, CTGF, and NOv (CCN1–CCN3) family
Ccn: cyclin
Ccnx–CDK*i*: type-x cyclin–type-*i* cyclin-dependent kinase dimer
CCPg: cell cycle progression protein
CCS: copper chaperone for superoxide dismutase
CCT: chaperonin containing T-complex protein
CCx: type-x chemokine CC (β)
CCR: chemokine CC motif receptor
CD: cluster determinant protein (cluster of differentiation)
CDase: ceramidase
CDC: cell division cycle protein
cDC: classical dendritic cell
CDH: CDC20 homolog
Cdh: cadherin
CDK: cyclin-dependent kinase
Cdm: caldesmon
CDO: cell adhesion molecule-related/downregulated by oncogenes
CE (CsE): cholesteryl esters
CEC: circulating endothelial cell
CELSR: cadherin, EGF-like, LAG-like, and seven-pass receptor
CenP: centromere protein
CEP: carboxyethylpyrrole
CeP: centrosomal protein
CEPC: circulating endothelial progenitor cell
Cer: ceramide
CerK: ceramide kinase
CerT: ceramide transfer protein
CETP: cholesterol ester transfer protein
CFD: computational fluid dynamics
CFLAR: caspase-8 and FADD-like apoptosis regulator
CFTR: cystic fibrosis transmembrane conductance regulator
CFU: colony-forming unit
CFUb: CFU basophil (basophil-committed stem cells)
CFUc: CFU in culture (granulocyte precursors, i.e., CFUgm)
CFUe: CFU erythroid
CFUeo: CFU eosinophil
CFUg: CFU granulocyte
CFUgm: CFU granulocyte–macrophage
CFUgemm: CFU granulocyte–erythroid–macrophage–megakaryocyte
CFUm: CFU macrophage
CFUmeg: CFU megakaryocyte
CFUs: colony-forming unit spleen (pluripotent stem cells)
CG: chromogranin
cGK: cGMP-dependent protein kinase (protein kinase G)
cGMP: cyclic guanosine monophosphate

CGN: cis-Golgi network
CGRP: calcitonin gene-related peptide
chanzyme: ion channel and enzyme
chemokine: chemoattractant cytokine
CHIP: C-terminus heat shock cognate-70-interacting protein
ChK: checkpoint kinase
CHK: CSK homologous kinase
CHOP: CCAAT/enhancer-binding protein homologous protein
CHREBP: carbohydrate-responsive element-binding protein
ChT: choline transporter
CI: cardiac index
CICR: calcium-induced calcium release
Cin: chronophin
CIP: CDC42-interacting protein
CIP2a: cancerous inhibitor of protein phosphatase-2A
CIPC: CLOCK-interacting protein, circadian
CIS: cytokine-inducible SH2-containing protein
CK: creatine kinase
CK: casein kinase
CKI: cyclin-dependent kinase inhibitor
CLAsP: CLiP-associated protein (microtubule binder)
ClASP: clathrin-associated sorting protein
CLC: cardiotrophin-like cytokine
ClC: voltage-gated chloride channel
ClCa: calcium-activated chloride channel
ClIC: chloride intracellular channel
CLINT: clathrin-interacting protein located in the trans-Golgi network
CLIP: corticotropin-like intermediate peptide
CLiP: cytoplasmic CAP-Gly domain-containing linker protein
iCliP: intramembrane-cleaving peptidase (that clips)
CLK: CDC-like kinase
ClNS: Cl^- channel nucleotide-sensitive
CLOCK: circadian locomotor output cycles kaput
CLP: common lymphoid progenitor
CLS: ciliary localization signal
Cmi: chylomicron
CMLP: common myeloid–lymphoid progenitor
CMP: common myeloid progenitor
CMC: cardiomyocyte
Col: collagen
CoLec: collectin
ColF: collagen fiber
CNG: cyclic nucleotide-gated channel
CNS: central nervous system
CNT: connecting tubule
CNTi: concentrative nucleoside transporter (SLC28ai)
CNTF: ciliary neurotrophic factor
CntnAP: contactin-associated protein
CO: cardiac output
CoBl: Cordon-bleu homolog (actin nucleator)
COLD: chronic obstructive lung disease
COOL: Cloned out of library (RhoGEF6/7)
coSMAD: common-mediator SMAD (SMAD4)
COx: cyclooxygenase
COx17: cytochrome-C oxidase copper chaperone
CoP: coat protein
CoP: constitutive photomorphogenic protein (Ub ligase)
COPD: chronic obstructive pulmonary disease
COUPTF: chicken ovalbumin upstream promoter transcription factor (NR2f1/2)
CP4H: collagen prolyl 4-hydroxylase
CPC: chromosomal passenger complex
CpG: cytidine-phosphate–guanosine oligodeoxynucleotide (motif)
cPKC: conventional protein kinase C
Cpx: complexin
CR: complement component receptor
Cr: creatine
cRABP: cellular retinoic acid-binding protein
CRAC: Ca^{++} release-activated Ca^{++} channel
CRACR: CRAC regulator
Crb: Crumbs homolog polarity complex
CRE: cAMP-responsive element

CREB: cAMP-responsive element-binding protein
CRF: corticotropin-releasing factor (family)
CRH: corticotropin-releasing hormone
CRIB: CDC42/Rac interactive-binding protein
CRIK: citron Rho-interacting, Ser/Thr kinase (STK21)
CRK: CT10 regulator of kinase
CRK: chicken tumor virus regulator of kinase
CRKL: V-CRK avian sarcoma virus CT10 homolog-like
CRL4: cullin-4A RING E3 ubiquitin ligase
CRLR: calcitonin receptor-like receptor
CRP: C-reactive protein
CRTC: CREB-regulated transcription coactivator
Cry: cryptochrome
Cs: cholesterol
CSBP: cytokine-suppressive anti-inflammatory drug-binding protein
CSE: cystathionine γ-lyase (H_2S production)
CSF: cerebrospinal fluid
CSF: colony-stimulating factor
CSF1: macrophage colony-stimulating factor (mCSF)
CSF2: granulocyte–macrophage colony-stimulating factors (gmCSF and sargramostim)
CSF3: granulocyte colony-stimulating factors (gCSF and filgrastim)
CSK: C-terminal Src kinase
Csk: cytoskeleton
Csq: calsequestrin
CSS: candidate sphingomyelin synthase
CT: cardiotrophin
CT: computed tomography
CTBP: C-terminal-binding protein
CTen: C-terminal tensin-like protein
CTF: C-terminal fragment
CTGF: connective tissue growth factor
CTL: cytotoxic T lymphocyte
CTLA: cytotoxic T-lymphocyte-associated protein
Ctn: catenin
CTr: copper transporter
CtR: calcitonin receptor
CTRC: CREB-regulated transcription coactivator
Cul: cullin
CUT: cryptic unstable transcript
CVI: chronic venous insufficiency
CVLM: caudal ventrolateral medulla
CVP: central venous pressure
CVS: cardiovascular system
Cx: connexin
CXCLi: type-i CXC (C-X-C motif; α) chemokine ligand
CXCRi: type-i CXC (C-X-C motif; α) chemokine receptor
CX3CLi: type-i CX3C (δ) chemokine ligand
CX3CRi: type-i CX3C (δ) chemokine receptor
cyCK: cytosolic creatine kinase
Cyld: cylindromatosis tumor suppressor protein (deubiquitinase USPL2)
CyP: member of the cytochrome-P450 superfamily
C3G: Crk SH3-binding GEF

D

D: dromotropy
D: vessel distensibility
$\mathcal{D}$: diffusion coefficient
$\mathcal{D}_T$: thermal diffusivity
D: deformation rate tensor
d: displacement vector
D: flexural rigidity
D: demobilization function (from proliferation to quiescence)
d: death rate, degradation rate
d: duration
Dab: Disabled homolog
DAD: delayed afterdepolarization
DAG: diacylglycerol
DAPC: dystrophin-associated protein complex
DAPK: death-associated protein kinase
DARC: Duffy antigen receptor for chemokine
DAT: dopamine active transporter

DAX: dosage-sensitive sex reversal, adrenal hypoplasia critical region on chromosome X (NR0b1)
DBC: deleted in breast cancer protein
DBP: albumin D-element binding protein (PAR/b–ZIP family)
DC: dendritic cell
DCA: directional coronary atherectomy
DCAF: DDB1- and Cul4-associated factor
DCC: deleted in colorectal carcinoma (netrin receptor)
DCT: distal convoluted tubule
Dctn: dynactin
DDAH: dimethylarginine dimethyl-aminohydrolase
DDB: damage-specific DNA-binding protein
DDEF: development and differentiation-enhancing factor (ArfGAP)
DDR: discoidin domain receptor
De: Dean number
DEC: differentially expressed in chondrocytes (DEC1 and DEC2 are a.k.a bHLHe40 and bHLHe41, bHLHb2 and bHLHb3, or HRT2 and HRT1)
DEC: deleted in esophageal cancer
DEG: delayed-early gene
DETC: dendritic epidermal $\gamma\delta$ T cell
DH: Dbl homology
DHET: dihydroxyeicosatrienoic acid
DHh: desert Hedgehog
Dia: Diaphanous
DICOM: digital imaging and communication for medicine
DICR: depolarization-induced Ca^{++} release
DISC: death-inducing signaling complex
Dkk: Dickkopf
DLg: Disc large homolog
DLL: Delta-like (Notch) ligand
DLx: distal-less homeobox protein
DM: double minute
DMM: DNA methylation modulator
DMPK: myotonic dystrophy-associated protein kinase
DMT: divalent metal transporter
DN1: double-negative-1 cell
DN2: double-negative-2 cell
DN3: double-negative-3 cell
DNA: deoxyribonucleic acid
DNAPK: DNA-dependent protein kinase
DoC2: double C2-like domain-containing protein
DOCK: dedicator of cytokinesis (GEF)
DOK: downstream of Tyr kinase docking protein
DOR: δ-opioid receptor
DPG: diphosphoglyceric acid
DRAM: damage-regulated modulator of autophagy
DRF: Diaphanous-related formin (for GTPase-triggered actin rearrangement)
Drl: Derailed
Dsc: desmocollin
Dsg: desmoglein
Dsh: Disheveled (Wnt-signaling mediator)
DSK: dual-specificity kinase
dsRNA: double-stranded RNA
Dst: dystonin
DUb: deubiquitinase
DUS: Doppler ultrasound
DUSP: dual-specificity phosphatase
DV: dead space volume
Dvl: Disheveled (cytoplasmic phosphoprotein; other alias Dsh)
DVT: deep-vein thrombosis
dynactin: dynein activator
DYRK: dual-specificity Tyr (Y) phosphorylation-regulated kinase

E

$\mathbf{E}$: strain tensor
$\mathbf{E}$: electric field
E: elastance modulus
E: elastance
$\mathcal{E}$: energy
$\{\hat{\mathbf{e}}_i\}_{i=1}^3$: basis
$\mathbf{e}$: strain vector
e: specific free energy
E-box: enhancer box sequence of DNA
E2: ubiquitin-conjugase
E3: ubiquitin-ligase

EAAT: excitatory amino acid (glutamatee–aspartate) transporter

EAD: early afterdepolarization

EAR: V-erbA-related nuclear receptor (NR2f6)

EB: end-binding protein

EBCT: electron beam CT

EBF: early B-cell factor

EC: endothelial cell

ECA: external carotid artery

ECF: extracellular fluid

ECG: electrocardiogram

ECM: extracellular matrix

ED1L: EGF-like repeat- and discoidin-1-like domain-containing protein

EDGR: endothelial differentiation gene receptor

EDHF: endothelial-derived hyperpolarizing factor

EDIL: EGF-like repeats and discoidin-1 (I)-like domain-containing protein

EDV: end-diastolic volume

EEA: early endosomal antigen

eEF: eukaryotic translation elongation factor

EEL: external elastic lamina

EET: epoxyeicosatrienoic acid

EFA6: exchange factor for ARF6 (ArfGEF)

EF-Tu: elongation factor Tu

EGF: epidermal growth factor

EGFL: EGF-like domain-containing protein

EGFR: epidermal growth factor receptor

EGR: early growth response transcription factor

EHD: C-terminal EGFR substrate-15 homology domain-containing protein

eIF: eukaryotic translation initiation factor

EL: endothelial lipase

ELAM: endothelial–leukocyte adhesion molecules

ELCA: excimer laser coronary angioplasty

ELK: ETS-like transcription factor (ternary complex factor [TCF] subfamily)

ElMo: engulfment and cell motility adaptor

Eln: elastin

ElnF: elastin fiber

ELP: early lymphoid progenitor

EMI: early mitotic inhibitor

EMR: EGF-like module containing, mucin-like, hormone receptor-like protein

EMT: epithelial–mesenchymal transition

ENA–VASP: Enabled homolog and vasoactive (vasodilator)-stimulated phosphoprotein family

ENaC: epithelial Na^+ channel

EnaH: Enabled homolog

endo-siRNA: endogenous small interfering RNA

ENPP: ectonucleotide pyrophosphatase–phosphodiesterase

Ens: endosulfine

ENT: equilibrative nucleoside transporter

ENTPD: ectonucleoside triphosphate diphosphohydrolase

EPAC: exchange protein activated by cAMP

EPAS: endothelial PAS domain protein

EPC: endothelial progenitor cell

EPCR: endothelial protein-C receptor

EPDC: epicardial-derived cell

Epgn: epigen (EGF superfamily member)

EPH: erythropoietin-producing hepatocyte receptor kinase or pseudokinase (EPHa10 and EPHb6)

ephrin: EPH receptor interactor

Epo: erythropoietin

EPS: epidermal growth factor receptor pathway substrate

ER: endoplasmic reticulum

ERx: type-x estrogen receptor (NR3a1/2)

eRas: embryonic stem cell-expressed Ras (or hRas2)
ErbB: erythroblastoma viral gene product B (HER)
ERE: estrogen response element (DNA sequence)
Ereg: epiregulin (EGF superfamily member)
eRF: eukaryotic release factor
ERGIC: endoplasmic reticulum–Golgi intermediate compartment
ERK: extracellular signal-regulated protein kinase
ERK1/2: usually refers to ERK1 and ERK2
ERM: ezrin–radixin–moesin
ERMES: endoplasmic reticulum–mitochondrion encounter structure
ERP: effective refractory period
ERR: estrogen-related receptor (NR3b1–NR3b3)
ESCRT: endosomal sorting complex required for transport
ESL: E-selectin ligand
ESRP: epithelial splicing regulatory protein
ESV: end-systolic volume
ET: endothelin
ETP: early thymocyte progenitor
ETR ($ET_{A/B}$): endothelin receptor
ETS: E-twenty six (transcription factor; erythroblastosis virus E26 proto-oncogene product homolog)
ETV: ETS-related translocation variant
EVAR: endovascular aneurysm repair
Exo: exocyst subunit
Ext: exostosin (glycosyltransferase)

F

$\mathbf{F}$: transformation gradient tensor
F: function fraction of proliferating cells
F: erythrocytic rouleau fragmentation rate
$\mathbf{f}$: surface force
$\hat{\mathbf{f}}$: fiber direction unit vector
f: binding frequency
f_{C}: cardiac frequency
f_{R}: breathing frequency
f: friction shape factor
f_{v}: head loss per unit length
f_{i}: molar fraction of gas component $\mathtt{i}$
FA: fatty acid
FABP: fatty acid-binding protein
FABP: filamentous actin-binding protein
FACAP: F-actin complex-associated protein
FAD: flavine adenine dinucleotide
FADD: Fas receptor-associated death domain
FAK: focal adhesion kinase
Fanc: Fanconi anemia protein
FAN: Fanconi anemia-associated nuclease
FAPP: phosphatidylinositol four-phosphate adaptor protein
Fas: death receptor (TNFRSF6a)
FasL: death ligand (TNFSF6)
FAST: Forkhead activin signal transdurcer
FB: fibroblast
Fbln (Fibl): fibulin
Fbn: fibrillin
FBS: F-box, Sec7 protein (ArfGEF)
FBX: F-box only protein (ArfGEF)
FC: fibrocyte
FCHO: FCH domain only protein
FcαR: Fc receptor for IgA
FcγR: Fc receptor for IgG
FcεR: Fc receptor for IgE
FDM: finite difference method
FEM: finite element method
FERM: four point-1, ezrin–radixin–moesin domain
Fer: Fes-related Tyr kinase
Fes: feline sarcoma kinase
FFA: free fatty acid
FGF: fibroblast growth factor
FGFR: fibroblast growth factor receptor
FGR: viral feline Gardner-Rasheed sarcoma oncogene homolog kinase
FHL: four-and-a-half LIM-only protein
FHoD: formin homology domain-containing protein (FmnL)
FIH: factor inhibiting HIF1α (asparaginyl hydroxylase)

FIP: focal adhesion kinase family-interacting protein
FIT: Fat-inducing transcript
FKBP: FK506-binding protein
FlIP: flice-inhibitory protein
FLK: fetal liver kinase
fMLP: N-formyl methionyl-leucyl-phenylalanine
FN: fibronectin
Fn: fibrin
Fng: fibrinogen
Fos: Finkel Biskis Jinkins murine osteosarcoma virus sarcoma proto-oncogene product
Fox: forkhead box transcription factor
Fpn: ferroportin
FR: flow ratio
FRK: Fyn-related kinase
FrmD: FERM domain-containing adaptor
FRNK: FAK-related non-kinase
FSH: follicle-stimulating hormone
FSI: fluid–structure interaction
FVM: finite volume method
FXR: farnesoid X receptor (NR1h4)
Fz: Frizzled (Wnt GPCR)

G

G: Green-Lagrange strain tensor
G: shear modulus
G': storage modulus
G'': loss modulus
$\mathcal{G}$: Gibbs function
G: conductance
G_{p}: pressure gradient
G_{e}: electrical conductivity
G_{h}: hydraulic conductivity
G_{T}: thermal conductivity
g: gravity acceleration
g: physical quantity
g: detachment frequency
g: free enthalpy
G protein: guanine nucleotide-binding protein (G$\alpha\beta\gamma$ trimer)
Gα: α subunit (signaling mediator) of G protein
Gα_i (Gi): inhibitory Gα subunit
Gα_s (Gs): stimulatory Gα subunit
Gα_t (Gt): transducin, Gα subunit of rhodopsin
Gs$_{XL}$: extra-large Gs protein
G$\alpha_{i/o}$ (Gi/o): Gα subunit class
G$\alpha_{q/11}$ (Gq/11): Gα subunit class
G$\alpha_{12/13}$ (G12/13): Gα subunit class
G$\beta\gamma$: dimeric subunit (signaling effector) of G protein
G$_{gust}$: gustducin, G protein α subunit (Gi/o) of taste receptor
G$_{olf}$: G protein α subunit (Gs) of olfactory receptor
GAB: GRB2-associated binder
GABA: γ-aminobutyric acid
GABA$_A$: GABA ionotropic receptor (Cl^- channel)
GABA$_B$: GABA metabotropic receptor (GPCR)
GABARAP: GABA$_A$ receptor-associated protein
GaBP: globular actin-binding protein
GADD: growth arrest and DNA-damage-induced protein
gadkin: γ1-adaptin and kinesin interactor
GAG: glycosaminoglycan
GAK: cyclin G-associated kinase
Gal: galanin
GAP: GTPase-activating protein
GAPDH: glyceraldehyde 3-phosphate dehydrogenase
GAS: growth arrest-specific non-coding, single-stranded RNA
GAT: γ-aminobutyric acid transporter
GATA: DNA sequence GATA-binding protein (TF)
GBF: Golgi-associated brefeldin-A-resistant guanine nucleotide-exchange factor
GCAP: guanylyl cyclase-activating protein
GCC: Golgi coiled-coil domain-containing protein
GCK: germinal center kinase
GCKR: GCK-related kinase
GCNF: germ cell nuclear factor (NR6a1)
GCN2: general control non-derepressible 2 (pseudokinase)

gCSF: granulocyte colony-stimulating factor (G-CSF; CSF3)
GD: disialoganglioside
GDP: guanosine diphosphate
GDF: growth differentiation factor
GDF: (Rab)GDI displacement (dissociation) factor
GDI: guanine nucleotide-dissociation inhibitor
GDNF: glial cell line-derived neurotrophic factor
GEF: guanine nucleotide (GDP-to-GTP)-exchange factor
GF: growth factor
GFAP: glial fibrillary acidic protein (intermediate filament)
GFL: GDNF family of ligands
GFP: geodesic front propagation
GGA: Golgi-localized γ-adaptin ear-containing Arf-binding protein
Ggust: (G protein) Gα subunit gustducin
GH: growth hormone
GHR: growth hormone receptor
GHRH: growth hormone-releasing hormone
GIP: GPCR-interacting protein
GIRK: G$\beta\gamma$-regulated inwardly rectifying K^+ channel
GIT: GPCR kinase-interacting protein
GKAP: guanylyl kinase-associated protein
GLK: GCK-like kinase
GluK: ionotropic glutamate receptor (kainate)
GluN: ionotropic glutamate receptor (NMDA)
GluR: ionotropic glutamate receptor (AMPA)
GluT: glucose transporter
GlyCAM: glycosylation-dependent cell adhesion molecule
GlyR: glycine receptor (channel)
GlyT: glycine transporter
GM: monosialoganglioside
gmCSF: granulocyte–monocyte colony-stimulating factor (GM-CSF; CSF2)
GMP: granulocyte–monocyte progenitor
GMP: guanosine monophosphate
GnRH: gonadotropin-releasing hormone
GP: glycoprotein
Gpc: glypican
GPI: glycosyl-phosphatidylinositol anchor
gpiAP: GPI-anchored protein
GPCR: G-protein-coupled receptor
GPx: glutathione peroxidase
GQ: quadrisialoganglioside
GR: glucocorticoid receptor (NR3c1)
GRAP: GRB2-related adaptor protein (or GAds)
GRB: growth factor receptor-bound protein
GRE: glucocorticoid response element (DNA sequence)
GRK: G-protein-coupled receptor kinase
GRP: G-protein-coupled receptor phosphatase
GSK: glycogen synthase kinase
GT: trisialoganglioside
GTF: general transcription factor
GTP: guanosine triphosphate
GTPase: guanosine triphosphatase
GuCy: guanylyl cyclase (CyG)
GWAS: genome-wide association study

H

H: height
$\mathcal{H}$: history function
$\mathtt{H}$: dissipation
$\mathtt{h}$: head loss
h: thickness
h: specific enthalpy
h_{T}: heat transfer coefficient
h_{m}: mass transfer coefficient
HA: hyaluronic acid
HAD: haloacid dehalogenase
HAP: huntingtin-associated protein
HAT: histone acetyltransferase
HAAT: heterodimeric amino acid transporter
HAND: heart and neural crest derivatives expressed protein
Hb: hemoglobin
Hb^{SNO}: ^{S}nitrosohemoglobin

HBEGF: heparin-binding EGF-like growth factor
HCK: hematopoietic cell kinase
HCLS: hematopoietic lineage cell-specific Lyn substrate protein
HCN: hyperpolarization-activated, cyclic nucleotide-gated K^+ channel
HCNP: hippocampal cholinergic neurostimulatory peptide
HCT: helical CT
HDAC: histone deacetylase complex
HDL: high-density lipoprotein
HDL–C: HDL–cholesterol
HDL–CE: HDL–cholesteryl ester
HDM: human double minute (E3 ubiquitin ligase)
HEET: hydroxyepoxyeicosatrienoic
hemin: heme oxygenase-1 inducer
HERG: human ether-a-go-go related gene
HER: human epidermal growth factor receptor (HER3: pseudokinase)
HES: Hairy enhancer of split
HETE: hydroxyeicosatrienoic acid
HEV: high endothelial venule
HGF: hepatocyte growth factor
HGFR: hepatocyte growth factor receptor
HGS: HGF-regulated Tyr kinase substrate (HRS)
HhIP: Hedgehog-interacting protein
HIF: hypoxia-inducible factor
HIP: huntingtin-interacting protein
HIP1R: HIP1-related protein
His: histamine
Hjv: hemojuvelin
HK: hexokinase
HL: hepatic lipase
HMG: high mobility group protein
HMGB: high mobility group box protein
HMGCoAR: hydroxy methylglutaryl coenzyme-A reductase
HMT: histone methyl transferase
HMWK: high-molecular weight kininogen
HNF: hepatocyte nuclear factor (NR2a1/2)
HNP: human neutrophil peptide
hnRNP: heterogeneous nuclear ribonucleoprotein
HODE: hydroxy octadecadienoic acid
HOP: HSP70–HSP90 complex-organizing protein
HotAIR: HOX antisense intergenic RNA (large intergenic non-coding RNA)
HOx: heme oxygenase
Hox: homeobox DNA sequence (encodes homeodomain-containing morphogens)
HPK: hematopoietic progenitor kinase (MAP4K)
hpRNA: long hairpin RNA
hRas: Harvey Ras
HRE: hormone response element (DNA sequence)
HRM: hypoxia-regulated microRNA
hRNP: heterogeneous ribonucleoprotein
HRS: hepatocyte growth factor-regulated Tyr kinase substrate
HRT: Hairy and enhancer of Split-related transcription factor
HS: heparan sulfate
HSC: hematopoietic stem cell
HSC: heat shock cognate
HSER: heat stable enterotoxin receptor (guanylyl cyclase 2C)
HSP: heat shock protein (chaperone)
HSPG: heparan sulfate proteoglycan
Ht: hematocrit
HTR: high temperature requirement endopeptidase

I

I: identity tensor
i: current
I: inotropy
IAP: inhibitor of apoptosis protein
IBABP: intestinal bile acid-binding protein
IC: isovolumetric contraction
ICA: internal carotid artery
ICAM: intercellular adhesion molecule (IgCAM member)
IgCAM: immunoglobulin-like cell adhesion molecule

ICF: intracellular fluid
ICliP: intramembrane-cleaving peptidase
ID: inhibitor of DNA binding
IDL: intermediate-density lipoprotein
IDmiR: immediately downregulated microRNA
IDOL: inducible degrader of LDL receptor (E3-Ub ligase)
IEG: immediate-early gene
IEL: internal elastic lamina
IEL: intra-epithelial lymphocyte
IfIH: interferon-induced with helicase-C domain-containing protein
Ifn: interferon
IFT: intraflagellar transport complex
Ig: immunoglobulin
IGF: insulin-like growth factor
IGFBP: IGF-binding protein
IgHC: immunoglobulin heavy chain
IgLC: immunoglobulin light chain
iGluR: ionotropic glutamate receptor
IH: intimal hyperplasia
IHh: Indian Hedgehog
IK: intermediate-conductance Ca^{++}-activated K^+ channel
IκB: inhibitor of NFκB
IKK: IκB kinase
IL: interleukin
ILC: innate lymphoid cell
iLBP: intracellular lipid-binding protein
ILK: integrin-linked (pseudo)kinase
IMM: inner mitochondrial membrane
IMP: Impedes mitogenic signal propagation
INADl: inactivation no after-potential D protein
InCenP: inner centromere protein
InF: inverted formin
INPP: inositol polyphosphate 5-phosphatase
InsIG: insulin-induced gene product (ER anchor)
InsL: insulin-like peptide
InsR (IR): insulin receptor
InsRR: insulin receptor-related receptor
IP: inositol phosphate
IP_3: inositol (1,4,5)-triphosphate
IP_3R: IP_3 receptor (IP_3-sensitive Ca^{++}-release channel)
IP_4: inositol (1,3,4,5)-tetrakisphosphate
IP_5: inositol pentakisphosphate
IP_6: inositol hexakisphosphate
IPCEF: interaction protein for cytohesin exchange factor
IPOD: (perivacuolar) insoluble protein deposit
IPP: ILK–PINCH–parvin complex
iPSC: induced pluripotent stem cell
IQGAP: IQ motif-containing GTPase-activating protein (IQ: first 2 amino acids of the motif: isoleucine [I; commonly] and glutamine [Q; invariably]).
IR: isovolumetric relaxation
IRAK: IL1 receptor-associated kinase (IRAK2: pseudokinase)
IRE: irreversible electroporation
IRES: internal ribosome entry site
IRF: interferon-regulatory protein (transcription factor)
IRFF: interferon-regulatory factor family
IRP: iron regulatory protein
IRS: insulin receptor substrate
ISA: intracranial saccular aneurysm
ISG: interferon-stimulated gene product
iSMAD: inhibitory SMAD (SMAD6 or SMAD7)
ITAM: immunoreceptor tyrosine-based activation motif
Itch: Itchy homolog (ubiquitin ligase)
ITIM: immunoreceptor tyrosine-based inhibitory motif
ITK: interleukin-2-inducible T-cell kinase
ITPK: inositol trisphosphate kinase
IVC: inferior vena cava
IVP: initial value problem
IVUS: intravascular ultrasound

J

J: flux
J_m: cell surface current density
JAM: junctional adhesion molecule
JaK: Janus (pseudo)kinase

JIP: JNK-interacting protein (MAPK8IP1 and -2)
JMy: junction-mediating and regulatory protein
JNK: Jun N-terminal kinase (MAPK8–MAPK10)
JNKBP: JNK-binding protein;
JNKK: JNK kinase
JSAP: JNK/SAPK-associated protein
Jun: avian sarcoma virus-17 proto-oncogene product (Japanese juunana: 17; TF)
JUNQ: juxtanuclear quality-control compartment

K

K: conductivity tensor
K: bending stiffness
K: reflection coefficient
K_d: dissociation constant
K_M: Michaelis constant (chemical reaction kinetics)
K_m: material compressibility
k: cross-section ellipticity
k_{ATP}: myosin ATPasic rate
k_B: Boltzmann constant (1.38×10^{-23} J/K)
k_c: spring stiffness
k_m: mass-transfer coefficient
k_P: Planck constant
K_R: resistance coefficient
KaP: karyopherin
K_{ATP}: ATP-sensitive K^+ channel
$K_{Ca}1.x$: BK channel
$K_{Ca}2/3/4.x$: SK channel
$K_{Ca}5.x$: IK channel
K_{IR}: inwardly rectifying K^+ channel
K_V: voltage-gated K^+ channel
KAP: kinesin (KIF)-associated protein
Kap: karyopherin
KAT: lysine (K) acetyltransferase
KCC: K^+–Cl^- cotransporter
KChAP: K^+ channel-associated protein
KChIP: K_V channel-interacting protein
KDELR: KDEL (Lys–Asp–Glu–Leu) endoplasmic reticulum retention receptor
KDR: kinase insert domain receptor
KHC: kinesin heavy chain
KIF: kinesin family
KIR: killer cell immunoglobulin-like receptor
KIT: cellular kinase in tyrosine (SCFR)
Kk: kallikrein
KLC: kinesin light chain
KLF: Krüppel-like factor
KLR: killer cell lectin-like receptor
Kn: Knudsen number
KOR: κ-opioid receptor
kRas: Kirsten Ras
Krt: keratin
KSR: kinase suppressor of Ras (adaptor; pseudokinase)

L

L: velocity gradient tensor
L: inertance
L: length
LA: left atrium
LAB: linker of activated B lymphocyte
LAd: LcK-associated adaptor
LANP: long-acting natriuretic peptide
LAP: leucine-rich repeat and PDZ domain-containing protein (4-member family)
LAP: latency-associated peptide (4 isoforms LAP1–LAP4)
LAP: nuclear lamina-associated polypeptide
LAR: leukocyte common-antigen-related receptor (PTPRF)
LAT: linker of activated T lymphocytes
LaTS: large tumor suppressor
LAX: linker of activated X cells (both B and T cells)
LBR: lamin B receptor
LCA: left coronary artery
LCAT: lysolecithin cholesterol acyltransferase
LCC: left coronary cusp
LCK: leukocyte-specific cytosolic (non-receptor) Tyr kinase
LCP: lymphocyte cytosolic protein (adaptor SLP76)
LDL: low-density lipoprotein
LDLR: low-density lipoprotein receptor

LDV: laser Doppler velocimetry
Le: entry length
LEF: lymphoid enhancer-binding transcription factor
LGalS: lectin, galactoside-binding, soluble cell adhesion molecule
LGIC: ligand-gated ion channel
LGL: lethal giant larva protein
LH: luteinizing hormone
LIF: leukemia-inhibitory factor
LIFR: leukemia-inhibitory factor receptor
LIMA: LIM domain and actin-binding protein
LIME: LcK-interacting molecule
LIMK: Lin1, Isl1, and Mec3 kinase
LIMS: LIM and senescent cell antigen-like-containing domain protein
LiNC: linker of nucleoskeleton and cytoskeleton
lincRNA: large intergenic non-coding RNA
LipC: hepatic lipase
LipD: lipoprotein lipase
LipE: hormone-sensitive lipase
LipG: endothelial lipase
LipH: lipase-H
liprin: LAR PTP-interacting protein
LIR: leukocyte immunoglobulin-like receptor
LIS: lissencephaly protein
LKB: liver kinase-B
LKLF: lung Krüppel-like factor
LLTC: large latent TGFβ complex
LMan: lectin, mannose-binding
LMO: LIM domain-only-7 protein
Lmod: leiomodin (actin nucleator)
LMPP: lymphoid-primed multipotent progenitor
LMR: laser myocardial revascularization
Ln: laminin
LOx: lipoxygenase
LP: lipoprotein
LPA: lysophosphatidic acid
LPase: lipoprotein lipase
lpDC: lamina propria dendritic cell
Lphn: latrophilin (adhesion-GPCR)
LPL: lysophospholipid
LPLase: lysophospholipase
LPP: lipid phosphate phosphatase
LPR: lipid phosphatase-related protein
LPS: lipopolysaccharide
LQTS: long-QT syndrome
LRAT: lecithin–retinol acyltransferase
LRH: liver receptor homolog (NR5a2)
LRO: lysosome-related organelle
LRP: LDL receptor-related protein
LRRTM: leucine-rich repeat-containing transmembrane protein
LSK: Lin−, SCA1+, KIT+ cell
LST: lethal with Sec-thirteen
LSV: long saphenous vein
LT (Lkt): leukotriene
LTBP: latent TGFβ-binding protein
LTCC: L-type Ca^{++} channel (Ca_V1)
LTK: leukocyte tyrosine kinase
LUbAC: Linear ubiquitin chain assembly complex
LV: left ventricle
LVAD: left ventricular assist device
LX: lipoxin
LXR: liver X receptor (NR1h2/3)
LyVE: lymphatic vessel endothelial hyaluronan receptor

M

M: molar mass
$\mathcal{M}$: moment
m: mass
Ma: Mach number
MACF: microtubule-actin crosslinking factor
mAChR: acetylcholine muscarinic receptor (metabotropic; GPCR)
MAD: mothers against decapentaplegic homolog
MAD: mitotic arrest-deficient protein
MAdCAM: mucosal vascular addressin cell adhesion molecule
MAF: V-maf musculoaponeurotic fibrosarcoma oncogene homolog (TF)
MAGI: membrane-associated guanylate kinase-related protein with inverted domain organization

MAGP: microfibril-associated glycoprotein
MAGuK: membrane-associated guanylyl kinase
MAIT: mucosal-associated invariant T lymphocyte
MALT: mucosa-associated lymphoid tissue
MAO: monoamine oxidase
MAP: microtubule-associated protein
MAP1LC3: microtubule-associated protein-1 light chain-3 (LC3)
mAP: mean arterial pressure
MAPK: mitogen-activated protein kinase
MAP2K: mitogen-activated protein kinase kinase
MAP3K: MAP kinase kinase kinase
MAPKAPK: MAPK-activated protein kinase
MARCKS: myristoylated alanine-rich C kinase substrate
MaRCo: macrophage receptor with collagenous structure (ScaRa2)
MARK: microtubule affinity-regulating kinase
MASTL: microtubule-associated Ser/Thr kinase-like protein
MAT: ménage à trois
MATK: megakaryocyte-associated Tyr kinase
MAVS: mitochondrial antiviral signaling protein
MBP: myosin-binding protein
MBP: myeloid–B-cell progenitor
MBTPSi: membrane-bound transcription factor peptidase site i
MCAK: mitotic centromere-associated kinesin
MCAM: melanoma cell adhesion molecule
MCL1: BCL2-related myeloid cell leukemia sequence protein-1
MCLC: stretch-gated Mid1-related chloride channel
MCM: minichromosome maintenance protein
MCP: monocyte chemoattractant protein
mCSF: macrophage colony-stimulating factor (M-CSF; CSF1)
MCT: monocarboxylate–proton cotransporter
mDC: myeloid dendritic cell
MDM: mitochondrial distribution and morphology protein
MDR: multiple drug resistance (ABC transporter)
MEF: myocyte enhancer factor
megCSF: megakaryocyte colony-stimulating factor
MEJ: myoendothelial junction
MELK: maternal embryonic leucine zipper kinase
MEP: megakaryocyte erythroid progenitor
MEP: myeloid–erythroid progenitor
MET: mesenchymal–epithelial transition factor (proto-oncogene; HGFR)
METC: mitochondrial electron transport chain
MGP: matrix Gla protein
mGluR: metabotropic glutamate receptor
MHC: major histocompatibility complex
MHC: myosin heavy chain
MyHC or MYH: myosin heavy chain gene
miCK: mitochondrial creatine kinase
MinK: misshapen-like kinase
miR: microRNA
miRNP: microribonucleoprotein
MiRP: MinK-related peptide
MIRR: multichain immune-recognition receptor
MIS: Müllerian inhibiting substance
MIS: mini-invasive surgery
MIS: mitochondrial intermembrane space
MIST: mast cell immunoreceptor signal transducer
MIT: mini-invasive therapy
MiV: mitral valve
MIZ: Myc-interacting zinc finger protein

MJD: Machado-Joseph disease protein domain-containing peptidase (DUb)
MKL: megakaryoblastic leukemia-1 fusion coactivator
MKP: mitogen-activated protein kinase phosphatase
MLC: myosin light chain
MLCK: myosin light-chain kinase
MLCP: myosin light-chain phosphatase
MLK: mixed lineage kinase
MLKL: mixed lineage kinase-like pseudokinase
MLL: mixed lineage [myeloid–lymphoid] leukemia factor
MLLT: mixed lineage leukemia translocated protein
MLP: muscle LIM protein
mmCK: myofibrillar creatine kinase
MME: membrane metalloendopeptidase
MMM: maintenance of mitochondrial morphology protein
MMP: matrix metallopeptidase
MO: mouse protein
Mo: monocyte
MOMP: mitochondrial outer membrane permeabilization
MOR: μ-opioid receptor
MP: MAPK partner
MPF: mitosis (maturation)-promoting factor (CcnB–CDK1 complex)
MPG: N-methylpurine (N-methyladenine)-DNA glycosylase
MPO: median preoptic nucleus
Mpo: myeloperoxidase
MP_P: membrane protein, palmitoylated
MPP: multipotent progenitor
MR: mineralocorticoid receptor (NR3c2)
mRas: muscle Ras (or rRas3)
MRCK: myotonic dystrophy kinase-related CDC42-binding kinase
MRI: (nuclear) magnetic resonance imaging
mRNA: messenger RNA
mRNP: messenger ribonucleoprotein
MRTF: myocardin-related transcription factor
MSC: mesenchymal stem cell
MSH: melanocyte-stimulating hormone
MSSCT: multi-slice spiral CT
MST: mammalian sterile-twenty-like kinase
MST1R: macrophage-stimulating-1 factor receptor
MT: metallothionein
MTM: myotubularin (myotubular myopathy-associated gene product)
mtMMP: membrane-type MMP (mt$i$MMP: type-i mtMMP)
MTMR: myotubularin-related phosphatase
MTOC: microtubule organizing center
MTP: myeloid–T-cell progenitor
MTP: microsomal triglyceride transfer protein
MuRF: muscle-specific RING finger (Ub ligase)
MuSK: muscle-specific kinase
MVB: multivesicular body
MVE: multivesicular endosome (MVB)
MVO2: myocardial oxygen consumption
MWSS: maximal wall shear stress
MyB: V-myb myeloblastosis viral oncogene homolog (TF)
MyC: V-myc myelocytomatosis viral oncogene homolog (TF)
MyHC: myosin heavy chain
MyLC or MYL: myosin light-chain gene
MyPT: myosin phosphatase targeting subunit
MyT: myelin transcription factor

N

N: sarcomere number
$\hat{\mathbf{n}}$: unit normal vector
n: mole number
n: PAM density with elongation x
n: myosin head density
$\mathcal{N}_A$: Avogadro number
N-terminus: amino (amine group NH_2)-terminus
NAADP: nicotinic acid adenine dinucleotide phosphate
nAChR: acetylcholine nicotinic receptor (ionotropic; LGIC)

NAD: nicotine adenine dinucleotide
NADPH: reduced form of nicotinamide adenine dinucleotide phosphate
NAd: noradrenaline
NAF: nutrient-deprivation autophagy factor
NALT: nasal-associated lymphoid tissu
NAmPT: nicotinamide phosphoribosyltransferase
Nanog: ever young (Gaelic)
NAP: NCK-associated protein (NCKAP)
NAT: noradrenaline transporter
Na_V voltage-gated Na^+ channel
NBC: Na^+–HCO_3^- cotransporters
NCC: non-coronary cusp
NCC: Na^+–Cl^- cotransporter
Ncdn: neurochondrin
NCK: non-catalytic region of Tyr kinase adaptor
NCoA: nuclear receptor coactivator
NCoR: nuclear receptor corepressor
NCR: natural cytotoxicity-triggering receptor
ncRNA: non-coding RNA
NCS: neuronal calcium sensor
NCKX: Na^+–Ca^{++}–K^+ exchanger
NCLX: Na^+–Ca^{++}–Li^+ exchanger
NCX: Na^+–Ca^{++} exchanger
NDCBE: Na^+-dependent Cl^-–HCO_3^- exchanger
NecL: nectin-like molecule
NEDD: neural precursor cell expressed, developmentally downregulated
NeK: never in mitosis gene-A (NIMA)-related kinase
NES: nuclear export signal
NESK: NIK-like embryo-specific kinase
nesprin: nuclear envelope spectrin repeat protein
NeuroD: neurogenic differentiation protein
NF: neurofilament protein (intermediate filament)
NF: neurofibromin (RasGAP)
NFAT: nuclear factor of activated T cells
NFe2: erythroid-derived nuclear factor-2
NFH: neurofilament, heavy polypeptide
NFκB: nuclear factor κ light chain-enhancer of activated B cells
NFL: neurofilament, light polypeptide
NFM: neurofilament, medium polypeptide
NGAL: neutrophil gelatinase-associated lipocalin
NGF: nerve growth factor
Ngn: neogenin (netrin receptor)
NHA: Na^+–H^+ antiporter
NHE: Na^+–H^+ exchanger
NHERF: NHE regulatory factor
NHR: nuclear hormone receptor
NIc: nucleoporin-interacting protein
NIK: NFκB-inducing kinase
NIK: NCK-interacting kinase
NIP: neointimal proliferation
NK: natural killer cell
NKCC: Na^+–Ka^+–$2Cl^-$ cotransporter
NKG: NK receptor group
NKT: natural killer T cell
NKx2: NK2 transcription factor-related homeobox protein
NLR: NOD-like receptor (nucleotide-binding oligomerization domain, Leu-rich repeat-containing)
NLS: nuclear localization signal
NMDAR: N-methyl D-aspartate receptor
NmU: neuromedin-U
NO: nitric oxide (nitrogen monoxide)
NonO: non-POU domain-containing octamer-binding protein
NOR: neuron-derived orphan receptor (NR4a3)
NOS: nitric oxide synthase
NOS1: neuronal NOS
NOS2: inducible NOS
NOS3: endothelial NOS
NOx: NAD(P)H oxidase
Noxa: damage (Latin)
NPAS: neuronal PAS domain-containing transcription factor
NPC: Niemann-Pick disease type-C protein
NPC1L: Niemann-Pick protein-C1-like
NPC: nuclear-pore complex
nPKC: novel protein kinase C
NPY: neuropeptide Y

NR: nuclear receptor
NRAP: nebulin-related actinin-binding protein
nRas: neuroblastoma Ras
NRBP: nuclear receptor-binding protein
NRF: nuclear factor erythroid-derived-2 (NF-E2)-related factor
NRF1: nuclear respiratory factor-1
Nrg: neuregulin (EGF superfamily member)
Nrgn: neuroligin
Nrp: neuropilin (VEGF-binding molecule; VEGFR coreceptor)
NRSTK: non-receptor Ser/Thr kinase
NRTK: non-receptor Tyr kinase
Nrxn: neurexin
NSCLC: non-small-cell lung cancer
NSF: N-ethylmaleimide-sensitive factor
NSLTP: non-specific lipid-transfer protein
NST: nucleus of the solitary tract
NT: neurotrophin
NT5E: ecto-5′-nucleotidase
NTCP: sodium–taurocholate cotransporter polypeptide
NTF: N-terminal fragment
NTRK: neurotrophic tyrosine receptor kinase (TRK)
NTP: nucleoside triphosphate
NTPase: nucleoside triphosphate hydrolase superfamily member
Nu: Nusselt number
NuAK: nuclear AMPK-related kinase
NuP: nucleoporin (nuclear-pore complex protein)
NuRD: nucleosome remodeling and histone deacetylase
NuRR: nuclear receptor-related factor (NR4a2)
nWASP: neuronal WASP

O

^{O}Glc^{N}Ac: β^{N}acetyl ^{D}glucosamine
OCRL: oculocerebrorenal syndrome of Lowe phosphatase
Oct: octamer-binding transcription factor
ODE: ordinary differential equation
OGA: ^{O}Glc^{N}Acase (β^{N}acetylglucosaminidase)
OMM: outer mitochondrial membrane
ORC: origin recognition complex
ORF: open reading frame
ORP: OSBP-related protein
OSA: obstructive sleep apnea
OSBP: oxysterol-binding protein
OSI: oscillatory shear index
OSM: oncostatin M
OSMR: oncostatin M receptor
OSR (OxSR): oxidative stress-responsive kinase
OTK: off-track (pseudo)kinase
OTU: ovarian tumor superfamily peptidase (deubiquitinase)
OTUB: otubain (Ub thioesterase of the OTU superfamily)
OVLT: organum vasculosum lamina terminalis

P

$\mathcal{P}$: permeability
P: power
P: cell division rate
p: pressure
$p_{\mathtt{i}}$: partial pressure of gas component $\mathtt{i}$
PA: phosphatidic acid
PAAT: proton–amino acid transporter
PAF: platelet-activating factor
PAFAH: platelet-activating factor acetylhydrolase
PAG: phosphoprotein associated with glycosphingolipid-enriched microdomains
PAH: polycyclic aromatic hydrocarbon
PAH: pulmonary arterial hypertension
PAI: plasminogen activator inhibitor
PAK: P21-activated kinase
PALR: promoter-associated long RNA
PALS: protein associated with Lin-7
PAMP: pathogen-associated molecular pattern
PAMP: proadrenomedullin peptide
PAR: poly^{ADP}ribose
PAR: promoter-associated, non-coding RNA

PARi: type-i peptidase-activated receptor
Par: partitioning defective protein
PARG: polyADPribosyl glycosidase
PARP: polyADPribose polymerase
PASR: promoter-associated short RNA
PATJ: protein (PALS1) associated to tight junctions
Pax: paxillin
Pax$i$: paired box protein-i (transcription regulator)
PBC: pre-Bötzinger complex (ventilation frequency)
PBIP: Polo box-interacting protein
PC: protein C
PCMRV: phase-contrast MR velocimetry
PCr: phosphocreatine
PCT: proximal convoluted tubule
PCTP: phosphatidylcholine-transfer protein
PD: pharmacodynamics
pDC: plasmacytoid dendritic cell
PdCD: programmed cell death protein
PdCD6IP: PdCD 6-interacting protein
PdCD1Lg: programmed cell death-1 ligand
PDE: phosphodiesterase
PDE: partial differential equation
PDGF: platelet-derived growth factor
PDGFR: platelet-derived growth factor receptor
PDI: protein disulfide isomerase
PDK: phosphoinositide-dependent kinase
Pe: Péclet number
PE: pulmonary embolism
PEBP: phosphatidylethanolamine binding protein
PECAM: platelet–endothelial cell adhesion molecule
PEDF: pigment epithelium-derived factor (serpin F1)
PEn2: presenilin enhancer-2
PEO: proepicardial organ
Per: Period homolog
PERK: protein kinase-like endoplasmic reticulum kinase
PERP: P53 apoptosis effector related to peripheral myelin protein PMP22
PET: positron emission tomography
Pex: peroxin
PF: platelet factor
PFK: phosphofructokinase
pFRG: parafacial respiratory group
PG: prostaglandin
PGC: PPARγ coactivator
pGC: particulate guanylyl cyclase
PGEA: prostaglandin ethanolamide
PGF: paracrine growth factor
PGG: prostaglandin glycerol ester
PGi2: prostacyclin
PGP: permeability glycoprotein
PH: pleckstrin homology domain
PHD: prolyl hydroxylase
PHLPP: PH domain and Leu-rich repeat protein phosphatase
PI: phosphoinositide (phosphorylated phosphatidylinositol)
PI(4)P: phosphatidylinositol 4-phosphate
PI(i)PiK: phosphatidylinositol i-phosphate i-kinase
PI(i,j)P_2: phosphatidylinositol (i,j)-bisphosphate (PIP_2)
PI(3,4,5)P_3: phosphatidylinositol (3,4,5)-trisphosphate (PIP_3)
PI3K: phosphatidylinositol 3-kinase
PI3KAP: PI3K adaptor protein
PIiK: phosphatidylinositol i-kinase
PIAS: protein inhibitor of activated STAT (SUMo E3 ligase)
PIC: pre-initiation complex
PICK: protein that interacts with C-kinase
PIDD: P53-induced protein with a death domain
PIKE: phosphoinositide 3-kinase enhancer (GTPase; ArfGAP)
PIKK: phosphatidylinositol 3-kinase-related kinase (pseudokinase)
PIM: provirus insertion of Molony murine leukemia virus gene product
PIN: protein peptidyl prolyl isomerase NIMA-interacting

PINCH: particularly interesting new Cys–His protein (or LIMS1)
PInK: PTen-induced kinase
PIP: phosphoinositide monophosphate
PIP*i*K: phosphatidylinositol phosphate *i*-kinase
PIP_2: phosphatidylinositol bisphosphate
PIP_3: phosphatidylinositol triphosphate
PIPP: proline-rich inositol polyphosphate 5-phosphatase
PIR: paired immunoglobulin-like receptor
piRNA: P-element-induced wimpy testis-interacting (PIWI) RNA
PIRT: phosphoinositide-interacting regulator of TRP channels
PITP: phosphatidylinositol-transfer protein
Pitx: pituitary (or paired-like) homeobox transcription factor
PIV: particle image velocimetry
PIX: P21-activated kinase (PAK)-interacting exchange factor (Rho(Arh)GEF6/7)
PK: pharmacokinetics
PK: protein kinase
PKA: protein kinase A
PKB: protein kinase B
PKC: protein kinase C
PKD: protein kinase D
PKG: protein kinase G
PKL: paxillin kinase linker
PKMYT (MYT): membrane-associated Tyr–Thr protein kinase
PKN: protein kinase novel
Pkp: plakophilin
PL: phospholipase
PLA2: phospholipase A2
PLC: phospholipase C
PLD: phospholipase D
PLb: phospholamban
PLd: phospholipid
PlGF: placental growth factor
PLK: Polo-like kinase
PLTP: phospholipid transfer protein
PMCA: plasma membrane Ca^{++} ATPase
PML: promyelocytic leukemia protein
PMR: percutaneous (laser) myocardial revascularization
PMRT: protein arginine methyltransferase
Pn: plasmin
Png: plasminogen
PoG: proteoglycan
PoM: pore membrane protein
Pon: paraoxonase
POPx: partner of PIX
POSH: scaffold plenty of SH3 domains
PP: protein phosphatase
PP3: protein phosphatase 3 (PP2b or calcineurin)
PPAR: peroxisome proliferator-activated receptor (NR1c1–3)
PPG: photoplethysmography
PPId: peptidyl prolyl isomerase-D
PPIP: monopyrophosphorylated inositol phosphate
$(PP)_2IP$: bispyrophosphorylated inositol phosphate
PPK: PIP kinase
PPM: protein phosphatase (magnesium-dependent)
PPR: pathogen-recognition receptor
PPRE: PPAR response element (DNA sequence)
PR: progesterone receptor (NR3c3)
PRC: protein regulator of cytokinesis
PRC: Polycomb repressive complex
Prdx: peroxiredoxin
pre-cDC: pre-classical dendritic cell
pre-miR: precursor microRNA
preBotC: pre-Bötzinger complex
preKk: prekallikrein
PREx: PIP_3-dependent Rac exchanger (RacGEF)
PRG: plasticity-related gene product
PRH: prolactin-releasing hormone
pri-miR: primary microRNA
Prl: prolactin
PrlR: prolactin receptor
PRMT: protein arginine (R) N-methyltransferase
Prompt: promoter upstream transcript
Protor: protein observed with Rictor
PROX: prospero homeobox gene

Prox: PROX gene product (transcription factor)
PrP: processing protein
PRPK: P53-related protein kinase
PRR: pattern recognition receptor
PRR: prorenin and renin receptor
PS: presenilin
PS: protein S
PSC: pluripotent stem cell
PSD: postsynaptic density adaptor
PsD: postsynaptic density
PSEF: pseudo-strain energy function
PSer: phosphatidylserine
PSGL: P-selectin glycoprotein ligand
PSKh: protein serine kinase H
Psm: proteasome subunit
PSTPIP: Pro–Ser–Thr phosphatase-interacting protein
PTA: plasma thromboplastin antecedent
Ptc: Patched receptor (Hedgehog signaling)
PTCA: percutaneous transluminal coronary angioplasty
PtcH: Patched Hedgehog receptor
PTCRA: PTC rotational burr atherectomy
PtdCho (PC): phosphatidylcholine
PtdEtn (PE): phosphatidylethanolamine
PtdSer (PS): phosphatidylserine
PtdIns (PI): phosphatidylinositol
PTen: phosphatase and tensin homolog deleted on chromosome ten (phosphatidylinositol 3-phosphatase)
PTFE: polytetrafluoroethylene
PTH: parathyroid hormone
PTHRP: parathyroid hormone-related protein
PTK: protein Tyr kinase
PTK7: pseudokinase (RTK)
PTP: protein Tyr phosphatase
PTPni: protein Tyr phosphatase non-receptor type i
PTPR: protein Tyr phosphatase receptor
PTRF: RNA polymerase-1 and transcript release factor
PUFA: polyunsaturated fatty acid
PUMA: P53-upregulated modulator of apoptosis
PuV: pulmonary valve
PVF: PDGF- and VEGF-related factor
PVNH: paraventricular nucleus of hypothalamus
PVR: pulmonary vascular resistance
PWS: pulse wave speed
Px: pannexin
PXR: pregnane X receptor (NR1i2)
PYK: proline-rich tyrosine kinase
P2X: purinergic ligand-gated channel
P53AIP: P53-regulated apoptosis-inducing protein
p75NtR: pan-neurotrophin receptor

Q

Q: material quantity
Q_e: electric current density
Q_T: thermal energy (heat)
q_T: transfer rate of thermal energy (power)
q: flow rate

R

R: resistance
$\mathcal{R}$: local reaction term
R_h: hydraulic radius
R_g: gas constant
R_R: respiratory quotient
R: recruitment function (from quiescence to proliferation)
r: cell renewal rate
r: radial coordinate
RA: right atrium
RAAS: renin–angiotensin–aldosterone system
Rab: Ras from brain
Rab11FIP: Rab11 family-interacting protein
Rac: Ras-related C3-botulinum toxin substrate
RACC: receptor-activated cation channel
RACK: receptor for activated C-kinase
RAD: recombination protein-A (RecA)-homolog DNA-repair protein

Rad: radiation sensitivity protein
Rag: Ras-related GTP-binding protein
Ral: Ras-related protein
RalGDS: Ral guanine nucleotide-dissociation stimulator
RAMP: (calcitonin receptor-like) receptor-activity-modifying protein
Ran: Ras-related nuclear protein
RANTES: regulated upon activation, normal T-cell expressed, and secreted product (CCL5)
RAP: receptor-associated protein
Rap: Ras-proximate (Ras-related) protein
Raptor: regulatory associated protein of TOR
RAR: retinoic acid receptor (NR1b2/3)
Ras: rat sarcoma viral oncogene homolog (small GTPase)
RasA: Ras p21 protein activator
rasiRNA: repeat-associated small interfering RNA (PIWI)
RASSF: Ras interaction/interference protein RIN1, afadin, and Ras association domain-containing family protein
RB: retinoblastoma protein
RBC: red blood cell (erythrocyte)
RBP: retinol-binding protein
RC: ryanodine calcium channel (RyR)
RCA: right coronary artery
RCan: regulator of calcineurin
RCC: right coronary cusp
RCC: regulator of chromosome condensation
Re: Reynolds number
REDD: regulated in development and DNA-damage response gene product
Rel: reticuloendotheliosis proto-oncogene product (TF; member of NFκB)
REP: Rab escort protein
ReR: renin receptor (PRR)
restin: Reed-Steinberg cell-expressed intermediate filament-associated protein (CLiP1)
ReT: rearranged during transfection (receptor Tyr kinase)
RevRE: reverse (Rev)-ErbA (NR1d1/2) response element (DNA sequence)
RFA: radiofrequency ablation
RGL: Ral guanine nucleotide dissociation stimulator-like protein (GEF)
RGS: regulator of G-protein signaling
RHEB: Ras homolog enriched in brain
RHS: equation right-hand side
Rho: Ras homologous
RIAM: Rap1-GTP-interacting adaptor
RIBP: RLK- and ITK-binding protein
RICH: RhoGAP interacting with CIP4 homolog
RICK: receptor for inactive C-kinase
Rictor: rapamycin-insensitive companion of TOR
RIF: Rho in filopodium
RIn: Ras and Rab interactor (RabGEF)
RIN: Ras-like protein expressed in neurons (GTPase)
RIP: regulated intramembrane proteolysis
RIPK: receptor-interacting protein kinase
RISC: RNA-induced silencing complex
RIT: Ras-like protein expressed in many tissues
RKIP: Raf kinase inhibitor protein
RlBP: retinaldehyde-binding protein
RLC: RISC-loading complex
RLK: resting lymphocyte kinase (TXK)
RNA: ribonucleic acid
RNABP: RNA-binding protein
RNase: ribonuclease
RnBP: renin-binding protein
RNF2: RING finger protein-2 (ubiquitin ligase)
RNP: ribonucleoprotein
Robo: roundabout
ROC: receptor-operated channel
RoCK: Rho-associated, coiled-coil-containing protein kinase
ROI: region of interest
ROMK: renal outer medullary potassium channel

ROR: RAR-related orphan receptor (NR1f1–NR1f3)
ROR(RTK): receptor Tyr kinase-like orphan receptor
ROS: reactive oxygen species
Ros: V-ros UR2 sarcoma virus proto-oncogene product (RTK)
RPIP: Rap2-interacting protein
RPS6: ribosomal protein S6
RPTP: receptor protein Tyr phosphatase
rRas: related Ras
rRNA: ribosomal RNA
RSA: respiratory sinus arrhythmia
RSE: rapid systolic ejection
RSK: P90 ribosomal S6 kinase
RSKL: ribosomal protein S6 kinase-like (pseudokinase)
rSMAD: receptor-regulated SMAD (SMAD1–SMAD3, SMAD5, and SMAD9)
RSMCS: robot-supported medical and surgical system
RSpo: R-spondin
RSTK: receptor Ser/Thr kinase
RTK: receptor Tyr kinase
RTN: retrotrapezoid nucleus
Rubicon: RUN domain and Cys-rich domain-containing, beclin-1-interacting protein
Runx: Runt-related transcription factor
RV: right ventricle
RVF: rapid ventricular filling
RVLM: rostral ventrolateral medulla
RVMM: rostral ventromedial medulla
RXR: retinoid X receptor (NR2b1–NR2b3)
RYK: receptor-like Tyr (Y) kinase (pseudokinase)
RyR: ryanodine receptor (ryanodine-sensitive Ca^{++}-release channel)

S

S: Cauchy-Green deformation tensor
$\mathfrak{s}$: entropy
s: sarcomere length
s: evolution speed
SAA: serum amyloid A
SAC: stretch-activated channel
SAc: suppressor of actin domain-containing 5-phosphatase
sAC: soluble adenylyl cyclase
SACM1L: suppressor of actin mutations 1-like
SAH: subarachnoid hemorrhage
SAN: sinoatrial node
SAP: SLAM-associated protein
SAP: stress-activated protein
SAPi: synapse-associated protein i
SAPK: stress-activated protein kinase (MAPK)
SAR: secretion-associated and Ras-related protein
Sc: Schmidt number
SCA: stem cell antigen
SCAMP: secretory carrier membrane protein
SCAP: SREBP cleavage-activating protein (SREBP escort)
SCAR: suppressor of cAMP receptor (WAVe)
ScaR: scavenger receptor
SCF: SKP1–Cul1–F-box Ub-ligase complex
SCF: stem cell factor
SCFR: stem cell factor receptor (KIT)
Scgb: secretoglobin
SCLC: small-cell lung cancer
scLC: squamous-cell lung cancer (NSCLC subtype)
SCN: suprachiasmatic nucleus
SCO: synthesis of cytochrome-C oxidase
Scp: stresscopin (urocortin 3)
Scrib: Scribble polarity protein
Sdc: syndecan
SDF: stromal cell derived factor
SDPR: serum deprivation protein response
SE: systolic ejection
SEF: strain-energy function
SEF: similar expression to FGF genes (inhibitor of RTK signaling)
SEK: SAPK/ERK kinase
Sema: semaphorin (Sema-, Ig-, transmembrane-, and short cytoplasmic domain)

SERCA: sarco(endo)plamic reticulum calcium ATPase
serpin: serine peptidase inhibitor
SerT: serotonin transporter
SF: steroidogenic factor (NR5a1)
SFK: SRC-family kinase
SFO: subfornical organ
sFRP: secreted Frizzled-related protein
SftP (SP): surfactant protein
sGC: soluble guanylyl cyclase
SGK: serum- and glucocorticoid-regulated kinase
SGlT: Na^+–glucose cotransporter (SLC5a)
Sgo: shugoshin (Japanese: guardian spirit)
SH: Src homology domain
Sh: Sherwood number
SH3P: Src homology-3 domain-containing adaptor protein
Shank: SH3- and multiple ankyrin repeat domain-containing protein
SHAX: SNF7 (VSP32) homolog associated with ALIX
SHB: Src homology-2 domain-containing adaptor
SHC: Src-homologous and collagen-like substrate
SHC: Src homology-2 domain-containing transforming protein
SHh: sonic Hedgehog
SHIP: SH-containing inositol phosphatase
SHP: SH-containing protein Tyr phosphatase (PTPn6/11)
SHP: small heterodimer partner (NR0b2)
shRNA: small (short) hairpin RNA
SIAH: Seven in absentia homolog (E3 ubiquitin ligase)
siglec: sialic acid-binding Ig-like lectin
SIK: salt-inducible kinase
SIn: stress-activated protein kinase-interacting protein
SIP: steroid receptor coactivator-interacting protein
siRNA: small interfering RNA
SiRP: signal-regulatory protein
SIRT: sirtuin (silent information regulator-2 [two]; histone deacetylase)
SIT: SHP2-interacting transmembrane adaptor
SK: small conductance Ca^{++}-activated K^+ channel
SKi: sphingosine kinase-i
SKIP: sphingosine kinase-1-interacting protein
SKIP: skeletal muscle and kidney-enriched inositol phosphatase
SKP: S-phase kinase-associated protein
SLA: Src-like adaptor
SLAM: signaling lymphocytic activation molecule
SLAMF: SLAM family member
SLAP: Src-like adaptor protein
SLC: solute carrier class member
SLK: Ste20-like kinase
Sln: sarcolipin
SLPI: secretory leukocyte peptidase inhibitor
SLTC: small latent TGFβ complex
SM: sphingomyelin
SMA: smooth muscle actin
SMAD: small mothers against decapentaplegic homolog
SmAP: Small ArfGAP protein
SMase: sphingomyelinase
SMC: smooth muscle cell
Smo: Smoothened
SMPD: sphingomyelin phosphodiesterase
SMRT: silencing mediator of retinoic acid and thyroid hormone receptor
SMS: sphingomyelin synthase
SMURF: SMAD ubiquitination regulatory factor
SNAAT: sodium-coupled neutral amino acid transporter
SNAP: soluble N-ethylmaleimide-sensitive factor-attachment protein
SnAP: synaptosomal-associated protein
SNARE: SNAP receptor
SNF7: sucrose non-fermenting (VPS32)
SNIP: SMAD nuclear-interacting protein

snoRNA: small nucleolar RNA
snoRNP: small nucleolar ribonucleoprotein
SNP: single-nucleotide polymorphism
snRNA: small nuclear RNA
snRNP: small nuclear ribonucleoprotein
SNx: sorting nexin
SOC: store-operated Ca^{++} channel
SOCE: store-operated Ca^{++} entry
SOCS: suppressor of cytokine signaling protein
SOD: superoxide dismutase
SorbS: sorbin and SH3 domain-containing adaptor
SOS: Son of sevenless (GEF)
Sost: sclerostin
SostDC: sclerostin domain-containing protein
SOX: sex-determining region Y (SRY)-box gene
Sox: SOX gene product (transcription factor)
SP1: specificity protein (transcription factor)
SPARC: secreted protein acidic and rich in cysteine
SPC: sphingosylphosphorylcholine
SPCA: secretory pathway Ca^{++} ATPase
SPECT: single photon emission CT
Sph: sphingosine
SphK: sphingosine kinase
SPI: spleen focus-forming virus (SFFV) proviral integration proto-oncogene product (transcription factor)
SPN: supernormal period
SPP: sphingosine phosphate phosphatase
SpRED: Sprouty-related protein with an EVH1 domain
SPURT: secretory protein in upper respiratory tract
SQTS: short-QT syndrome
SR: sarcoplasmic reticulum
SR: Arg/Ser domain-containing protein (alternative splicing)
SRA: steroid receptor RNA activator
SRC: steroid receptor coactivator
Src: sarcoma-associated (Schmidt-Ruppin A2 viral oncogene homolog) kinase
SREBP: sterol regulatory element-binding protein
SRF: serum response factor
SRM/SMRS: Src-related kinase lacking regulatory and myristylation sites
SRP: stresscopin-related peptide (urocortin 2)
SRPK: splicing factor RS domain-containing protein kinase
SRY: sex-determining region Y
SSAC: shear stress-activated channel
SSE: slow systolic ejection
SSH: slingshot homolog protein
SSI: STAT-induced STAT inhibitor
ssRNA: single-stranded RNA
Sst: somatostatin
SSV: short saphenous vein
St: Strouhal number
STAM: signal-transducing adaptor molecule
STAMBP: STAM-binding protein (ubiquitin isopeptidase)
StAR: steroidogenic acute regulatory protein
StART: StAR-related lipid transfer protein
STAT: signal transducer and activator of transduction
STEAP: six transmembrane epithelial antigen of the prostate
STICK: substrate that interacts with C-kinase
StIM: stromal interaction molecule
STK: protein Ser/Thr kinase
STK1: stem cell protein Tyr kinase receptor
STLK: Ser/Thr kinase-like (pseudo)kinase
Sto: Stokes number
StRAd: STe20-related adaptor
StRAP: stress-responsive activator of P300
Stx: syntaxin ($SNARE^Q$)
SUMo: small ubiquitin-related modifier
SUn: Sad1 and Unc84 homology protein
SUR: sulfonylurea receptor

SUT: stable unannotated transcript
SV: stroke volume
SVC: superior vena cava
SVCT: sodium-dependent vitamin-C transporter
SVF: slow ventricular filling
SVP: synaptic vesicle precursor
SVR: systemic vascular resistance
SW: stroke work
SwAP70: 70-kDa switch-associated protein (RacGEF)
SYK: spleen tyrosine kinase
Synj: synaptojanin
Syp: synaptophysin
Syt: synaptotagmin
S1P: sphingosine 1-phosphate
S6K: P70 ribosomal S6 kinase ($P70^{RSK}$)

T

T: extrastress tensor
T: transition rate from a cell cycle phase to the next
T: temperature
T_C: cytotoxic T lymphocyte (CD8+ effector T cell; CTL)
T_{C1}: type-1 cytotoxic T lymphocyte
T_{C2}: type-2 cytotoxic T lymphocyte
T_{CM}: central memory T lymphocyte
T_{Conv}: conventional T lymphocyte
T_{Eff}: effector T lymphocyte
T_{EM}: effector memory T lymphocyte
T_{FH}: follicular helper T lymphocyte
T_H: helper T lymphocyte (CD4+ effector T cell)
T_{Hi}: type-i helper T lymphocyte ($i = 1/2/9/17/22$)
T_{H3}: TGFβ-secreting T_{Reg} lymphocyte
T_L: lung transfer capacity (alveolocapillary membrane)
T_{R1}: type-1, IL10-secreting, regulatory T lymphocyte
T_{Reg}: regulatory T lymphocyte
aT_{Reg}: CD45RA−, FoxP3^{hi}, activated T_{Reg} cell
iT_{Reg}: inducible T_{Reg} lymphocyte
nT_{Reg}: naturally occurring (natural) T_{Reg} lymphocyte
rT_{Reg}: CD45RA+, FoxP3^{low}, resting T_{Reg} cell
$\hat{\mathbf{t}}$: unit tangent
t: time
TβR(1/2): TGFβ receptor
TAA: thoracic aortic aneurysm
TAB: TAK1-binding protein
TACE: tumor-necrosis factor-α-converting enzyme
TACE: transarterial chemoembolization
TAF: TBP-associated factor
TAK: TGFβ-activated kinase (MAP3K7)
TALK: TWIK-related alkaline pH-activated K^+ channel
TANK: TRAF family member-associated NFκB activator
TASK: TWIK-related acid-sensitive K^+ channel
TASR: terminus-associated short RNA
TAP: transporter associated with antigen processing (ABC transporter)
Taz: taffazin
TBC1D: Tre2 (or USP6), BUB2, CDC16 domain-containg RabGAP
TBCK: tubulin-binding cofactor kinase (pseudokinase)
TBK: TANK-binding kinase
TBP: TATA box-binding protein (subclass-4F transcription factor)
TBx: T-box transcription factor
TC: thrombocyte (platelet)
TCA: tricarboxylic acid cycle
TCF: T-cell factor
TCF: ternary complex factor
TcFi: type-i transcription factor
TCP: T-complex protein
TCR: T-cell receptor
TEA: transluminal extraction atherectomy
TEC: Tyr kinase expressed in hepatocellular carcinoma
TEF: thyrotroph embryonic factor (PAR/b–ZIP family)
TEK: Tyr endothelial kinase
TEM: transendothelial migration
Ten: tenascin
TF: transcription factor

Tf: transferrin
TFPI: tissue factor pathway inhibitor
TfR: transferrin receptor
TG: triglyceride (triacylglycerol)
TGF: transforming growth factor
TGFBR: TGFβ receptor gene
TGN: trans-Golgi network
THETE: trihydroxyeicosatrienoic acid
THIK: tandem pore-domain halothane-inhibited K$^+$ channel
THR: thyroid hormone receptor (NR1a1/2)
TIAM: T-lymphoma invasion and metastasis-inducing protein (RacGEF)
TICE: transintestinal cholesterol efflux
TIE: Tyr kinase with Ig and EGF homology domains (angiopoietin receptor)
TIEG: TGFβ-inducible early gene product
TIGAR: TP53-inducible glycolysis and apoptosis regulator
TIM: T-cell immunoglobulin and mucin domain-containing protein
Tim: timeless homolog
TIMM: translocase of inner mitochondrial membrane
TIMP: tissue inhibitor of metallopeptidase
TIRAP: Toll–IL1R domain-containing adaptor protein
tiRNA: transcription initiation RNA
TJ: tight junction
TKR: Tyr kinase receptor
TLC: total lung capacity
TLR: Toll-like receptor
TLT: TREM-like transcript
TLX: tailless receptor (NR2e1)
TM: thrombomodulin
TMC: twisting magnetocytometry
TMePAI: transmembrane prostate androgen-induced protein
TMy: tropomyosin
Tnn (TN): troponin
Tn: thrombin
TNF: tumor-necrosis factor
TNFαIP: tumor-necrosis factor-α-induced protein
TNFR: tumor-necrosis factor receptor
TNFRSF: tumor-necrosis factor receptor superfamily member
TNFSF: tumor-necrosis factor superfamily member
TNK: Tyr kinase inhitor of NFκB
Tns: tensin
TOR: target of rapamycin
TORC: target of rapamycin complex
TORC: transducer of regulated CREB activity (a.k.a. CRTC)
TP: thromboxane A2 Gq/11 coupled receptor
TP53I: tumor protein P53-inducible protein
tPA: tissue plasminogen activator
Tpo: thrombopoietin
TPPP: tubulin polymerization-promoting protein
TPST: tyrosylprotein sulftotransferase
TR: testicular receptor (NR2c1/2)
TRAAK: TWIK-related arachidonic acid-stimulated K$^+$ channel
TRADD: tumor-necrosis factor receptor-associated death domain adaptor
TRAF: tumor-necrosis factor receptor-associated factor
TRAM: TRIF-related adaptor molecule
transceptor: transporter-related receptor
TraPP: transport protein particle
TRAT: T-cell receptor-associated transmembrane adaptor
Trb: Tribbles homolog (pseudokinase)
TRE: trapped in endoderm
TREK: TWIK-related K$^+$ channel
TREM: triggering receptor expressed on myeloid cells
TRESK: TWIK-related spinal cord K$^+$ channel
TRF: TBP-related factor
TRH: thyrotropin releasing hormone
TRIF: Toll–IL1R domain-containing adaptor inducing Ifnβ
TRIM: T-cell receptor interacting molecule
TRK: tropomyosin receptor kinase (NTRK)

tRNA: transfer RNA
TRP: transient receptor potential channel
TRPA: ankyrin-like transient receptor potential channel
TRPC: canonical transient receptor potential channel
TRPM: melastatin-related transient receptor potential channel
TRPML: mucolipin-related transient receptor potential channel
TRPN: no mechanoreceptor potential C
TRPP: polycystin-related transient receptor potential channel
TRPV: vanilloid transient receptor potential channel
TrrAP: transactivation (transformation)/transcription domain-associated protein (pseudokinase)
TrV: tricuspid valve
TRx: thioredoxin
TRxIP: thioredoxin-interacting protein
TSC: tuberous sclerosis complex
TSH: thyroid-stimulating hormone
TSLP: thymic stromal lymphopoietin
Tsp: thrombospondin
Tspan: tetraspanin
TsPO: translocator protein of the outer mitochondrial membrane
tSNARE: target SNARE
tsRNA: tRNA-derived small RNA
tssaRNA: transcription start site-associated RNA
Ttn: titin (pseudokinase)
TUT: terminal uridine transferase
TWIK: tandem of P domains in a weak inwardly rectifying K^+ channel
TxA2: thromboxane A2 (thromboxane)
TxB2: thromboxane B2 (thromboxane metabolite)
TXK: Tyr kinase mutated in X-linked agammaglobulinemia
TyK: tyrosine kinase
T_3: tri-iodothyronine
T_4: thyroxine
$^+$TP: plus-end-tracking proteins

U

U: right stretch tensor
u: displacement vector
u: electrochemical command
u: specific internal energy
Ub: ubiquitin
UbC: ubiquitin-conjugating enzyme
UbE2: E2 ubiquitin conjugase
UbE3: E3 ubiquitin ligase
UbL: ubiquitin-like protein
UCH: ubiquitin C-terminal hydrolase (DUb)
Ucn: urocortin
UCP: uncoupling protein
UDP: uridine diphosphate-glucose
UK: urokinase
ULK: uncoordinated-51-like kinase (pseudokinase)
Unc: uncoordinated receptor
uPA: urokinase-type plasminogen activator (urokinase)
uPAR: uPA receptor
uPARAP: uPAR-associated protein (CLec13e)
UPR: unfolded protein response
UPS: ubiquitin-proteasome system
UP4A: uridine adenosine tetraphosphate
Uro: urodilatin
US: ultrasound
USC: unipotential stem cell
USF: upstream stimulatory factor
USI: ultrasound imaging
USP: ubiquitin-specific peptidase (deubiquitinase)
UTP: uridine triphosphate
UTR: untranslated region
UVRAG: ultraviolet wave resistance-associated gene product

V

V: left stretch tensor
V: volume
V_q: cross-sectional average velocity
V_s: specific volume
v: velocity vector
v: recovery variable
V1(2)R: type-1(2) vomeronasal receptor

$V_{1A/1B/2}$: type-1a/1b/2 arginine vasopressin receptor
VAAC: volume-activated anion channel
VACamKL: vesicle-associated CamK-like (pseudokinase)
VAChT: vesicular acetylcholine transporter
VAMP: vesicle-associated membrane protein (synaptobrevin)
VanGL: Van Gogh (Strabismus)-like protein
VAP: VAMP-associated protein
VASP: vasoactive stimulatory phosphoprotein
VAT: vesicular amine transporter
vATPase: vesicular-type H^+ ATPase
VAV: ventriculoarterial valve
Vav: GEF named from Hebrew sixth letter
VC: vital capacity
VCAM: vascular cell adhesion molecule
VCt: vasoconstriction
VDAC: voltage-dependent anion channel (porin)
VDCC: voltage-dependent calcium channel
VDP: vesicle docking protein
VDt: vasodilation
VEGF: vascular endothelial growth factor
VEGFR: vascular endothelial growth factor receptor
VF: ventricular fibrillation
VF: ventricular filling
VGAT: vesicular GABA transporter
VGC: voltage-gated channel
VgL: Vestigial-like protein
VGluT: vesicular glutamate transporter
VHL: von Hippel-Lindau protein (E3 ubiquitin ligase)
VIP: vasoactive intestinal peptide
VLDL: very-low-density lipoprotein
VLDLR: very-low-density lipoprotein receptor
VMAT: vesicular monoamine transporter
VN: vitronectin
VPS: vacuolar protein sorting-associated kinase
VR: venous return
VRAC: volume-regulated anion channel
VRC: ventral respiratory column
VRK: vaccinia-related kinase
VS: vasostatin
vSMC: vascular smooth muscle cell
vSNARE: vesicular SNAP receptor (SNARE)
VSOR: volume-sensitive outwardly rectifying anion channel
VSP: voltage-sensing phosphatase
VVO: vesiculo-vacuolar organelle
vWF: von Willebrand factor

W

$\mathbf{W}$: vorticity tensor
$\mathcal{W}$: strain energy density
W: work, deformation energy
$\mathtt{w}$: weight
$\mathbf{w}$: grid velocity
WASH: WASP and SCAR homolog
WASP: Wiskott-Aldrich syndrome protein
WAT: white adipose tissue
WAVe: WASP-family verprolin homolog
WBC: white blood cell
WDR: WD repeat-containing protein
Wee: small (Scottish)
WHAMM: WASP homolog associated with actin, membranes, and microtubules
WIP: WASP-interacting protein
WIPF: WASP-interacting protein family protein
WIPI: WD repeat domain-containing phosphoinositide-interacting protein
WNK: with no K (Lys) kinase
Wnt: wingless-type
WPWS: Wolff-Parkinson-White syndrome
WSB: WD-repeat and SOCS box-containing protein (Ub ligase)
WSS: wall shear stress
WSSTG: WSS transverse gradient
WWTR: WW domain-containing transcription regulator

X

$\mathcal{X}$: trajectory
$\mathtt{X}$: reactance
X: Lagrangian position vector
x: position vector
$\{x, y, z\}$: Cartesian coordinates
XBP: X-box-binding protein (transcription factor)
XIAP: X-linked inhibitor of apoptosis (Ub ligase)

Y

$\mathtt{Y}$: admittance coefficient
YAP: Yes-associated protein
YBP: Y-box-binding protein (transcription factor)
YY: yin yang (transcriptional repressor)

Z

$\mathtt{Z}$: impedance
ZAP70: ζ-associated protein 70
ZBTB: zinc finger and BTB (Broad complex, Tramtrack, and bric-à-brac) domain-containing transcription factor
ZnF: zinc finger protein
ZO: zonula occludens

Miscellaneous

2-5A: 5′-triphosphorylated, (2′,5′)-phosphodiester-linked oligoadenylate
2AG: 2-arachidonyl glycerol
3DR: three-dimensional reconstruction
3BP2: Abl Src homology-3 domain-binding adaptor
4eBP1: inhibitory eIF4e-binding protein
5HT: serotonin
7TMR: 7-transmembrane receptor (GPCR)

Complementary Lists of Notations

Greek Letters

α: volumic fraction
α: convergence/divergence angle
α: attenuation coefficient
α_{k}: kinetic energy coefficient
α_{m}: momentum coefficient
β: inclination angle
$\{\beta_i\}_1^2$: myocyte parameters
β_T: coefficient of thermal expansion
Γ: domain boundary
Γ_{L}: local reflection coefficient
Γ_{G}: global reflection coefficient
γ: heat capacity ratio
γ: activation factor
γ_{g}: amplitude ratio (modulation rate) of g
γ_s: surface tension
$\dot{\gamma}$: shear rate
δ: boundary layer thickness
ϵ_T: emissivity (thermal energy radiation)
ϵ_e: electric permittivity
ϵ: strain
ε: small quantity
ζ: singular head loss coefficient
ζ: transmural coordinate
$\{\zeta_j\}_1^3$: local coordinate
η: azimuthal spheroidal coordinate
θ: circonferential polar coordinate
θ: $(\hat{\mathbf{e}}_x, \hat{\mathbf{t}})$ angle
κ: wall curvature
κ_{c}: curvature ratio
κ_{d}: drag coefficient
κ_{h}: hindrance coefficient
κ_{o}: osmotic coefficient
κ_{s}: size ratio
$\{\kappa_k\}_{k=1}^9$: tube law coefficients
κ_{e}: correction factor
Λ: head loss coefficient
λ_{L}: Lamé coefficient
λ: stretch ratio
λ: wavelength
λ_A: area ratio
λ_a: acceleration ratio
λ_L: length ratio
λ_q: flow rate ratio
λ_t: time ratio
λ_v: velocity ratio
μ: dynamic viscosity
μ_{L}: Lamé coefficient
ν: kinematic viscosity
ν_{P}: Poisson ratio
Π: osmotic pressure
ρ: mass density
τ: time constant
Φ: potential
$\phi(t)$: creep function
φ: phase
χ: Lagrangian label
chi_i: molar fraction of species i
χ_i: wetted perimeter
$\psi(t)$: relaxation function
Ψ: porosity
ω: angular frequency
Ω: computational domain

Dual Notations

Bφ: basophil
Eφ: eosinophil
Lφ: lymphocyte
Mφ: macrophage
aaMφ: alternatively activated macrophage
caMφ: classically activated macrophage
Nφ: neutrophil
Σc: sympathetic
pΣc: parasympathetic

Subscripts

$_{A}$: alveolar, atrial
$_{Ao}$: aortic
$_{a}$: arterial
$_{app}$: apparent
$_{atm}$: atmospheric
$_{b}$: blood
$_{\mathbf{c}}$: contractile
$_{c}$: center
$_{c}$: point-contact
$_{D}$: Darcy (filtration)
$_{\mathrm{d}}$: diastolic
$_{dyn}$: dynamic
$_{E}$: expiration, Eulerian
$_{e}$: external
$_{\mathrm{e}}$: extremum
$_{eff}$: effective
$_{f}$: fluid
$_{g}$: grid
$_{I}$: inspiration
$_{i}$: internal
$_{inc}$: incremental
$_{L}$: Lagrangian
$_{l}$: limit
$_{\ell}$: line-contact
$_{M}$: macroscopic
$_{m}$: mean
$_{max}$: maximum
$_{m}$: muscular, mouth
$_{met}$: metabolic
$_{\mu}$: microscopic
$_{P}$: pulmonary
$_{\mathbf{p}}$: parallel
$_{p}$: particle
$_{q}$: quasi-ovalization
$_{r}$: radial
$_{rel}$: relative
$_{S}$: systemic
$_{s}$: solute
$_{\mathbf{s}}$: serial
$_{\mathrm{s}}$: systolic
$_{t}$: stream division
$_{T}$: total
$_{t}$: turbulence
$_{t}$: time derivative of order 1
$_{tt}$: time derivative of order 2
$_{tis}$: tissue
$_{V}$: ventricular
$_{v}$: venous
$_{w}$: wall
$_{\mathrm{w}}$: water (solvent)
$_{\Gamma}$: boundary
$_{\theta}$: azimuthal
$_{+}$: positive command
$_{-}$: negative command
$_{*}$: at interface
$_{0}$: reference state ($\cdot_0$: unstressed or low shear rate)
$_{\infty}$: high shear rate

Superscripts

^{a}: active state
^{e}: elastic
^{f}: fluid
^{h}: hypertensive
^{n}: normotensive
^{p}: passive state
^{P}: power
^{s}: solid
^{T}: transpose
^{v}: viscoelastic
*: scale
*: complex variable
$\cdot'$: first component of complex elastic and shear moduli
$\cdot''$: second component of complex elastic and shear moduli
$^{\natural}$: static, stationary, steady variable

Mathematical Notations

T: bold face capital letter means tensor
v: bold face lower case letter means vector

S, s: upper or lower case letter means scalar
$\Delta\bullet$: difference
$\delta\bullet$: increment
$d\bullet/dt$: time gradient
∂_t: first-order time partial derivative
∂_{tt}: second-order time partial derivative
∂_i: first-order space partial derivative with respect to spatial coordinate x_i
∇: gradient operator
$\nabla \mathbf{u}$: displacement gradient tensor
$\nabla \mathbf{v}$: velocity gradient tensor
$\nabla\cdot$: divergence operator
∇^2: Laplace operator
$|\ |_+$: positive part
$|\ |_-$: negative part
$\dot{\bullet}$: time derivative
$\bar{\bullet}$: time mean
$\breve{\bullet}$: space averaged
$\langle\bullet\rangle$: ensemble averaged
$\tilde{\bullet}$: dimensionless
$\bullet^+$: normalized ($\in [0, 1]$)
$\hat{\bullet}$: peak value
$\bullet_\sim$: modulation amplitude
det($\bullet$): determinant
cof($\bullet$): cofactor
tr($\bullet$): trace

Cranial Nerves

I: olfactory nerve (sensory)
II: optic nerve (sensory)
III: oculomotor nerve (mainly motor)
IV: trochlear nerve (mainly motor)
V: trigeminal nerve (sensory and motor)
VI: abducens nerve (mainly motor)
VII: facial nerve (sensory and motor)
VIII: vestibulocochlear (auditory-vestibular) nerve (mainly sensory)
IX: glossopharyngeal nerve (sensory and motor)
X: vagus nerve (sensory and motor)
XI: cranial accessory nerve (mainly motor)
XII: hypoglossal nerve (mainly motor)

Chemical Notations

[X]: concentration of X species
X: upper and lower case letters correspond to gene and corresponding protein or conversely (i.e., Fes, FES, and fes designate protein, a proto-oncogene product that acts as a kinase, and corresponding gene and oncogene product, respectively)
X_i: receptor isoform i of ligand X (i: integer)
XRi: receptor isoform i of ligand X (i: integer)
X+: molecule X expressed (X-positive)
X^+: cation; also intermediate product X of oxidation (loss of electron) from a reductant (or reducer) by an oxidant (electron acceptor that removes electrons from a reductant)
X−: molecule X absent (X-negative)
X^-: anion; also intermediate product X of reduction (gain of electron) from an oxidant (or oxidizer) by a reductant (electron donor that transfers electrons to an oxidant)
small $GTPase^{GTP(GDP)}$: active (inactive) form of small (monomeric), regulatory guanosine triphosphatase
$X^{GTP(GDP)}$: GTP (GDP)-loaded molecule X
X^M: methylated molecule X
X^P: phosphorylated molecule X
pAA: phosphorylated amino acid (pSer, pThr, and pTyr)
X^S: soluble form
X^{SNO}: ^{S}nitrosylated molecule X
X^U: ubiquitinated protein X
X_{alt}: alternative splice variant
$X_{h(l)}$: high (low)-molecular-weight isotype
$X_{L(S)}$: long (short) isoform (splice variants)
X_C: C-terminal fragment (after proteolytic cleavage)
X_c: catalytic subunit
X_N: N-terminal fragment
X_P: palmitoylated molecule X

X_i: number of molecule or atom (i: integer, often 2 or 3)
$(X_1–X_2)_i$: oligomer made of i complexes constituted of molecules X_1 and X_2 (e.g., histones)
$^{D(L)}$X: D (L)-stereoisomer of amino acids and carbohydrates (chirality prefixes for dextro- [dexter: right] and levorotation [lævus: left]), i.e., dextro(levo)rotatory enantiomer
$^{F(G)}$actin: polymeric, filamentous (monomeric, globular) actin
$_t$X: truncated isoform
a, c, nX: atypical, conventional, novel molecule X (e.g., PKC)
acX: acetylated molecule X (e.g., acLDL)
al, ac, nX: alkaline, acidic, neutral molecule X (e.g., sphingomyelinase)
asX: alternatively spliced molecule X (e.g., asTF)
cX: cellular, cytosolic, constitutive (e.g., cNOS), or cyclic (e.g., cAMP and cGMP) molecule X
caX: cardiomyocyte isoform (e.g., caMLCK)
dX: deoxyX
eX: endothelial isoform (e.g., eNOS and eMLCK)
hX: human form (ortholog); heart type (e.g., hFABP); hormone-like isoform (FGF)
iX: inhibitory mediator (e.g., iSMAD) or intracellular (e.g., iFGF) or inducible (e.g., iNOS) isoform
kX: renal type (kidney) molecule X
ksX: kidney-specific isoform of molecule X
lX: lysosomal molecule X
l,acX: lysosomal, acidic molecule X
mX: mammalian species or membrane-associated molecule X (e.g., mTGFβ)
mtX: mitochondrial type of molecule X
nX: neutral X; neuronal type (e.g., nWASP)
oxX: oxidized molecule X (e.g., oxLDL)
plX: plasmalemmal type of molecule X
rX: receptor-associated mediator or receptor-like enzyme; also regulatory type of molecular species (e.g., rSMAD)
sX: secreted, soluble form of molecule X
s,acX: secreted, acidic molecule X
skX: skeletal myocyte isoform (e.g., skMLCK)
smcX: smooth muscle cell isoform (e.g., smcMLCK)
tX: target type of X (e.g., tSNARE); tissue type (e.g., tPA)
vX: vesicle-associated (e.g., vSNARE) or vacuolar (e.g., vATPase) type of X
GPX: glycoprotein (X: molecule abbreviation or assigned numeral)
Xx: (x: single letter) splice variants
X1: human form (ortholog)
Xi: isoform type i (paralog or splice variant; i: integer)
Xi/j: (i,j: integers) refers to either both isoforms (i.e., Xi and Xj, such as ERK1/2) or heterodimer (i.e., Xi–Xj, such as ARP2/3)
X1/X2: molecular homologs or commonly used aliases (e.g., contactin-1/F3)
PI(i)P, PI(i,j)P_2, PI(i,j,k)P_3: i,j,k (integers): position(s) of phosphorylated OH groups of the inositol ring of phosphatidylinositol mono-, bis-, and trisphosphates

Amino Acids

Ala (A): alanine
Arg (R): arginine
Asn (N): asparagine
Asp (D): aspartic acid
CysH (C): cysteine
Cys: cystine
Gln (Q): glutamine
Glu (E): glutamic acid
Gly (G): glycine
His (H): histidine
Iso, Ile (I): isoleucine
Leu (L): leucine
Lys (K): lysine
Met (M): methionine
Phe (F): phenylalanine
Pro (P): proline

Ser (S): serine
Thr (T): threonine
Trp (W): tryptophan
Tyr (Y): tyrosine
Val (V): valine

Ions

Asp^-: aspartate (carboxylate anion of aspartic acid)
ATP^{4-}: ATP anion
Ca^{++}: calcium cation
Cl^-: chloride anion
Co^{++}: cobalt cation
Cu^+: copper monovalent cation
Cu^{++}: copper divalent cation
Fe^{++}: ferrous iron cation
Fe^{3+}: ferric iron cation
Glu^-: glutamate (carboxylate anion of glutamic acid)
H^+: hydrogen cation (proton)
H_3O^+: hydronium (oxonium or hydroxonium) cation
HCO_3^-: bicarbonate anion
HPO_4^{2-}: hydrogen phosphate anion
K^+: potassium cation
Mg^{++}: magnesium cation
$MgATP^{2-}$: ATP anion
Mn^{++}: manganese cation
Na^+: sodium cation
Ni^{++}: nickel cation (common oxidation state)
OH^-: hydroxide anion
PO_4^{3-}: phosphate anion
SO_4^{2-}: sulfate anion
Zn^{++}: zinc cation (common oxidation state)

Inhaled and Signaling Gas

CO: carbon monoxide (or carbonic oxide; signaling gas and pollutant)
CO_2: carbon dioxide (cell waste)
H_2S: hydrogen sulfide (signaling gas)
He: helium (inert monatomic gas)
N_2: nitrogen (inert diatomic gas)
NO: nitric oxide (or nitrogen monoxide; signaling gas and pollutant)
NO_2: nitrogen dioxide (air pollutant)
O_2: oxygen (cell energy producer)
SO_2: sulfur dioxide (air pollutant)

Nitric Oxide Derivatives

$NO^\star$: free radical form
NO^+: nitrosonium (nitrosyl) cation
NO^-: nitroxyl or hyponitrite anion (inodilator)
HNO: nitroxyl (protonated nitroxyl anion)
NO_2^-: nitrite anion
NO_3^-: nitrate anion

Reactive Oxygen Species

H_2O_2: hydrogen peroxide
O_2^-: superoxide
OH^-: hydroxyl radical, hydroxide
$ONOO^-$: peroxynitrite

Time Units

d: day
h: hour
mn: minute
s: second
wk: week

SI-Based and Non-SI Units of Quantity

mmol, nmol, µmol: milli-, nano-, micromoles
(amount of a chemical species, one mole containing about $6.02214078 \times 10^{23}$ molecules)
mosm: milliosmole
(osm: number of moles of a osmotically active chemical compound)
kDa: kiloDalton
(Da: atomic or molecular mass unit)
ppm: parts per million
l: liter

Index

D

E

O

P

R

S

T